Graduate Texts in Contemporary Physics

Springer

New York
Berlin
Heidelberg
Hong Kong
London
Milan
Paris
Tokyo

Graduate Texts in Contemporary Physics

S.T. Ali, J.P. Antoine, and J.P. Gazeau: **Coherent States, Wavelets and Their Generalizations**

A. Auerbach: **Interacting Electrons and Quantum Magnetism**

T.S. Chow: **Mesoscopic Physics of Complex Materials**

B. Felsager: **Geometry, Particles, and Fields**

P. Di Francesco, P. Mathieu, and D. Sénéchal: **Conformal Field Theories**

A. Gonis and W.H. Butler: **Multiple Scattering in Solids**

K. Gottfried and T.-M. Yan: **Quantum Mechanics: Fundamentals, 2nd Edition**

K.T. Hecht: **Quantum Mechanics**

J.H. Hinken: **Superconductor Electronics: Fundamentals and Microwave Applications**

J. Hladik: **Spinors in Physics**

Yu.M. Ivanchenko and A.A. Lisyansky: **Physics of Critical Fluctuations**

M. Kaku: **Introduction to Superstrings and M-Theory, 2nd Edition**

M. Kaku: **Strings, Conformal Fields, and M-Theory, 2nd Edition**

H.V. Klapdor (ed.): **Neutrinos**

R.L. Liboff (ed): **Kinetic Theory: Classical, Quantum, and Relativistic Descriptions, 3rd Edition**

J.W. Lynn (ed.): **High-Temperature Superconductivity**

H.J. Metcalf and P. van der Straten: **Laser Cooling and Trapping**

R.N. Mohapatra: **Unification and Supersymmetry: The Frontiers of Quark-Lepton Physics, 3rd Edition**

R.G. Newton: **Quantum Physics: A Text for Graduate Students**

H. Oberhummer: **Nuclei in the Cosmos**

G.D.J. Phillies: **Elementary Lectures in Statistical Mechanics**

R.E. Prange and S.M. Girvin (eds.): **The Quantum Hall Effect**

S.R.A. Salinas: **Introduction to Statistical Physics**

B.M. Smirnov: **Clusters and Small Particles: In Gases and Plasmas**

(continued after index)

Richard L. Liboff

Kinetic Theory
Classical, Quantum,
and Relativistic Descriptions

Third Edition

With 112 Illustrations

Springer

Richard L. Liboff
School of Electrical Engineering
Cornell University
Ithaca, NY 14853
richard@ee.cornell.edu

Series Editors

R. Stephen Berry
Department of Chemistry
University of Chicago
Chicago, IL 60637
USA

Joseph L. Birman
Department of Physics
City College of CUNY
New York, NY 10031
USA

Mark P. Silverman
Department of Physics
Trinity College
Hartford, CT 06106
USA

H. Eugene Stanley
Center For Polymer Studies
Physics Department
Boston University
Boston, MA 02215
USA

Mikhail Voloshin
Theoretical Physics Institute
Tate Laboratory of Physics 424
The University of Minnesota
Minneapolis, MN 55455
USA

Library of Congress Cataloging-in-Publication Data
Liboff, Richard L., 1931–
    Kinetic theory: classical, quantum, and relativistic descriptions / Richard L. Liboff
    3rd ed.
        p. cm.—(Graduate texts in contemporary physics)
    Includes bibliographical references and index.
    ISBN 0-387-95551-8 (alk. paper)
    1. Kinetic theory of matter. I. Title. II. Series.
QC174.9 L54 2003
530.13′6—dc21   2002026658

ISBN 0-387-95551-8   Printed on acid-free paper.

Printed in the United States of America.

9 8 7 6 5 4 3 2 1   SPIN 10887268

www.springer-ny.com

Springer-Verlag   New York Berlin Heidelberg
A member of BertelsmannSpringer Science+Business Media GmbH

To the Memory of

Harold Grad

Teacher and Friend

I learned much from my teachers, more from my colleagues and most of all from my students

Rabbi Judah ha-Nasi (ca. 135–C.E.)
Babylonian Talmud (Makkot, 10a)

Ludwig Boltzmann (University of Vienna, courtesy AIP Emilio Segrè Visual Archives)

Preface to the Third Edition

Since the first edition of this work, kinetic theory has maintained its position as a cornerstone of a number of disciplines in science and mathematics. In physics, such is the case for quantum and relativistic kinetic theory. Quantum kinetic theory finds application in the transport of particles and radiation through material media, as well as the non-stationary quantum–many-body problem. Relativistic kinetic theory is relevant to controlled thermonuclear fusion and to a number of problems in astrophysics. In applied mathematics, kinetic theory relates to the phenomena of localization, percolation, and hopping, relevant to transport properties in porous media. Classical kinetic theory is the foundation of fluid dynamics and thus is important to aerospace, mechanical, and chemical engineering. Important to the study of transport in metals is the Lorentz–Legendre expansion, which in this new edition appears in an appendix. A new section in Chapter 1 was included in this new edition that addresses constants of motion and symmetry. A number of small but important revisions were likewise made in this new edition. A more complete description of the contents of the text follows.

The text comprises seven chapters. In Chapter 1, the transformation theory of classical mechanics is developed for the purpose of deriving Liouville's theorem and the Liouville equation. Four distinct interpretations of the solution to this equation are presented. The fourth interpretation addresses Gibbs's notion of a distribution function that is the connecting link between the Liouville equation and experimental observation. The notion of a Markov process is discussed, and the central-limit theorem is derived and applied to the random walk problem.

In Chapter 2, the very significant BBKGY hierarchy is obtained from the Liouville equation, and the first two equations of this sequence are applied in the derivation of conservation of energy for a gas of interacting particles. In nondimensionalizing this sequence, parameters emerge that differentiate between weakly and strongly coupled fluids. Correlation functions are introduced through the Mayer expansions. Examining a weakly coupled fluid composed of particles interacting under long-range interaction leads to the Vlasov equation and the closely allied concept of a self-consistent solution. Prigogine's diagrammatic technique and related operator formalism for examining the Liouville equation are described. The Bogoliubov *ansatz* concerning the equilibration of a gas, as well as the Klimontovich formulation of kinetic theory, are also included in this chapter.

The Boltzmann equation is derived in Chapter 3 and applied to the derivation of fluid dynamic equations and the $\mathcal{H}$ theorem. Poincaré's recurrence theorem is proved and is discussed relative to Boltzmann's $\mathcal{H}$ theorem. Transport coefficients are defined, and the Chapman–Enskog expansion is developed. Results of this technique of solution to the Boltzmann equation are compared with experimental data and are found to be in good agreement for various molecular samples. Grad's method of solution of the Boltzmann equation involving expansion in tensor Hermite polynomials is described. The chapter continues with a derivation of the Druyvesteyn distribution relevant to a current carrying plasma in a dc electric field. In the last section of the chapter, the topic of irreversibility is revisited. Ergodic and mixing flows are discussed. Action-angle variables are introduced, and the notions of classical degeneracy and resonant domains in phase space are described in relation to the chaotic behavior of classical systems. A statement of the closely allied KAM theorem is also given.

In the first half of Chapter 4, the Vlasov equation is applied to linear wave theory for a two-component plasma composed of electrons and heavy ions. Landau damping and the Nyquist criterion for wave instabilities are described. The chapter continues with derivations of other important kinetic equations: Krook–Bhatnager–Gross (KBG), Fokker–Planck, Landau, and Balescu–Lenard equations. A table is included describing the interrelation of the classical kinetic equations discussed in the text. The chapter concludes with a description of the widely used Monte Carlo numerical analysis in kinetic theory.

Quantum kinetic theory is developed in Chapter 5. A brief review of basic principles leads to a description of the density matrix, the Pauli equation, and the closely related Wigner distribution. Various equivalent forms of the Wigner–Moyal equation are derived. A quantum modified KBG equation is applied to photon transport and electron propagation in solids. Thomas–Fermi screening and the Mott transition are also discussed. The Uehling–Uhlenbeck quantum modified Boltzmann equation is developed and applied to a Fermi liquid. The chapter continues with an overview of classical and quantum hierarchies of equations connecting reduced distributions. A table of hierarchies is included where the reader is easily able to view distinctions among these sets of equations. The Kubo formula, described previously in Chapter 3, is revisited and applied to the derivation of a quantum expression for electrical conductivity. The chapter concludes with an introduction to Green's function analysis and related diagrammatic representations.

Chapter 6 addresses relativistic kinetic theory. The discussion begins with elementary concepts, including a statement of Hamilton's equations in covariant form. Stemming from a covariant distribution function in four space, together with Maxwell's equations in covariant form, a relativistic Vlasov equation is derived for a plasma in an electromagnetic field. An important component of this chapter is the derivation and compilation of a table of Lorentz invariants in kinetic theory. The chapter continues with a derivation of the relativistic Maxwellian and concludes with a brief description of relativity in non-Cartesian coordinates.

Chapter 7 has been added in this new edition, and addresses kinetic and thermal properties of metals and amorphous media. The first component of the chapter begins with a review of the notion of thermopower, and the Wiedermann–Franz law is derived. The discussion continues with a formulation for electrical and thermal conductivity in metals (encountered previously in Chapter 5), stemming from the quantum Boltzmann equation, in which Bloch's classic low-temperature T^5 dependence of metallic resistivity and canonical high-temperature linear T dependence are derived. In addition, the formalism yields a residual resistivity at 0 K. The chapter continues with a discussion of properties of amorphous media and related processes of localization, hopping, and percolation. Bloch waves and the notion of extended states are reviewed. Anderson's parameter of the ratio of the spread-of-states to the band width is introduced. Localization occurs at some critical value of this parameter. At smaller values of the parameter, energies of localized and extended states are separated at the "mobility edge." Transition of the Fermi energy from the domain of extended states to the domain of localized states represents the Mott metal–insulator transition. Mechanisms of electrical conduction are discussed in three temperature intervals in which the notions of thermally assisted and variable-range hopping emerge. The chapter continues with the concepts of bond and site percolation. A number of percolation scaling laws are discussed. The chapter concludes with a review of localization in second quantization. Throughout the chapter, many discussions related to material science are included.

Each chapter is preceded by a brief introductory statement of the subject matter contained in the chapter. Problems appear at the end of each chapter, many of which carry solutions. A number of problems include self-contained descriptions of closely allied topics. In such cases, these are listed in the chapter table of contents under the heading, *Topical Problems*. In addition to references cited in the text, a comprehensive list of references is included in Appendix E. Assorted mathematical formulas are included in Appendices A and B, including a list of properties of Laguerre and Hermite polynomials (B4). Appendix D, addressing the Lorentz–Legendre expansion in kinetic theory, is new to this edition.

Stemming from the observation that science and society are inextricably entwined, a time chart is included (Appendix E) listing early contributors to science and technology of the classical Greek and Roman eras. The reader will note that a central figure in this display is the Greek philosopher, Democritus, who, at about 400 BCE, was the first to propose an atomic theory of matter. Readers of my earlier work [*Introduction to the Theory of Kinetic Equations*, Wiley, New York (1969)] will recall that it, too, included a time chart describing contributions to dynamics from the fifteenth to the nineteenth centuries. The appendix on Mathematical Formulas has been expanded in this new edition to include a list of properties of Laguene and Hermite polynomials (Appendix B4).

Many individuals have contributed to the development of this work. I remain indebted to these kind colleagues and would like here to express my sincere gratitude for their encouragement, support, and constructive criticism: Sidney Leibovich, Terrence Fine, Robert Pay, Christof Litwin, Kenneth Gardner, Neal Maresca, K. C. Liu, Danny Heffernan, Edwin Dorchek, Philip Bagwell, Ronald Kline, Steve Seidman, S. Ramakrishna, G. George, Timir Datta, William Morrell, Wayne Scales, Daniel Koury, Erich Kunhardt, Marvin Silver, Hercules Neves, James Hartle, Kenneth Andrews, Clifford Pollock, Veit Elser, Chuntong Ying, Michael Parker, Jack Freed, Richard Zallen, Abner Shimony, Philip Holmes, Lloyd Hillman, Arthur Ruoff, L. Pearce Williams, Lloyd Motz, John Guckenheimer, Isaac Rabinowitz, Gregory Schenter, and Ilya Prigogine.

Some of these individuals are former students. It is due to my association with these gifted and talented colleagues that the talmudic inscription for this work is motivated.

Peace,

RICHARD L. LIBOFF

תושלב"ע.

Contents

Preface to the Third Edition   ix

1   The Liouville Equation   1

     1.1   Elements of Classical Mechanics   2

           1.1.1   Generalized Coordinates and the Lagrangian . . .   2

           1.1.2   Hamilton's Equations   4

           1.1.3   Constants of the Motion   6

           1.1.4 Γ-Space   7

           1.1.5   Dynamic Reversibility   9

           1.1.6   Equation of Motion for Dynamical Variables . . .   10

     1.2   Canonical Transformations   11

           1.2.1   Generating Functions   11

           1.2.2   Generating Other Transformations   14

           1.2.3   Canonical Invariants   15

           1.2.4   Group Property of Canonical Transformations . .   15

           1.2.5   Constants of Motion and Symmetry   16

     1.3   Liouville Theorem   17

           1.3.1   Proof   17

           1.3.2   Geometric Significance   18

           1.3.3   Action Generates the Motion   19

     1.4   Liouville Equation   20

           1.4.1   The Ensemble: Density in Phase Space   20

           1.4.2   First Interpretation of $D(q, p, t)$ 20

           1.4.3   Most General Solution: Second Interpretation of

                 $D(q, p, t)$ 22

           1.4.4   Incompressible Ensemble   23

           1.4.5   Method of Characteristics   24

	1.4.6	Solutions to the Initial-Value Problem	25
	1.4.7	Liouville Operator	27
1.5		Eigenfunction Expansions and the Resolvent	29
	1.5.1	Liouville Equation Integrating Factor	29
	1.5.2	Example: The Ideal Gas	30
	1.5.3	Free-Particle Propagator	32
	1.5.4	The Resolvent	33
1.6		Distribution Functions	35
	1.6.1	Third Interpretation of $D(q, p, t)$	35
	1.6.2	Joint-Probability Distribution	36
	1.6.3	Reduced Distributions	37
	1.6.4	Conditional Distribution	37
	1.6.5	s-Tuple Distribution	38
	1.6.6	Symmetric Properties of Distributions	39
1.7		Markov Process	41
	1.7.1	Two-Time Distributions	41
	1.7.2	Chapman–Kolmogorov Equation	41
	1.7.3	Homogeneous Processes in Time	43
	1.7.4	Master Equation	43
	1.7.5	Application to Random Walk	44
1.8		Central-Limit Theorem	47
	1.8.1	Random Variables and the Characteristic Function	47
	1.8.2	Expectation, Variance, and the Characteristic Function	47
	1.8.3	Sums of Random Variables	49
	1.8.4	Application to Random Walk	50
	1.8.5	Large n Limit: Central-Limit Theorem	54
	1.8.6	Random Walk in Large n Limit	55
	1.8.7	Poisson and Gaussian Distributions	57
	1.8.8	Covariance and Autocorrelation Function	59
Problems			60
2	**Analyses of the Liouville Equation**		**77**
2.1		BBKGY Hierarchy	78
	2.1.1	Liouville, Kinetic Energy, and Remainder Operators	78
	2.1.2	Reduction of the Liouville Equation	80
	2.1.3	Further Symmetry Reductions	81
	2.1.4	Conservation of Energy from BY_1 and BY_2	83
2.2		Correlation Expansions: The Vlasov Limit	86
	2.2.1	Nondimensionalization	86
	2.2.2	Correlation Functions	89
	2.2.3	The Vlasov Limit	90
	2.2.4	The Vlasov Equation: Self-Consistent Solution	92
	2.2.5	Debye Distance and the Vlasov Limit	95

	2.2.6	Radial Distribution Function	96
2.3	Diagrams: Prigogine Analysis	97
	2.3.1	Perturbation Liouville Operator	98
	2.3.2	Generalized Fourier Series	99
	2.3.3	Interpretation of $\mathbf{a}_l$ Coefficients	99
	2.3.4	Equations of Motion for $\mathbf{a}_{(\mathbf{k})}$ Coefficients	102
	2.3.5	Selection Rules for Matrix Elements	103
	2.3.6	Properties of Diagrams	104
	2.3.7	Long-Time Diagrams: The Boltzmann Equation .	107
	2.3.8	Reduction of Time Integrals	108
	2.3.9	Application of Diagrams to Plasmas	113
2.4	Bogoliubov Hypothesis	115
	2.4.1	Time and Length Intervals	115
	2.4.2	The Three Temporal Stages	115
	2.4.3	Bogoliubov Distributions	117
	2.4.4	Density Expansion	117
	2.4.5	Construction of $F_2^{(0)}$	119
	2.4.6	Derivation of the Boltzmann Equation	121
2.5	Klimontovich Picture	124
	2.5.1	Phase Densities	124
	2.5.2	Phase Density Averages	125
	2.5.3	Relation to Correlation Functions	126
	2.5.4	Equation of Motion	127
2.6	Grad's Analysis	127
	2.6.1	Liouville Equation Revisited	127
	2.6.2	Truncated Distributions	128
	2.6.3	Grad's First and Second Equations	129
	2.6.4	The Boltzmann Equation	130
Problems	. . . .	132

3 The Boltzmann Equation, Fluid Dynamics, and Irreversibility   137
3.1	Scattering Concepts	138
	3.1.1	Separation of the Hamiltonian	138
	3.1.2	Scattering Angle	140
	3.1.3	Cross Section	143
	3.1.4	Kinematics	146
3.2	The Boltzmann Equation	149
	3.2.1	Collisional Parameters and Derivation	149
	3.2.2	Multicomponent Gas	153
	3.2.3	Representation of Collision Integral for Rigid Spheres	153
3.3	Fluid Dynamic Equations and the Boltzmann $\mathcal{H}$ Theorem	155
	3.3.1	Collisional Invariants	155
	3.3.2	Macroscopic Variables and Conservative Equations	157

	3.3.3	Conservation Equations and the Boltzmann Equation	159
	3.3.4	Temperance: Variance of the Velocity Distribution	161
	3.3.5	Irreversibility	162
	3.3.6	Poincaré Recurrence Theorem	163
	3.3.7	Boltzmann and Gibbs Entropies	164
	3.3.8	Boltzmann's $\mathcal{H}$ Theorem	166
	3.3.9	Statistical Balance	167
	3.3.10	The Maxwellian	168
	3.3.11	The Barometer Formula	170
	3.3.12	Central-Limit Theorem Revisited	172
3.4		Transport Coefficients	174
	3.4.1	Response to Gradient Perturbations	174
	3.4.2	Elementary Mean-Free-Path Estimates	178
	3.4.3	Diffusion and Random Walk	186
	3.4.4	Autocorrelation Functions, Transport Coefficients, and Kubo Formula	188
3.5		The Chapman–Enskog Expansion	194
	3.5.1	Collision Frequency	194
	3.5.2	The Expansion	195
	3.5.3	Second-Order Solution	199
	3.5.4	Thermal Conductivity and Stress Tensor	202
	3.5.5	Sonine Polynomials	203
	3.5.6	Application to Rigid Spheres	207
	3.5.7	Diffusion and Electrical Conductivity	208
	3.5.8	Expressions of $\Omega^{(l,q)}$ for Inverse Power Interaction Forces	209
	3.5.9	Interaction Models and Experimental Values	211
	3.5.10	The Method of Moments	213
3.6		The Linear Boltzmann Collision Operator	217
	3.6.1	Symmetry of the Kernel	217
	3.6.2	Negative Eigenvalues	218
	3.6.3	Comparison of Boltzmann and Liouville Operators	219
	3.6.4	Hard and Soft Potentials	220
	3.6.5	Maxwell Molecule Spectrum	220
	3.6.6	Further Spectral Properties	222
3.7		The Druyvesteyn Distribution	225
	3.7.1	Basic Parameters and Starting Equations	225
	3.7.2	Legendre Polynomial Expansion	227
	3.7.3	Reduction of $\hat{J}_0(f_0 \mid F)$	232
	3.7.4	Evaluation of I_0 and I_1 Integrals	233
	3.7.5	Druyvesteyn Equation	236
	3.7.6	Normalization, Velocity Shift, and Electrical Conductivity	237

3.8   Further Remarks on Irreversibility   240
    3.8.1   Ergodic Flow   240
    3.8.2   Mixing Flow and Coarse Graining   243
    3.8.3   Action-Angle Variables   245
    3.8.4   Hamilton–Jacobi Equation   248
    3.8.5   Conditionally Periodic Motion and Classical
         Degeneracy   249
    3.8.6   Bruns's Theorem   253
    3.8.7   Anharmonic Oscillator   253
    3.8.8   Resonant Domains and the KAM Theorem   255
Problems   258

**4   Assorted Kinetic Equations with Applications to Plasmas and
Neutral Fluids**   **278**
4.1   Application of the Vlasov Equation to a Plasma   279
    4.1.1   Debye Potential and Dielectric Constant   279
    4.1.2   Waves, Instabilities, and Damping   285
    4.1.3   Landau Damping   288
    4.1.4   Nyquist Criterion   292
4.2   Further Kinetic Equations of Plasmas and Neutral Fluids .   294
    4.2.1   Krook–Bhatnager–Gross Equation   295
    4.2.2   KBG Analysis of Shock Waves   298
    4.2.3   The Fokker–Planck Equation   301
    4.2.4   The Landau Equation   307
    4.2.5   The Balescu–Lenard Equation   309
    4.2.6   Convergent Kinetic Equation   315
    4.2.7   Fokker–Planck Equation Revisited   317
4.3   Monte Carlo Analysis in Kinetic Theory   319
    4.3.1   Master Equation   319
    4.3.2   Equivalence of Master and Integral Equations . .   320
    4.3.3   Interpretation of Terms   321
    4.3.4   Application of Random Numbers   323
    4.3.5   Program for Evaluation of Distribution Function .   324
Problems   325

5   Elements of Quantum Kinetic Theory   **329**
5.1   Basic Principles   330
    5.1.1   The Wave Function and Its Properties   330
    5.1.2   Commutators and Measurement   332
    5.1.3   Representations   334
    5.1.4   Coordinate and Momentum Representations . . .   335
    5.1.5   Superposition Principle   337
    5.1.6   Statistics and the Pauli Principle   339
    5.1.7   Heisenberg Picture   339
5.2   The Density Matrix   341

		5.2.1	The Density Operator	342
		5.2.2	The Pauli Equation	346
		5.2.3	The Wigner Distribution	351
		5.2.4	Weyl Correspondence	354
		5.2.5	Wigner–Moyal Equation	357
		5.2.6	Homogeneous Limit: Pauli Equation Revisited	359
	5.3	Application of the KBG Equation to Quantum Systems	363
		5.3.1	Equilibrium Distributions	363
		5.3.2	Photon Kinetic Equation	366
		5.3.3	Electron Transport in Metals	369
		5.3.4	Thomas–Fermi Screening	376
		5.3.5	Mott Transition	378
		5.3.6	Relaxation Time for Charge-Carrier Phonon Interaction	379
	5.4	Quantum Modifications of the Boltzmann Equation	385
		5.4.1	Quasi-Classical Boltzmann Equation	385
		5.4.2	Kinetic Theory for Excitations in a Fermi Liquid	387
		5.4.3	H Theorem for Quasi-Classical Distribution	392
	5.5	Overview of Classical and Quantum Hierarchies	394
		5.5.1	Second Quantization and Fock Space	394
		5.5.2	Classical and Quantum Distribution Functions	395
		5.5.3	Equations of Motion	399
		5.5.4	Generalized Hierarchies	400
	5.6	Kubo Formula Revisited	403
		5.6.1	Charge Density and Current	403
		5.6.2	Identifications for Kubo Formula	404
		5.6.3	Electrical Conductivity	405
		5.6.4	Reduction to Drude Conductivity	407
	5.7	Elements of the Green's Function Formalism	408
		5.7.1	Schrödinger Equation Green's Function	408
		5.7.2	The s-Body Green's Function	410
		5.7.3	Averages and the Green's Function	410
		5.7.4	The Quasi-Free Particle	413
		5.7.5	One-Body Green's Function	414
		5.7.6	Retarded and Advanced Green's Functions	415
		5.7.7	Coupled Green's Function Equations	415
		5.7.8	Diagrams and Expansion Techniques	416
	5.8	Spectral Function for Electron–Phonon Interactions	421
		5.8.1	Hamiltonian	421
		5.8.2	Green's Function Equations of Motion	423
		5.8.3	The Spectral Function	425
		5.8.4	Lorentzian Form	426
		5.8.5	Lifetime and Energy of a Quasi-Free Particle	428
	Problems			429

6 Relativistic Kinetic Theory   **449**

6.1 Preliminaries 450
   6.1.1 Postulates 450
   6.1.2 Events, World Lines, and the Light Cone 450
   6.1.3 Four-Vectors 451
   6.1.4 Lorentz Transformation 452
   6.1.5 Length Contraction, Time Dilation, and Proper Time 454
   6.1.6 Covariance, Hamiltonian, and Hamilton's Equations 454
   6.1.7 Criterion for Relativistic Analysis 455

6.2 Covariant Kinetic Formulation 456
   6.2.1 Distribution Function 456
   6.2.2 One-Particle Liouville Equation 457
   6.2.3 Covariant Electrodynamics 458
   6.2.4 Vlasov Equation 459
   6.2.5 Covariant Drude Formulation of Ohm's Law . . . 460
   6.2.6 Lorentz Invariants in Kinetic Theory 461
   6.2.7 Relativistic Electron Gas and Darwin Lagrangian 464

6.3 The Relativistic Maxwellian 467
   6.3.1 Normalization 467
   6.3.2 The Nonrelativistic Domain 469

6.4 Non-Cartesian Coordinates 470
   6.4.1 Covariant and Contravariant Vectors 470
   6.4.2 Metric Tensor 471
   6.4.3 Lagrange's Equations 472
   6.4.4 The Christoffel Symbol 473
   6.4.5 Liouville Equation 474

Problems 474

7 Kinetic Properties of Metals and Amorphous Media   **480**

7.1 Metallic Electrical and Thermal Conduction 482
   7.1.1 Background 482
   7.1.2 Thermopower 484
   7.1.3 Electron–Phonon Scattering Matrix Elements . . 487
   7.1.4 Quantum Boltzmann Equation 493
   7.1.5 Perturbation Distribution 496
   7.1.6 Electrical Resistivity 498
   7.1.7 Scale Parameters of ρ_0 503
   7.1.8 Electron Distribution Function 504
   7.1.9 Thermal Conductivity 504

7.2 Amorphous Media 505
   7.2.1 Background 505
   7.2.2 Localization 508
   7.2.3 Conduction Mechanisms and Hopping 511

	7.2.4	Percolation Phenomena	514
	7.2.5	Pair-Connectedness Function and Scaling	517
	7.2.6	Localization in Second Quantization	520
Problems			524

Location of Key Equations — **527**

List of Symbols — **529**

Appendices

A Vector Formulas and Tensor Notation — **537**
  A.1 Definitions — 537
  A.2 Vector Formulas and Tensor Equivalents — 539

B Mathematical Formulas — **541**
  B.1 Tensor Integrals and Unit Vector Products — 541
  B.2 Exponential Integral, Γ Function, ζ Function, and Error Function — 542
    B.2.1 Exponential Integral — 542
  B.3 Other Useful Integrals — 545
  B.4 Hermite and Laguerre Polynomials and the Hypergeometric Function — 546

C Physical Constants — **550**

D Lorentz–Legendre Expansion — **552**

E Additional References — **554**
  E.1 Early Works — 554
  E.2 Recent Contributions to Kinetic Theory and Allied Topics — 558

F Science and Society in the Classical Greek and Roman Eras — **563**

Index — **565**

CHAPTER 1

The Liouville Equation

Introduction

This chapter begins with a review of the basic principles and equations of classical mechanics. These lead naturally to the Liouville theorem, which together with the notions of ensemble and phase space, give rise to the Liouville equation. This equation is perhaps the most significant relation in all of classical kinetic theory, and in the remainder of the chapter various techniques of solution to this equation, as well as various interpretations of its solution, are presented. Thus, for example, it is found that the Liouville equation is an equation of motion for the N-particle joint-probability distribution, where N denotes the number of particles in the system. Integration of this distribution over subdomains in phase space gives reduced distributions whose equations of motion (BBKGY Eqs.) are developed in Chapter 2.

A discussion is included of the Chapman–Kolmogorov equation and the closely allied master equation, which is applied to the random-walk problem. The master equation emerges again in Chapter 5 in analysis of the quantum mechanical density matrix.

The chapter concludes with a brief description of probability theory and a derivation of the central-limit theorem. This theorem is applied to obtain the long-time displacement of the random walk and yields the Gaussian distribution. This distribution is fundamental to diffusion theory and is encountered again in Chapter 3. The central-limit theorem is likewise again encountered in Chapter 3 in a derivation of the distribution of center-of-mass momenta of a stationary gas of molecules. The discrete Poisson distribution is obtained from random-walk results and applied to small shot noise.

Classical mechanics and probability theory are the foundations of kinetic theory. Consequently, a firm grasp of the basic principles and relations of these formalisms will lead to a richer working knowledge of kinetic theory.

1.1 Elements of Classical Mechanics

1.1.1 Generalized Coordinates and the Lagrangian

The number of *degrees of freedom* a given system has is equal to the minimum number of independent parameters necessary to uniquely determine the location and orientation of the system in physical space. Such independent parameters are called *generalized coordinates*. Thus, for example, a rigid dumbbell molecule free to move in three-space has five degrees of freedom: three to locate the center of mass of the molecule and two (angles) to determine the orientation of the molecule. A fluid of N such molecules has $5N$ degrees of freedom. A rigid triangular molecule has six degrees of freedom: five to locate an edge of the triangle and an additional angle to fix the orientation of the plane of the molecule about this axis.

For a system with N degrees of freedom, generalized coordinates are labeled $q_1, q_2, \ldots, q_N$. These parameters may be taken to comprise the components of an N-dimensional vector,

$$q = (q_1, q_2, \ldots, q_N) \tag{1.1}$$

The corresponding *generalized velocities* also comprise components of an N-dimensional vector, and we write

$$\dot{q} = (\dot{q}_1, \dot{q}_2, \ldots, \dot{q}_N) \tag{1.2}$$

The composite vector $(q, \dot{q})$ determines the *state* of the system.

Let the system at hand exist in a conservative force field with kinetic energy $T(q, \dot{q})$ and potential $V(q)$. The *Lagrangian* of the system is then given by

$$L(q, \dot{q}) = T(q, \dot{q}) - V(q) \tag{1.3}$$

The dynamics of the system may be formulated in terms of *Hamilton's principle*. This principle states that the motion of the system between two fixed points, $(q, t)_1$ and $(q, t)_2$, renders the action integral:

$$S = \int_1^2 L(q, \dot{q}, t)\, dt \tag{1.4}$$

an extremum.[1] That is,

$$\delta \int_1^2 L(q, \dot{q}, t)\, dt = 0 \tag{1.5}$$

---

[1] More precisely, a minimum.

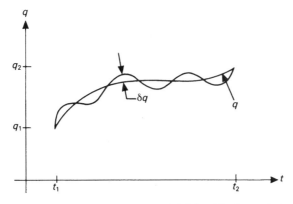

FIGURE 1.1. End points q_1 and q_2 are held fixed in the variation (1.5).

where δ denotes a variation about the motion of the system.[2] See Fig. 1.1

Lagrange's equations

The dynamical principle (1.5) may be employed to obtain a differential equation for $L(q, \dot{q}, t)$. That is, let $q(t)$ be the motion that renders S an extremum.

Effecting the variation (1.5) and Taylor-series expanding the integrand gives

$$\delta S = \int_1^2 L(q + \delta q, \dot{q} + \delta \dot{q}, t)\, dt - \int_1^2 L(q, \dot{q}, t)\, dt$$

$$= \int_1^2 \left(L + \frac{\delta L}{\delta q} \cdot \delta q + \frac{\delta L}{\delta \dot{q}} \cdot \delta \dot{q} \right) dt - \int_1^2 L(q, \dot{q}, t)\, dt \qquad (1.6)$$

Note that

$$\frac{\partial L}{\partial \dot{q}} \frac{d}{dt}(\delta q) = \frac{d}{dt}\left(\frac{\partial L}{\partial \dot{q}} \cdot \delta q) \right) - \frac{d}{dt}\left(\frac{\partial L}{\partial \dot{q}} \right) \cdot \delta q$$

Inserting this expression into the preceding equation gives

$$\left[\frac{\partial L}{\partial \dot{q}} \cdot \delta q \right]_1^2 + \int_1^2 \left[\frac{\partial L}{\partial q} - \frac{d}{dt}\left(\frac{\partial L}{\partial \dot{q}} \right) \right] \cdot \delta q\, dt = 0$$

Since the end points of the trajectory are held fixed in the variation, the first bracketed "surface terms" vanish. Insofar as the remaining integral must vanish for any arbitrary, infinitesimal variation δq, it is necessary that

$$\boxed{\frac{d}{dt}\left(\frac{\partial L}{\partial \dot{q}_l} \right) - \frac{\partial L}{\partial q_l} = 0 \qquad (l = 1, 2, \ldots, N)} \qquad (1.7)$$

---

[2]Furthermore, the variation is a *virtual* displacement, that is, on that occurs in zero time.

Here we have returned to the component representation of $(q, \dot{q})$. Equations (1.7) are called Lagrange's equations.

1.1.2 Hamilton's Equations

When written in the form

$$\frac{d}{dt}\left(\frac{\partial L}{\partial \dot{q}_l}\right) = \frac{\partial L}{\partial q_l} \tag{1.8}$$

Lagrange's equations (1.7) reveal the following important observation: If the coordinate q_l is missing in the Lagrangian, the quantity $\partial L/\partial \dot{q}_l$ is constant in time. In this event, we say that the coordinate q_1 is *cyclic* or *ignorable*.

To better exhibit this symmetry principle of mechanics, we effect the following transformation:

$$(q_l, \dot{q}_l) \rightarrow (q_l, p_l) \tag{1.9}$$

where

$$p_l \equiv \frac{\partial L}{\partial \dot{q}_l} \tag{1.10}$$

The variable p_l, so defined, is called the *canonical momentum* or *momentum conjugate* to the coordinate q_l.

The transformation (1.9) is accomplished through a Legendre transformation (familiar to thermodynamics). This transformation carries the Lagrangian $L(q, \dot{q})$ to the *Hamiltonian* $H(q, p)$ by the following recipe:

$$H = \sum_l \frac{\partial L}{\partial \dot{q}_l} \dot{q}_l - L \tag{1.11}$$

Equations of motion in terms of the Hamiltonian are obtained by forming the differential of the latter equation.

$$dH(q, p, t) = \sum_l \left[d\left(\frac{\partial L}{\partial \dot{q}_l}\right)\dot{q}_l + \frac{\partial L}{\partial \dot{q}_l} d\dot{q}_l - \frac{\partial L}{\partial \dot{q}_l} d\dot{q}_l - \frac{\partial L}{\partial q_l} dq_l - \frac{\partial L}{\partial t} dt \right]$$

With Lagrange's equations (1.7) and the definition (1.10), we obtain

$$dH(q, p, t) = \sum_l (\dot{q}_l \, dp_l - \dot{p}_l \, dq_l) - \frac{\partial L}{\partial t} dt$$

Expanding the left side of this equation and identifying terms gives

$$\boxed{\dot{p}_l = -\frac{\partial H}{\partial q_l}, \qquad \dot{q}_l = \frac{\partial H}{\partial p_l}} \tag{1.12}$$

together with $\partial H/\partial t = -\partial L/\partial t$. Equations (1.12) are called Hamilton's equations.

The Hamiltonian and the energy

An important theorem concerning the Hamiltonian is as follows. Consider a system of particles with radii vectors $\{\mathbf{r}_i\}$. Consider that the transformation to the set of generalized coordinates,

$$\mathbf{r}_i = \mathbf{r}_i(q_1, \ldots, q_N) \tag{1.13}$$

is independent of time and that the Lagrangian of the system is not an explicit function of time. Then, with Lagrange's equations (1.7), we obtain

$$\frac{dL}{dt} = \sum_l \frac{\partial L}{\partial q_l} \dot{q}_l + \frac{\partial L}{\partial \dot{q}_l} \frac{d\dot{q}_l}{dt}$$

$$= \sum_l \left[\frac{d}{dt}\left(\frac{\partial L}{\partial \dot{q}_l}\right)\dot{q}_l + \left(\frac{\partial L}{\partial \dot{q}_l}\right)\frac{d\dot{q}_l}{dt}\right]$$

$$= \sum_l \frac{d}{dt}\left(\frac{\partial L}{\partial \dot{q}_l}\dot{q}_l\right)$$

It follows that

$$\frac{d}{dt}\left(\sum \frac{\partial L}{\partial \dot{q}_l}\dot{q}_l - L\right) = 0 = \frac{d}{dt}H(q, p) \tag{1.14}$$

We may conclude that for such systems H is constant in time. This constant may further be identified with the energy, E (see Problem 1.32).

$$H = E = \text{constant} \tag{1.15}$$

Electrodynamic and relativistic Hamiltonian

Two important Hamiltonians that emerge in physics are as follows. The Hamiltonian of a particle of charge e and mass m that moves in an electromagnetic field with vector potential $\mathbf{A}$ and scalar potential Φ is given by (see Problem (1.22)

$$H = \frac{1}{2m}\left(\mathbf{p} - \frac{e}{c}\mathbf{A}\right)^2 + e\Phi$$

$$\mathbf{p} = m\mathbf{v} + \frac{e}{c}\mathbf{A} \tag{1.16}$$

The electric $\mathcal{E}$ and magnetic field $\mathbf{B}$ are related to the fields through the relations

$$\mathcal{E} = -\nabla\Phi - \frac{1}{c}\frac{\partial \mathbf{A}}{\partial t}$$

$$\mathbf{B} = \nabla \times \mathbf{A} \tag{1.17}$$

Hamilton's equations applied to the preceding Hamiltonian returns the Lorentz force law

$$\mathbf{F} = e\left(\mathcal{E} + \frac{\mathbf{v}}{c} \times \mathbf{B}\right) \tag{1.18}$$

The second example addresses a relativistic particle with rest mass m. It is given by[3,4]

$$H = \sqrt{p^2c^2 + m^2c^4} + V(\mathbf{r}) \qquad (1.19)$$

The relativistic momentum has the form

$$\mathbf{p} = \gamma m \mathbf{v}, \qquad \gamma^2 \equiv \frac{1}{1 - \beta^2}, \qquad \beta \equiv \frac{v}{c} \qquad (1.19a)$$

In the preceding expressions, c is the speed of light.

A transformation of the potentials Φ and $\mathbf{A}$ in which the fields $\mathcal{E}$, $\mathbf{B}$ are invariant is called a 'gauge transformation.' Consider, for example, the transformation

$$\Phi \to \Phi' = \Phi - \frac{1}{c}\frac{\partial \Psi}{\partial t} \qquad \mathbf{A} \to \mathbf{A}' = \mathbf{A} + \nabla \Psi \qquad (1.19b)$$

Substitution of these expressions into (1.17) indicates that the fields $\mathcal{E}$, $\mathbf{B}$ remain the same. (See Problem 1.22).

1.1.3   Constants of the Motion

If a dynamical function

$$W = W(q_1, \ldots, q_N, p_1, \ldots, p_N, t)$$

remains constant as the motion of the system unfolds in time, W is a *constant of the motion* or an *integral of the system*.

How many such constants are there? For a system with N degrees of freedom, integration of Hamilton's equations (1.12) gives

$$\dot{p}_l = -\frac{\partial H}{\partial q_1} \to p_l(t) = p_l(0) - \int_0^t \frac{\partial H}{\partial q_1} dt$$

$$\dot{q}_l = \frac{\partial H}{\partial p_l} \to q_l(t) = q_l(0) + \int_0^t \frac{\partial H}{\partial p_l} dt \qquad (1.20)$$

Thus we find that the state of the system is specified by $2N$ constants, $\{q_l(0), p_l(0)\}$.[5] This conclusions may also be inferred geometrically.

In $2N + 1$-dimensional space, $2N$ surfaces define a curve. For example, in three-space, the two surfaces

$$f(x, y, z) = C_1$$
$$g(x, y, z) = C_2$$

---

[3] An alternative expression for H written in covariant form is given in Chapter 6.
[4] The Hamiltonian of an aggregate of N interacting particles is given by (4.27).
[5] Although initial values are constant, subsequent motion may grow chaotic. These topics are further discussed in Section 3.8

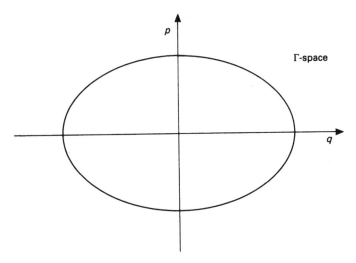

FIGURE 1.2. For a harmonic oscillator, the state of the system lies on an ellipse in Γ-space.

define a curve. Only one of three variables x, y, z is independent. A curve is a one-dimensional locus. Thus the intersection of the two surfaces written above may be written in parametric form as

$$x = x(t), \qquad y = y(t), \qquad z = z(t)$$

Before proceeding further with this geometric construction, we define Γ-space.

1.1.4   Γ-Space

For a system with N degrees of freedom. Γ-space is a $2N$ dimensional Cartesian space whose axes are the $\{q_l, p_l\}$ variables. Thus the state of the system at any given instant $(q_1, \ldots, q_N, p_1, \ldots, p_N)$ is a single point in $2N$-dimensional Γ-space.

Let the system be a single harmonic oscillator for which Γ-space is two-dimensional. The orbit of the oscillator may be written

$$q = a \cos \omega t, \qquad p = b \sin \omega t$$

from which we find

$$\left(\frac{q}{a}\right)^2 + \left(\frac{p}{b}\right)^2 = 1$$

Thus the state of the system (q, p) lies on an ellipse in Γ-space. See Fig. 1.2.

This example illustrates that time is suppressed in Γ-space. All we know from the curve shown in Fig. 1.2 is that the particle has q, p values that lie somewhere on the ellipse. We do not know what these values are at a specific time.

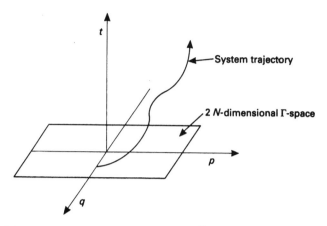

FIGURE 1.3. The system trajectory is a curve in $\tilde{\Gamma}$-space. At any instant of time, the curve intersects Γ-space at a single point (q, p).

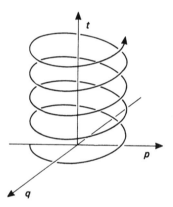

FIGURE 1.4. Dynamical trajectory of a harmonic oscillator in three-dimensional $(2 + 1)$ $\tilde{\Gamma}$-space. Projection onto Γ-space is an ellipse.

Time is exhibited explicitly in $2N + 1$ $\tilde{\Gamma}$-space. This Cartesian space is comprised of Γ-space and an additional orthogonal time axis. The *system trajectory* or *dysfunctional orbit*,

$$[q_1(t), \ldots, q_N(t); p_1(t), \ldots, p_N(t)]$$

is a curve in $\tilde{\Gamma}$-space. See Fig. 1.3. The system trajectory for the harmonic oscillator is shown in Fig. 1.4.

A constant of the motion

$$W_1(q, p, t) = C_1$$

is constant on the system trajectory. But $W = C_1$ is a surface in $\tilde{\Gamma}$-space (that is, a $2N$-dimensional locus). The intersection of $2N$ such independent surfaces

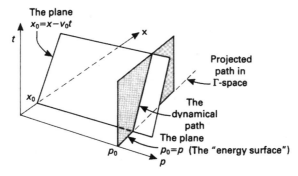

FIGURE 1.5. Dynamical path of a free particle in three-dimensional $\tilde{\Gamma}$-space.

$(W_1, W_2, \ldots, W_N)$ determines a curve (one-dimensional locus) in $\tilde{\Gamma}$-space.[6] This curve may be written

$$q_l = q_l(\eta), \quad p_l = p_l(\eta), \quad t = t(\eta), \quad l = 1, \ldots, N \qquad (1.21)$$

which is the system trajectory. So again we find that the dynamical orbit is determined by $2N$ constants of the motion. In practice, we are often more interested in the projected motion in Γ-space. This motion, we recall, is determined by $2N - 1$ constants. See also Section 3.8.6.

These concepts are well illustrated by the simple example of a free particle moving in one dimension. Since this system has one degree of freedom, it has only two constants of motion:

$$x_0 = x - vt$$
$$p_0 = p$$

Intersection of these two surfaces in three-dimensional $\tilde{\Gamma}$-space reveals the orbit

$$x = x_0 + \frac{p_0}{m}t$$

See Fig. 1.5. These topics are returned to in Section 3.8.

1.1.5 Dynamic Reversibility

Consider a system whose Hamiltonian is invariant under time reversal. That is, $H(t) = H(-t)$. Suppose $\{q(t), p(t)\}$ is a trajectory for this system. Then another solution is $\{q(-t), -p(-t)\}$. To establish this property first, set $t' = -t$

---

[6]Such constants that intersect the energy surface are called *isolating integrals*. Constants such as $f(H)$, where f is any smooth function, are evidently constant on the energy surface and therefore have no intersection with it.

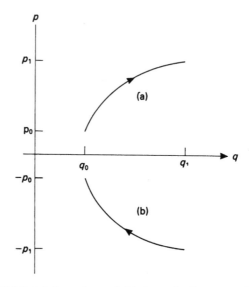

FIGURE 1.6. (a) A motion and (b) dynamically reversed motion.

in Hamilton's equations (1.12). There results

$$-\frac{dp}{dt'} = -\frac{\partial H}{\partial q}, \qquad -\frac{dq}{dt'} = \frac{\partial H}{\partial p}$$

So if we set $p' = -p$, $q' = q$, Hamilton's equations (1.12) are returned. It follows that, if $\{q(t), p(t)\}$ is a solution to Hamilton's equations, then $\{q(-t), -p(-t)\}$ is also a solution. This situation is depicted in Fig. 1.6.

1.1.6 Equation of Motion for Dynamical Variables

Let $u(q, p, t)$ denote a dynamical variable. Then we may write

$$\frac{du}{dt} = \sum_l \left(\frac{\partial u}{\partial q_l} \dot{q}_l + \frac{\partial u}{\partial p_l} \dot{p}_l \right) + \frac{\partial u}{\partial t} \tag{1.22}$$

With Hamilton's equations, the preceding equation may be rewritten

$$\frac{du}{dt} = \sum_t \left(\frac{\partial u}{\partial q_l} \frac{\partial H}{\partial p_l} - \frac{\partial H}{\partial q_l} \frac{\partial u}{\partial p_l} \right) + \frac{\partial u}{\partial t} \tag{1.23}$$

Let us introduce the following *Poisson bracket* notation relevant to two dynamical variables A and B:

$$[A, B] = \sum_l \left(\frac{\partial A}{\partial q_l} \frac{\partial B}{\partial p_l} - \frac{\partial B}{\partial q_l} \frac{\partial A}{\partial p_l} \right) \tag{1.24}$$

Then (1.23) may be written

$$\boxed{\frac{du}{dt} = [u, H] + \frac{\partial u}{\partial t}} \qquad (1.25)$$

This is the *equation of motion* for the dynamical variable $u(q, p, t)$. The following properties of Poisson brackets are easily verified:

$$[A, B] = -[B, A] \qquad (1.26a)$$

$$[A, c] = 0 \quad (c \text{ is constant}) \qquad (1.26b)$$

$$[A + B, F] = A[B, F] + [B, F] \qquad (1.26c)$$

$$[AB, F] = A[B, F] + [A, F]B \qquad (1.26d)$$

$$[A, [B, F]] + [B[F, A]] + [F, [A, B]] = 0 \qquad (1.26e)$$

$$\frac{\partial}{\partial t}[A, B] = \left[\frac{\partial A}{\partial t}, B\right] + \left[A, \frac{\partial B}{\partial t}\right] \qquad (1.26f)$$

$$[A, B(A)] = 0 \qquad (1.26g)$$

The relation (1.26e) is called *Jacob's identity*. In (1.26g), the function $B(A)$ is assumed to have a Taylor series expansion.

In the event that a dynamical variable u is not explicitly dependent on time, then $\partial u / \partial t = 0$ and (1.25) becomes

$$\frac{du}{dt} = [u, H] \qquad (1.27)$$

If, further, $u = u(H)$, then with the property (1.26e) we may write

$$\frac{du}{dt} = 0 \qquad (1.28)$$

So we may conclude that any dynamical variable that is function only of the Hamiltonian is a constant of the motion.

With (1.25), Hamilton's equations (1.12) may be rewritten

$$\dot{q} = [q, H] \qquad (1.29a)$$

$$\dot{p} = [p, H] \qquad (1.29b)$$

Note also the *fundamental* Poisson bracket relations:

$$[q_l, q_{l'}] = [p_l, p_{l'}] = 0, \qquad [q_l, p_{l'}] = \delta_{ll'} \qquad (1.30)$$

1.2 Canonical Transformations

1.2.1 Generating Functions

Consider the transformation of variables

$$q, p \rightarrow q', p' \qquad (2.1)$$

If these two sets of variables are to describe the dynamics of the same system, then when

$$\delta \int_{t_1}^{t_2} L(q, \dot{q}) \, dt = 0 \tag{2.2}$$

we must also find

$$\delta \int_{t_1}^{t_2} L'(q', \dot{q}') \, dt = 0 \tag{2.3}$$

When this agreement is obeyed, the transformation (2.1) is said to be *canonical*.

The variational equality (2.3) will follow from (2.2) provided a function $G_1(q, q', t)$ exists such that

$$L(q, \dot{q}) = L'(q', \dot{q}') + \frac{dG_1(q, q', t)}{dt} \tag{2.4}$$

Substituting this relation into (2.2) gives

$$\delta \int L \, dt = 0 = \delta \int L'(q', \dot{q}') \, dt + \delta\{G_1[q, q'(q), t]_{t_2} - G_1[q, q'(q), t]_{t_1}\}$$

$$= \delta \int L'(q', \dot{q}') \, dt \tag{2.5}$$

We may conclude that the existence of a function $G_1(q, q', t)$ that satisfies (2.4) guarantees that the transformation (2.1) is canonical.

The specifics of this transformation are obtained by casting (2.4) in terms of the Hamiltonian of the system (deleting summational indexes):

$$\sum p\dot{q} - H(p, q) = \sum p'\dot{q}' - H'(q', p') + \frac{dG_1(q, q', t)}{dt} \tag{2.6}$$

Transposing terms gives

$$\sum p \, dq - \sum p' \, dq' + (H' - H) \, dt = dG_1(q, q', t) \tag{2.7}$$

With the left side of (2.7) written for the expansion of $dG(q, q', t)$, we may conclude that

$$\boxed{p = \frac{\partial G_1(q, q')}{\partial q}, \qquad p' = -\frac{\partial G_1(q, q')}{\partial q'}} \tag{2.8}$$

Here we have assumed that G_1 is not explicitly dependent on time. Otherwise, (2.7) further implies that

$$\frac{\partial G_1}{\partial t} = H' - H \tag{2.9}$$

Equations (2.8) indicate the manner in which $G_1(q, q')$ "generates" the transformation (2.1). For G_1 non-time dependent, (2.7) gives

$$\sum p \, dq - \sum p' \, dq' = dG_1(q, q') \tag{2.10}$$

From the development leading to this equation, we conclude that *if the differences of differential forms on the left side of (2.10) are equal to a total differential, then the transformation is canonical.*

We wish to examine more carefully the nature of the generating function $G_1(q, q')$. For a system with N degrees of freedom, (2.8) comprises $2N$ equations. Thus consider that (2.8) refers specifically to the components (q_s, p_s) and q'_s, p'_s. The left equation of (2.8) is an implicit relation for obtaining $q'_s = q'_s(q, p)$. Such inversion is possible if and only if[7]

$$\det \left| \frac{\partial^2 G_1}{\partial q_s \partial q'_r} \right| \neq 0 \tag{2.11}$$

Once having obtained $q'_s = q'_s(q, p)$, the right equation of (2.8) gives $p'_s = p'_s(q, p)$, thereby completing the transformation (2.1).

As an example of a continuous function of q, q' that does *not* generate a canonical transformation, consider

$$G_1(q, q') = f(q) + h(q')$$

where $f(q)$ and $h(q')$ are arbitrary continuous functions. This function does not satisfy (2.11) and does not give a canonical transformation. Thus, when it is stated that the relations (2.10) guarantees the transformation to be canonical, it is tacitly assumed that $G_1(q, q')$ obeys (2.11).

Exchange transformation

An instructive example of the relation (2.8) is given by the generating function

$$G_1(q, q') = -qq'$$

There results

$$p' = q, \qquad q' = -p$$

Save for a sign reversal, the roles of p and q are reversed. Accordingly, this transformation is called the *exchange transformation*.

Hamiltonian criterion

In the preceding description it was stated that, for a transformation to be canonical, Hamilton's principle must be obeyed in both coordinate frames [that is, (2.2) and (2.3)]. Equivalently, we say that the transformation $(q, p) \rightarrow (q', p')$ is canonical if and only if there exists a function $H'(q', p')$ such that the equations of motion in the new frame maintain the Hamiltonian form (1.12):

$$\dot{p}'_l = -\frac{\partial H'}{\partial q'_l}, \qquad \dot{q}'_l = \frac{\partial H'}{\partial p'_l} \tag{2.12}$$

---

[7]For further discussion, see E. C. G. Sudarshan and N. Mukunda, *Classical Dynamics: A Modern Perspective*, Wiley, New York (1974).

1.2.2   Generating Other Transformations

It is possible through Legendre transformations of (2.10) to effect generating functions that are functions of parts of "new" and "old" variables other than q, q'. Thus, for example, rewriting (2.10) as

$$\sum p\,dq - d\left(\sum p'q'\right) + \sum q'dp' = dG_1(q, q')$$

gives

$$\sum p\,dq + \sum q'dp' = d\left(G_1 + \sum p'q'\right)$$

It is evident from the differentials on the left side of this equation that the variables of the functions on the right are q, p'. Thus we may write

$$\sum p\,dq + \sum q'dp' = dG_2(q, p') \tag{2.13}$$

so that

$$\boxed{p = \frac{\partial G_2(q, p')}{\partial q}, \qquad q' = \frac{\partial G_2(q, p')}{\partial p'}} \tag{2.14}$$

where again, G_2 obeys (2.11) with respect to variables q, p'.

An instructive example of this transformation is given by the ordinary transformation of coordinates:

$$q'_l = f_l(a_1, \dots, q_N)$$

This transformation may be effected by the generating function

$$G_2(q, p') = \sum_l p'_l f_l(q) \tag{2.15}$$

With (2.14) we find

$$q'_k = \frac{\partial}{\partial p'_k} \sum_l p'_l f_l(q)$$

$$q'_k = f_k(q)$$

$$p_k = \frac{\partial}{\partial q_k} \sum_l p'_l f_l(q) = \sum_l p'_l \frac{\partial}{\partial q_k} f_l(q)$$

The second of these three equations returns (2.15), whereas the third completes the total canonical transformation:

$$(q, p) \rightarrow (q', p')$$

We return to the generating relations (2.14) in our derivation of Liouville's theorem (Section 1.3).

1.2.3 Canonical Invariants

A canonical invariant is a dynamical quantity that remains invariant under a canonical transformation. An important example of a canonical invariant is the Poisson brackets of two dynamical functions A, B. That is, if

$$A(q, p) \xrightarrow{C} A'(q', p')$$
$$B(q, p) \xrightarrow{C} B'(q', p')$$

then

$$[A, B]_{qp} \xrightarrow{C} [A', B']_{q',p'} = [A, B]_{q,p} \tag{2.16}$$

A simple proof of this statement can be constructed if we keep in mind that a canonical transformation is a change in variables related to a given system. Consider that B is the Hamiltonian of the system at hand. Then by (1.27) (for time-independent B)

$$\frac{dA}{dt} = [A, B]_{qp}$$

But the time rate of change of A can only depend on the properties of motion of the system and not on a particular choice of variables. So dA/dt is the same in all coordinate frames and (2.16) follows.

1.2.4 Group Property of Canonical Transformations

Let the canonical transformation

$$(q, p) \xrightarrow{C_1} (q', p') \tag{2.17}$$

be associated with the generating function $G(q, q')$. That is,

$$\sum p\,dq - \sum p'dq' = dG(q, q') \tag{2.18}$$

Similarly, let

$$(q', p') \xrightarrow{C_2} (q'', p'') \tag{2.19}$$

be related to the generating function $K(q', q'')$ so that

$$\sum p'dq' - \sum p''dq'' = dK(q', q'') \tag{2.20}$$

Adding dG to dK gives

$$\sum p\,dq - \sum p''dq'' = d\Phi(q, q'') \tag{2.21}$$

where we have set

$$d\Phi = d[G + K]$$

The relation (2.21) implies that the transformation

$$(q, p) \xrightarrow{C_3} (q'', p'') \tag{2.22}$$

is also canonical. Since the transformation C_3 is obtained by first effecting C_1 and *then* effecting C_2, we may say that C_3 is the *product* of C_1 and C_2. This is written as

$$C_3 = C_1 \otimes C_2 \qquad (2.23)$$

Thus we find that canonical transformations obey the *group property*; that is, the product of any two canonical transformations is itself canonical.

1.2.5   Constants of Motion and Symmetry

Constants of motion of a system may be associated with symmetries of a system. Consider a system described by the coordinates: $(q_1, q_2, \cdots, q_n)$ and suppose that $\partial H/\partial q_j = 0$. With (1.2) we conclude that $p_j = $ constant. It follows that if a system is symmetric with respect to displacements of a given coordinate, the momentum conjugate to that displacement is constant. As noted in Problem 1.2, such coordinates are called *cyclic*. Conservation principles follow from this symmetry rule.

For example, consider an aggregate of N particles. The coordinates of the system are $(\mathbf{R}, q_1', q_2' \cdots q_{3N}')$, where $\{q_i'\}$ are particle Cartesian coordinates relative to the center-of-mass and $\mathbf{R}$, $\mathbf{P}$ are coordinates and momenta of the center-of-mass, respectively. The Hamiltonian for this system is obtained from the kinetic energy form given in Problem 1.7, namely,

$$H = \frac{P^2}{2M} + \sum_i p_i'^2/2m_i + \sum_{i \neq j} V(r_{ij}') \qquad (2.24)$$

$$r_{ij}' \equiv |\mathbf{r}_i' - \mathbf{r}_j'|$$

where M is mass of the center-of-mass and $V(r_{ij}')$ is interparticle potential. Thus, with homogeneity of space, H is independent of $\mathbf{R}$ and one may conclude that the momentum of the center-of-mass is constant.

In the study of conservation of angular momentum we write the angular momentum of the system as [see (P1.9)]

$$\mathbf{J} = \mathbf{J}_{CM} + \{\mathbf{j}_1' + \mathbf{j}_2' + \cdots + \mathbf{j}_N'\} \qquad (2.25)$$

where $\mathbf{J}_{CM} = \mathbf{R} \times \mathbf{P}$ is the angular momentum of the center-of-mass and $\mathbf{j}_i'$ is the angular momentum of the ith particle relative to the center-of-mass. We note that rotation of all particles in the system about a given origin that is not coincident with the center-of-mass, is equivalent to rotation of the center-of-mass about this same origin. As space is isotropic if follows that the Hamiltonian of the system is invariant to this rotation. The related conservation rule may be obtained in the study of infinitesimal variation. Thus, consider the change in angular momentum of the center-of-mass due to infinitesimal displacement of the center-of-mass in the time interval, δt,

$$\delta \mathbf{J}_{CM} = \mathbf{R} \times M\dot{\mathbf{R}}\delta t \propto \mathbf{R} \times \delta \mathbf{R} = \mathbf{R} \times R\delta\theta \qquad (2.26)$$

where $\delta\theta$ is the angle swept out by $\delta\mathbf{R}$. The vectors $(\delta\mathbf{J}_{CM}, \mathbf{R}, R\delta\theta)$ form an orthogonal triad. Again, as space is isotropic, system dynamics are invariant to changes in θ. With (2.26), we conclude that the angular momentum of the center-of-mass is constant. (See also Problem 1.2)

Invariance of a Hamiltonian to a given variable may be examined through Hamilton's equations (1.12) or the Poisson-bracket equations (1.25). For the latter case, if H is independent of a variable u that is not explicitly time dependent, then

$$\frac{du}{dt} = [u, H] = 0$$

and $u = $ constant [see (1.27).] A similar situation occurs in quantum mechanics. If u and H commute, (the Poisson brackets become the commutator) then H is independent of u. In this event if u is not an explicit function of time, then the expectation of u is constant. (See Section 5.1.)

1.3   Liouville Theorem

1.3.1   Proof

The Liouville theorem states that the Jacobian of a canonical transformation is unity. For a system with N degrees of freedom, we write (in various notations)

$$J\left(\frac{q', p'}{q, p}\right) = \frac{\partial(q', p')}{\partial(q, p)} = \begin{vmatrix} \dfrac{\partial q_1'}{\partial q_1} & \dfrac{\partial q_2'}{\partial q_1} & \cdots & \dfrac{\partial p_1'}{\partial q_1} & \cdots & \dfrac{\partial p_N'}{\partial q_1} \\ \vdots & & & & & \vdots \\ \dfrac{\partial q_1'}{\partial p_N} & \dfrac{\partial q_2'}{\partial p_N} & \cdots & \dfrac{\partial p_1'}{\partial p_N} & \cdots & \dfrac{\partial p_N'}{\partial p_N} \end{vmatrix} = 1$$

(3.1)

This is a $2N \times 2N$ determinant. To prove the Liouville theorem, we note the following:

1. Jacobians may be treated like fractions so that

$$J = \frac{[\partial(q', p')]}{[\partial(q, p')]} \bigg/ \frac{[\partial(q, p)]}{[\partial(q, p')]}$$

(3.2)

2. When the same quantities appear in both partial derivatives, the Jacobian reduces to one in fewer variables with repeated variables taken as constant. It follows that (3.2) may be rewritten

$$J = \frac{\partial q'/\partial q|_{p'}}{\partial p/\partial p'|_q} \equiv \frac{J_n}{J_d}$$

(3.3)

The subscript p' on the "numerator" term J_n denotes that all N of the p' variables are held fixed in q differentiation. The ik element of J_n is

$$J_n^{ik} = \frac{\partial q_i'}{\partial q_k}$$

Since $(q, p) \rightarrow (q'p')$ is a canonical transformation, we may employ the generating equations (2.14) to obtain

$$J_n^{ik} = \frac{\partial^2 G_2}{\partial p_1' \partial q_k}$$

In a similar manner, the denominator term may be written

$$J_d^{ik} = \frac{\partial^2 G_2}{\partial q_i \partial p_k'}$$

So we find

$$J_n^{ik} = J_d^{ki}$$

and we may conclude that J_n differs from J_d by an interchange of rows and columns, which as may be recalled, has no effect on a determinant. Thus (3.3) gives

$$J = \frac{J_n}{J_d} = 1 \tag{3.4}$$

which establishes Liouville's theorem.

1.3.2   Geometric Significance

This theorem has an important geometrical consequence. Under an arbitrary transformation $(q, p) \rightarrow (q', p')$, a volume integral in Γ-space transforms as follows:

$$\iint_\Omega dq\, dp = \iint_{\Omega'} J\left(\frac{qp}{q'p'}\right) dq'\, dp' \tag{3.5}$$

In this expression, Ω denotes volume in phase space.

If the transformation is canonical, then (3.5) becomes

$$\iint_\Omega dq\, dp = \iint_{\Omega'} dq'\, dp' \tag{3.6}$$

We may conclude that volume elements in Γ-space are canonical invariants. See Fig. 1.7.

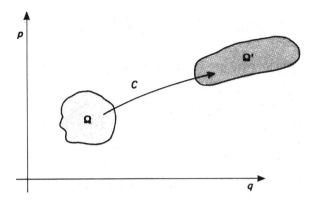

FIGURE 1.7. Volume elements in Γ-space are canonical invariants. $\Omega = \Omega'$.

1.3.3   Action Generates the Motion

We now come to an important relation that is a central point in connecting dynamics to kinetic theory.

Consider the action integral corresponding to motion in the interval from t to $t + T$.

$$S(t, T) = \int_t^{t+T} L(q, \dot{q}) \, d\tau$$

$$= \int_0^{t+T} \left[\sum p\dot{q} - H \right] d\tau - \int_0^t \left[\sum p\dot{q} - H \right] d\tau \qquad (3.7)$$

Differentiating, we obtain

$$\frac{dS}{dt} = \sum p'\dot{q}' - \sum p\dot{q} + (H - H') \qquad (3.8)$$

Here we have labeled

$$p' \equiv p(t + T), \qquad q' \equiv q(t + T) \qquad (3.9)$$

Assuming that H is time independent, (3.8) gives

$$dS = \sum p' dq' - \sum p \, dq \qquad (3.10)$$

Three important conclusions are evident from this relation:

1. The action S is a generating function for the actual physical motion in time.
2. The differential motion in time is a canonical transformation.
3. Because of the group property of canonical transformations, (2.23), the extended motion in time is a canonical transformation in time.

These properties will come into play in our first derivation of the Liouville equation to follow.

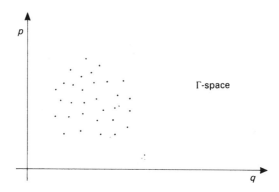

FIGURE 1.8. The ensemble comprises $\mathcal{N}$ points in Γ-space.

1.4 Liouville Equation

1.4.1 The Ensemble: Density in Phase Space

As mentioned previously, the state of a system is a single point in Γ-space. As time evolves, the system point moves on the system trajectory: $[q_1(t), \ldots, q_N(t); p_1(t), \ldots, p_N(t)]$.

Now imagine a large number of independent replicas of the same system. Such an abstract collection of identical systems is called an *ensemble*. Suppose there are $\mathcal{N}$ systems in the ensemble. Then at $t = 0$ the state of the ensemble is $\mathcal{N}$ points in Γ-space. See Fig. 1.8.

1.4.2 First Interpretation of $D(q, p, t)$

Let us introduce the function $D(q, p, t)$, which is defined as follows. The product

$$D(q, p, t)\, dq\, dp \equiv D(q, p, t)\, d\Omega \qquad (4.1a)$$

represents the number of system points in the phase volume $d\Omega$ about the point (q, p) at the time t. We may write

$$D(q, p, t) = \frac{d\mathcal{N}}{d\Omega} \qquad (4.1b)$$

An important property of the ensemble is that trajectories of the ensemble never cross in Γ-space. This follows from the fact that for a system with N degrees of freedom the system trajectory [given by (1.2)] is uniquely specified by $2N$ pieces of data: $[q(0), p(0)]$. See Fig. 1.9.

Consider that, at a given instant of time, ensemble points in a differential volume of Γ-space are contained within a continuously closed surface. As time evolves, with the property that trajectories do not cross, we conclude that points interior to the surface remain interior and that surface points remain surface points. See Fig. 1.10.

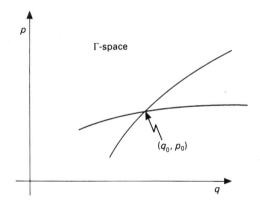

FIGURE 1.9. An impossible situation in Γ-space. Only one system trajectory passes through (q_0, p_0).

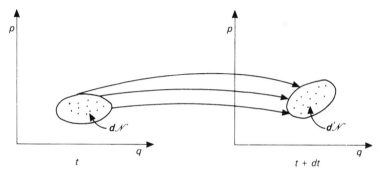

FIGURE 1.10. Interior points remain interior as time evolves. $d\mathcal{N} = d\mathcal{N}'$.

If $d\mathcal{N}$ represents the number of enclosed system points, then we may conclude that in the evolution of the system

$$d\mathcal{N} \to d\mathcal{N}' = d\mathcal{N} \qquad (4.2)$$

Let $d\Omega$ denote the volume occupied by these ensemble points. As established by (3.10), the motion of these points comprises a canonical transformation. Consequently, with the Liouville theorem (3.6), we may write

$$d\Omega \to d\Omega' = d\Omega \qquad (4.3)$$

Combining these relations, we find

$$\frac{d\mathcal{N}}{d\Omega} \to \left(\frac{d\mathcal{N}}{d\Omega}\right) = \frac{d\mathcal{N}}{d\Omega} \qquad (4.4)$$

Recalling the definition (4.1),

$$D = \frac{d\mathcal{N}}{d\Omega}$$

we may conclude that $D(q, p, t)$ *is constant on a system trajectory* or, equivalently, that

$$\frac{dD}{dt} = 0 \qquad (4.5)$$

on a system trajectory. This is the *Liouville equation*. It may be rewritten in Poisson bracket form. We have found previously (1.25) that if $u(q, p, t)$ is any dynamical variable, then

$$\frac{du}{dt} = [u, H] + \frac{\partial u}{\partial t} \qquad (4.6)$$

With (4.5) we may then write the Liouville equation in the form

$$\boxed{\frac{\partial D}{\partial t} + [D, H] = 0} \qquad (4.7)$$

1.4.3   Most General Solution: Second Interpretation of $D(q, p, t)$

We wish to establish the following important theorem: $g(q, p, t)$ is a solution of the Liouville equation if and only if g is a constant of the motion.

1. Let g be a constant of the motion. Then,

$$\frac{dg}{dt} = 0$$

But

$$\frac{dg}{dt} = \frac{\partial g}{\partial t} + [g, H]$$

Therefore,

$$\frac{\partial g}{\partial t} + [g, H] = 0$$

and g is a solution to the Liouville equation.

2. Let g be a solution to the Liouville equation. Then

$$\frac{\partial g}{\partial t} + [g, H] = 0$$

But

$$\frac{dg}{dt} = \frac{\partial g}{\partial t} + [g, H]$$

so that

$$\frac{dg}{dt} = 0$$

and we may conclude that g is a constant of the motion.

We may conclude that the *most general solution to the Liouville equation is an arbitrary function of all the constants of the motion.* We may write this most general solution in the form

$$D = D(g_1, g_2, \ldots, g_{2N}) \tag{4.8}$$

Thus, knowledge of the most general solution of the Liouville equation is equivalent to knowledge of all the constants of the motion:

$$g_1 = g_1(q, p, t)$$
$$g_2 = g_2(q, p, t)$$
$$\vdots \tag{4.9}$$
$$g_{2N} = g_{2N}(q, p, t)$$

Inverting these equations gives the dynamical orbits:

$$q_1 = q_1(g_1, \ldots, g_{2N}, t)$$
$$\vdots$$
$$q_N = q_N(g_1, \ldots, g_{2N}, t)$$
$$p_1 = p_1(g_1, \ldots, g_{2N}, t)$$
$$\vdots$$
$$p_N = p_N(g_1, \ldots, g_{2N}, t)$$

Thus we may conclude that knowledge of the most general solution to the Liouville equation is equivalent to knowledge of all the orbits of the system. This is our second interpretation of $D(q, p, t)$.

1.4.4 Incompressible Ensemble

A second derivation of the Liouville equation stems from a fluid-dynamic interpretation of ensemble flow in Γ-space.

Since system points in an ensemble are neither created nor destroyed, the rate of change of the number of system points in the volume Ω, $\int_\Omega D \, d\Omega$ is equal to the net flux of points that pass through the closed surface S that bounds Ω. Let $\mathbf{u}$ denote the velocity of system points in the neighborhood of the element of surface $d\mathbf{S}$, where, in general,

$$\mathbf{u} = (\dot{q}_1, \dot{q}_2, \ldots, \dot{q}_N; \dot{p}_1, \dot{p}_2, \ldots, \dot{p}_N) \tag{4.10}$$

Then the net flux out of the volume through the closed surface S is $\oiint \mathbf{u} \cdot D \, d\mathbf{S}$. We conclude that

$$\frac{\partial}{\partial t} \int_\Omega D \, d\Omega = -\oiint_S \mathbf{u} \cdot D \, d\mathbf{S} = -\int_\Omega \nabla \cdot \mathbf{u} D \, d\Omega$$

$$\int_{\Omega} d\Omega \left[\frac{\partial}{\partial t} D + \nabla \cdot (\mathbf{u}D) \right] = 0$$

Passing to the limit $\Omega \rightarrow 0$, together with the mean-value theorem, gives the equation

$$\frac{\partial D}{\partial t} + \nabla \cdot \mathbf{u}D = 0 \qquad (4.11)$$

Consider the term $\nabla \cdot \mathbf{u}$. With (4.10) and Hamilton's equation (1.12), we find

$$\nabla \cdot \mathbf{u} = \sum_{l=1}^{N} \left(\frac{\partial \dot{q}_l}{\partial q_l} + \frac{\partial \dot{p}_l}{\partial p_l} \right)$$

$$= \sum \left(\frac{\partial^2 H}{\partial p_l \partial q_l} - \frac{\partial^2 H}{\partial q_l \partial p_l} \right) = 0 \qquad (4.12)$$

Thus the fluid of system points is *incompressible*. Combining the latter two equations returns the Liouville equation, which now appears as

$$\frac{\partial D}{\partial t} + \mathbf{u} \cdot \nabla D = 0 \qquad (4.13)$$

Note that

$$\mathbf{u} \cdot \nabla = \sum_l \left(\dot{q}_l \frac{\partial}{\partial q_l} + \dot{p}_l \frac{\partial}{\partial p_l} \right)$$

$$= \sum_l \left(\frac{\partial H}{\partial p_l} \frac{\partial}{\partial q_i} - \frac{\partial H}{\partial q_l} \frac{\partial}{\partial p_l} \right) \qquad (4.14)$$

which renders (4.13) in the Poisson-bracket form (4.7).

1.4.5   Method of Characteristics

The property (4.8) that the most general solution to the Liouville equation is an arbitrary function of the $2N$ constants of motion may be seen in another way. In $2N + 1$ dimensional $\tilde{\Gamma}$-space, the gradient of $D(q, p, t)$ has components

$$\nabla D = \left(\frac{\partial D}{\partial t}, \frac{\partial D}{\partial q_1}, \ldots, \frac{\partial D}{\partial p_N} \right) \qquad (4.15)$$

In the notation, the Liouville equation (4.7) appears as

$$\mathbf{V} \cdot \nabla D = 0 \qquad (4.16)$$

where

$$\mathbf{V} \equiv \left(1, \frac{\partial H}{\partial p_1}, \ldots, -\frac{\partial H}{\partial q_N} \right) \qquad (4.16a)$$

The relation (4.16) says that the gradient of D is normal to the vector $\mathbf{V}$ in $\tilde{\Gamma}$-space. This will be the case if D is a function of orbits that are tangent at every point to the vector **V**. With (4.16a), we see that such orbits are given by

$$\frac{dt}{1} = \frac{dq_1}{\partial H/\partial p_1} = \frac{dq_2}{\partial H/\partial p_2} = \cdots = -\frac{dp_N}{\partial H/\partial q_N} \qquad (4.17)$$

These are evidently Hamilton's equations (1.12), whose solutions in the present context are called *characteristic curves*.

In general, the gradient of a function is normal to surfaces on which the function is constant. If D is a function of all the constants of the motion, then all orbits lie on a surface of constant D. Tangents to this surface therefore have components in $\tilde{\Gamma}$-space given by (4.17).[8]

So, once again, we find that solution to the Liouville equation is an arbitrary function of the orbits of the system or, equivalently, the constants of motion of the system.

1.4.6   Solutions to the Initial-Value Problem

In what follows, we present four approaches to solving the initial-value problem for the Liouville equation.

Taylor series expansion: Case 1

Let the initial distribution be

$$D(q, p, 0) \equiv D_0(q, p) \qquad (4.18)$$

Expanding $D(q, p, t)$ about $t = 0$ at fixed values of (q, p) gives

$$D(q, p, \Delta t) = D(q, p, 0) + \left.\frac{\partial D}{\partial t}\right|_0 \Delta t + \frac{1}{2}\left.\frac{\partial^2 D}{\partial t^2}\right|_0 (\Delta t)^2 + \cdots \qquad (4.19)$$

Employing the Liouville equation (4.7), we find

$$\frac{\partial D}{\partial t} = [H, D]$$

$$\frac{\partial^2 D}{\partial t^2} = \frac{\partial}{\partial t}[H, D] = \left[H, \frac{\partial D}{\partial t}\right]$$

$$= [H, [H, D]]$$

Inserting these relations into the expansion (4.19) gives the solution

$$D(q, p, \Delta t) = \{1 + \Delta t[H, \quad] + \frac{(\Delta t)^2}{2}[H, [H, \quad]] + \cdots\}D_0(q, p) \quad (4.20)$$

This solution may be viewed geometrically as the evolution of D in time at a point in the hyper (q, p) Γ-plane in $\tilde{\Gamma}$-space. See Fig. 1.11.

---

[8]The method of characteristics play a key role in the analysis of the wave equation. For further discussion, see R. Courant and K. O. Friedrichs, *Supersonic Flow and Shock Waves*, vol. I, Wiley-Interscience, New York (1956).

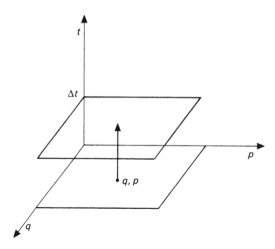

FIGURE 1.11. The series solution (4.20) gives $D(q, p, \Delta T)$ along a line of constant (q, p) in $\tilde{\Gamma}$-space.

Solution from trajectories: Case 2.

In this method of solution, it is assumed that the orbits

$$q = q_0 + \tilde{q}(t)$$
$$p = p_0 + \tilde{p}(t) \tag{4.21}$$

as well as the initial distribution (4.15) are known. The functions $\tilde{q}(t)$ and $\tilde{p}(t)$ vanish at $t = 0$, so

$$q(0) = q_0, \qquad p(0) = p_0$$

Solution in this case is derived based on the property that $D(q, p, t)$ is constant along system trajectories. Thus, we write

$$D(q, p, t) = D_0[q - \tilde{q}(t), p - \tilde{p}(t)] \tag{4.22}$$

This function has the following properties:

1. At $t = 0$,

$$D(q, p, 0) = D_0(q, p) \tag{4.22a}$$

which is the correct initial value.

2. For values of q, p on the system trajectory,

$$q - \tilde{q}(t) = q_0$$
$$p - \tilde{p}(t) = p_0 \tag{4.22b}$$

It follows that on the system trajectory in $\tilde{\Gamma}$-space

$$D(q, p, t) = D_0(q_0, p_0) = \text{constant} \tag{4.22c}$$

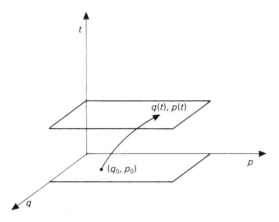

FIGURE 1.12. Graphical display of the solution (4.22) with $D(q, p, t)$ constant along a system trajectory.

We may conclude that (4.22) is the solution to the Liouville equation corresponding to the initial value (4.18). This solution may be viewed geometrically as depicted in Fig. 1.12.

1.4.7 Liouville Operator

The *Liouville operator* emerges in writing the Liouville equation in the form of the Schroedinger equation:

$$i\frac{\partial D}{\partial t} = i[H, D] \equiv \hat{\Lambda} D \tag{4.23}$$

Thus

$$\hat{\Lambda} = i[H, \quad] = i \sum \left(\frac{\partial H}{\partial q} \frac{\partial}{\partial p} - \frac{\partial H}{\partial p} \frac{\partial}{\partial q} \right) \tag{4.23a}$$

Properties of $\hat{\Lambda}$.

We wish to show that $\hat{\Lambda}$ is Hermitian in the space $\mathcal{L}_{2N}$. A function $\psi(q, p)$ is an element of $\mathcal{L}_{2N}$ if and only if the norm of ψ,

$$\|\psi\|^2 = \langle \psi \mid \psi \rangle = \int \psi^* \psi \, dq \, dp < \infty$$

An operator $\hat{\Lambda}$ is Hermitian providing

$$\hat{\Lambda} = \hat{\Lambda}^\dagger \tag{4.24}$$

where $\hat{\Lambda}^\dagger$ is the Hermitian adjoint of $\hat{\Lambda}$. The quantity $\hat{\Lambda}^\dagger$ may be defined in terms of its matrix elements in $\mathcal{L}_{2N}$. Let

$$\Lambda_{kl} = \int u_k^* \hat{\Lambda} u_l \, dq \, dp \tag{4.25}$$

where u_k and u_l are elements of a set that spans $\mathcal{L}_{2N}$. The matrix representation of (4.24) is given by

$$\Lambda_{kl} = (\Lambda_{lk})^* \tag{4.26}$$

or, equivalently,

$$\int u_k^* \hat{\Lambda} u_l \, dq \, dp = \int u_l \hat{\Lambda}^* u_k^* \, dq \, dp \tag{4.26a}$$

We consider an N-body Hamiltonian

$$H = \sum_{l=1}^{N} \frac{p_l^2}{2m} + \sum_{i<j}^{N} \Phi(\mathbf{x}_i, \mathbf{x}_j) \tag{4.27}$$

where Φ is potential of interaction between particles. With (4.23a) and (4.27), we obtain

$$\Lambda_{kl} = i \int u_k^* \sum \left(\frac{\partial H}{\partial q} \frac{\partial u_l}{\partial p} - \frac{\partial u_l}{\partial q} \frac{\partial H}{\partial p} \right) dq \, dp \tag{4.28}$$

Writing

$$u_k^* \frac{\partial u_l}{\partial p} = \frac{\partial}{\partial p}(u_k^* u_l) - u_l \frac{\partial u_k^*}{\partial p}$$

in the first term in (4.28) and

$$u_k^* \frac{\partial u_l}{\partial q} = \frac{\partial}{\partial q}(u_k^* u_l) - u_l \frac{\partial u_k^*}{\partial q}$$

in the second term and dropping surface terms gives[9]

$$
\begin{aligned}
\Lambda_{kl} &= i \int u_l \sum \left(-\frac{\partial H}{\partial q} \frac{\partial u_k^*}{\partial p} + \frac{\partial u_k^*}{\partial q} \frac{\partial H}{\partial p} \right) dq \, dp \\
&= \left[i \int u_l^* \sum \left(\frac{\partial H}{\partial q} \frac{\partial u_k}{\partial q} - \frac{\partial u_k}{\partial q} \frac{\partial H}{\partial p} \right) dq \, dp \right]^* \\
&= \Lambda_{lk}^*
\end{aligned}
\tag{4.29}
$$

We may conclude that $\hat{\Lambda}$ is a Hermitian operator. An immediate consequence of this property is that (see Problem 1.15):

1. Eigenvalues of $\hat{\Lambda}$ are real.
2. Eigenfunctions of $\hat{\Lambda}$ are orthogonal.

These properties will be employed in Section 1.5 in construction of an eigenfunction expansion for $D(q, p, t)$.

---

[9]Here we are assuming that $\partial H / \partial p$ is independent of q.

Two points should now be made concerning the preceding derivation. First, note that the property (4.24) is quite general. That is, in its derivation nothing was said of the particulars of the interaction between molecules. We merely assumed two-body conservative interactions. The second point is somewhat more pragmatic. Thus, if eigenvalues of $\hat{\Lambda}$ are real, then eigenvalues of $-i\hat{\Lambda} = [H, \quad]$ are purely imaginary. With the structure of the Liouville equation (4.7), it follows that there are solutions to this equation that oscillate in time. The reader may witness a common phenomenon that stems from this property, sound propagation. Sound waves carry pressure-density disturbances. The manner in which these fluid-dynamic variables are related to the Liouville equation is fully discussed in Chapter 2.

1.5   Eigenfunction Expansions and the Resolvent

1.5.1   Liouville Equation Integrating Factor

With the Liouville equation written as (4.20), our third (case 3) solution to the initial-value problem may be obtained as follows. Multiplying (4.20) through by the integrating factor (assuming $\hat{\Lambda}$ to be time-dependent)

$$\exp\left(i\int_0^t dt'\,\hat{\Lambda}\right)$$

permits the equation to be rewritten

$$\frac{\partial}{\partial t}\left[\left(\exp i\int_0^t dt'\,\hat{\Lambda}\right)D(t)\right] = 0$$

Integration gives the solution

$$D(q, p, t) = e^{-i\int_0^t dt'\,\hat{\Lambda}}\,D(q, p, 0) \tag{5.1}$$

For short time intervals we may write

$$\int_0^{\Delta t} dt'\,\hat{\Lambda} \simeq \Delta t\,\hat{\Lambda}$$

Expanding the exponential in (5.1) about $\Delta t = 0$ then gives

$$D(q, p, t) = \left[1 - i\Delta t\,\hat{\Lambda} + \frac{1}{2}(-i\Delta t\,\hat{\Lambda})^2 + \cdots\right]D(q, p, 0) \tag{5.2}$$

With (4.20) this expansion may be written

$$D(q, p, t) = \left\{1 + \Delta t[\hat{H}, \quad] + \frac{(\Delta t)^2}{2}[H, [H, \quad]] + \cdots\right\}D(q, p, 0) \tag{5.3}$$

This relation was previously obtained (4.20) through Taylor-series expansion of $D(q, p, t)$.

For longer time intervals, we revert to eigenproperties of $\hat{\Lambda}$ discussed in the preceding section. Again, we consider the initial-value problem with $D(q, p, 0)$ as given by (4.22a). Furthermore, it is assumed that the (orthogonal) eigenfunctions and (real) eigenvalues of $\hat{\Lambda}$ are known. Here we are taking $\hat{\Lambda}$ to be time independent. Thus we may write

$$\hat{\Lambda}\psi_n = \omega_n \psi_n \tag{5.4}$$

Assuming that the set $\{\psi_n\}$ spans the Hilbert space containing $D(q, p, 0)$, we may write

$$D_0(q, p) = \sum_{\forall n} D_n \psi_n \tag{5.5}$$

By virtue of the orthogonality of the functions ψ_n, the coefficients D_n are given by

$$D_n = \langle \psi_n \mid D_0(q, p) \rangle \tag{5.6}$$

Substituting the series (5.5) into (5.1), we obtain

$$D(q, p, t) = e^{-it\hat{\Lambda}} \sum_{\forall n} D_n \psi_n$$

which gives the solution[10]

$$\boxed{D(q, p, t) = \sum_{\forall n} D_n e^{-it\omega_n} \psi_n} \tag{5.7}$$

This completes our third method of solution for $D(q, p, t)$.

1.5.2   Example: The Ideal Gas

As an example of the application of (5.7), we will construct the solution to the Liouville equation for a collection of N noninteracting molecules (that is, an ideal gas). The gas is confined to a cubical box of edge length L and of sufficiently large size. The Hamiltonian is purely kinetic and is given by

$$H = \sum_{s=1}^{N} \frac{p_s^2}{2m}, \qquad 0 \le x_s^{(i)} \le L \tag{5.8}$$

where $x_s^{(i)}$ denotes any Cartesian component of $\mathbf{x}_s$. Let us call the $\hat{\Lambda}$ operator for this case $\hat{\Lambda}_0$. Thus

$$\hat{\Lambda}_0 = -i \sum_s \frac{\partial H}{\partial \mathbf{p}_s} \cdot \frac{\partial}{\partial \mathbf{x}_s} = -i \sum_s \mathbf{v}_s \cdot \frac{\partial}{\partial \mathbf{x}_s} \tag{5.9}$$

---

[10]Here we recall that, with $f(x)$ denoting any continuous function, $f(\hat{\Lambda})\psi_n = f(\omega_n)\psi_n$.

Here we have reverted to velocity $\mathbf{v}_s = \mathbf{p}_s/m$. Eigenfunctions of $\hat{\Lambda}_0$ satisfy the eigenvalue equation

$$\hat{\Lambda}_0 \psi_{(\mathbf{k})} = \omega_{(\mathbf{k})} \psi_{(\mathbf{k})} \tag{5.10}$$

where $(\mathbf{k})$ represents a sequence of wave-vectors:

$$(\mathbf{k}) \equiv (\mathbf{k}_1.\mathbf{k}_2, \ldots, \mathbf{k}_N) \tag{5.10a}$$

With (5.9), (5.10) becomes

$$-i \sum_s \mathbf{v}_s \cdot \frac{\partial}{\partial \mathbf{x}_s} \psi_{(\mathbf{k})} = \omega_{(\mathbf{k})} \psi_{(\mathbf{k})} \tag{5.11}$$

Substituting the trial solution

$$\psi_{(\mathbf{k})} = A \exp\left(i \sum_s \mathbf{k}_s \cdot \mathbf{x}_s \right) \tag{5.12}$$

into (5.11) gives the eigenvalues

$$\omega_{(\mathbf{k})} = \sum_{s=1}^{N} \mathbf{v}_s \cdot \mathbf{k}_s \tag{5.13}$$

Assuming that $\psi_{(\mathbf{k})}$ satisfies periodic boundary conditions gives (see Problem 1.17)

$$\mathbf{k}_s = \frac{2\pi}{L} \mathbf{n}_s \tag{5.14}$$

where the components of the vector $\mathbf{n}_s$ are integers. The constant A in (5.12) is fixed by normalization. We obtain

$$\psi_{(\mathbf{k})} = \frac{1}{L^{3N/2}} \exp\left(i \sum_s \mathbf{k}_s \cdot \mathbf{x}_s \right) \tag{5.15}$$

Substituting these eigenfunctions in the expansion (5.7) gives the explicit form:

$$D(\mathbf{x}^N, \mathbf{p}^N, t) = \sum_{(\mathbf{k})} D_{(\mathbf{k})}(\mathbf{p}^N) \psi_{(\mathbf{k})}(\mathbf{x}^N) e^{-i\omega_{(\mathbf{k})}t} \tag{5.16}$$

where $\mathbf{x}^N$ and $\mathbf{p}^N$ represent $3N$-dimensional vectors. To employ the initial data (4.22a), we examine (5.16) at the time $t = 0$.

$$D_0(\mathbf{x}^N, \mathbf{p}^N) = \sum_{(\mathbf{k})} D_{(\mathbf{k})}(\mathbf{p}^N) \psi_{(\mathbf{k})}(\mathbf{x}^N) \tag{5.17}$$

With the orthogonality of the eigenfunctions $\psi_{(\mathbf{k})}$, (5.17) gives

$$D_{(\mathbf{k})}(\mathbf{p}^N) = \frac{1}{L^{3N/2}} \int d\mathbf{x}^N \left[\exp\left(-i \sum_s \mathbf{k}_s \cdot \mathbf{x}_s \right) \right] D_0(\mathbf{x}^N, \mathbf{p}^N) \tag{5.18}$$

Having obtained the coefficients $D_{(\mathbf{k})}$, the solution (5.16) is complete. Substituting (5.13) for $\omega_{(\mathbf{k})}$ and (5.15) for $\psi_{(\mathbf{k})}$ into (5.16) permits the solution to be more explicitly written:

$$D(\mathbf{x}^N, \mathbf{p}^N, t) = \frac{1}{L^{3N/2}} \sum_{(\mathbf{k})} D_{(\mathbf{k})}(\mathbf{p}^N) e^{i \sum_s \mathbf{k}_s \cdot (\mathbf{x}_s - \mathbf{v}_s t)} \qquad (5.19)$$

Thus we find that our eigenfunction solution to the initial-value problem for the Liouville equation is a function of the $2 \times (3N)$ constants of the motion:

$$(\mathbf{p}_1, \mathbf{p}_2, \ldots, \mathbf{p}_N, \mathbf{x}_1 - \mathbf{v}_1 t, \mathbf{x}_2 - \mathbf{v}_2 t, \ldots, \mathbf{x}_N - \mathbf{v}_N t)$$

1.5.3   Free-Particle Propagator

Again consider the free-particle Hamiltonian (5.8) and related operator $\hat{\Lambda}_0$ given by (5.9). The formal solution to the initial-value problem is given by (5.1), which in the present case we write as

$$D(\mathbf{x}_1, \mathbf{x}_2, \ldots, \mathbf{v}_1, \ldots, \mathbf{v}_N, t) = e^{-it\hat{\Lambda}_0} D(\mathbf{x}_1, \ldots, \mathbf{v}_N, 0) \qquad (5.20)$$

$$-it\hat{\Lambda}_0 = -\sum_{s=1}^{N} t\mathbf{v}_s \cdot \frac{\partial}{\partial \mathbf{x}_s} \qquad (5.20a)$$

As particle variables are mutually independent, the factors in $\exp(-it\hat{\Lambda}_0)$ commute. Thus we may examine the evolution of the coordinates and momenta of particle 1.

$$D(\mathbf{x}_1, \mathbf{v}_1, t) = \exp\left(-t\mathbf{v}_1 \cdot \frac{\partial}{\partial \mathbf{x}_1}\right) D(\mathbf{x}_1, \mathbf{v}_1, 0)$$

Expanding the exponential gives

$$\exp\left(-t\mathbf{v}_1 \cdot \frac{\partial}{\partial \mathbf{x}_1}\right) D(\mathbf{x}_1, \mathbf{v}_1, 0)$$

$$= \left[1 - t\mathbf{v}_1 \cdot \frac{\partial}{\partial \mathbf{x}_1} + \frac{1}{2}\left[t\mathbf{v}_1 \cdot \frac{\partial}{\partial \mathbf{x}_1}\right]^2 + \cdots\right] D(\mathbf{x}_1, \mathbf{v}_1, 0)$$

$$= D(\mathbf{x}_1, -t\mathbf{v}_1, \mathbf{v}_1, 0) \qquad (5.21)$$

Repeating this construction for all N particles indicates that the value of $D(\mathbf{x}^N, \mathbf{v}^N, t)$ is the value it had t seconds earlier on a free-particle trajectory. Note also that

$$e^{-it\hat{\Lambda}_0} f(\mathbf{x}^N, \mathbf{p}^N, 0) = f(\mathbf{x}^N - \mathbf{v}^N t, \mathbf{p}^N, 0) = f(\mathbf{x}^N, \mathbf{p}^N, t) \qquad (5.22)$$

so that $\exp(-it\hat{\Lambda}_0)$ propagates variables backward in time.[11] For this reason, $\exp(-it\hat{\Lambda}_0)$ is called the free-particle propagator.

---

[11] This operator occurs again in Section 2.4, where we label $\exp(-it\hat{\Lambda}) \equiv \hat{\Lambda}_{-t}$.

1.5.4   The Resolvent

Our last technique of solution (case 4) to the initial-value problem for the Liouville equation stems from the Laplace transform defined by

$$\tilde{D}(s) = \int_0^\infty dt e^{-st} D(t) \qquad (5.23)$$

To obtain an equation for $\tilde{D}(s)$, we operate on the Liouville equation (4.20) as follows

$$\int_0^\infty dt e^{-st} \left(i \frac{\partial D}{\partial t} = \hat{\Lambda} D \right)$$

Substituting the expansion

$$e^{-st} \frac{\partial D}{\partial t} = \frac{\partial}{\partial t} (D e^{-st}) + s e^{-st} D$$

into the preceding equation gives

$$i \int_0^\infty \left(\frac{\partial}{\partial t} D e^{-st} \right) dt + is \int_0^\infty e^{-st} D \, dt = \hat{\Lambda} \int_0^\infty e^{-st} D \, dt$$

$$i D e^{-st} \Big|_0^\infty + is \tilde{D}(s) = \hat{\Lambda} \tilde{D}(s)$$

With Re $s > 0$, we obtain

$$- i D(0) = (\hat{\Lambda} - is) \tilde{D}(s) \qquad (5.24)$$

or, equivalently,

$$\tilde{D}(s) = -i(\hat{\Lambda} - is)^{-1} D(0)$$
$$\tilde{D}(s) = -i \hat{R} D(0) \qquad (5.25)$$

where

$$\hat{R} = (\hat{\Lambda} - is)^{-1}$$

represents the *resolvent* operator.

The inverse of (5.23) is written[12]

$$D(t) = \frac{1}{2\pi i} \int_{\gamma - i\infty}^{\gamma + i\infty} ds e^{ts} \tilde{D}(s)$$

where the line $s = \gamma$ in the complex s-plane lies to the right of all the singularities of $\hat{R}$. See Fig. 1.13.

As noted previously, since $\hat{\Lambda}$ is Hermitian, its eigenvalues are real. In this event we may conclude that the singularities of $\hat{R}$ occur for s purely imaginary.

---

[12]The Laplace transform comes into play again in discussion of Landau damping (Section 4.1.3).

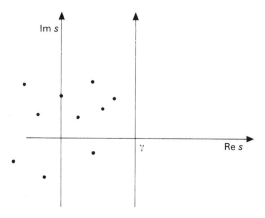

FIGURE 1.13. Path of integration in inverse Laplace transform (5.22) lies to the right of all singularities of $\hat{R}$.

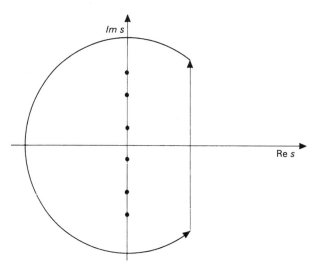

FIGURE 1.14. Closing the path of the inverse Laplace transform for singularities of $\hat{R}$ corresponding to s pure imaginary and finite in number.

This permits us to close the path of (5.22) as shown in Fig. 1.14. Here we are assuming that $\hat{R}$ has a finite number of singularities and that $\hat{R}$ is a bounded operator for all nonreal s.[13]

---

[13]The operator $\hat{R}$ is bounded in $\mathcal{L}_{2N}$, if for any element ψ of $\mathcal{L}_{2N}$ with finite norm, $\|\psi\|$, there exists a finite constant M such that $\|\hat{R}\psi\| < m\|\psi\|$.

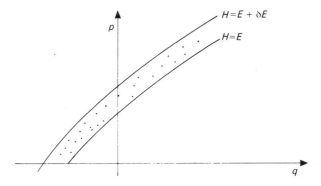

FIGURE 1.15. System points lie in the energy shell in Γ-space.

1.6 Distribution Functions

1.6.1 Third Interpretation of $D(q, p, t)$

To this point in our discourse, we have discovered two interpretations of $D(q, p, t)$: (1) A density of system points in Γ-space, and (2) a function whose general form for a given system implies a complete description of the state of the system in time.

In this section we come to yet another interpretation of $D(q, p, t)$. We will find that, apart from a multiplicative constant, it is the joint probability density for the system.

The energy shell

Consider an isolated system with N degrees of freedom with Hamiltonian

$$H(q, p) = E = \text{constant}$$

The system lies on this surface in Γ-space. Quantum mechanics prescribes an uncertainty to E, so we consider a small spread of energies about the mean value of E.[14] This generates an *energy shell* in Γ-space. System points of the ensemble lie in the energy shell, which we take to be of volume Ω. See Fig. 1.15.

As previously noted (Section 1.4), we call the number of system points in the ensemble $\mathcal{N}$ and $D(q, p, t)$ the ensemble density function. Since an individual member of the ensemble is at a given point q, p in Γ-space, and since such points may be identified with the classical state of the system, we may write

$$\mathcal{N} = \int_{\Omega} D \, dq \, dp = \text{total number of occupied states in } \Omega \qquad (6.1)$$

---

[14]Thus, for example, the smallest volume in six-dimensional phase space for which the state of a single particle can be defined is h^3, where h is Planck's constant. This concept is further discussed in Section 5.3.2.

It follows that in the volume $\Delta\Omega \subset \Omega$,

$$\Delta\mathcal{N} = \int_{\Delta\Omega} D\, dq\, dp = \text{number of occupied states in } \Delta\Omega \qquad (6.1a)$$

Dividing these two expressions, we find

$$\frac{\Delta\mathcal{N}}{\mathcal{N}} = \begin{array}{l} \text{ratio of number of states occupied in } \Delta\Omega \text{ to the} \\ \text{total number of occupied states of the ensemble} \end{array} \qquad (6.2)$$

We may equate (6.2) to the *to the probability that an unspecified state is occupied in $\Delta\Omega$.*[15] We label this entity $\int_{\Delta\Omega} f_N\, dq\, dp$ (each member of the ensemble is comprised of N particles) so that

$$\int_{\Delta\Omega} f_N\, dq\, dp = \frac{\int_{\Delta\Omega} D\, dq\, dp}{\int_{\Omega} D\, dq\, dp} \qquad (6.3)$$

Taking $\Delta\Omega$ to be infinitesimal (6.3) gives

$$f_N\, dq\, dp = \frac{D\, dq\, dp}{\int_{\Omega} D\, dq\, dp} \qquad (6.4)$$

We may conclude that

$$f_N(q, p, t) = CD(q, p, t) \qquad (6.5)$$

where C is a constant. It follows that $f_N(q, p, t)$ satisfies the Liouville equation

$$\frac{\partial f_N}{\partial t} - [f_N, H] = 0 \qquad (6.6)$$

1.6.2   Joint-Probability Distribution

What does it mean to say that $f_N\, d\Omega$ is the probability that a state is occupied in the volume $d\Omega$? We have been writing (q, p) for $(\mathbf{x}_1, \mathbf{x}_2, \ldots, \mathbf{x}_N; \mathbf{p}_1, \ldots, \mathbf{p}_N)$. So if the state (q, p) is occupied, then particle 1 is in the state $\mathbf{x}_1, \mathbf{p}_1$, particle 2 is in the $\mathbf{x}_2, \mathbf{p}_2$, and so on. It follows that $f_N(q, p, t)\, dq\, dp$ is the probability that particle 1 is in the volume $d\mathbf{x}_1\, d\mathbf{p}_1$ about the point $\mathbf{x}_1, \mathbf{p}_1$ *and* particle 2 is in the volume $d\mathbf{x}_2\, d\mathbf{p}_2$ about the point $\mathbf{x}_2, \mathbf{p}_2$, and so on, at the time t. Therefore, we may label f_N the *N-body joint-probability density* for the N-body system. Thus, whereas $D(q, p, t)$ is relevant to an ensemble of system points, $f_N(q, p, t)$ addresses a single system.

Note that with (6.4), f_N enjoys the proper probability-density normalization

$$\int_{\Omega} f_N\, dq\, dp = 1 \qquad (6.7)$$

The integration is over the entire space accessible to the system. That is, the energy shell.

---

[15]This interpretation is due to J. W. Gibbs. See *Collected Works of J. Willard Gibbs*, Dover, New York (1960).

If $G(q, p, t)$ is any dynamical variable, then with f_N a probability density we may write the following for the *expectation* or *average* of G.

$$\langle G \rangle = \int f_N G \, dq \, dp \tag{6.8}$$

The experimental meaning of $\langle G \rangle$ is given by

$$\langle G \rangle = \lim_{N \to \infty} \frac{1}{N} \sum_{i=1}^{N} G_i \tag{6.9}$$

In this expression, G_i represents the observed value of G in the ith experimental run, each run performed under identical circumstances.[16]

The last equations say that if we solve the Liouville equation for f_N and with (6.8) calculate $\langle G \rangle$, the same value will be found if we perform the experimental evaluation (6.9).

1.6.3   Reduced Distributions

Let us introduce the notation

$$d1 \equiv d\mathbf{x}_1 \, d\mathbf{p}_1, \; d2 \equiv d\mathbf{x}_2 \, d\mathbf{p}_2, \ldots$$

relevant, respectively, to particle 1, 2, ...

Consider a system of N identical particles. We direct our attention to the subsystem comprised of $s < N$ particles. The probability of finding this subsystem in the phase volume $d1 \, d2 \cdots ds$ about the state $(1, 2, \ldots, s)$ is

$$f_s(1, \ldots, s) \, d1 \cdots ds$$

To construct f_s from f_N, we must integrate out information on the state of particles $s + 1, \ldots, N$ from $f_N(1, 2, \ldots, N)$. That is,

$$\boxed{f_s(1, \ldots, s) = \int f_N(1, \ldots, N) \, d(s + 1) \cdots dN} \tag{6.10}$$

This relation is returned to in Chapter 2 in the derivation of the equations of motion for the reduced distributions.

1.6.4   Conditional Distribution

In addition to the joint-probability distribution $f_N(1, 2, \ldots, N, t)$, we also encounter conditional distributions. Thus, for example, the product

$$h_1^{(N)}(1 \mid 2, 3, \ldots, N) \, d1$$

---

[16]Phase averages, as relevant to the ergodic theorem, are further discussed in Section 3.8.1.

represents the probability that particle 1 is in the phase volume $d1$ about the phase point 1, *granted* that particles $2, \ldots, N$ are in the volume $d2 \ldots dN$ about the point $2, \ldots, N$. If particles are statistically independent, then

$$h_1^{(N)}(1 \mid 2, 3, \ldots, N) = f_1(1) \tag{6.11}$$

More generally, Bayes's formula relates $h^{(N)}$ to f_N as follows:[17]

$$h_1^{(N)}(1 \mid 2, \ldots, N) = \frac{f_N(1, \ldots, N)}{f_{N-1}(2, \ldots, N)} \tag{6.12a}$$

$$h_2^{(N)}(1, 2 \mid 3, \ldots, N) = \frac{f_N(1, \ldots, N)}{f_{N-2}(3, \ldots, N)} \tag{6.12b}$$

and so forth. These distributions are returned to later in our discussion of a Markov process.

1.6.5   s-Tuple Distribution

The s-tuple distribution, $F_s(1, \ldots, s)$, is defined as follows: The product

$$F_s(1, \ldots, s) \, d1 \cdots ds$$

represents the probable number of s-tuples of particles such that one of the particles is in the phase volume $d1$ about the point 1, another in $d2$ about 2, and so on, at a given time. Thus, for example, $F_1(1) \, d1$ is the probable number of particles in the phase element $d1$ about 1 at a specific time. The product $F_2 \, d1 \, d2$ is the probable number of pairs of particles such that in each pair one particle is in the state $d1$ about 1 and the other is in $d2$ about 2 at a specific time.

The relation between the joint-probability s-particle distribution function f_s and the s-tuple distribution function F_s is as follows: The function f_s relates to the s-particle state of a specific group of s particles. The function F_s refers to the same s-particle state but is independent of the specific particles occupying this state and independent of the manner in which these particles occupy this state. Thus the function $f_2(1, 2)$ is a property of the configuration where particle 1 occupies the state 1, and particle 2, the state 2. The function $F_2(1, 2)$ is a measure of the number of pairs of particles that occupy these states. To find the relation between F_s and f_s, we first note that the number of ways of choosing s particles from the total number N is given by

$$\binom{N}{s} = \frac{N!}{s!(N-s)!}$$

---

[17]Explicit time dependence is tacitly assumed in all distributions.

Assuming identical particles, each such choice gives the same value for f_s. Thus

$$\bar{F}_s = \binom{N}{s} f_s \tag{6.13}$$

Here we have used a bar over F_s to denote that a given s-tuple state has only been counted once in (6.13). The number of ways of distributing each subgroup of s particles while still obtaining the same s-tuple state is $s!$ This gives the desired relation

$$F_s = s! \binom{N}{s} f_s = \frac{N!}{(N-s)!} f_s \tag{6.14}$$

Thus, for example,

$$F_1 = N f_1 \tag{6.14a}$$
$$F_2 = N(N-1) f_2 \tag{6.14b}$$

Normalization is given by

$$\int F_s \, d1 \cdots ds = s! \binom{N}{s} \int f_s d1 \cdots ds = \frac{N!}{(N-s)!} \tag{6.15}$$

The average kinetic and potential energies of a collection of particles are best written in terms of the F_s functions. Thus, for example,

$$E_k = \frac{1}{2} \int F_1(\mathbf{x}, \mathbf{p}, t) \frac{p^2}{m} d\mathbf{x} \, d\mathbf{p}$$
$$E_\Phi = \frac{1}{2} \int F_2(\mathbf{x}, \mathbf{x}', \mathbf{p}, \mathbf{p}', t) \Phi(\mathbf{x}, \mathbf{x}') \, d\mathbf{x} \, d\mathbf{p} \, d\mathbf{x}' \, d\mathbf{p}' \tag{6.16}$$

where $\Phi(\mathbf{x}, \mathbf{x}')$ is the two-particle interaction potential. These relations are returned to in Chapter 2 in our proof of the conservation of total kinetic and potential energy, E_K and E_Φ, respectively, from the equations of motion for F_1 and F_2.

1.6.6   Symmetric Properties of Distributions

As the s-tuple distribution F_s does not address the state of individual particles, it must be a symmetric function of its arguments. Thus, in the relations (6.14) and following, it is assumed that f_s is properly symmetrized. Suppose f_s is not a symmetric function of its arguments; then a symmetric function is constructed as follows:

$$f_s(1, 2, \ldots, s) = \frac{1}{s!} \sum_{P(1,\ldots,s)} \tilde{f}_s(1, \ldots, s) \tag{6.17}$$

where the sum is over the permutations P of the numbers, $1 \ldots s$ and $\tilde{f}_s$ is the given asymmetric distribution.

Example

Consider the example that one molecule in an ideal gas of N identical molecules is moving with velocity $\mathbf{v}^0$ and is at the origin at time $t = 0$. The remaining molecules are stationary at given positions, $(\mathbf{x}_2^0, \ldots, \mathbf{x}_N^0)$. Let us construct a symmetric distribution f_N that describes this state of the system. From this distribution, let us then obtain f_1 and the number density $n(\mathbf{x}, t)$.

The symmetric distribution is given by

$$f_N(1, \ldots, N) = \frac{1}{N!} \sum_{P(1,\ldots,N)} \delta(\mathbf{v}_1 - \mathbf{v}^0)\delta(\mathbf{x}_1 - \mathbf{v}^0 t)$$

$$\times \, \delta(\mathbf{v}_2) \cdots \delta(\mathbf{v}_N)\delta(\mathbf{x}_2 - \mathbf{x}_2^0) \cdots \delta(\mathbf{x}_N - \mathbf{x}_N^0) \quad (6.18)$$

The permutation operator permutes the phase variables $(\mathbf{z}_1, \ldots, \mathbf{z}_N)$, where $\mathbf{z}_i \equiv (\mathbf{x}_i, \mathbf{v}_i)$. Note, in particular, that (6.18) may be written in determinant form. Setting

$$D_{i1} \equiv \delta(\mathbf{v}_i - \mathbf{v}_1^0)\delta(\mathbf{x}_i - \mathbf{v}_1^0 t)$$

$$D_{ij} \equiv \delta(\mathbf{v}_i)\delta(\mathbf{x}_i - \mathbf{x}_j^0), \qquad j \neq 1 \quad (6.19)$$

we find

$$f_N(1, \ldots, N) = \frac{1}{N!} \begin{vmatrix} D_{11} & D_{21} & D_{31} & \cdots & D_{N1} \\ D_{12} & D_{22} & D_{32} & \cdots & D_{N2} \\ D_{31} & D_{23} & D_{33} & \cdots & D_{N3} \\ \vdots & & & & \\ D_{1N} & \cdot & \cdot & \cdots & D_{NN} \end{vmatrix}_{\oplus} \quad (6.20)$$

The symbol $\oplus$ reminds us that only positive signs are employed in the expansion of the determinant. To obtain f_1, we must evaluate

$$f_1(1) = \int f_N(1, \ldots, N) \, d2 \cdots dN$$

All terms in (6.20) integrate to unity except for terms in the first column. The minor of each such term is of dimension $(N - 1)$. There results

$$f_1(1) = \frac{1}{N!}(N - 1)! \sum_{k=1}^{N} D_{1k}$$

$$f_1(1) = \frac{1}{N}[\delta(\mathbf{v}_1 - \mathbf{v}_1^0)\delta(\mathbf{x}_1 - \mathbf{v}_1^0 t) + \delta(\mathbf{v}_1)\delta(\mathbf{x}_1 - \mathbf{x}_2^0) \quad (6.21)$$

$$+ \cdots + \delta(\mathbf{v}_1)\delta(\mathbf{x}_1 - \mathbf{x}_N^0)]$$

This result is closely akin to a superposition state in quantum mechanics where the particle in question has equal probability of being in one of a number of states.[18]

The macroscopic variable number density, $n(\mathbf{x}, t)$, is such that $n(\mathbf{x}, t) \, d\mathbf{x}$ is the number of molecules in the volume $d\mathbf{x}$ about $\mathbf{x}$ at time t. It is obtained from $f_1(1)$ as follows.[19]

$$n(\mathbf{x}, t) = N \int f_1(1) \, d\mathbf{v}_1 \qquad (6.22)$$

Returning to our example, inserting $f_1(1)$ as given by (6.21) into the preceding formula gives

$$n(\mathbf{x}, t) = \delta(\mathbf{x} - \mathbf{v}_1^0 t) + \sum_{i=2}^{N} \delta(\mathbf{x} - \mathbf{x}_i^0) \qquad (6.23)$$

The number density is peaked along the trajectory $\mathbf{x} = \mathbf{v}_1^0 t$ and at the fixed sites $\mathbf{x}_i^0$. Furthermore, note that

$$\int n(\mathbf{x}, t) \, d\mathbf{x} = N \qquad (6.24)$$

which is the correct normalization for the number density.

1.7   Markov Process

1.7.1   Two-Time Distributions

Let us introduce the phase vector

$$z = (1, 2, \ldots, N)$$

In terms of this variable, the two-time, or conditional, distribution functions may be written $\prod(z, t \mid z_0, t_0)$. The product

$$\prod(z, t \mid z_0, t_0) \, dz \qquad (7.1)$$

represents the probability of finding the system in the state dz about the point z at time t, granted that it was in the state z_0 at t_0.

1.7.2   Chapman–Kolmogorov Equation

The two-time distribution has three fundamental properties:

---

[18]The superposition principle is discussed in Section 5.1.4.
[19]Macroscopic variables are more fully described in Chapter 3.

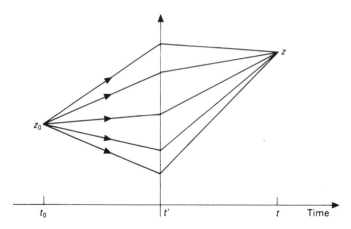

FIGURE 1.16. Summing over intermediate states in the CK equation.

1. The system undergoes a transition to some state in the interval $(t - t_0)$.

$$\int_{\forall_z \text{ all } z} \prod(z, t \mid z_0, t_0)\, dz = 1 \qquad (7.2)$$

2. The system does not change its state in zero time.

$$\prod(z, t_0 \mid z_0, t_0) = \delta(z - z_0) \qquad (7.3)$$

3. The $\prod$ function obeys the Chapman–Kolmogorov (CK) equation:

$$\prod(z, t \mid z_0, t_0) = \int \prod(z, t \mid z', t') \prod(z', t' \mid z_0, t_0)\, dz' \qquad (7.4)$$

The integral in (7.4) represents a sum over intermediate states z', as depicted in Fig. 1.16. Note that the integral in (7.4) is independent of the arbitrary intermediate time, t'.

Careful examination of (7.4) indicates that it contains an approximation. That is, the integrand function should more accurately be written

$$\prod(z, t \mid z', t'; z_0, t_0) \qquad (7.5)$$

That is, this term should include the information that the system was in the state z_0 at t_0. The approximate form of the function (7.5) as written in (7.4) is appropriate to systems with "short memories". A process so characterized is called a *Markov process*. In this sense property 3 is not an exact relation, as opposed to the precise properties 1 and 2. Note further that in the situation depicted in Fig. 1.16 it is uncertain which orbit passes from z_0 to z. In the event that $\prod(z \mid z_0)$ is equivalent to the joint-probability function (6.4), which satisfies the Liouville equation (4.7), then as we have found previously there is only one orbit through z_0. We may conclude that the CK equation is more relevant to systems governed by probabilistic laws, such as occur in random processes and quantum systems. Thus, the equivalent differential equation to (6.6), derived below for discrete processes [see (7.20)], is first applied to the random-walk problem. It is returned to in Chapter 4 in the derivation of the Fokker–Planck equation and in Chapter 5, where it is applied to quantum processes.

1.7.3 Homogeneous Processes in Time

For systems whose behavior is homogenous in time, processes can only depend on time intervals, not on initial or final times. In such cases we write

$$\prod(z_2, t_2 \mid z_1, t_1) \rightarrow \prod(z_1 \mid z_1; |t_2 - t_1|) \tag{7.6}$$

With $|t_2 - t_1| \equiv \Delta t$ and setting $t_0 = 0$, the CK equation (7.4) becomes

$$\prod(z \mid z_0; t + \Delta t) = \int \prod(z \mid z'; \Delta t) \prod(z' \mid z_0; t) \, dz' \tag{7.7}$$

1.7.4 Master Equation

Our aim at this point is to obtain a differential equation for $\prod$ appropriate to the case where phase variables go over to a countable set:

$$\{z\} \rightarrow \{l\}, \qquad l = 0, 1, 2, \ldots \tag{7.8}$$

The integral equation (7.7) then becomes

$$\prod(l \mid l_0; t + \Delta t) = \sum_{\forall j} \prod(l \mid j; \Delta t) \prod(j \mid l_0; t) \tag{7.9}$$

Conditions 1 and 2 become

$$\sum_{\forall l} \prod(l \mid l_0; t) = 1 \tag{7.10}$$

$$\prod(l \mid l_0; 0) = \delta_{ll_0} \tag{7.11}$$

For $l \neq j$, we note that

$$\prod(l \mid j; \Delta t) \rightarrow 0 \quad \text{as } \Delta t \rightarrow 0 \tag{7.12}$$

so we may write

$$\prod(l \mid j; \Delta t) = w_{jl} \Delta t + \cdots, \qquad l \neq j \tag{7.13}$$

From the property

$$\prod(l \mid l; \Delta t) \rightarrow 1 \quad \text{as } \Delta t \rightarrow 0 \tag{7.14}$$

we may write

$$\prod(l \mid l; \Delta t) = 1 - \text{(probability that there is a transition out of } l \text{ in } \Delta t) \tag{7.15}$$

Equivalently,

$$\prod(l \mid l; \Delta t) = 1 - \Delta t \sum_{l' \neq l} w_{ll'} \qquad (7.16)$$

Equations (7.13) and (7.16) may be combined to give (note that the summation now includes *all* l' values)

$$\prod(l \mid j; \Delta t) = w_{jl} \Delta t + \delta_{lj} \left(1 - \Delta t \sum_{\forall l'} w_{ll'} \right) \qquad (7.17)$$

Let us check this equation. For $j = l$,

$$\prod(l \mid l; \Delta t) = w_{ll} \Delta t + 1 - \Delta t \sum_{\forall l} w_{ll'} = 1 - \Delta t \sum_{l' \neq l} w_{ll'}$$

which is (7.17).

Substituting the key result (7.17) into the CK equation (7.9) gives

$$\prod(l \mid l_0; t + \Delta t) = \sum_{j} \left\{ w_{jl} \Delta t + \delta_{lj} \left[1 - \Delta t \sum_{\forall l'} w_{ll'} \right] \right\} \prod(j \mid l_0; t) \qquad (7.18)$$

Summing the δ_{lj} factor gives

$$\prod(l \mid l_0, t + \Delta t)$$
$$= \Delta t \sum_{j} w_{jl} \prod(j \mid l_0; t) + \prod(l \mid l_0; t) - \Delta t \sum_{\forall l'} w_{ll'} \prod(l \mid l_0; t)$$

Changing the dummy variable l' to j in the second sum gives

$$\frac{\prod(l \mid l_0; t + \Delta t) - \prod(l \mid l_0; t)}{\Delta t} = \sum_{j} \left[w_{jl} \prod(j \mid l_0; t) - w_{lj} \prod(l \mid l_0; t) \right] \qquad (7.19)$$

Passing to the limit $\Delta t \to 0$, we obtain

$$\boxed{\frac{\partial \prod(l \mid l_0; t)}{\partial t} = \sum_{j} [w_{jl} \prod(j \mid l_0; t) - w_{lj} \prod(l \mid l_0; t)]} \qquad (7.20)$$

which is a canonical form of the *master equation*. Its meaning is best revealed again through a diagram. See Fig. 1.17. The w_{jl} coefficient in (7.20) represents the probable rate at which transitions from j to l occur. The master equation in a form closely allied to (7.20) comes into play in a number of instances in our discussion of application of kinetic theory to quantum systems in Chapter 5.

1.7.5   Application to Random Walk

We wish to apply the master equation (7.20) to the random-walk problem in one dimension. For this problem, time is replaced by the total number of steps,

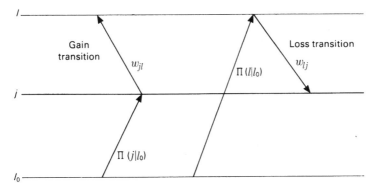

FIGURE 1.17. Geometrical description of terms in the master equation (7.20).

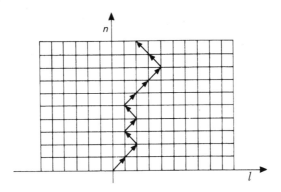

FIGURE 1.18. The random walk in one dimension (at $n = 10, l = 2$).

$n = t/\Delta t$, with $\Delta n = 1$. Furthermore, we set

$$P(l, n) \equiv \prod(l + l_0 \mid l_0; n\Delta t)$$

which denotes the probability that the particle has moved through the displacement l in n steps. (See Fig. 1.18.) Setting $\bar{w}_{jl} = w_{jl}\Delta t$ and with reference to (7.19), we write

$$P(l, n) - P(l, n - 1) = \sum_{j=-\infty}^{\infty} [\bar{w}_{jl} P(j, n - 1) - \bar{w}_{lj} P(l, n - 1)] \quad (7.21)$$

The transition probability $\bar{w}_{jn}$ has the value

$$\bar{w}_{jl} = \frac{1}{2}[\delta_{j,l+1} + \delta_{j,l-1}] \quad (7.22)$$

Inserting this form into (7.21) and summing over j gives

$$P(l, n) - P(l, n - 1) = \frac{1}{2}[P(l + 1, n - 1) + P(l - 1, n - 1)] - P(l, n - 1)$$

There results

$$P(l, n) = \frac{1}{2}[P(l+1, n-1) + P(l-1, n-1)] \qquad (7.23)$$

This equation states that step n follows step $n-1$ and that the displacement l comes from $l \pm 1$ with equal probability. We wish to establish the following key result for the random-walk problem:

$$\langle l^2 \rangle \propto n \qquad (7.24)$$

To this end, we define the moments:

$$M_1(n) = \sum_l l P(l, n) = \langle l \rangle$$
$$M_2(n) = \sum_l l^2 P(l, n) = \langle l^2 \rangle \qquad (7.25)$$

Stemming from the property $P(l, 0) = \delta_{l0}$ it is readily shown that

$$M_1(n) = 0 \qquad (7.26)$$

For M_2, with (7.23), we write

$$M_2(n) = \frac{1}{2} \sum_l [l^2 P(l-1, n-1) + l^2 P(l+1, n-1)] \qquad (7.27)$$

which may be rewritten

$$M_2(n) = \frac{1}{2} \sum_l [(l-1)^2 P(l-1, n-1) - (1 - 2l) P(l-1, n-1)$$
$$+ (l+1)^2 P(l+1, n-1) - (1 + 2l) P(l+1, n-1)]$$

Setting $k \equiv l - 1$, $\tilde{k} \equiv l + 1$ allows the preceding to be written

$$M_2(n) = \frac{1}{2} \left\{ \sum_k [k^2 P(k, n-1) + 2k P(k, n-1) + P(k, n-1)] \right.$$
$$\left. + \sum_k [\tilde{k}^2 P(\tilde{k}, n-1) - 2\tilde{k} P(\tilde{k}, n-1) + P(\tilde{k}, n-1)] \right\}$$

We note that

$$\sum_k 2k P(k, n-1) = 2M_1(n-1) = 0$$

and

$$\sum_k P(k, n-1) = 1$$

There results

$$M_2(n) = \sum_k k^2 P(k, n-1) + 1 \tag{7.28}$$

$$M_2(n) = M_2(n-1) + 1$$

Since $M_2(0) = 0$, we find

$$M_2(n) = n \tag{7.29}$$

which was to be shown. This relation, or equivalently (7.24), plays an important role in the theory of diffusion and will be returned to in our discussion of transport coefficients (Chapter 3).

1.8   Central-Limit Theorem

1.8.1   Random Variables and the Characteristic Function

In the preceding example we did not obtain an expression for the probability $P(l, n)$ relevant to the random-walk problem. That is, we did not obtain a solution to the discrete equation (7.23).

This result is readily obtained from elementary notions of probability theory, whereas the long-time behavior of $P(l, n)$ may be found with aid of the central-limit theorem. To obtain these results, we first describe some basic language of probability theory.

A *random variable* (rv) is a mapping, ξ, from a *sample space* to the real line. The sample space is a space of objects or outcomes of experiments. Thus, for example, in a coin flip the sample space consists of two results, which we may label H and T. The mapping ξ maps H and T onto, say, the values 1, 2. For a sample space comprised of elements of the alphabet, ξ maps A, B, C, ... onto, say, the numbers 1, 2, 3, ... See Fig. 1.19. These are examples of discrete rv's, in which case ξ maps the sample space onto integers. For a sample space comprised of a continuum of elements, ξ maps the sample space onto the continuous real line.

We may associate a probability with such mappings, which we write $P(\xi)$, or $P(\xi = x)$ (where x are elements of the real line) or, more simply, $P(x)$. Thus

$$\sum_{-\infty}^{\infty} P(x) = 1 \qquad \text{or} \qquad \int_{-\infty}^{\infty} dx\, P(x) = 1 \tag{8.1}$$

1.8.2   Expectation, Variance, and the Characteristic Function

The *expectation* or average of ξ is written

$$\mathcal{E}(\xi) = \langle \xi \rangle \tag{8.2a}$$

(a) Coin Flip

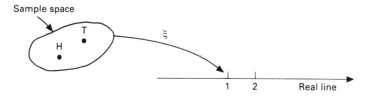

(b) Choosing letters from the alphabet.

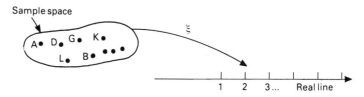

FIGURE 1.19. Examples of sample spaces.

whereas the *variance* is written

$$D(\xi) = \mathcal{E}[(\xi - \mathcal{E}(\xi))^2]$$
$$= \langle (\xi - \langle\xi\rangle)^2\rangle$$
$$= \sum_{\forall x}(x - \langle x\rangle)^2 P(x) \qquad (8.2b)$$

The *characteristic function* $\phi(a)$ of the probability function $P(x)$ is given by

$$\phi(a) = \mathcal{E}(e^{ia}) = \sum_{\forall x} P(x)e^{iax} \qquad (8.3)$$

Thus we see that the characteristic function is the Fourier transform of $P(x)$. Inverting (8.3) gives

$$P(x) = \frac{1}{2\pi} \int_{-\pi}^{\pi} da\phi(a)e^{-iax} \qquad (8.4)$$

From (8.3), we find

$$\mathcal{E}(\xi) = -i\left(\frac{d\ln\phi}{da}\right)_{a=0}$$

$$D(\xi) = -\left(\frac{d^2\ln\phi}{da^2}\right)_{a=0} \qquad (8.5)$$

Consider the expression

$$\ln\phi(a) = \ln\sum P(x)e^{iax}$$

$$\ln\phi(0) = \ln\sum P(x) = \ln 1 = 0 \qquad (8.6)$$

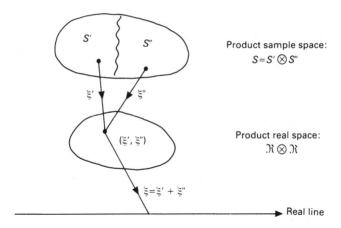

FIGURE 1.20. The random variable ξ maps pairs of elements from the sample spaces S' and S'' onto the real line.

Continuing in this manner, we obtain the series expansion

$$\ln \phi(a) = 0 + i\mathcal{E}(\xi)a - \frac{1}{2}D(\xi)a^2 + \cdots \qquad (8.7)$$

This expansion will come into play in the derivation of the central-limit theorem in Section 8.5.

1.8.3   Sums of Random Variables

Let ξ' have the probability distribution $P'(x)$ and ξ'' the distribution $P''(x)$. Let $\xi = \xi' + \xi''$. Then ξ is a valid rv because it maps pairs of elements from a sample space S onto the real line. See Fig. 1.20. Let us obtain $P(\xi = x)$. If $\xi' = k$, then $\xi = k$, providing $\xi'' = x - k$, $[\xi' + \xi''] = \xi = x]$. It follows that

$$P(x) = \sum_{\forall k} P_{\text{joint}}(\xi' = k, \xi'' = x - k) \qquad (8.8)$$

where P_{joint} represents a joint-probability distribution. If ξ' and ξ'' are statistically independent, then

$$P_{\text{joint}}(\xi', \xi'') = P'(\xi')P''(\xi'') \qquad (8.9)$$

Consider the characteristic function

$$\phi(a) = \mathcal{E}(e^{ia\xi}) = \mathcal{E}[e^{ia(\xi'+\xi'')}]$$

$$\phi(a) = \mathcal{E}[(e^{ia\xi'})(e^{ia\xi''})] \qquad (8.10)$$

If ξ' and ξ'' are statistically independent, then (8.10) reduces to

$$\phi(a) = \mathcal{E}(e^{ia\xi'})\mathcal{E}(e^{ia\xi''}) = \phi'(a)\phi''(a) \qquad (8.11)$$

So the characteristic function of a sum of statistically independent rv's is equal to the product of characteristic functions related respectively to individual rv's.

More generally, we may write, for n statistically independent rv's.

$$\phi(a) = \prod_{r=1}^{n} \phi_r(a), \qquad \xi_n = \sum_{r=1}^{n} \xi_r \tag{8.12}$$

The related probability that measurement finds the sum $\xi_n = x$ is given by (8.4).

Thus, for example, consider the following problem involving three slot machines. The first can show the numbers 1, 3, 5, and 7, the second the numbers 1, 2, and 3, and the third the numbers 0, 8, and 9. What is the probability that the sum of readings is 4 in any single turn of the machines? To answer this question, we note first that the respective probability distributions for the three machines are $P'(x) = \frac{1}{4}$, $P''(x) = \frac{1}{3}$, and $P'''(x) = \frac{1}{3}$. Thus

$$\phi'(a) = \sum_x P'(x)e^{iax} = \frac{1}{4}(e^{ia} + e^{3ia} + e^{5ia} + e^{7ia})$$

$$\phi''(a) = \sum_x P''(x)e^{iax} = \frac{1}{3}(e^{ia} + e^{2ia} + e^{3ia})$$

$$\phi'''(a) = \sum_x P'''(x)e^{iax} = \frac{1}{3}(e^{8ia} + e^{9ia} + e^{0ia})$$

The answer to our problem is then given by

$$P(4) = \frac{1}{4} \times \frac{1}{3} \times \frac{1}{3} \times \frac{1}{2\pi} \int_{-\pi}^{\pi} 2e^{0ia} da = \frac{1}{18}$$

We turn next to application of the notion of a summed rv to the random-walk problem discussed previously in Section 7.4.

1.8.4 Application to Random Walk

Each step in the random walk is considered an independent event. The rv of the rth step, ξ_r, maps the sample space (L, R) onto the numbers $-1, +1$. The probability distribution of ξ_r is

$$P(\xi_r = +1) = P(1) = \mathrm{p}$$
$$P(\xi_r = -1) = P(-1) = \mathrm{q} \tag{8.13}$$
$$P(\xi_r \neq \pm 1) = 0$$

It follows that in any single step

$$\sum_{\forall x} P(x) = \mathrm{p} + \mathrm{q} = 1 \tag{8.14}$$

Here we are assuming that the random walk is biased so that p and q need not be equal. For an unbiased random walk, $\mathrm{p} = \mathrm{q} = \frac{1}{2}$. The characteristic function for this process is

$$\phi_r(a) = \sum_{\forall x} P(x)e^{iax} = \mathrm{q}e^{-ia} + \mathrm{p}e^{ia} \tag{8.15}$$

so that

$$\mathcal{E}(\xi_r) = -i \left(\frac{d \ln \phi_r}{da} \right)_{a=0} = \mathfrak{p} - \mathfrak{q}$$

(8.16)

$$D(\xi_r) = -\left(\frac{d^2 \ln \phi_r}{da^2} \right)_{a=0} = 4\mathfrak{p}\mathfrak{q}$$

The probability that the net displacement to the right is l after n total steps is written $P(l, n)$ [as in (7.23)]. The displacement rv after n steps, ξ, is given by (8.12). As ξ_r are statistically independent rv's, and all $\phi_r(a)$ are equal, with (8.12) we write

$$\phi(a) = [\phi_r(a)]^n = [\mathfrak{p}e^{ia} + \mathfrak{q}e^{-ia}]^n$$

(8.17)

Recalling (8.4), we write

$$P(l, n) = \frac{1}{2\pi} \int_{-\pi}^{\pi} da\phi(a)e^{-ial}$$

(8.18)

From the binomial expansion, we may write

$$[\mathfrak{p}e^{ia} + \mathfrak{q}e^{-ia}]^n = \sum_m \binom{n}{m} \mathfrak{q}^m e^{-iam} \mathfrak{p}^{n-m} e^{i(n-m)a}$$

(8.19)

Inserting this result into (8.18) indicates that the integrand contains the factor $\exp ia(n - 2m - l)$ so that integration gives a nonzero result only when $m = (n - l)/2$. There results

$$P(l, n) = \frac{n!}{[(1/2)(n + l)]![(1/2)(n - l)]!} \mathfrak{p}^{(n+l)/2} \mathfrak{q}^{(n-l)/2}$$

(8.20)

The meaning of the $\mathfrak{p}$ and $\mathfrak{q}$ factors are as follows. The net displacement to the right is l. Remaining steps are $n - l$. Of these, $(n - l)/2$ must be to the right and $(n - l)/2$ must be to the left. The total number of steps to right is $l + (n - l)/2 = (n + l)/2$). For an unbiased random walk, $\mathfrak{p} = \mathfrak{q} = \frac{1}{2}$. In general, (8.20) is called the binomial distribution.

Simple substitution indicates that the distribution (8.20) satisfies (7.23) rewritten for a biased random walk:

$$P(l, n) = \mathfrak{q}P(l + 1, n - l) + \mathfrak{p}P(l - 1, n + 1)$$

(7.23a)

See Problems 1.26 and 1.27.

Poisson distribution

An important distribution relevant to many applications may be obtained directly from (8.20). Let

$$r \equiv \frac{n + l}{2}$$

denote the total number of steps to the right. Then (8.20) gives

$$P(l, n) = \frac{n!}{r!(n-r)!} p^r (1-p)^{n-r} \qquad (8.21)$$

We may interpret this form to give the probability that an event with probability p occurs r times in n trials.

Let us consider the limit $n \to \infty$, $p \to 0$, with $np \equiv \lambda \ll n$, constant. The meaning of this assumption is best illustrated by viewing the given process as one in time so that n is the total number of subintervals (say, seconds) in a total time interval. The probability p of an event occurring in one of the subintervals will decrease if the number of these subintervals is increased. It is evident, however, that in this process, the average, $\lambda = np = \langle r \rangle$, remains fixed.

With the preceding assumption, we write

$$\ln(1 - p) \simeq -p$$
$$1 - p = e^{-p}$$

Thus we obtain

$$(1 - p)^{n-r} \simeq e^{-p(n-r)} \simeq e^{-pn} = e^{-\lambda}$$

Furthermore,

$$\frac{n!}{(n-r)!} = n(n-1)\ldots[n-(r-1)]$$
$$\simeq n^r$$

Inserting these results into (8.21) gives

$$P\left(r, \frac{\lambda}{p}\right) = \frac{\lambda^r}{r!} e^{-\lambda} \equiv P(r) \qquad (8.22)$$

This distribution gives the probability of finding a total number of events r in λ/p trials, where p is the probability of an event.

The distribution (8.22) is called the *Poisson distribution*. As it stems from (8.20), it is relevant to spontaneous events with no memory, like raindrops on a roof or particles emitted by a radioactive sample or electrons by a hot cathode.

Note that (8.22) is properly normalized. That is,

$$\sum_r P(r) = e^{-\lambda} \sum_r \frac{\lambda^r}{r!} = 1$$

Furthermore,

$$\mathcal{E}(r) = \langle r \rangle = e^{-\lambda} \sum_r \frac{r\lambda^r}{r!} = \lambda$$

which agrees with our previous interpretation of λ. In addition, the variance,

$$D(r) = \langle (r - \langle r \rangle)^2 \rangle = \lambda$$

so that

$$D(r) = \mathcal{E}(r) = \lambda \qquad (8.23)$$

in a Poisson distribution.

Small shot noise

As an elementary example of these results, consider that a hot cathode emits electrons to a collecting plate in an elementary dc circuit. We label

$\langle I \rangle$ = average emission current

$\langle Q \rangle = \tau \langle I \rangle$ = total charge collected by anode in time τ

Thus the average number of electrons collected in time τ (which we identify with $\langle r \rangle$) is

$$\frac{1}{e}\langle Q \rangle = \frac{\tau}{e}\langle I \rangle = \langle r \rangle$$

$$\langle I \rangle = \frac{e}{\tau}\langle r \rangle$$

where e is electronic charge. If we assume electrons are emitted in a Poisson distribution, then with (8.23) we obtain[20]

$$\delta I = \sqrt{\frac{e}{\tau}}\sqrt{\langle I \rangle} \qquad (8.24)$$

Under ordinary circumstances,

$$\frac{\delta I}{\langle I \rangle} = \frac{\sqrt{e/\tau}}{\sqrt{\langle I \rangle}} \ll 1$$

and δI is not readily observable. However, with careful observation, Hull and Williams in 1925[21] were able to employ (8.24) to infer the value of electronic charge e. They found

$$e = \frac{\tau(\delta I)^2}{\langle I \rangle} = 1.585 \times 10^{-19} \text{ C}$$

which is seen to differ from present-day values,

$$e = 1.602 \times 10^{-19} \text{ C}$$

by only 1%.

---

[20]Noise stemming from fluctuations in the number of particles (or charges, as in the present case) is termed *shot noise*. Noise stemming from fluctuations in velocities is called *Johnson noise* (see Section 3.4.4).

[21]A. W. Hull and N. W. Williams, *Phys. Rev.* 25, 147 (1925).

1.8.5 Large n Limit: Central-Limit Theorem

The central-limit theorem addresses a sum of n statistically independent rv's, $\{\xi_r\}$, which all have the same probability distributions, $P(\xi_r)$, and same expectations, $\mathcal{E}(\xi_r)$, and same variances, $D(\xi_r)$. Furthermore, it is assumed that $D(\xi_r) < \infty$ for all r. The theorem addresses the asymptotic domain $n \to \infty$ relevant to the displacement rv and related characteristic function given by (8.12). We find that

$$\mathcal{E}(\xi_n) = -i \left(\frac{d \ln \phi}{da}\right)_{a=0} = -i \left(\frac{d \ln \prod \phi_r}{da}\right)_{a=0}$$

$$= -i \sum_{r=1}^{n} \left(\frac{d \ln \phi_r}{da}\right)_{a=0} = \sum \mathcal{E}(\xi_r)$$

Likewise,

$$D(\xi_n) = -\left(\frac{d^2 \ln \phi}{da^2}\right)_{a=0} = -\sum \left(\frac{d^2 \ln \phi_r}{da^2}\right)_{a=0} = \sum D(\xi_r)$$

It follows that if all $\mathcal{E}(\xi_r)$ are equal and all $D(\xi_r)$ are equal, as in the stated assumptions above, then

$$\mathcal{E}(\xi_n) = n\mathcal{E}(\xi_r) \tag{8.25a}$$

$$D(\xi_n) = nD(\xi_r) \tag{8.25b}$$

where ξ_r denotes any of the rv's of the process. The result (8.25a) reflects the fact that ξ_n is a sum of n statistically independent rv's. To understand (8.25b), we recall that $D(\xi_n)$ [see (8.2b)] is a measure of the deviation from the mean of ξ_n, which, we expect, should also grow with n. From the definition of the characteristic function, we write

$$P(\xi_n = l) \equiv P(l, n) = \frac{1}{2\pi} \int_{-\pi}^{\pi} da e^{-ial} \phi(a) \tag{8.26}$$

With our previous result (8.7), the preceding integral becomes

$$P(l, n) = \frac{1}{2\pi} \int_{-\pi}^{\pi} da e^{-ial} \exp\left[ia\mathcal{E}(\xi_n) - \frac{1}{2}D(\xi_n)a^2 + \cdots\right] \tag{8.27}$$

We have found [see (8.25)] that $D(\xi_n)$ increases with n under the stated assumptions. Thus, as $n \to \infty$, only the $a \simeq 0$ values contribute to the integral (8.27). In this event, the limits of integration in (8.27) may be replaced by $(-\infty, +\infty)$ without incurring gross error in the result. Setting

$$i(l - \mathcal{E}) = u$$

permits (8.27) to be rewritten

$$P(l, n) = \frac{1}{2\pi} \int_{-\infty}^{\infty} da \exp\left\{-\frac{D}{2}\left[\left(a + \frac{u}{D}\right)^2 - \left(\frac{u}{D}\right)^2\right]\right\}$$

$$= \frac{1}{2\pi} e^{u^2/2D} \sqrt{\frac{2}{D}} \int_{-\infty}^{\infty} d\lambda e^{-\lambda^2} = \frac{1}{\sqrt{2\pi D}} e^{u^2/2D}$$
(8.28)

where λ is the dummy variable as implied. Thus we obtain the key result of the central-limit theorem:

$$P(l, n) = \frac{1}{[2\pi D(\xi_n)^{1/2}]} \exp\left\{\frac{-[l - \mathcal{E}(\xi_n)]^2}{2D(\xi_n)}\right\}$$
(8.29a)

With (8.25), this result may be written

$$P(l, n) = \frac{1}{[2\pi n D(\xi)]^{1/2}} \exp\left\{\frac{-[l - n\mathcal{E}(\xi)]^2}{2n D(\xi)}\right\}$$
(8.29b)

Here we have written ξ for any of the n statistically independent rv's, $\{\xi_r\}$. The central-limit theorem is employed in the following to obtain the asymptotic displacement in the random-walk problem. It comes into play again in Chapter 3 in a derivation of the distribution for center-of-mass momentum of a fluid.

1.8.6 Random Walk in Large n Limit

The result (8.29) is the widely employed relation of the central-limit theorem. It gives the probability that the rv, ξ, has the value l after n steps (with $n \gg 1$) of the given process.

In applying this result to the random-walk process, we recall (8.16)

$$\mathcal{E}(\xi_r) = \mathfrak{p} - \mathfrak{q}$$
$$D(\xi_r) = 4\mathfrak{p}\mathfrak{q}$$

Combining these results with (8.25) and substituting into the central-limit formula (8.29), we find

$$P(l, n) \simeq \frac{1}{[8\pi \mathfrak{p}\mathfrak{q}n]^{1/2}} \exp\left\{\frac{-[l - n(\mathfrak{p} - \mathfrak{q})]^2}{8n\mathfrak{p}\mathfrak{q}}\right\}$$
(8.30)

For an unbiased random walk, we set $\mathfrak{p} = \mathfrak{q} = \frac{1}{2}$ and (8.30) becomes

$$P(l, n) \simeq \frac{1}{(2\pi n)^{1/2}} \exp\left(-\frac{l^2}{2n}\right)$$
(8.31)

The continuum limit

For large n, we may assume that the displacement l grows large compared to the unit step interval, which we label δ. In this limit, it is consistent to introduce a continuous displacement variable x in place of the discrete variable l so that $x = l\delta$. Let Δx be an interval that contains many δ intervals, but sufficiently small so that $P(l, n)$ does not change appreciably over Δx. Then the probability that the net displacement lies in the interval $(x, x + \Delta x)$ at the nth step is

$$P(x, n)\Delta x = \sum_{\forall l \in \Delta x} P(l, n) = P(l, n)\frac{\Delta x}{\delta}$$

where the integer $\Delta x/\delta$ is the number of intervals in the displacement Δx and therefore represents the number of terms in the partial sum. Substituting (8.31) into this equation gives

$$P(x, n) = \frac{1}{\delta(2\pi n)^{1/2}} \exp\left(-\frac{x^2}{2\delta^2 n}\right) \tag{8.32}$$

Associating n with the time interval of the walk, $n = t$, and setting $\delta^2 \equiv A$ gives the probability distribution

$$P(x, t) = \frac{1}{(2\pi At)^{1/2}} \exp\left(-\frac{x^2}{2At}\right) \tag{8.33}$$

The function $P(x, t)$ at any instant of time is a well-known distribution. That is, with

$$\sigma^2 \equiv At$$

(8.33) becomes

$$P(x) = \frac{1}{\sqrt{2\pi}\sigma} e^{-x^2/2\sigma^2} \tag{8.34}$$

which we recognize to be the *Gaussian* distribution. Note in particular that $P(x)$ as given by (8.34) maintains normalization.

$$\int_{-\infty}^{\infty} P(x)\,dx = 1$$

Furthermore,

$$D(x) = \langle x^2 \rangle - \langle x \rangle^2 = \langle x^2 \rangle$$

$$\langle x^2 \rangle = \int_{-\infty}^{\infty} x^2 P(x)\,dx = \sigma^2 \tag{8.35}$$

$$\langle x \rangle = 0$$

so that we may associate σ^2 with the *variance* of the Gaussian (and σ with *standard deviation*). Note that the preceding result returns the fundamental finding (7.24), now written as

$$\langle x^2 \rangle \propto t \tag{8.36}$$

This result, as well as the Gaussian distribution (8.34), will be encountered again in our discussion of diffusion (see Section 3.4.3).

Stemming from the central-limit theorem, the Gaussian distribution (8.34), as well as the Poisson distribution (8.22), is relevant to independent events with no memory.

Asymptotic values

Let us consider application of the Gaussian (8.34) to the problem of evaluating the probability that observation finds x in the domain $|x| < x_0$, where x_0 is some constant value. This probability is given by

$$P(|x| < x_0) = 2 \int_0^{x_0} P(x)\,dx = \frac{2}{\sqrt{\pi}} \int_0^{x_0/\sigma\sqrt{2}} e^{-y^2}\,dy \equiv \mathrm{erf}\,\frac{x_0}{\sigma\sqrt{2}} \quad (8.37)$$

where erf is written for the *error function*. Thus, at $x_0 = \sigma$, we find

$$P(|x| < \sigma) = \mathrm{erf}\,\frac{1}{\sqrt{2}} \simeq 0.68 \quad (8.38)$$

There is $\simeq 70\%$ probability of finding $|x| < \sigma$.

For large z, we note (see Appendix B, Section B.2)

$$\mathrm{erf}\,z \simeq 1 - \frac{e^{-z^2}}{z\sqrt{\pi}}\left(1 - \frac{1}{2z^2} + \cdots\right)$$

Thus, in the large x_0 domain, we find

$$P(|x| > x_0) = 1 - \mathrm{erf}\,\frac{x_0}{\sigma\sqrt{2}} \simeq \sqrt{\frac{2}{\pi}}\,\frac{e^{-\frac{1}{2}(x_0/\sigma)^2}}{x_0/\sigma}\left(1 - \frac{1}{(x_0/\sigma)^2} + \cdots\right) \quad (8.39)$$

This result will come into play in Section 3.3.12 in application of the central-limit theorem to a fluid.

1.8.7   Poisson and Gaussian Distributions

The Poisson distribution (8.22) is relevant to discrete events with $r = 0, 1, 2,$ $\ldots$, whereas the Gaussian (8.34) is a continuous function of x. Furthermore, the Poisson distribution gives the main value $\langle r \rangle = \lambda > 0$, whereas the Gaussian gives $\langle x \rangle = 0$. See Fig. 1.21.

We wish to illustrate a limiting transformation in which the Poisson distribution goes to the Gaussian distribution with unit variance, $\sigma = 1$. The transformation is

$$x = \frac{r - \lambda}{\sqrt{\lambda}} \quad (8.40)$$

in the limit that λ becomes large. Note that in this limit the new variable x grows continuous. Two approximations come into play in this transformation: The first is Stirling's approximation for large r,

$$r! \sim \sqrt{2\pi r}\left(\frac{r}{e}\right)^r$$

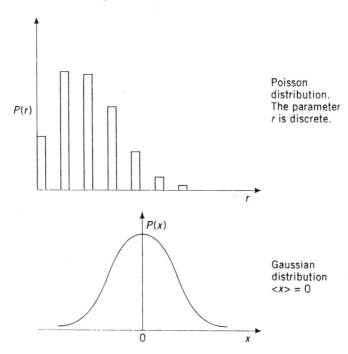

Poisson distribution. The parameter r is discrete.

Gaussian distribution $\langle x \rangle = 0$

FIGURE 1.21. Collapse of the Poisson to Gaussian distribution under the transformation $x = (r - \lambda)/\sqrt{\lambda}$, $\lambda \to \infty$.

or, equivalently,

$$\ln r! \sim \frac{1}{2}\ln(2\pi r) + r(\ln r - 1) \tag{8.41a}$$

The second approximation in the expansion

$$\ln(1 + x) = x - \frac{x^2}{2} + \cdots \tag{8.41b}$$

for $x \ll 1$.

We recall the Poisson distribution (8.22).

$$P(r) = \frac{\lambda^r e^{-\lambda}}{r!}$$

We wish to transform this distribution under (8.40):

$$r = x\sqrt{\lambda} + \lambda \tag{8.42}$$

Let us call the new distribution $P(x)$ so that

$$P(r)\,dr = P(x)\,dx$$

$$P(x) = P(r)\frac{dr}{dx} = P(r)\sqrt{\lambda} \tag{8.43}$$

$$P(x) = \frac{\lambda^{(r+\frac{1}{2})}e^{-\lambda}}{r!}$$

with $r = r(x)$, as given by (8.42). Taking the ln of (8.43) gives

$$\ln P(x) = \left(r + \frac{1}{2}\right)\ln \lambda - \lambda - \ln r! \qquad (8.44)$$

With (8.42) and (8.41), we write

$$\ln r = \ln \lambda + \frac{x}{\sqrt{\lambda}} - \frac{x^2}{2\lambda} + \cdots$$

and

$$\ln r! = \ln \sqrt{2\pi} + \left(r + \frac{1}{2}\right)\left[\ln \lambda + \frac{x}{\sqrt{\lambda}} - \frac{x^2}{2\lambda}\right] - r$$

Substituting these results into (8.44) and neglecting terms of $0(x/\sqrt{\lambda})$ gives

$$P(x) = \frac{e^{-x^2/2}}{\sqrt{2\pi}} \qquad (8.45)$$

which agrees with the Gaussian (8.34) for $\sigma = 1$.

1.8.8   Covariance and Autocorrelation Function

Another important function entering in probability theory is the *covariance* of two rv's written $\mathrm{cov}(\xi', \xi'')$. It is defined as

$$\mathrm{cov}(\xi', \xi'') = \mathcal{E}[(\xi - \mathcal{E}(\xi'))(\xi'' - \mathcal{E}(\xi''))]$$

More concisely, we may write

$$\mathrm{cov}(\xi', \xi'') = \langle(\xi' - \langle\xi'\rangle)(\xi'' - \langle\xi''\rangle)\rangle$$
$$= \langle\xi'\xi''\rangle - \langle\xi'\rangle\langle\xi''\rangle \qquad (8.46)$$

Note in particular that if ξ' and ξ'' are statistically independent or *uncorrelated* then

$$\mathrm{cov}(\xi', \xi'') = 0 \qquad (8.47)$$

The *correlation function* of two rv's is given by the dimensionless ratio

$$C(\xi', \xi'') = \frac{\mathrm{cov}(\xi', \xi'')}{\sqrt{D(\xi')D(\xi'')}} \qquad (8.48)$$

where $D(\xi)$ is the variance (8.2b).

The random variable of a *stochastic* process is a function of time. Thus, for example, in a stochastic process one writes

$$\mathcal{E}[\xi(t)] = \langle\xi(t)\rangle$$
$$D[\xi(t)] = \langle\xi^2(t)\rangle - \langle\xi(t)\rangle^2 \qquad (8.49)$$

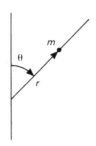

FIGURE 1.22. The coordinates for Problem 1.1.

The *autocorrelation function* of a stochastic process is the image of the covariance and we write

$$\text{cov}[\xi(t_1), \xi(t_2)] = \langle \xi(t_1)\xi(t_2) \rangle - \langle \xi(t_1) \rangle \langle \xi(t_2) \rangle \qquad (8.50)$$

If $\langle \xi(t) \rangle = 0$, we may set $t_2 = t_1 + \tau$ and the preceding becomes

$$\text{cov}[\xi(t_1), \xi(t_2)] = \langle \xi(t_1)\xi(t_1 + \tau) \rangle = \langle \xi(t)\xi(t + \tau) \rangle \qquad (8.51)$$

In the limit of no correlation between the distinct stochastic processes,

$$\langle \xi(t)\xi(t + \tau) \rangle = \langle \xi(t) \rangle \langle \xi(t + \tau) \rangle \qquad (8.52)$$

which vanishes by previous assumption.

The autocorrelation function is returned to in Section 3.4.4 concerning analysis of transport coefficients and in Section 4.2.5 in derivation of the Balescu–Lenard equation. In the quantum domain it may be reformulated in terms of the Green's function. See Problem 5.42.

Problems

1.1. A weightless rigid rod with a frictionless movable bead on it is constrained to rotate in a fixed plane with one end fixed. The plane of rotation is parallel to the gravity field. The bead has mass m.

(a) Choose a set of generalized coordinates and write down the Lagrangian for the system.

(b) What are Lagrange's equations for this system?

Answer

(a) Generalized coordinates are r, θ (see Fig. 1.22).

(b) The r-equation

$$\frac{\partial L}{\partial r} = m r \dot{\theta}^2 - mg \cos \theta$$

$$\frac{d}{dt} \left(\frac{\partial L}{\partial \dot{r}} \right) = \frac{d}{dt}(m\dot{r})$$

This equation gives

$$\frac{d}{dt}(m\dot{r}) = mr\dot{\theta}^2 - mg\cos\theta$$

The first term on the right side represents centripetal force, whereas the second term is the gravity force.

The θ equation:

$$\frac{\partial L}{\partial \theta} = mgr\sin\theta, \qquad \frac{\partial L}{\partial \dot{\theta}} = mr^2\dot{\theta}$$

$$\frac{d}{dt}(mr^2\dot{\theta}) = mgr\sin\theta$$

That is, the rate of change of angular momentum is equal to the applied torque.

1.2. **(a)** What is the Hamiltonian for a free particle expressed in spherical coordinates?

(b) What are the constant momenta of the particle in this representation?

(c) What are the constant momenta in Cartesian coordinates?

Answer

(a) The Hamiltonian is given by

$$H = \frac{p_r^2}{2m} + \frac{p_\theta^2}{2mr^2} + \frac{p_\phi^2}{2mr^2\sin^2\theta}$$

which may be rewritten

$$H = \frac{p_r^2}{2m} + \frac{L^2}{2mr^2}$$

where $\mathbf{L}$ denotes the total angular momentum of the particle.

(b) Since ϕ is a cyclic coordinate, p_ϕ is constant. With

$$\dot{\phi} = \frac{\partial H}{\partial p_\phi}$$

we find

$$p_\phi = mr^2\dot{\phi}\sin^2\theta = \text{constant}$$

Two remaining constant are H and L^2.

(c) In Cartesian coordinates

$$H = \frac{p_x^2 + p_y^2 + p_z^2}{2m}$$

Since x, y, z are cyclic,

$$\mathbf{p} = \text{constant}$$

which again comprises three independent constants.

1.3. Show that the Lorentz force law (1.18) follows from the electrodynamic Hamiltonian (1.16).

1.4. An inductively coupled double-circuit network has the Lagrangian

$$L = \frac{1}{2} \sum_{i=1}^{2} \left[L_i \dot{I}_i^2 - \frac{I_i^2}{2C_i} + \dot{V}_i(t)I_i \right] + \sum_{j \neq k}^{2} \sum^{2} M_{jk} \dot{I}_j \dot{I}_k$$

In this expression, I is current, V is voltage, L is inductance, C is capacitance, and M_{jk} is mutual inductance.

(a) What are Lagrange's equations for this system?
(b) What is the Hamiltonian for this system?

1.5. Establish (a) Jacobi's identity (1.26e), and (b) the relation (1.26g).

Answer (partial)

(b) Let $B(A)$ have the Taylor-series expansion

$$B(A) = \sum_{\forall n} b_n A^n$$

Then (1.26g) appears as

$$\sum b_n [A, A^n] = \sum b_n A [A, A^{n-1}]$$
$$= \sum b_n A^2 [A, A^{n-2}] = \cdots$$
$$= \sum b_n A^{(n-1)} [A, A] = 0$$

Here we have used the expansion (1.26d).

1.6. Show that $[q, p]$ is a canonical invariant.

Answer

$$[q, p] = \frac{\partial q}{\partial q} \frac{\partial p}{\partial p} - \frac{\partial p}{\partial q} \frac{\partial q}{\partial p} = 1$$

In the primed frame, we have

$$[q, p]_{q'p'} = \frac{\partial q}{\partial q'} \frac{\partial p}{\partial p'} - \frac{\partial p}{\partial q'} \frac{\partial q}{\partial p'} = J \left(\frac{pq}{q'p'} \right) = 1$$

which proves the statement.

1.7. Consider a collection of N mass points each of mass m. Let radii vectors to these points be written $\mathbf{r}_i$. Then the radius to the center of mass is given by

$$\mathbf{R} = \frac{\sum_i m_i \mathbf{r}_i}{\sum_i m_i} = \frac{\sum m_i \mathbf{x}_i}{M}$$

The summation runs over all N particles. Let

$$\mathbf{r}_i' \equiv \mathbf{r}_i - \mathbf{R}$$

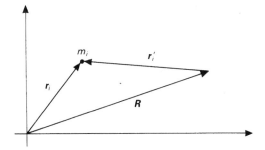

FIGURE 1.23. Coordinates relative to the center of mass.

designate the particle radius vector relative to the center of mass. See Fig. 1.23. Show that the kinetic energy of the system may be written

$$2T = M\dot{R}^2 + \sum_i m_i \dot{\mathbf{r}}_i'^2$$

1.8. A rigid weightless wheel of radius a carries a dumbbell of radius $2a$. The dumbbell rotates with its end points constrained to move on the rim of the wheel. The arm of the dumbbell is weightless, but its end points each have mass m. The system moves in three-space.

(a) How many degrees of freedom does this system have?
(b) Choose generalized coordinates for this system that include the coordinates of the center of mass, X, Y, Z.
(c) The system moves free of gravity. What is the Lagrangian of this system in the coordinates you have chosen? *Hint*: See Problem 1.7.
(d) Which momenta are conserved for this system in the coordinate frame you have chosen?

1.9. (a) With the notation of Problem 1.7, show that the total angular momentum of a system of N particles may be written

$$\mathbf{J} = \mathbf{R} \times \mathbf{P} + \sum_{i=1}^{N} \mathbf{r}_i' \times \mathbf{p}_i' \tag{P9.1}$$

In this expression

$$\mathbf{P} = M\dot{\mathbf{R}}$$

is the total linear momentum of the system, and

$$\mathbf{p}_i' = m_i \dot{\mathbf{r}}_i'$$

is the linear momentum of the ith particle relative to the center of mass.

(b) For a finite rigid body with number density $n(\mathbf{r})$, show that (P9.1) may be written

$$\mathbf{J} = \mathbf{L} + \mathbf{S}$$

where $\mathbf{L} = \mathbf{R} \times \mathbf{P}$ and

$$\mathbf{S} = \int d\mathbf{r}' n(\mathbf{r}') \mathbf{r}' \times \mathbf{p}'(\mathbf{r}')$$

1.10. Show that the Poisson bracket relation

$$[p, q] = -1$$

is invariant under the transformation generated by

$$G_2(q, p') = q \sin p'$$

Answer

We find

$$p = \sin p', \qquad q = \frac{q}{\cos p'}$$

Thus

$$[p, q] \to \left[\sin p', \frac{q'}{\cos p} \right]'$$

$$= \frac{\partial}{\partial q'} \sin p' \frac{\partial}{\partial p'} \left(\frac{q'}{\cos p'} \right) - \frac{\partial}{\partial q'} \left(\frac{q'}{\cos p'} \right) \frac{\partial}{\partial p'} (\sin p') = -1$$

1.11. Is the transformation generated by

$$G_2(q, p') = A \ln[f(q)h(p')]$$

canonical? The parameter A is constant and f and h are arbitrary but continuous functions.

1.12. Show that the transformation

$$q' = \ln \left(\frac{1}{q} \sin p \right), \qquad p' = q \cot p$$

is canonical.

1.13. Use the generating function

$$G_1 = \frac{m\omega q^2}{2} \cot q'$$

to obtain the orbits of the Hamiltonian

$$H = \frac{p^2}{2m} + \frac{kq^2}{2}$$

1.14. **(a)** Show that the equality

$$p_x dx + p_y dy + p_z dz = p_r dr + p_\theta d\theta + p_\phi d\phi$$

follows from the transformation equations from Cartesian to spherical coordinates.

(b) Argue that this transformation is canonical.

1.15. Show that if $\hat{\Lambda}$ is a Hermitian operator, then:

    (a) Eigenvalues of $\hat{\Lambda}$ are real.
    (b) Eigenfunctions of $\hat{\Lambda}$ are orthogonal.
    (c) What additional assumption comes into play in your answer to (b)?

1.16. Show that if

$$\hat{\Lambda} \equiv i\hat{L}$$

is Hermitian then:

    (a) Eigenvalues of $\hat{L}$ are purely imaginary.
    (b) Eigenstates of $\hat{L}$ are orthogonal.
    (c) $\hat{\Lambda}$ and $\hat{L}$ have common eigenstates.
    (d) What type of operator is $\hat{L}$?

1.17. Assuming that $\psi_{(k)}$ as given by (5.12) satisfies periodic boundary conditions, show that wave vectors have the values

$$\mathbf{k}_s = \frac{2\pi}{L}\mathbf{n}_s$$

where the components of n_s are integers and L^3 is the volume of cubical confinement.

1.18. Let a surface be generated by rotating a curve that passes through the points (x_1, y_1); (x_2, y_2) about the y axis. Show that the curve that makes the surface area a minimum is $y = a \cosh^{-1}(x.a) + b$.

1.19. If u and v are two dynamical functions that are both constants of the motion for a given system, show that $[u, v]$ is also a constant of the motion.

1.20. Show that the Poisson brackets of two Cartesian components of the angular momentum $\mathbf{L}$ of a particle about a specified origin obey the relation

$$[L_i, L_j] = L_k$$

where (i, j, k) are the variables (x, y, z) in cyclic order.

1.21. Constants of the motion for a free particle moving in one dimension are

$$g_1 = p$$
$$g_2 = \frac{pt}{m} - x$$

Show that

$$\frac{\partial g_2}{\partial t} + [g_2, H] = 0$$

1.22. This problem addresses *generalized potentials*, which come into play for certain nonconservative systems. Consider that the force F_s in the direction of the generalized coordinate q_s may be written

$$F_s = -\frac{\partial U}{\partial q_s} + \frac{d}{dt}\frac{\partial U}{\partial q_s}$$

where U is the generalized potential. If this condition is satisfied, then the dynamics of the system is still given in terms of Lagrange's or Hamilton's equations with

$$L = T - U$$

$$p_s = \frac{\partial L}{\partial \dot{q}_s}$$

$$H = \sum p_i \dot{q}_i - L$$

Consider a particle of mass m and charge e that moves in an electromagnetic field. The force on the particle is given by (cgs)

$$\mathbf{F} = e\left[\mathcal{E} + \frac{1}{c}(\mathbf{v} \times \mathbf{B})\right]$$

where c is the speed of light. Working with the potentials $\mathbf{A}$ and Φ given by

$$\mathcal{E} = -\nabla\Phi - \frac{1}{c}\frac{\partial \mathbf{A}}{\partial t}$$

$$\mathbf{B} = \nabla \times \mathbf{A}$$

$$\nabla \cdot \mathbf{A} + \frac{1}{c}\frac{\partial \Phi}{\partial t} = 0 \qquad \text{(Lorentz gauge)}$$

(a) Show that the generalized potential for the charged particle is

$$U = e\left(\Phi - \frac{1}{c}\mathbf{v} \cdot \mathbf{A}\right)$$

(b) Show that the canonical momentum of the particle is

$$\mathbf{p} = m\mathbf{v} + \frac{e}{c}\mathbf{A}$$

(c) Show that the Hamiltonian for the particle is given by(1.16).
(d) Show that the preceding defining equations for $\mathbf{A}$ and Φ remain invariant under the transformation $\Phi \rightarrow \Phi' = \Phi - \partial\psi/c\partial t$; $\mathbf{A} \rightarrow \mathbf{A}' + \nabla\psi$.
(e) What equations of motion for $\mathbf{A}$ and Φ, respectively, do Maxwell's equations give?

1.23. Three identical particles move in one dimension and are confined to the interval $(0, L)$. At a given instant of time, the joint-probability distribution for this system is known to be

$$f_3(1, 2, 3) = [\exp -a(p_1^2 + p_2^2 + p_3^2)]$$
$$\times [\sin bx_1 \sin bx_2 \sin bx_3 \cos c(x_1x_2 + x_2x_3 + x_1x_3)]$$

where a, b, and c are constants.

(a) Is this distribution symmetric under exchange of particles?
(b) What is the value of the constant b
(c) What is the pair distribution function $F_2(x, x')$ for this system?
(d) What is the conditional distribution function $F_2(x, x')$ for this system?

In your answer to (c) and (d), integral expressions will suffice.

1.24. A collection of four identical particles moving in one dimension are known to be in the following state at a given time $t > 0$. One particle is moving with velocity v_1^0 and another with v_2^0. Both these particles were at the origin $x = 0$ at $t = 0$. The remaining two particles are stationary at x_3^0 and x_4^0, respectively.

(a) Write down a determinantal joint-probability distribution that describes this state.

(b) Obtain an expression for $f_1(x, v)$ from your answer to (a).

1.25. A student argues the following: It has been stated that the relation

$$\sum p \, dq = \sum p' \, dq'$$

establishes the transformation $(q, p) \rightarrow (q', p')$ to be canonical. He argues that this is in violation of the criterion (2.11). Punch a hole in the student's argument.

Answer

Assuming that the given transformation stems from $G_1(q, q')$ leads to inconsistencies. For with q, q' as independent variables, the transformation gives $p = p' = 0$ On the other hand, $G_2(q, p')$ may be shown to be consistent with the given transformation. Consider, for example (for one degree of freedom),

$$G_2(q, p') = p' f(q)$$

The given relation becomes

$$p \, dq - p' \, dq' = p' \frac{df}{dq} dq - p' \, df = 0$$

which is seen to be consistent.

1.26. Show that the probability distribution (8.20), relevant to the random walk, satisfies (7.23a).

1.27. Show that the difference equation (7.23) gives the *diffusion equation*[22]

$$\frac{\partial P(x, t)}{\partial t} = D \frac{\partial^2}{\partial x^2} P(x, t) \tag{P27.1}$$

where time t and displacement x are related to the integers n and l through the interval constants δ_1 and δ_x as

$$n = \frac{t}{\delta_1}, \qquad l = \frac{x}{\delta_x}$$

and

$$D \equiv \lim_{\delta_t, \delta_x \to 0} \left[\frac{\delta_x^2}{2\delta_t} \right]$$

---

[22] Suggested by G. K. Schenter.

is assumed to be a finite parameter.

Answer

Rewrite (7.23) as

$$P(l, n) - P(l, n-1) = \frac{1}{2}[P(l+1, n-1) - 2P(l, n-1) + P(l-1, n-1)]$$
(P27.2)

Now note that

$$\Delta P(n) = P(n) - P(n-1)$$
$$\Delta[\Delta P(l)] = \Delta[P(l+1) - P(l)]$$
$$= P(l+1) - 2P(l) + P(l-1)$$

Thus, with $\Delta n = \Delta l = 1$, (P27.2) becomes

$$\frac{\Delta P}{\Delta n} = \frac{1}{2}\frac{\Delta(\Delta P/\Delta l)}{(\Delta l)^2}$$

or, equivalently,

$$\delta_t \frac{\Delta P}{\Delta t} = \frac{\delta_x^2}{2}\frac{\Delta^2 P}{\Delta x^2}$$

which, in the said limit, gives the diffusion equation (P27.1). Note in particular that, with correspondence between the difference equation (7.23) and the diffusion equation (P27.1) established, we may conclude that the diffusion equation describes a Markov process.

1.28. **(a)** What is the equation of motion in Γ space for the ensemble density function D that corresponds to the fact that system points are neither created nor destroyed?

(b) How does this equation differ from the Liouville equation?

1.29. This problem concerns the *integral invariants of Poincaré*.

(a) Consider a dynamical system with N degrees of freedom. Prove Liouville's theorem for the subsystem comprised of $s < N$ particles. That is show that

$$J\left(\frac{q', p'}{q, p}\right) = 1$$

where $(q, p) \to (q', p')$ is a canonical transformation for the s-particle subsystem.

(b) What is the geometrical significance of this statement in Γ-space?

(c) If we attempt a derivation of Liouville'e equation for the subsystem following the first derivation given in the text, where does the argument collapse?

1.30. Consider an N-body system whose Hamiltonian is given by

$$H = H_0 + \sum_{i=1}^{N} \lambda(\mathbf{x}_i - a\mathbf{p}_i)^2$$

The function H_0 is given by (4.24) and λ and a are constants.

(a) What are Hamilton's equations for this system?
(b) Is the Liouville operator (4.20a) Hermitian for this Hamiltonian?

1.31. (a) Construct a generating function $G_2(q^N, p'^N)$ that gives the transformation

$$q_1' = \alpha q_1, \qquad q_n' = q_n$$
$$p_1' = \frac{1}{\alpha} p1, \qquad p_n' = p_n$$

where $2 \leq n \leq N$.
(b) Show explicitly that the Jacobian of this transformation is unity.

1.32. Show that the Hamiltonian of an N-particle isolated system with conservative forces may be identified with the energy of the system. *Hint*: To establish this result, recall Euler's theorem which states that, if $F(x_1, x_2, \ldots, x_N)$ is a linear combination of products of x_i variables with the sum of exponents of each product equal to n, then

$$\sum_{i=1}^{N} x_i \frac{\partial F}{\partial x_i} = nF$$

and employ the Hamiltonian–Lagrangian relation.

1.33. Show that the Gaussian distribution in three dimensions

$$n(\mathbf{r}, t) = \frac{N_0}{(4\pi Dt)^{3/2}} \exp\left(-\frac{r^2}{4Dt}\right)$$

has the following properties:

(a) $\int n(\mathbf{r}, t) \, d\mathbf{r} = N_0$
(b) $\left(\frac{\partial}{\partial t} - D\nabla^2\right) n(\mathbf{r}, t) = 0$
(c) $n(\mathbf{r}, 0) = N_0 \delta(\mathbf{r})$

1.34. If $p(n)$ is a discrete probability density for the rv ξ, and $\phi(a)$ is its characteristic function, then show that

(a) $\phi(0) = 1$
(b) $\mathcal{E}(\xi) = -i \left(\dfrac{d\phi}{da}\right)_{a=0} = -i \left(\dfrac{d \ln \phi}{da}\right)_{a=0}$
(c) $|\phi(a)| \leq 1$
(d) $D(\xi) = -\left[\dfrac{d^2(\ln \phi)}{da^2}\right]_{a=0}$

1.35. Consider the Poisson distribution for the random variable r,

$$P(r) = \frac{\lambda^r e^{-\lambda}}{r!}$$

(a) Construct the characteristic function for this distribution as a sum over r.
(b) Use your answer to obtain $\mathcal{E}(r)$ and $D(r)$. (Compare your findings with values given in the text.)

Answer (partial)

(a) $\ln \phi(a) = -\lambda(1 - e^{ia})$

1.36. The faces of a six-sided cubical die are numbered 1 to 6, and the faces of a four-sided tetrahedral die are numbered 1 to 4. Consider the numbers that appear on the bottom faces of the dice in a random throw.

    (a) What are the respective characteristic functions $\phi'(a)$ and $\phi''(a)$ corresponding to throws of the individual dice?

    (b) What is the probability that the sum of numbers on the bottom faces adds to 8 in a throw of the dice?

1.37. Do the canonical transformations on a system comprise a group? Explain you answer.

1.38. Show that the discrete random-walk probability $P(l, n)$ given by (8.20) goes over to the asymptotic form (8.30) for large n.

1.39. This problem addresses classical field dynamics and the Sine–Gordon equation. Consider a field $\phi(x, t)$ with time and space derivatives denoted by ϕ_t and ϕ_x, respectively, which has the Lagrangian

$$L = \int \left[\frac{1}{2}(\phi_t^2 - k_1\phi_x^2) - k_2 \cos \phi \right] dx$$

The field coordinate and its conjugate momentum are (ϕ, Π), where

$$\Pi(x, t) = \frac{\delta L}{\delta \phi_t}$$

Variation on the right denotes functional differentiation.

    (a) What is the form of Π for the given Lagrangian?

    (b) What is the Hamiltonian for this field?

    (c) Write down Hamilton's equations for a field (in functional form).

    (d) From the relations found in part (c) obtain an equation of motion for the field $\phi(x, t)$.

Answer

(a) $\Pi(x, t) = \dfrac{\delta L}{\delta \phi_t} = \dfrac{\frac{1}{2}\int [(\phi_t + \delta\phi_t)^2 - B]dx' - \frac{1}{2}\int [(\phi_t^2 - B]dx'}{\delta \phi_t}$

where B represents constant terms in the variation. There results

$$\Pi(x, t) = \int \phi_t(x') \left[\frac{\delta \phi_t(x')}{\delta \phi_t(x)} \right] dx'$$

$$= \int \phi_t(x')\delta(x - x') dx' = \phi_t(x, t)$$

Thus

$$\Pi(x, t) = \phi_t(x, t)$$

(b) $H = \int \Pi \phi_t dx - L$

$$H = \int \left[\frac{1}{2}(\Pi^2 + k_1 \phi_x^2) + k_2 \cos \phi \right] dx$$

(c) $\dfrac{\delta H}{\delta \Pi} = \phi_t$

$\dfrac{\delta H}{\delta \phi} = -\Pi_t$

(d) We must evaluate

$$H(\phi + \delta \phi) - H(\phi)$$
$$= \int \left[\frac{1}{2}(\Pi^2 + k_1(\phi + \delta \phi)_x^2) + k_2 \cos(\phi + \delta \phi) \right] dx$$
$$- \int \left[\frac{1}{2}(\Pi^2 + k_1 \phi_x^2) + k_2 \cos \phi \right] dx$$
$$= \int (k_1 \phi_{xx} + k_2 \sin \phi) \delta \phi \, dx$$

There results

$$\frac{\delta H}{\delta \phi} = -(k_1 \phi_{xx} + k_2 \sin \phi)$$

Combining this finding with preceding results gives

$$\phi_{tt} - k_1 \phi_{xx} - k_2 \sin \phi = 0$$

which is the celebrated Sine–Gordon (nonlinear partial differential) equation. *Note:*[23] The classical field $\phi(x, t)$ may be viewed as appropriate to a system with an infinite number of degrees of freedom. Thus, for example, the Lagrangian given in the statement of this problem is related to an infinite set of freely rotating pendula coupled to each other by springs with force constant k_1. The function $\phi(x, t)$ is related to the angle of a pendulum at x, t.

1.40. Chebyshev's inequality states that, for an arbitrary probability distribution with $\mathcal{E}(x) = \bar{x}$ and $D(x) = \sigma^2$,

$$P(x - \bar{x}| > \lambda \sigma) \leq \frac{1}{\lambda^2}$$

Show that this inequality is valid for a Gaussian distribution.

1.41. A system with N degrees of freedom has coordinates $\{x_i\}$ and momenta $\{p_i\}$, $i = 1, \ldots, N$. Consider that $\{x_i\}$ is written as a column vector $\mathbf{x}$. A

---

[23] For further discussion, see P. G. Drazin, *Solitons*, Cambridge University Press, New York (1983). The phase "Sine–Gordon" was coined as a pun on the Klein–Gordon equation of relativistic quantum mechanics.

transformation to new coordinates $\bar{\mathbf{x}}$ is given by

$$\bar{\mathbf{x}} = \bar{\bar{M}}\bar{\mathbf{x}}$$

where the elements of the $N \times N$ matrix, $\bar{\bar{M}}$, are constants.

(a) If we write

$$\bar{\mathbf{p}} = \bar{\bar{Q}}\mathbf{p}$$

what is the relation between $\bar{\bar{Q}}$ and $\bar{\bar{M}}$ that ensures that the transformation is canonical?

(b) Under what conditions will the transformation be canonical if $\bar{\bar{Q}} = \bar{\bar{M}}$?

Answer

(a) For a canonical transformation, in matrix notation, we write

$$\bar{\mathbf{x}}^{\dagger} d\bar{\mathbf{p}} = \mathbf{x}^{\dagger}\, d\mathbf{p}$$

where, in general, $\bar{\bar{A}}^{\dagger}$ is written for the Hermitian adjoint of $\bar{\bar{A}}$. [Recall, $(\bar{\bar{A}}^{\dagger})_{ij} = A^{*}_{ji}$.] Inserting the given transformations into the preceding equation gives

$$(\bar{\bar{M}}\mathbf{x})^{\dagger}\, d(\bar{\bar{Q}}\mathbf{p}) = \mathbf{x}^{\dagger}\, d\mathbf{p}$$
$$\mathbf{x}^{\dagger}\bar{\bar{M}}^{\dagger}\,\bar{\bar{Q}}\, d\mathbf{p} = \mathbf{x}^{\dagger}\, d\mathbf{p}$$

which implies that

$$\bar{\bar{M}}^{\dagger}\,\bar{\bar{Q}} = \bar{\bar{I}}$$

or, equivalently,

$$\bar{\bar{Q}}^{-1} = M^{\dagger}$$

(b) If $\bar{\bar{Q}} = \bar{\bar{M}}$, the preceding condition becomes

$$\bar{\bar{M}}^{\dagger}\,\bar{\bar{M}} = 1$$

That is, $\bar{\bar{M}}$ must be unitary.

1.42. If $f_N(1, 2, \ldots, N)$ is translationally invariant,

$$f_N(\mathbf{x}_1 + \mathbf{a}, \mathbf{p}_1; \mathbf{x}_2 + \mathbf{a}, \mathbf{p}_2; \ldots; \mathbf{x}_N + \mathbf{a}, \mathbf{p}_N) = f_N(\mathbf{x}_1, \mathbf{p}_1; \mathbf{x}_2, \mathbf{p}_2; \ldots; \mathbf{x}_N, \mathbf{p}_N)$$

and rotationally invariant,

$$f_N(\mathbf{x}_1, \mathbf{p}_1; \mathbf{x}_2, \mathbf{p}_2; \ldots) = f_N(\mathbf{x}_1 + \varepsilon \times \mathbf{x}_1, \mathbf{p}_1; \mathbf{x}_2 + \varepsilon \times \mathbf{x}_2, \mathbf{p}_2, \ldots)$$

where ε is infinitesimal, then show that

(a) $f_1(\mathbf{x}_1, \mathbf{p}_1) = g(\mathbf{p}_1)$
and

(b) $f_2(\mathbf{x}_1, \mathbf{p}_1; \mathbf{x}_2, \mathbf{p}_2) = h(|\mathbf{x}_1 - \mathbf{x}_2|, \mathbf{p}_1, \mathbf{p}_2)$
The functions g and h are arbitrary.

Answer

Consider an arbitrary transformation from variables $(\mathbf{x}, \mathbf{p})$ to variables $(\bar{\mathbf{x}}, \bar{\mathbf{p}})$ such that (as in this problem)

$$f_N(\mathbf{x}_1, \ldots, \mathbf{x}_N; \mathbf{p}_1, \ldots, \mathbf{p}_N) = f_N(\bar{\mathbf{x}}_1, \ldots, \bar{\mathbf{x}}_N; \bar{\mathbf{p}}_1, \ldots, \bar{\mathbf{p}}_N)$$

Suppose further that the Jacobian of the transformation is unity, as in this problem (show this). Then we have the following (for $s \le N$);

$$f_s(1, \ldots, s) = \int d(s+1) \ldots dN f_N(1, \ldots, N)$$

$$= \int d(s+1) \ldots dN f_N(\bar{1}, \ldots, \bar{N})$$

$$= \int d\overline{(s+10} \ldots d\bar{N} f_N(\bar{1}, \ldots, \bar{N}) = f_s(\bar{1}, \ldots, \bar{s})$$

Thus the translational and rotational invariance of f_N is obeyed by reduced distributions as well, and we may write

$$f_1(\mathbf{x}_1, \mathbf{p}_1) = f_1(\mathbf{x}_1 + \mathbf{a}, \mathbf{p}_1)$$

which implies that

$$\frac{\partial f_1}{\partial \mathbf{x}_1} = 0$$

Thus

$$f_1(\mathbf{x}_1, \mathbf{p}_1) = g(\mathbf{p}_1)$$

which is property (a).
To establish (b), first we write

$$f_2(\mathbf{x}_1, \mathbf{x}_2, \mathbf{p}_1, \mathbf{p}_2) = h(\mathbf{x}_1 + \mathbf{x}_2, \mathbf{x}_2 - \mathbf{x}_2, \mathbf{p}_1, \mathbf{p}_2)$$

Translational invariance of f_2 then implies, with $\mathbf{y} \equiv \mathbf{x}_1 + \mathbf{x}_2$, that

$$h(\mathbf{y}, \mathbf{x}_1 - \mathbf{x}_2, \mathbf{p}_1, \mathbf{p}_2) = h(\mathbf{y} + 2\mathbf{a}, \mathbf{x}_1 - \mathbf{x}_2, \mathbf{p}_1, \mathbf{p}_2)$$

Thus

$$\frac{\partial h}{\partial \mathbf{y}} = 0$$

and the preceding becomes, with $\mathbf{w} \equiv \mathbf{x}_1 - \mathbf{x}_2$,

$$f_2(\mathbf{x}_1, \mathbf{x}_2, \mathbf{p}_1, \mathbf{p}_2) = (\mathbf{w}, \mathbf{p}_1, \mathbf{p}_2)$$

Let $\hat{R}$ denote the infinitesimal rotation

$$\mathbf{x} \rightarrow \mathbf{x}' = \hat{R}\mathbf{x} = \mathbf{x} + \varepsilon \times \mathbf{x}$$

Then rotational invariance of f_2 gives

$$h(\mathbf{w}, \mathbf{p}_1, \mathbf{p}_2) = h(\hat{R}\mathbf{w}, \mathbf{p}_1, \mathbf{p}_2)$$

Thus h is independent of the direction of $\mathbf{w}$, whence it can only depend on $|\mathbf{w}|$, and we conclude

$$f_2(\mathbf{x}_1, \mathbf{x}_2, \mathbf{p}_1, \mathbf{p}_2) = h(|\mathbf{x}_1 - \mathbf{x}_2|\mathbf{p}_1, \mathbf{p}_2)$$

which is property (b).

1.43. Show that for nonvelocity-dependent potentials, canonical momenta are dependent only on the kinetic energy of the system at hand. Thus, in general, canonical momentum may be considered a kinematic variable. (An example where this property does not hold is the case of a charged particle in an electromagnetic environment, for which the related "generalized potential" is velocity dependent. See Problem 1.22).

1.44. A particle of mass m moves on a two-dimensional plane surface.

(a) Transforming from Cartesian to polar coordinates, obtain expressions for new momenta (p_r, p_θ) employing the generating function G_2 given beneath (2.15). State explicitly for p_r and p_θ directly from the Lagrangian, L, written in polar coordinates. State explicitly what the arguments of G_2 are for this problem.

(b) Obtain expressions for p_r and p_θ directly from the Lagrangian, L, written in polar coordinates. State explicitly what the arguments of L are.

(c) Do your answers to the preceding change if the particle moves in a potential field? Explain.

1.45. A pendulum consists of a weightless rod of length a and a mass m at one end. The other end is fixed and the pendulum moves in a fixed plane whose normal is normal to the gravity field.

(a) Write down the Hamiltonian of the pendulum in polar coordinates (θ, p_θ) (with $\theta = 0$ corresponding to the direction of gravity).

(b) Solve for the motion $\theta(t)$ in quadrature.

(c) Sketch the orbits of the pendulum in (θ, p_θ) phase space. Show that orbits divide into sets: closed (vibration) curves and open (rotation) curves. The orbit that separates these classes of curves is called the *separatrix*.

(d) Show that the separatrix [in $\theta = (-\pi, +\pi)$] includes one *fixed point* $(p_\theta = 0)$. Such orbits are called *homoclinic* as opposed to *heteroclinic* orbits, which include more than one fixed point.

(e) Show that for motion on the separatrix it takes infinite time to reach the fixed point.

1.46. Given that $dL(q, \dot{q}, t)/dt = 0$, show that the null variation of

$$\int_{t_1}^{t_2} L \, dt \qquad \text{and} \qquad \int_{t_1}^{t_2} f(L) \, dt$$

give the same Lagrange's equations, where $f(L)$ is any continuous function of L.

1.47. **(a)** Suppose $\hat{M}$ is an arbitrary square matrix. If there exists a matrix $\hat{C}$ such that

$$\hat{\Lambda} = \hat{C}^{-1}\hat{M}\hat{C}$$

is diagonal, describe the matrices $\hat{C}$ and $\hat{\Lambda}$.

(b) If $\hat{A}$, $\hat{B}$ are two square matrices and $\hat{A}$ is nonsingular, then show that

$$\exp \hat{A}\hat{B}\hat{A}^{-1} = \hat{A}(\exp \hat{B})\hat{A}^{-1}$$

(c) Consider the N-dimensional dynamical equation

$$\dot{\mathbf{v}} = \hat{M}\mathbf{v}$$

where $\hat{M}$ is a time-dependent $N \times N$ matrix, a dot represents time differentiation, and $\mathbf{v}$ is an N-dimensional column vector. Write the solution of this equation in terms of $(\exp \hat{\Lambda}t)$ and $\mathbf{v}(0)$.

(d) If $\hat{M}$ is Hermitian, how is your answer to (c) simplified?

Answers (in part)

(a) The matrix $\hat{C}$ is comprised of columns that are eigenvectors of $\hat{M}$, whereas $\hat{\Lambda}$ is comprised of the eigenvalues of $\hat{M}$.

(c) $\mathbf{v}(t) = e^{t\hat{M}}\mathbf{v}(0)$
From (a), $\hat{M} = C\hat{\Lambda}C^{-1}$ so that

$$\mathbf{v}(t) = [\exp t(C\hat{\Lambda}C^{-1})]\mathbf{v}(0)$$
$$= [\exp(Ct\hat{\Lambda}C^{-1})]\mathbf{v}(0)$$

with (b) we may then write

$$\mathbf{v}(t) = \hat{C}e^{t\hat{\Lambda}}\hat{C}^{-1}\mathbf{v}(0)$$

(d) The preceding form involves $\hat{C}^{-1}$, which may not be trivial to calculate. However, if $\hat{M}$ is Hermitian, then $\hat{\Lambda}$ is real and $\hat{C}$ is unitary. In this event, if $\hat{C}$ is known, so is $\hat{C}^{-1} = \hat{C}^{\dagger}$.

1.48. The initial value of an ensemble density is given on the spherical surface in Γ-space, $\sum p_i^2 + \sum q_i^2 = a^2$, to be

$$D_0(q, o) = \sum q_i^2 \exp[-\sum p_i^2]$$

System points of the ensemble are comprised of N independent harmonic oscillators with potential $V = q^2/2$ and unit mass. What is the solution $D(q, p, t)$ to the Liouville equation?

1.49. A canonical set of coordinates and momenta obey the fundamental Poisson bracket relations (1.12). Expressions for the Poisson brackets of the components of angular momentum are given in Problem 1.20.

(a) Are the components of angular momentum valid canonical momenta?

(b) The energy of a rigid molecule, free to rotate about its center of mass, is given by

$$E = \frac{L_x^2 + L_y^2}{2I_1} + \frac{L_z^2}{2I_3}$$

where moments of inertia, (I_1, I_2, I_3) are evaluated in principal axis with $I_1 + I_2$ and the origin at the center of mass. Is the preceding expression a valid Hamiltonian form?

(c) Write down the proper Hamiltonian for this molecule. *Hint*: Evaluate $[L_2, L^2]$.

Answers (partial)

(b) As L_x, L_y, L_z are not canonical momenta, the given form is not a valid Hamiltonian.

(c) The appropriate Hamiltonian is given by

$$H = \frac{L^2 - L_z^2}{2I_1} + \frac{L_z^2}{2I_3}$$

CHAPTER 2

Analyses of the Liouville Equation

Introduction

This chapter begins with a derivation of the sequence of equations called the BBKGY equations (also called the *hierarchy*). These are equations for the reduced distributions encountered in Section 1.6.3 and play a key role in the kinetic theory of gases and fluids. The first two equations of this sequence are employed in the derivation of the conservation of energy of a gas of interacting particles. This description serves as a good example of basic techniques used in kinetic-theory problems. A nondimensionalization procedure is introduced in Section 2.2, which leads to the notions of strongly and weakly coupled fluids. Correlation functions are defined through the Mayer–Mayer expansion, and the Vlasov kinetic equation is found to result in the limit of weak coupling and long-range interaction. This equation leads to the notion of a self-consistent solution. The section concludes with a brief account of spatial correlation functions important to the theory of the equilibrium structure of fluids. These are the radial distribution and total correlations functions.

In the following two sections, techniques are described that attempt to develop solutions to the hierarchy through perturbation-expansion techniques. A diagrammatic representation of integrals occurs in the Prigogine technique. Summing diagrams corresponding to the rare-gas limit gives the Boltzmann equation. Diagrams relevant to a plasma are also discussed.

Temporal intervals are introduced in the Bogoliubov analysis, relevant to the equilibrium of a fluid. Expanding in powers of inverse specific volume and keeping leading terms again give rise to the Boltzmann equation.

The chapter continues with a description of the Klimontovich picture of kinetic theory. This formalism involves phase densities that are delta-function expressions of the state of the system. Moments of these phase densities are found to be related to multiparticle distribution functions. The discussion concludes with rederivation of the first equation in the hierarchy.

In the concluding section of the chapter, Grad's derivation of the Boltzmann equation from the Liouville equation is presented. It is concluded that all three techniques are given by Prigogine, Bogoliubov, and Grad contain procedures for obtaining higher-order corrections to the Boltzmann equation appropriate to denser fluids.

In Chapter 3 the very important Boltzmann equation is rederived more directly and studied in detail.

2.1   BBKGY Hierarchy

The reduced joint-probability distributions $f_s(1, \ldots, s)$, $s < N$, were introduced in Section 1.6.3. It was noted that f_1 and f_2 determine the kinetic and potential energy of an aggregate of particles. In Chapter 3 the extreme relevance of these first two distributions to all fluid dynamics will be discussed. Thus it is important to obtain equations of motion for these distributions. With the definition of f_s, as given by (1.6.10), we see that an equation of motion for f_s is obtained by integrating the Liouville equation over the phase volume $d(s + 1) \ldots dN$. Prior to so proceeding, we introduce a few key operators important to obtaining the desired equations.

2.1.1   Liouville, Kinetic Energy, and Remainder Operators

First we rewrite the Liouville equation (1.6.6)

$$\frac{\partial f_N}{\partial t} + [f_N, H] = 0 \tag{1.1}$$

in terms of a slightly modified Liouville operator

$$\frac{\partial f_N}{\partial t} - \hat{L}_N f_N = 0 \tag{1.2}$$

Comparison of $\hat{L}_N$ so defined with $\hat{\Lambda}$ given by (1.4.23) indicates that $i\hat{L}_N = \hat{\Lambda}$. With Problem 1.16, we find that $\hat{L}_N$ has purely imaginary eigenvalues corresponding to oscillatory solutions of (1.2). [Recall (1.5.7)].

With (1.1) and (1.2), we find

$$\hat{L}_N = [H, \quad] \tag{1.3}$$

which for an aggregate of N particles gives

$$\hat{L}_N = \sum_{l=1}^{N} \left(\frac{\partial H}{\partial \mathbf{x}_l} \cdot \frac{\partial}{\partial \mathbf{p}_l} - \frac{\partial H}{\partial \mathbf{p}_l} \cdot \frac{\partial}{\partial \mathbf{x}_l} \right) \tag{1.4}$$

Again, the Hamiltonian, H, is given by (1.4.27):

$$H = \sum_{i=1}^{N} \frac{p_i^2}{2m} + \sum_{i<j}^{N} \sum \Phi_{ij} \tag{1.5}$$

where

$$\Phi_{ij} \equiv \Phi(|\mathbf{x}_i - \mathbf{x}_j|) \tag{1.5a}$$

is the two-particle interaction potential. We obtain

$$\hat{L}_N = -\sum_{l=1}^{N} \frac{\mathbf{p}_l}{m} \cdot \frac{\partial}{\partial \mathbf{x}_l} + \sum_{i<j}^{N} \sum \sum_{l=1}^{N} \frac{\partial}{\partial \mathbf{x}_l} \Phi_{ij} \cdot \frac{\partial}{\partial \mathbf{p}_l} \tag{1.6}$$

Consider the operator

$$\hat{O}_{ij} \equiv \sum_{l} \frac{\partial}{\partial \mathbf{x}_l} \Phi_{ij} \cdot \frac{\partial}{\partial \mathbf{p}_l}$$

Terms in the sum are nonzero only when $l = i$ or $l = j$. Thus

$$\hat{O}_{ij} = \frac{\partial}{\partial \mathbf{x}_i} \Phi_{ij} \cdot \frac{\partial}{\partial \mathbf{p}_i} + \frac{\partial}{\partial \mathbf{x}_j} \Phi_{ij} \cdot \frac{\partial}{\partial \mathbf{p}_j}$$

With the equality

$$\frac{\partial}{\partial \mathbf{x}_i} \Phi_{ij} = -\frac{\partial}{\partial \mathbf{x}_j} \Phi_{ij}$$

we obtain

$$\hat{O}_{ij} = -\mathbf{G}_{ij} \cdot \left(\frac{\partial}{\partial \mathbf{p}_i} - \frac{\partial}{\partial \mathbf{p}_j} \right) \tag{1.7}$$

Here we have written

$$\mathbf{G}_{ij} = -\frac{\partial}{\partial \mathbf{x}_i} \Phi_{ij} \tag{1.7a}$$

for the force on the ith particle due to the jth particle. With (1.7), (1.6) may be written

$$\hat{L}_N = -\sum_{l=1}^{N} \frac{\mathbf{p}_l}{m} \cdot \frac{\partial}{\partial \mathbf{x}_l} + \sum_{i<j}^{N} \sum \hat{O}_{ij} \tag{1.8}$$

or, equivalently,

$$L_N = -\sum_{l} \hat{K}_l + \sum_{i<j} \sum \hat{O}_{ij} \tag{1.9}$$

where $\hat{K}_l$, the kinetic energy operator, is as implied. As we are interested in an equation for f_s, we partition $\hat{L}_N$ as follows:

$$\hat{L}_N = \hat{L}_s + \hat{L}_{N,s+1} \tag{1.10}$$

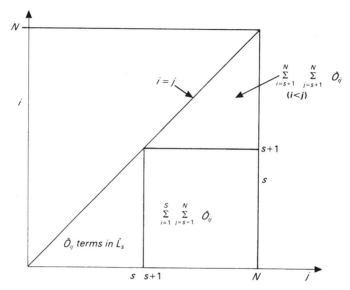

FIGURE 2.1. Enumeration of terms in the remainder operator $\hat{R}_{N,s+1}$.

where, the s-particle Liouville operator, $\hat{L}_s$, is given by

$$\hat{L}_s = -\sum_{l=1}^{s} \hat{K}_l + \sum \sum_{i<j} \hat{O}_{ij} \qquad (1.11)$$

The *remainder operator*, $\hat{L}_{N,s+1}$, is given by

$$\hat{L}_{N,s+1} = -\sum_{l=s+1}^{N} \hat{K}_l + \hat{R}_{N,s+1} \qquad (1.12)$$

To delineate the $\hat{O}_{ij}$ terms in $\hat{R}_{N,s+1}$, we refer to Fig. 2.1. Thus we find

$$\hat{R}_{N,s+1} = \sum_{i=1}^{s}\sum_{j=s+1}^{N} \hat{O}_{ij} + \sum_{\substack{i=s+1 \\ (i<j)}}^{N}\sum_{j=s+1}^{N} \hat{O}_{ij} \qquad (1.13)$$

2.1.2 Reduction of the Liouville Equation

With these operators at hand, we are prepared to carry out the integration of (1.2) as described above. Specifically with (1.10) and (1.12) substituted into the Liouville equation (1.2), integration gives

$$\left(\frac{\partial}{\partial t} - \hat{L}_s\right) f_s = \int d(s+1)\ldots dN \hat{L}_{N,s+1} f_n$$

$$= \int d(s+1)\ldots dN \left(-\sum_{l=s+1}^{N} \hat{K}_l + \hat{R}_{N,s+1}\right) f_N \quad (1.14)$$

Inserting (1.13) gives

$$
\text{RHS}(1.14) = \int d(s+1)\dots dN
$$

$$
\cdot\left(-\sum_{l=s+1}^{N}\frac{\mathbf{p}_l}{m}\cdot\frac{\partial}{\partial\mathbf{x}_l} + \sum_{i=1}^{s}\sum_{j=s+1}^{N}\hat{O}_{ij} + \sum_{i=s+1}^{N}\sum_{j=s+1}^{N}\hat{O}_{ij}\right) f_N
$$

The first and third terms in this expression contribute only surface terms and vanish. Thus we obtain

$$
\left(\frac{\partial}{\partial t} - \hat{L}_s\right) f_s = \int d(s+1)\dots dN \sum_{i=1}^{s}\sum_{j=s+1}^{N}\hat{O}_{ij} f_N \tag{1.15}
$$

Inserting the representation (1.7) for $\hat{O}_{ij}$ gives

$$
\text{RHS}(1.15) = -\int d(s+1)\dots dN \sum_{i=1}^{s}\sum_{j=s+1}^{N} \mathbf{G}_{ij}\cdot\left(\frac{\partial}{\partial\mathbf{p}_i} - \frac{\partial}{\partial\mathbf{p}_j}\right) f_N
$$

As the $\mathbf{p}_j$ derivatives run over $j \geq s+1$, they too are surface terms and vanish. We are left with

$$
\left(\frac{\partial}{\partial t} - \hat{L}_s\right) f_s = -\sum_{i=1}^{s}\frac{\partial}{\partial\mathbf{p}_i}\cdot\int d(s+1)\dots dN \sum_{j=s+1}^{N} \mathbf{G}_{ij} f_N \tag{1.16}
$$

This is a key equation in the present development for the following reasons. First, we note that up to this point the mass of particles in our aggregate, written m, many have been written with an index, such as m_i in the expression for the Hamiltonian (1.5). That is, in obtaining (1.16) it was not necessary to assume that particles are of equal mass. Second, it is apparent from (1.16) that the dynamics of the fluid partitions so that the interaction of the subgroup of s particles, described by $f_s(1,\dots,s)$, with the remaining particles in the fluid is given by the right side of (1.16).

2.1.3 Further Symmetry Reductions

To further reduce (1.16), it is necessary to assume that the constituent particles in the fluid are identical and that $f_s(1, 2, \dots, s)$ is symmetric under interchange of particle phase numbers. Thus, for example, in $f_s(1, 3, 2)$, particle 2 is in state 3 and particle 3 is in state 2. With the preceding assumptions, we may write

$$
f_3(1, 2, 3) = f_3(1, 3, 2) \tag{1.17}
$$

To examine the equivalence of the integrals under the j sum in (1.16), we must show, for example, that

$$
\int d2\,d3\dots\mathbf{G}_{12} f_N(1, 2, 3, \dots) = \int d2\,d3\dots\mathbf{G}_{13} f_N(1, 2, 3, \dots)
$$

or, equivalently, that

$$\int d2\, d3 \mathbf{G}_{12} f_3(1, 2, 3) = \int d2\, d3 \mathbf{G}_{13} f_3(1, 2, 3) \qquad (1.18)$$

Integrating the left side over $d3$ and the right side over $d2$ gives

$$\int d2 \mathbf{G}_{12} f_2(1, 2) = \int d3 \mathbf{G}_{13} f_2(1, 3)$$

Changing the dummy variable from 3 to 2 in the right integral establishes the equality. This result may also be argued geometrically, as demonstrated in Fig. 2.2. Thus, in (1.16) each term in the $(N-1)j$ summation gives identical terms and we obtain

$$\left(\frac{\partial}{\partial t} - \hat{L}_s \right) f_s + (N - s) \sum_{i=1}^{s} \frac{\partial}{\partial \mathbf{p}_i} \cdot \int \mathbf{G}_{i,s+1} f_N d(s + 1) \ldots dN = 0 \quad (1.19)$$

Here we have set j in $\mathbf{G}_{i,j}$ equal to $s+1$ to facilitate further reduction. Thus, for example, with this choice, integration of (1.19) over the volume $d(s+2)\ldots dN$ leaves f_{s+1} in the integrand and we obtain

$$\boxed{\left(\frac{\partial}{\partial t} - \hat{L}_s \right) f_s + (N - s) \sum_{i=1}^{s} \frac{\partial}{\partial \mathbf{p}_i} \cdot \int \mathbf{G}_{i,s+1} f_{s+1}\, d(s + 1) = 0} \quad (1.20)$$

$$1 \leq s \leq N$$

The coupled N equations given by (1.20) are called the BBKGY equations. This abbreviation is written for N. N. Bogoliubov, M. Born, G. Kirkwood, H. S. Green, and J. Yvon. These equations are also called the *hierarchy*. We will use the shorthand notation BY_s to denote the sth equation in the hierarchy.

We first note three basic properties of these equations:

1. They are a coupled sequence of N equations, the Nth of which is the Liouville equation for f_N.
2. Writing

$$\frac{Df_s}{Dt} \equiv \frac{\partial f_s}{\partial t} - \hat{L}_s f_s \qquad (1.21)$$

relevant to a subgroup of $s < N$ particles, with (1.20) we note that

$$\frac{Df_s}{Dt} \neq 0$$

This indicates that $f_s(1, 2, \ldots, s)$ is not constant along the subsystem trajectory in $\tilde{\Gamma}$-space. The term Df_s/Dt does not vanish due to interaction of the s subgroup of particles with the remaining aggregate. As noted previously, this interaction is represented by the sum of integrals in (1.20).

3. The third property is of special significance to the present work. It concerns the first equation in the sequence (1.20), BY_1. This equation is the generic form for all kinetic equations. A *kinetic equation* is a closed equation of

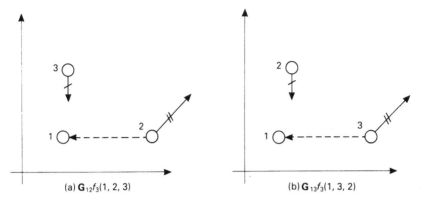

(a) $\mathbf{G}_{12}f_3(1, 2, 3)$ (b) $\mathbf{G}_{13}f_3(1, 3, 2)$

FIGURE 2.2. Graphical representation of force terms in (1.16). Arrows represent momenta; dashed arrows, interaction; and circles, position. Integrating over the phase variables of 2 and 3 gives equal results.

motion for $f_1(\mathbf{x}, \mathbf{p}, t)$. The manner in which this occurs is as follows. Let us rewrite BY_1 as

$$\frac{\partial f_1}{\partial t} + \frac{\mathbf{p}_1}{m} \cdot \frac{\partial}{\partial \mathbf{x}_1} f_1 = -\hat{A}_1 f_2(1, 2) \tag{1.22}$$

where $\hat{A}_1$ is the operator implied by (1.20) with $s = 1$. A kinetic equation results if we are able to effect the transformation

$$\hat{A}_1 f_2(1, 2) = \hat{J}(f_1) \tag{1.23}$$

where $\hat{J}$, typically called collision integral, maps functions onto functions. The simplest way to effect this mapping is merely to write $f_2(1, 2)$ as some functional of $f_1(1)$. Thus, for example in the Vlasov approximation to be considered below, we set $f_2(1, 2) = f_1(1)f_1(2)$. In the Bogoliubov ansatz (Section 2.4), we assume that $f_2(1, 2)$ may be written in the general functional form $f_s(1, 2) = f_2[1, 2, f_1]$. Both these assumptions serve to close BY_1 and produce kinetic equations. The first assumption leads to the Vlasov equation and the second leads to the Boltzmann equation.

2.1.4   Conservation of Energy from BY_1 and BY_2

We turn now to the explicit form of BY_1 and BY_2:

$$BY_1 : \frac{\partial f_1}{\partial t} + \frac{\mathbf{p}_1}{m} \cdot \frac{\partial f_1}{\partial \mathbf{x}_1} + (N - 1)\frac{\partial}{\partial \mathbf{p}_1} \cdot \int \mathbf{G}_{12} f_2(1, 2) \, d2 = 0 \tag{1.24}$$

$$BY_2 : \frac{\partial f_2}{\partial t} + \frac{1}{m} \left(\mathbf{p}_1 \cdot \frac{\partial}{\partial \mathbf{x}_1} + \mathbf{p}_2 \cdot \frac{\partial}{\partial \mathbf{x}_2} \right) f_2 + \mathbf{G}_{12} \cdot \left(\frac{\partial}{\partial \mathbf{p}_1} - \frac{\partial}{\partial \mathbf{p}_2} \right) f_2$$

$$+ (N - 2) \left\{ \frac{\partial}{\partial \mathbf{p}_1} \cdot \int \mathbf{G}_{13} f_3 \, d3 + \frac{\partial}{\partial \mathbf{p}_2} \cdot \int \mathbf{G}_{23} f_3 \, d3 \right\} = 0 \tag{1.25}$$

We wish to apply these equations to establish conservation of energy for an isolated collection of N interacting particles.

Equations (1.6.16) relate kinetic energy density E_K and potential energy density E_Φ, respectively, to the single and pair distributions F_1 and F_2 [see (1.6.14)]. In the present discussion it is more convenient to work with velocities in place of momenta. Thus we write (for a fluid whose mean macroscopic velocity is zero)

$$E_K(\mathbf{x}, t) = \frac{1}{2} \int mv^2 F_1 \, d\mathbf{v} \tag{1.26}$$

$$E_\Phi(\mathbf{x}, t) = \frac{1}{2} \int \Phi(\mathbf{x}, \mathbf{x}') F_2 \, d\mathbf{v} \, d\mathbf{v}' \, d\mathbf{x}' \tag{1.27}$$

To obtain BY_1 for F_1, we multiply (1.24) by N and remove particle labels. To obtain BY_2 for F_2, we multiply (1.25) by $N(N-1)$ and again remove labels. There results

$$\frac{\partial F_1}{\partial t} + \mathbf{v} \cdot \frac{\partial}{\partial \mathbf{x}} F_1 - \frac{1}{m} \frac{\partial}{\partial \mathbf{v}} \cdot \int \left(\frac{\partial}{\partial \mathbf{x}} \Phi \right) F_2 \, d\mathbf{x}' \, d\mathbf{v}' = 0 \tag{1.28}$$

$$\frac{\partial F_2}{\partial t} + \left(\mathbf{v} \cdot \frac{\partial}{\partial \mathbf{x}} + \mathbf{v}' \cdot \frac{\partial}{\partial \mathbf{x}'} \right) F_2 + \frac{\partial}{\partial \mathbf{v}} \cdot \mathbf{X} + \frac{\partial}{\partial \mathbf{v}'} \cdot \mathbf{Y} = 0 \tag{1.29}$$

where $\mathbf{X}$ and $\mathbf{Y}$ are implied interaction terms (which contain F_3). To find the equation for E_K, we we operate (1.28) with $\int d\mathbf{v}(mv^2/2)$. There results

$$\frac{\partial}{\partial t} E_K + \frac{\partial}{\partial \mathbf{x}} \cdot \int \frac{1}{2} mv^2 \mathbf{v} F_1 \, d\mathbf{v} - \frac{1}{2} \int d\mathbf{v} v^2 \frac{\partial}{\partial \mathbf{v}} \cdot \int \left(\frac{\partial}{\partial \mathbf{x}} \Phi \right) F_2 \, d\mathbf{x}' \, d\mathbf{v}'$$

Now note that

$$\frac{1}{2} v^2 \frac{\partial}{\partial \mathbf{v}} \cdot \mathbf{G} = \frac{\partial}{\partial \mathbf{v}} \cdot \left(\frac{v^2}{2} \mathbf{G} \right) - \mathbf{G} \cdot \mathbf{v}$$

Inserting this relation in the preceding equation and dropping the surface terms gives

$$\frac{\partial}{\partial t} E_K + \frac{\partial}{\partial \mathbf{x}} \cdot \int \frac{1}{2} mv^2 \mathbf{v} F_1 \, d\mathbf{v} + \int d\mathbf{v} \, d\mathbf{v}' \, d\mathbf{x}' \mathbf{v} \cdot \left(\frac{\partial \Phi}{\partial \mathbf{x}} \right) F_2 = 0 \tag{1.30}$$

Operating on BY_2 as giving by (1.29) with

$$\frac{1}{2} \int d\mathbf{x}' \, d\mathbf{v} \, d\mathbf{v}' \Phi(\mathbf{x}, \mathbf{x}')$$

and dropping the $\mathbf{X}$ and $\mathbf{Y}$ surface terms, we find

$$\frac{\partial E_\Phi}{\partial t} + \frac{1}{2} \int d\mathbf{v} \, d\mathbf{v}' \, d\mathbf{x}' \Phi(\mathbf{x}, \mathbf{x}') \left(\mathbf{v} \cdot \frac{\partial}{\partial \mathbf{x}} + \mathbf{v}' \cdot \frac{\partial}{\partial \mathbf{x}'} \right) F_2 = 0 \tag{1.31}$$

Now note that

$$\Phi \frac{\partial}{\partial \mathbf{x}} F_2 = \frac{\partial}{\partial \mathbf{x}} (\Phi F_2) - F_2 \frac{\partial}{\partial \mathbf{x}} \Phi \tag{1.32a}$$

$$\Phi \frac{\partial}{\partial \mathbf{x'}} F_2 = \frac{\partial}{\partial \mathbf{x'}}(\Phi F_2) - F_2 \frac{\partial}{\partial \mathbf{x'}} \Phi \qquad (1.32b)$$

When the first of these relations is substituted into (1.31), the $\partial(\Phi F_2)/\partial \mathbf{x}$ term remains, as $\mathbf{x}$ is not integrated. However, the corresponding term in (1.32b) yields a surface term, and we obtain

$$\frac{\partial}{\partial t} E_\Phi + \frac{1}{2} \frac{\partial}{\partial \mathbf{x}} \cdot \int d\mathbf{v} \, d\mathbf{v'} \, d\mathbf{x'} \mathbf{v} \Phi F_2$$

$$- \frac{1}{2} \int d\mathbf{v} \, d\mathbf{v'} \, d\mathbf{x'} F_2 \left(\mathbf{v} \cdot \frac{\partial}{\partial \mathbf{x}} \Phi + \mathbf{v'} \cdot \frac{\partial}{\partial \mathbf{x'}} \Phi \right) = 0 \qquad (1.33)$$

Equations (1.31) and (1.33) have the form

$$\frac{\partial E_K}{\partial t} + \nabla \cdot \mathbf{Q}_K + \int d\mathbf{x'} \, d\mathbf{v} \, d\mathbf{v'} F_2 \mathbf{v} \Phi F_2 \mathbf{v} \cdot \frac{\partial}{\partial \mathbf{x}} \Phi = 0 \qquad (1.34a)$$

$$\frac{\partial E_\Phi}{\partial t} + \nabla \cdot \mathbf{Q}_\Phi - \frac{1}{2} \int d\mathbf{x'} \, d\mathbf{v} \, d\mathbf{v'} F_2 \left(\mathbf{v} \cdot \frac{\partial}{\partial \mathbf{x}} \Phi + \mathbf{v'} \cdot \frac{\partial}{\partial \mathbf{x}} \Phi \right) = 0 \qquad (1.34b)$$

where the flow vectors $\mathbf{Q}_K$ and $\mathbf{Q}_\Phi$ are as implied. We label the total energies

$$\bar{E}_K(t) = \int E_K(\mathbf{x}, t) \, d\mathbf{x}$$
$$\bar{E}_\Phi(t) = \int E_\Phi(\mathbf{x}, t) \, d\mathbf{x} \qquad (1.35)$$

Thus, operating on (1.34a,b) with $\int d\mathbf{x}$ and adding the resulting equations gives

$$\frac{d}{dt}(\bar{E}_K + \bar{E}_\Phi) = \frac{1}{2} \int d\mathbf{x} \, d\mathbf{x'} \, d\mathbf{v} \, d\mathbf{v'} F_2 \left(\mathbf{v} \cdot \frac{\partial \Phi}{\partial \mathbf{x}} + \mathbf{v'} \cdot \frac{\partial \Phi}{\partial \mathbf{x'}} \right)$$
$$- \int d\mathbf{x} \, d\mathbf{x'} \, d\mathbf{v} \, d\mathbf{v'} F_2 \mathbf{v} \cdot \frac{\partial \Phi}{\partial \mathbf{x}} \qquad (1.36)$$

Owing to the symmetry relations $F_2(\mathbf{z}, \mathbf{z'}) = F_2(\mathbf{z'}, \mathbf{z})$, where $\mathbf{z} \equiv (\mathbf{x}, \mathbf{y})$, and $\Phi(\mathbf{x}, \mathbf{x'}) = \Phi(|\mathbf{x} - \mathbf{x'}|) = \Phi(\mathbf{x'}, \mathbf{x})$, we find that all three integrals on the right side of (1.36) have the same value. Thus we obtain the desired result:

$$\frac{d}{dt}(\bar{E}_K + \bar{E}_\Phi) = 0$$
$$\bar{E}_{\text{total}} = \bar{E}_K + \bar{E}_\Phi = \text{constant} \qquad (1.37)$$

A more precise statement of conservation of energy in classical kinetic theory may be obtained directly from the dynamic equation (1.1.25). Identifying the Hamiltonian $H(\mathbf{x}^N, \mathbf{p}^N)$ with the energy of the system, we find

$$\frac{dH}{dt} = [H, H] + \frac{\partial H}{\partial t} = 0 \qquad (1.38)$$

Here we have observed that H is not an explicit function of the time. The finding (1.38) is a more exact result than (1.37) in that (1.38) identifies the total precise energy (1.5) as being constant as opposed to (1.37), which identifies merely the average of total energy as being constant.

2.2   Correlation Expansions: The Vlasov Limit

2.2.1   Nondimensionalization

In this section we introduce the correlation function representation of distribution functions. To motivate this representation, it proves valuable to nondimensionalize BY_s as given by (1.20), which, with $N \gg s$, we first rewrite as

$$\left(\frac{\partial}{\partial t} - \hat{L}_s\right) f_s = \hat{I}_s f_{s+1} \tag{2.1}$$

where

$$\hat{I}_s \equiv -N \sum_{i=1}^{s} \frac{\partial}{\partial \mathbf{p}_i} \cdot \int d(s+1) \mathbf{G}_{i,s+1} \tag{2.2}$$

With (1.11), (2.1) is more explicitly written

$$\left(\frac{\partial}{\partial t} + \sum_{l-1}^{s} \hat{K}_l - \sum_{i<j}^{s} \hat{O}_{ij}\right) f_s = \hat{I}_s f_{s+1} \tag{2.3}$$

where $\hat{O}_{ij}$ is the momentum operator (1.7).

The fluid being considered is comprised of N particles, which we now assume occupies a volume V. This permits us to introduce a characteristic number density

$$n_0 \equiv \frac{N}{V} \tag{2.4a}$$

We further assume that we may assign a mean thermal speed, C, and corresponding temperature, T, to the fluid, which are related as

$$mC^2 \equiv k_B T \tag{2.4b}$$

where k_B is Boltzmann's constant.

The strength of potential Φ_0 and characteristic scale length r_0 are defined by

$$G_{ij} = \frac{\Phi_0}{r_0} \bar{G}_{ij} \tag{2.5c}$$

where $\bar{G}_{ij}$ is nondimensional.

Finally, we renormalize f_s so that[1]

$$F_s \equiv V^s f_s \tag{2.6}$$

For a spatially homogeneous fluid,

$$f_1(\mathbf{x}, \mathbf{p}) = \frac{1}{V} F_1(\mathbf{p})$$

In general, number density, $n(\mathbf{x}, t)$, is given by

$$n(\mathbf{x}, t) = N \int f_1 \, d\mathbf{p} = n_0 \int F_1 \, d\mathbf{p} \tag{2.6a}$$

To convert (2.3) to an equation for F_s, we multiply both sides by V^s to obtain

$$\left(\frac{\partial}{\partial t} + \sum_l \hat{K}_l - \sum_{i<j} \sum \hat{O}_{ij} \right) F_s = \frac{1}{V} \hat{I}_s F_{s+1} \tag{2.7}$$

The nondimensionalization scheme we introduce is given by the following relations (barred variables are nondimensional):

$$
\begin{aligned}
x &= r_0 \bar{x}, & p &= mC \bar{p} \\
t &= \frac{r_0}{C} \bar{t}, & F_s &= (mC)^{-3s} \bar{F}_s
\end{aligned}
\tag{2.8}
$$

Note that a number of choices for characteristic length enter this analysis. We may set r_0 equal to the range of interaction or we may set

$$r_0 = n_0^{-1/3}$$

or we may set

$$r_0 = V^{1/3}$$

These choices are noted by the following scheme:

$$
r_0 \left\{
\begin{array}{l}
\text{range of interaction} \\
V^{1/3} \\
n_0^{-1/3}
\end{array}
\right.
\tag{2.9}
$$

Substituting the relations (2.8) into (2.7) gives

$$\left(\frac{\partial}{\partial \bar{t}} + \sum_l \hat{K}_l - \frac{\Phi_0}{mC^2} \sum_{i<j} \sum \hat{O}_{ij} \right) \bar{F}_s = (n_0 r_0^3) \left(\frac{\Phi_0}{mC^2} \right) \hat{I}_s \bar{F}_{s+1} \tag{2.10}$$

The factor r_0^3 stems from nondimensionalization of $d\mathbf{x}$, whereas the V^{-1} factor in n_0 stems from the definition of F_s. It is evident that this equation suggest

---

[1]This F_s distribution should not be confused with the s-tuple distribution introduced in Chapter 1, which was also labeled F_s.

we identify the parameters

$$\alpha \equiv \frac{\Phi_0}{k_B T}, \qquad \gamma^{-1} \equiv n_0 r_0^3 \qquad (2.11)$$

Dropping bars over dimensionless variables, (2.10) becomes

$$\left(\frac{\partial}{\partial t} + \sum_{l=1}^{s} \hat{K}_l - \alpha \sum_{i<j}^{s} \sum \hat{O}_{ij} \right) F_s = \frac{\alpha}{\gamma} \hat{I}_s F_{s+1} \qquad (2.12)$$

Consider the coefficient

$$\frac{\alpha}{\gamma} = \frac{n_0 r_0^3 \Phi_0}{k_B T} \qquad (2.13)$$

With the first choice for r_o in (2.9), we obtain

$$\mathcal{N}_0 \equiv n_0 r_0^3$$

which is the number of particles in a range sphere. Inserting this result into (2.13) gives

$$\frac{\alpha}{\gamma} = \frac{\mathcal{N}_0^2 \Phi_0}{\mathcal{N}_0 k_B T} \simeq \frac{\langle E_\Phi \rangle}{\langle E_\kappa \rangle} \qquad (2.14)$$

The numerator of this expression represents pair-interaction energy per range volume, and the denominator represents thermal energy per range volume. We take this ratio to be a measure of the ratio of average potential to kinetic energy in the fluid. Note that if $\mathcal{N}_0$ is small than few particles in the fluid interact, and $\langle E_\Phi \rangle$ must likewise be small.

With (2.14), we may term the fluid *strongly coupled* when $\alpha/\gamma \gtrsim 1$, and *weakly coupled* when $a/\gamma \ll 1$. Thus we write

$$\frac{\alpha}{\gamma} \begin{cases} \alpha/\gamma \gtrsim 1, & \text{strongly coupled} \\ \\ \alpha/\gamma \ll 1, & \text{weakly coupled} \end{cases} \qquad (2.15)$$

A representation that well exploits this partitioning is that of correlation functions discussed below.

First let us delineate coupling domains of the fluid generated by varying the magnitudes of α and γ in the coupling constant α/γ. These domains are shown in Fig 2.3.[2]

---

[2]The ε-ordering diagram appeared previously in R. L. Liboff, *Introduction to the Theory of Kinetic Equations*, Wiley, New York, (1969). More recently, it was employed in an analysis of galaxy correlations: M. A. Guillen and R. L. Liboff, *Mon. Not. Roy. Astr. Soc. 231*, 957 (1988), *234*, 1119 (1988).

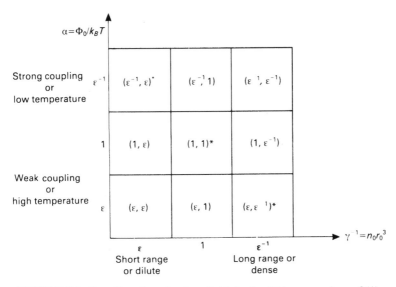

FIGURE 2.3. Coupling domains in a fluid. In the (*) boxes, $\alpha/\gamma = 0(1)$.

2.2.2 Correlation Functions

If particles in a gas are statistically independent, the conditional probability
(see Section 1.6.4)

$$h_1^{(N)}(1|2, \ldots, N) = \frac{f_N(1, \ldots, N)}{f_{N-1}(2, \ldots, N)} = f_1(1) \qquad (2.16)$$

This follows since the state of particle 1 is in no way dependent on the state
of the remaining particles in the fluid. For (2.16) to hold for all particles in the
fluid, we must have, in general,

$$f_s(1, 2 \ldots, s) = \prod_{i=1}^{s} f_1(i), \qquad 1 \le s \le N \qquad (2.17)$$

Another way to say that particles in the fluid are statistically independent is to
say that they are *uncorrelated*. Let us define the correlation function between
two particles, $C_2(1, 2)$, according to the relation

$$f_2(1, 2) = f_1(1)f_1(2) + C_2(1, 2) \qquad (2.18)$$

Thus, when $C_2 = 0$, particles are uncorrelated and the preceding returns (2.17).
Continuing in this way gives a transformation of distributions:

$$(f_1, f_2, f_3, \ldots, f_N) \rightarrow (f_1, C_2, C_3, \ldots, C_N)$$

These relations are given by [repeating (2.18)][3]

$$f_2(1, 2) = f_1(1)f_1(2) + C_2(1, 2)$$

$$f_3(1, 2, 3) = f_1(1)f_1(2)f_1(3) + \sum_{P(1,2,3)} f_1(1)C_2(2, 3) + C_3(1, 2, 3) \quad (2.19)$$

$$\vdots$$

where $P(1, 2, 3)$ denotes permutations of 1, 2, 3.

We may represent these equations with the aid of diagrams. Thus, the equation for f_5, say has the representation

$$f_5 = \text{o o o o o} + \sum_P \text{o o-o-o-o} + \sum_P \text{o-o o-o-o} + \sum_P \text{o o o-o-o}$$

$$+ \sum_P \text{o-o o o-o} + \sum_P \text{o-o o o o} + \text{o-o-o-o-o} \quad (2.20)$$

where, for example, o–o represents C_2 and o o represents $f_1 f_1$ and P is written for $P(1, 2, 3, 4, 5)$.

2.2.3   The Vlasov Limit

We wish to consider the limit $\Phi_0/k_B T \ll 1$ with long-range interaction so that $n_0 r_0^3 \gg 1$. We may describe this limit by setting $\alpha \to \varepsilon\alpha$ and $\gamma^{-1} \to (1/\varepsilon)\gamma^{-1}$, where ε is a parameter of smallness (see Fig. 2.3). Note that α/γ is of order unity. As $k_B T \gg \Phi_0$, we suspect that correlation between particles in the fluid is small. This condition may be described by attaching an ε factor with each coupling bar in the corresponding diagram equation. In this manner we find

$$f_2 = f_1 f_1 + \varepsilon C_2$$

$$f_3 = f_1 f_1 f_1 + \varepsilon \sum_P f_1 C_2 + \varepsilon^2 C_3 \quad (2.21)$$

Substituting these equations into (2.12) and keeping terms of $O(\varepsilon)$, we obtain [reverting to F_S distributions defined by (2.6)][4]

$$\left(\frac{\partial}{\partial t} + \hat{\kappa}_1\right) F_1 = \frac{\alpha}{\gamma}\hat{I}_1[F_1(1)F_1(2) + \varepsilon C_2(1, 2)]$$

$$\left(\frac{\partial}{\partial t} + \hat{\kappa}_2 - \varepsilon\alpha\hat{O}_{12}\right)[F_1(1)F_1(2) + \varepsilon C_2(1, 2)]$$

$$= \frac{\alpha}{\gamma}\hat{I}_2[F_1(1)F_1(2)F_1(3) + \varepsilon F_1(1)C_2(2, 3)$$

$$+ \varepsilon F_1(2)C_2(3, 1) + \varepsilon F_1(3)C_2(1, 2)]$$

$$\vdots$$

$$(2.22)$$

---

[3]J. E. Mayer and M. G. Mayer, *Statistical Mechanics*, Wiley, New York (1940).

[4]Here we have assumed that (2.21) is written in nondimensional form, in which case $\bar{f}_s = \bar{F}_s$.

where

$$\hat{\kappa}_s \equiv \sum_{i=1}^{s} \hat{K}_i \tag{2.22a}$$

Keeping lowest-order terms in (2.22) gives

$$\left(\frac{\partial}{\partial t} + \hat{\kappa}_1\right) F_1(1) = \frac{\alpha}{\gamma} \hat{I}_1 F_1(1) F_1(2) \tag{2.23a}$$

$$\left(\frac{\partial}{\partial t} + \hat{\kappa}_2\right) F_1(1) F_2(2) = \frac{\alpha}{\gamma} \hat{I}_2 F_1(1) F_1(2) F_1(3) \tag{2.23b}$$

$$\vdots$$

$$\left(\frac{\partial}{\partial t} + \hat{\kappa}_s\right) \prod_{i=1}^{s} F_1(i) = \frac{\alpha}{\gamma} \hat{I}_s \prod_{i=1}^{s+1} F_1(i) \tag{2.23c}$$

$$\vdots$$

These equations compromise N equations for the single unknown F_1. For this sequence to be logically consistent, it follows that these N equations must be redundant.[5] To establish this redundancy for the sequence (2.23), we note the following two equalities:

$$\left(\frac{\partial}{\partial t} + \kappa_s\right) \prod_{i=1}^{s} F_1(i) = \sum_{l=1}^{s} F_1(1) \dots F_1(l-1)$$

$$\times F_1(l+1) \dots F_1(s) \left(\frac{\partial}{\partial t} + \hat{K}_l\right) F_1(l) \tag{2.24}$$

$$\hat{I}_s \prod_{i=1}^{s+1} F_1(i) = \sum_{l=1}^{s} F_1(1) \dots F_1(l-1)$$

$$\times F_1(l+1) \dots F_1(s) \hat{I}_s^l F_1(l) F_1(s+1)$$

where we have written

$$\hat{I}_s^l \equiv -\frac{\partial}{\partial \mathbf{p}_l} \cdot \int d(s+1) \mathbf{G}_{i,s+1} \tag{2.25}$$

Substituting (2.24) into (2.23c) gives

$$\sum_{l=1}^{s} F_1(1) \cdots F_1(l-1) F_1(l+1) \cdots F_1(s) \left(\frac{\partial}{\partial t} + \hat{K}_l\right) F_1(l)$$

$$= -\frac{\alpha}{\gamma} \sum_{l=1}^{2} F_1(1) \cdots F_1(l-1) F_1(l+1) \cdots F_1(s) \hat{I}_s^l F_1(l) F_1(s+1)$$

---

[5]The problem of redundancy in BBKGY expansions was critically examined by R. L. Liboff and G. E. Perona, *J. Math. Phys.* 8, 2001 (1967).

or, equivalently

$$\sum \prod F_1 \left[\left(\frac{\partial}{\partial t} + \hat{K}_l \right) F_1(l) - \frac{\alpha}{\gamma} \hat{I}_s F_1(l) F_1(s+1) \right] = 0 \qquad (2.26)$$

With reference to (2.23a), we see that the bracketed term vanishes. Thus, if $F_l(i)$ satisfies (2.23a), it also satisfies (2.23c), and we may conclude that the sequence (2.23 is consistent.

2.2.4 The Vlasov Equation: Self-Consistent Solution

We have found that the equation that emerges from the BBKGY sequence in the limit $\alpha = 0(\varepsilon)$, $\gamma = 0(1/\varepsilon)$ is (2.23a), which we now write in fully dimensional form (deleting the subscript on F_1):

$$\left(\frac{\partial}{\partial t} + \mathbf{v} \cdot \frac{\partial}{\partial x} \right) F(\mathbf{x}, \mathbf{v}, t) = -\frac{n_0}{m} \frac{\partial}{\partial \mathbf{v}} \cdot \int d\mathbf{x}' \, d\mathbf{v}' \mathbf{G}(\mathbf{x}, \mathbf{x}') F(\mathbf{x}, \mathbf{v}, t) F(\mathbf{x}', \mathbf{v}', t)$$

$$(2.27)$$

[Here we have recalled the equality $F(\mathbf{p}) \, d\mathbf{p} = F(\mathbf{v}) \, d\mathbf{v}$.] Equation (2.27) is the *Vlasov equation*. Its physical significance is revealed when cast in terms of the number density [see (2.6a)],

$$n(\mathbf{x}', t) = n_0 \int F(\mathbf{x}', \mathbf{v}', t) \, d\mathbf{v}' \qquad (2.28a)$$

and the mean force field

$$\mathbf{G}(\mathbf{x}, t) = \int n(\mathbf{x}', t) \mathbf{G}(\mathbf{x}, \mathbf{x}') \, d\mathbf{x}' \qquad (2.28b)$$

The integral in (2.27) then becomes

$$\frac{\partial}{\partial \mathbf{v}} F_1(\mathbf{x}, \mathbf{v}, t) \cdot n_0 \int F_1(\mathbf{x}', \mathbf{v}', t) \, d\mathbf{v}' \int \mathbf{G}(\mathbf{x}, \mathbf{x}') \, d\mathbf{x}'$$

$$= \frac{\partial}{\partial \mathbf{v}} F_1(\mathbf{x}, \mathbf{v}, t) \cdot \int n(\mathbf{x}', t) \mathbf{G}(\mathbf{x}, \mathbf{x}') \, d\mathbf{x}'$$

$$= \frac{\partial}{\partial \mathbf{v}} F_1(\mathbf{x}, \mathbf{v}, t) \cdot \mathbf{G}(\mathbf{x}, t) \qquad (2.29)$$

Thus (2.27) may be written

$$\left(\frac{\partial}{\partial t} + \mathbf{v} \cdot \frac{\partial}{\partial \mathbf{x}} + \frac{\mathbf{G}}{m} \cdot \frac{\partial}{\partial \mathbf{v}} \right) F(\mathbf{x}, \mathbf{v}, t) = 0 \qquad (2.30)$$

where $\mathbf{G}$ is given by (2.28). The preceding equation indicates the following: $F(\mathbf{x}, \mathbf{v}, t)$ develops under a force field that is the instantaneous average of all two-particle forces in the fluid, as given by (2.28). This force field is sometimes called a "snapshot" field. As this cumulative force changes slowly, we may except particle trajectories in a *Vlasov fluid* to likewise change smoothly. See Fig 2.4.

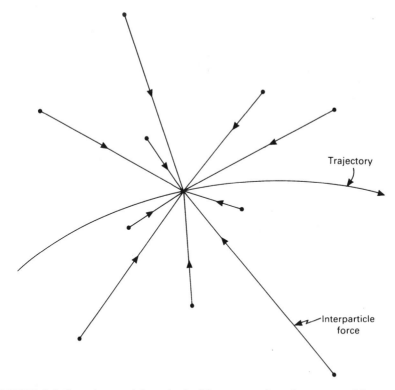

FIGURE 2.4. Snapshot total force in the Vlasov equation gives a smoothly varying trajectory.

The Vlasov equation (2.30) bears a striking similarity to the *one-particle Liouville equation*, which is relevant to a single particle moving in an externally supported force field. The Hamiltonian for this system is given by

$$H = \frac{p^2}{2m} + \Phi(\mathbf{x})$$

where

$$\tilde{\mathbf{G}}(\mathbf{x}) = -\frac{\partial}{\partial \mathbf{x}}\Phi(\mathbf{x})$$

is the externally supported force field. The Liouville equation

$$\frac{\partial F}{\partial t} + [F, H] = 0 \tag{2.31}$$

for this system returns (2.30) with $\mathbf{G}$ replaced by $\tilde{\mathbf{G}}$. That is,

$$\left(\frac{\partial}{\partial t} + \mathbf{v} \cdot \frac{\partial}{\partial \mathbf{x}} + \frac{\tilde{\mathbf{G}}}{m} \cdot \frac{\partial}{\partial \mathbf{v}} \right) F(\mathbf{x}, \mathbf{v}, t) = 0 \tag{2.30a}$$

The solution to both (2.30) and (2.30a) is obtained by solving the characteristic equations

$$dt = \frac{d\mathbf{x}}{\mathbf{v}} = \frac{m\,d\mathbf{v}}{\mathbf{G}} \tag{2.32}$$

(where $\mathbf{a}/\mathbf{b} = a_x/b_x = a_y/b_y = a_z/b_z$). These are equivalent to the orbit equations

$$\mathbf{v} = \frac{d\mathbf{x}}{dt}, \qquad \mathbf{G} = m\frac{d\mathbf{v}}{dt}$$

which give

$$m\frac{d^2\mathbf{x}}{dt^2} = \mathbf{G}(\mathbf{x}) = -\frac{\partial\Phi}{\partial\mathbf{x}} \tag{2.33}$$

These equations have the integral

$$E = \frac{m}{2}\left(\frac{d\mathbf{x}}{dt}\right)^2 + \Phi(\mathbf{x}) \tag{2.34}$$

Thus any function of the form

$$F = F\left[\frac{p^2}{2m} + \Phi(\mathbf{x})\right] \tag{2.35}$$

is a solution to the one-particle Liouville equation (2.30a). In this case, $\Phi(\mathbf{x})$ is a given known function.

Returning to the Vlasov equation, we see that for (2.35) to be a solution to this equation $\Phi(\mathbf{x})$ must satisfy (2.28):

$$\frac{\partial\Phi}{\partial\mathbf{x}} = -n_0 \int F(\mathbf{x}', \mathbf{v}', t)\mathbf{G}(\mathbf{x}, \mathbf{x}')\,d\mathbf{x}'\,d\mathbf{v}' \tag{2.36}$$

So in this case knowledge of Φ demands knowledge of the distribution function. An additional requirement for (2.35) to be a solution to the Vlasov equation is that any explicit time dependence of Φ be negligible. Consider, for example, the force equation (2.33). Multiplying this equation by $\dot{\mathbf{x}}$ gives

$$m\dot{\mathbf{x}} \cdot \ddot{\mathbf{x}} + \dot{\mathbf{x}} \cdot \frac{\partial}{\partial\mathbf{x}}\Phi = 0 \tag{2.37}$$

If $\Phi = \Phi(\mathbf{x}, t)$, then

$$\frac{d\Phi}{dt} = \dot{\mathbf{x}} \cdot \frac{\partial}{\partial\mathbf{x}}\Phi + \frac{\partial\Phi}{\partial t} \tag{2.38}$$

which does not allow (2.37) to be integrated. However, for

$$\frac{\partial\Phi}{\partial t} \ll \dot{\mathbf{x}} \cdot \frac{\partial}{\partial\mathbf{x}}\Phi \tag{2.39}$$

(2.37) returns the integral (2.34).

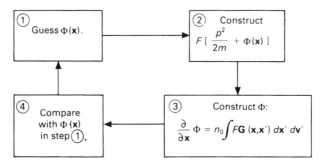

FIGURE 2.5. Flow chart illustrating the notion of a self-consistent solution to the Vlasov equation.

Another important distinction between the Vlasov and one-particle Liouville equations is that the former is nonlinear. That is, when writing the Vlasov equation in the form (2.30), we must keep in mind that the force field $\mathbf{G}$ is a functional of the distribution F. Thus we may write

$$\mathbf{G}(\mathbf{x}, t) = \mathbf{G}[F(\mathbf{x}, \mathbf{v}, t)]$$

which exposes the nonlinearity of (2.30).

Let us return to the condition (2.36). This is a statement of *self-consistency*, which is demonstrated by the flow chart for the construction of a solution to the Vlasov equation shown in Fig 2.5.

The Vlasov equation is returned to in Chapter 4 where applications of this equation are made to a plasma.

2.2.5 Debye Distance and the Vlasov Limit

We have found that the Vlasov equation results in the limit

$$\gamma^{-1} = n_0 r_0^3 = O(1/\varepsilon) \tag{2.40a}$$

$$\alpha = \frac{\Phi_0}{k_B T} = O(\varepsilon) \tag{2.40b}$$

$$\alpha \gamma^{-1} = O(1) \tag{2.40c}$$

In a plasma comprised of electrons in a neutralizing, fixed, positive background, the strength of the interaction potential between electrons is the Coulomb form

$$\Phi_0 = \frac{e^2}{\lambda_d} \tag{2.41}$$

where we have set the range of interaction r_0 equal to the Debye distance λ_d. For γ^{-1}, we write

$$\gamma^{-1} = 4\pi n_0 \lambda_d^3 \tag{2.42}$$

Note that, in the present case, weak coupling corresponds to the limit $4\pi n_0 \lambda_d^3 \gg 1$. Substituting the latter two relations into (2.40c) and solving for λ_d, we find

$$4\pi \lambda_d^2 = \frac{k_B T}{e^2 n_0} \tag{2.43}$$

This characteristic distance comes into play a number of times in the text.[6]

2.2.6 Radial Distribution Function

In concluding this section we return to the theory of correlations introduced in Section 2.2. An important correlation function that often comes into play in the theory of the equilibrium structure of liquids is the radial distribution function.[7] This is a two-particle spatial correlation function that, for a fluid in equilibrium, is defined by

$$f_2(1, 2) = f_0(p_1) f_0(p_2) g(r) \tag{2.44}$$

where $g(r)$ is the radial distribution function.

In (2.44), we have labeled

$$r \equiv |\mathbf{x}_2 - \mathbf{x}_1|$$

and have introduced the Maxwellian distribution[8]

$$f_0(p) = \frac{1}{V(2\pi m^2 C^2)^{3/2}} \exp\left(-\frac{p^2}{2m^2 C^2}\right) \tag{2.45}$$

where V is the volume occupied by the fluid, and the thermal speed C is defined by (2.4b). The Maxwellian (2.45 has the normalization

$$\iint f_0(p) \, d\mathbf{p} \, d\mathbf{x} = 1 \tag{2.46}$$

The *total correlation function* $h(r)$ is defined by

$$h(r) = g(r) - 1 \tag{2.47}$$

which, with the first to the sequence (2.19), gives the relation

$$C_2(1, 2) = f_0(p_1) f_0(p_2) h(r) \tag{2.48}$$

Thus $g(r)$ is directly related to $f_2(1, 2)$, whereas $h(r)$ is directly related to $C_2(1, 2)$. For a fluid with no spatial correlation between molecules, $h(r) = 0$

---

[6]Thus, for example, the Debye distance is encountered again in Section 3.9 in application of the diagrammatic approach to a plasma and in Chapter 4 in application of the KBG equation to a plasma. Numerical values of λ_d as a function of T/n are given in (4.1.37). *Note:* With λ_d^2 and C^2/λ_d^2 taken as constant, (2.40) et seq. is referred to as the Rosenbluth-Rostoker limit. (See *Phys. Fluids*, **3**, 1, 1960.)

[7]For further discussion on this topic, see D. L. Goodstein, *States of Matter*, Prentice-Hall, Englewood Cliffs, N.J. (1975).

[8]The derivation of this one-particle equilibrium distribution is given in Section 3.3.

and $g(r) = 1$. As two molecules in a fluid that are separated by a distance large compared to their interaction length may be assumed not to be correlated, we write

$$g(\infty) = 1 \tag{2.49}$$

Furthermore, as two molecules cannot occupy the same position in space, $f_2(r = 0) = 0$, which, with (2.44) gives

$$g(0) = 0 \tag{2.50}$$

These latter two statements are standard boundary conditions in studies of the radial distribution function.

An important property of these correlation functions is their Fourier transform, the *structure factor*:

$$S(k) = 1 + n \int h(r)e^{i\mathbf{k}\cdot\mathbf{r}} \, d\mathbf{r} \tag{2.51}$$

where $\mathbf{k}$ is the Fourier transform wave vector and n is the number density of molecules in the fluid. The function $S(k)$ serves to relate data from scattering experiments to the correlation functions. For an ideal gas, there is no interaction between particles, and (2.44) gives $g(r) = 1$, which, with (2.51), gives the value $S(k) = 1$.

Two key relations involving the radial distribution function, which may be obtained either from the partition function of equilibrium statistical mechanics[9] or BY_1 (see Problem 3.44), are as follows:

$$\frac{E}{nk_BT} = \frac{3}{2} + \frac{n}{2k_BT} \int_0^\infty \Phi(r)g(r)4\pi r^2 \, dr \tag{2.52}$$

$$\frac{P}{nk_BT} = 1 - \frac{n}{6k_BT} \int_0^\infty \Phi'(r)g(r)4\pi r^2 \, dr \tag{2.53}$$

The first of these gives the energy density of the fluid. The second gives the equation of state of the fluid, where P is scalar pressure.[10] Note that the integral in (2.52) gives the interaction potential contribution to fluid energy, whereas the integral in (2.53) gives the interaction potential contribution to pressure.

2.3   Diagrams: Prigogine Analysis

In this section we offer an introduction to a perturbation technique of solution to the Liouville equation. A diagrammatic representation of interaction integrals

---

[9] D. A. McQuarrie, *Statistical Mechanics*, Harper & Row, New York (1973). See also T. M. Reed and K. E. Gubbins, *Applied Statistical Mechanics*. McGraw-Hill, New York (1973). See also Problem 2.12.

[10] The notion of scalar pressure, as well as other dynamic variables, is discussed in Section 3.4.

occurring in the perturbation expansion is given. It is found possible to associate the topology of these diagrams with the order of interaction between particles (that is, two particle, three particle, an so on). Application of this formalism is made in derivation of the Boltzmann equation and the temporal evolution of a plasma.

2.3.1 Perturbation Liouville Operator

Let us recall the Liouville equation (1.2):

$$\frac{\partial f_N}{\partial t} = \hat{L}_N f_N \tag{3.1}$$

where $\hat{L}_N$ is given by (1.3).

$$\hat{L}_N = [H, \quad] \tag{3.2}$$

The operator $\hat{L}_N$ has the explicit representation [see (1.9)]

$$L_N = \sum_{l=1}^{N} \hat{K}_l + \sum_{i<j}^{N} \hat{O}_{ij} \tag{3.3}$$

We rewrite this operator in this form

$$\hat{L}_N = \hat{L}_0 + \delta\hat{L} \tag{3.4}$$

where $\hat{L}_0$ is the free-particle, kinetic-energy operator and

$$\delta\hat{L} = \sum_{i<j}^{N} \frac{\partial}{\partial \mathbf{x}_i} \Phi_{ij} \cdot \left[\frac{\partial}{\partial \mathbf{p}_i} - \frac{\partial}{\partial \mathbf{p}_j} \right] \tag{3.5}$$

represents the perturbation to $\hat{L}_0$.

Eigenfunctions of $\hat{L}_0$ are given by (see Section 1.5.2)

$$\hat{L}_0 \psi_{(\mathbf{k})} = -i\omega_{(\mathbf{k})} \psi_{(\mathbf{k})}$$

$$\psi_{(\mathbf{k})} = L^{-3N/2} \exp\left[i \sum \mathbf{k}_l \cdot \mathbf{x}_l \right] \equiv |(\mathbf{k})\rangle \tag{3.6}$$

$$\omega_{(\mathbf{k})} = \sum \mathbf{k}_l \cdot \mathbf{v}_l$$

where $\mathbf{v}_l = \mathbf{p}_l/m$. Here, as in (1.5.10a), we are writing $(\mathbf{k})$ for the sequence

$$(\mathbf{k}) = (\mathbf{k}_1, \mathbf{k}_2, \ldots, \mathbf{k}_N) \tag{3.7}$$

$$\mathbf{k}_i = \frac{2\pi}{L} \mathbf{n}_i \tag{3.7a}$$

where, with (1.5.14), the components of $\mathbf{n}_i$ are integers.

We wish to employ the basis functions (3.6) to solve the Liouville equation (3.1), with $\hat{L}_N$ given by the general form (3.3). Following (1.15.16) relevant to the free-particle case, we write

$$f_N(1, \ldots, N) = \sum_{(\mathbf{k})} a_{(\mathbf{k})}(\mathbf{p}^N, t)\psi_{(\mathbf{k})}(\mathbf{x}^N)e^{-i\omega_{(\mathbf{k})}t} \tag{3.8}$$

Terms in this series may be regrouped so that coefficients in this new series bear special physical significance.

2.3.2 Generalized Fourier Series

The regrouped series is given by

$$\begin{aligned}
f_N = \frac{1}{V^N}\Bigg[a_0(\mathbf{p}^N, t) &+ \frac{1}{\bar{V}}\sum_{j=1}^{N}\sum_{\mathbf{k}_j} a_1(\mathbf{k}_j, \mathbf{p}^N, t)e^{i\mathbf{k}_j \cdot \mathbf{x}_j}e^{-i\omega_j t} \\
&+ \frac{1}{\bar{V}^2}\sum_{j<l}\sum_{\mathbf{k}_j \mathbf{k}_l}([1 - \delta_{\mathbf{k}_j + \mathbf{k}_l}] + \delta_{\mathbf{k}_j + \mathbf{k}_l}) \\
&\times a_2(\mathbf{k}_j, \mathbf{k}_l, \mathbf{p}^N, t)e^{i[\mathbf{k}_j \cdot \mathbf{x}_j + \mathbf{k}_l \cdot \mathbf{x}_l]}e^{-i\omega_{jl}t} + \cdots \Bigg]
\end{aligned} \tag{3.9}$$

The coefficients in this series have the property that $a_n(\mathbf{k}^n, \mathbf{p}^N, t)$, for example, contains only n nonzero $\mathbf{k}$ vectors. Thus the coefficient a_2 contains only two nonvanishing $\mathbf{k}$ vectors. The Kronecker symbol

$$\begin{aligned}
\delta_{\mathbf{k}} &= 1, \qquad \mathbf{k} = 0 \\
\delta_{\mathbf{k}} &= 0, \qquad \mathbf{k} \neq 0
\end{aligned}$$

Thus the bracketed terms in the second sum in (3.9) contains no contributions with $\mathbf{k}_j + \mathbf{k}_l = 0$. The volume $\bar{V} \equiv V/(2\pi)^3$.

Terms in the second summation of (3.9) with $\mathbf{k}_j + \mathbf{k}_l = 0$ are relevant to the *homogeneous limit*. This may be seen as follows. If the system is homogeneous, then f_N is invariant under translation of coordinates, $(\mathbf{x}_l) \to (\mathbf{x}_l') = (\mathbf{x}_l + \mathbf{b})$. Thus, for a typical term in (3.9), we have

$$\sum_{(\mathbf{k})} a_{(\mathbf{k})}e^{i\sum \mathbf{k}_l \cdot \mathbf{x}_l} = \sum_{(\mathbf{k})} a_{(\mathbf{k})}e^{i\sum \mathbf{k}_l \cdot \mathbf{x}_l}e^{i\mathbf{b}\cdot\sum \mathbf{k}_l}$$

This equality is satisfied for all $\mathbf{k}_l$ and $\mathbf{x}_l$, providing $\sum \mathbf{k}_l = 0$ for all $(\mathbf{k})$ sequences.

2.3.3 Interpretation of $\mathbf{a}_l$ Coefficients

Integration of (3.9) overall $\mathbf{x}^N$ gives

$$\int f_N \, d\mathbf{x}^N = \int \frac{d\mathbf{x}^N}{V^N}\Bigg[a_0(\mathbf{p}^N, t) + \frac{1}{\bar{V}}\sum_j\sum_{\mathbf{k}_j} a_1(\mathbf{k}_j, \mathbf{p}^N, t)e^{i\mathbf{k}_j \cdot \mathbf{x}_j}e^{-i\omega_j t} + \cdots \Bigg] \tag{3.10}$$

The meaning of the double sum in this equation is that for each value of j the vector $\mathbf{k}_j$ is summed over all values given by (3.7a).

In the limit of large volume,

$$\frac{1}{V} \sum_{\mathbf{k}} \to \int d\mathbf{k} \qquad (3.11a)$$

Furthermore, in this same limit

$$\frac{1}{(2\pi)^3} \int e^{i\mathbf{k}\cdot\mathbf{x}} \, d\mathbf{x} = \delta(\mathbf{k}) \qquad (3.11b)$$

so that the second term in (3.10) gains a $\delta(\mathbf{k})$ factor upon spatial integration. Thus the only terms that enter this sum are those for which $\mathbf{k}_j = 0$. But $a_1(0, \mathbf{p}^N, t) = 0$ by definition. It follows that

$$a_0(\mathbf{p}^N, t) = \int f_N \, d\mathbf{x}^N \qquad (3.12a)$$

$$\int a_0(\mathbf{p}^N, t) \, d\mathbf{p}^N = 1 \qquad (3.12b)$$

represents the N-particle momentum distribution.

Let us introduce the momentum integrals of a_l:

$$\bar{a}_1(\mathbf{k}_j, t) \equiv \int d\mathbf{p}^N a_1(\mathbf{k}_j, \mathbf{p}^N, t) e^{-i\omega_j t}$$

$$\bar{a}_2(\mathbf{k}_j, \mathbf{k}_l, t) \equiv \int d\mathbf{p}^N a_2(\mathbf{k}_j, \mathbf{k}_l, \mathbf{p}^N, t) e^{-i\omega_{jl} t} \qquad (3.13)$$

$$\vdots$$

The number density $n_1(\mathbf{x})$ is given by

$$n_1(\mathbf{x}) = n_1(\mathbf{x}_l) = N \int f_N \, d\mathbf{p}^N \, d\mathbf{x}_1 \, d\mathbf{x}_2 \cdots d\mathbf{x}_{l-1} \, d\mathbf{x}_{l+1} \cdots d\mathbf{x}_N \qquad (3.14)$$

whereas the pair distribution is given by

$$n_2(\mathbf{x}, \mathbf{x}') = n_2(\mathbf{x}_s, \mathbf{x}_n) \qquad (3.15)$$
$$= \frac{N(N-1)}{2} \int f_N \, d\mathbf{p}^N \, d\mathbf{x}_1 \ldots d\mathbf{x}_{s-1} \, d\mathbf{x}_{s+1} \ldots d\mathbf{x}_{n-1} \, d\mathbf{x}_{n+1} \ldots d\mathbf{x}_N$$

We turn first to evaluation of $n_1(\mathbf{x})$. Integrating (3.9) in accord with (3.14) gives

$$n_1(\mathbf{x}_s) = \frac{N}{V^N} \int d\mathbf{p}^N \prod_{i \neq s} d\mathbf{x}_i$$

$$\cdot \left[a_0(\mathbf{p}^N, t) + \frac{1}{V} \sum_{j=1}^{N} \sum_{\mathbf{k}_j} a_1(\mathbf{k}_j, \mathbf{p}^N, t) e^{i(\mathbf{k}_j \cdot \mathbf{x}_i - \omega_j t)} + \cdots \right] \qquad (3.16)$$

In the limit of large volume with (3.11) and (3.12), we see that all terms in the j-sum except for the $j = s$ term give $a_1(0, \mathbf{p}^N, t)$, which vanishes by definition.

Each term in the sum over $a_2(\mathbf{k}_j, \mathbf{k}_l, \mathbf{p}^N, t)$ contains either a $\delta(\mathbf{k}_j)$ or $\delta(\mathbf{k}_l)$ factor, which stem respectively from $\int ds_j$ or $\int d\mathbf{k}_l$ integrations. It follows that all terms in this sum vanish since

$$a_2(0, \mathbf{k}_l) = a_2(\mathbf{k}_j, 0) = 0$$

again by definition. The same argument applies to all higher-order terms. There results

$$n_1(\mathbf{x}) = \frac{N}{V}\left[1 + \int \bar{a}_1(\mathbf{k}, t)e^{i\mathbf{k}\cdot\mathbf{x}}\,d\mathbf{k}\right] \tag{3.17}$$

In the limit of spatial homogeneity, $\bar{a}_1 = 0$ and

$$n_1 = \frac{N}{V} \qquad \text{(homogeneous case)}$$

Next we turn to evaluation of $n_2(\mathbf{x}, \mathbf{x}')$. Integrating (3.9) over

$$\frac{N(N-1)}{2}\int d\mathbf{p}_N \int \prod_{j\neq s,n} d\mathbf{x}_j$$

gives

$$n_2(\mathbf{x}_s, \mathbf{x}_n) = \frac{N(N-1)}{2}\left\{\frac{1}{V}\left[\frac{n_1(\mathbf{x}_s, t)}{N} + \frac{n_1(\mathbf{x}_n, t)}{N} - \frac{1}{V}\right]\right.$$
$$+ \frac{1}{V^2}\left[\int' \bar{a}_2(\mathbf{k}_s, \mathbf{k}_n, t)e^{i[\mathbf{k}_s\cdot\mathbf{x}_s+\mathbf{k}_n\cdot\mathbf{x}_n]}\,d\mathbf{k}_s\,d\mathbf{k}_n\right.$$
$$\left.\left. + \int \bar{a}_2(\mathbf{k}_s, -\mathbf{k}_s, t)e^{i\mathbf{k}_s\cdot(\mathbf{x}_s-\mathbf{x}_n)}\,d\mathbf{k}_s\right]\right\} \tag{3.18}$$

The prime on the first integral stipulates that the value $\mathbf{k}_j + \mathbf{k}_l$ has been deleted.

Returning to (3.18) and passing to the homogeneous limit gives

$$n_2(\mathbf{x}, \mathbf{x}') = \frac{N(N-1)}{2V^2}\left[1 + \int \bar{a}_2(\mathbf{k}, -\mathbf{k}, t)e^{i\mathbf{k}\cdot(\mathbf{x}-\mathbf{x}')}\,d\mathbf{k}\right] \tag{3.19}$$

The integral term represents the deviation from the state of no two-particle correlation and is seen to be a function of $|\mathbf{x} - \mathbf{x}'|$. In the state of no two-particle correlation, $\bar{a}_2 = 0$ and

$$n_2(\mathbf{x}, \mathbf{x}') = \frac{N(N-1)}{2V^2} \tag{3.20}$$

which has the normalization

$$\iint n_2(\mathbf{x}, \mathbf{x}')\,d\mathbf{x}\,d\mathbf{x}' = \frac{N(N-1)}{2} \tag{3.21}$$

This is the total number of pairs of particles in the system.

Having obtained the key relations [(3.17), and following] connecting $\bar{a}_1$ and $\bar{a}_2$ to the fluid variables n_1 and n_2, we return to formulation of equations of motion for the general $a_{(\mathbf{k})}$ coefficients.

2.3.4 Equations of Motion for $\mathbf{a}_{(\mathbf{k})}$ Coefficients

Substituting the series (3.9) into the Liouville equation (3.1) gives

$$\frac{\partial}{\partial t} \sum_{(\mathbf{k}')} a_{(\mathbf{k}')} e^{-i\omega_{(\mathbf{k}')}t} |(\mathbf{k}')\rangle = (\hat{L}_0 + \delta\hat{L}) \sum |(\mathbf{k}')\rangle a_{(\mathbf{k}')} e^{-i\omega_{(\mathbf{k}')}t} \tag{3.22}$$

Differentiating the left side and then operating from the left with $\langle(\mathbf{k})|$ and noting the orthogonality of these eigenstates and the eigenvalue relation (3.6), we obtain

$$\frac{\partial}{\partial t} a_{(\mathbf{k})} = \sum_{(\mathbf{k}')} e^{i\omega_{(\mathbf{k})}t} \langle(\mathbf{k})|\delta\hat{L}|(\mathbf{k}')\rangle (e^{i\omega_{(\mathbf{k}')}t} a_{(\mathbf{k}')}) \tag{3.23}$$

The operator $\delta\hat{L}$ is given by (3.5). Fourier expanding the interparticle potential we obtain

$$\Phi_{ln} = \frac{1}{\bar{V}} \sum_{\mathbf{K}} \Phi_K e^{i\mathbf{K}\cdot(\mathbf{x}_l - \mathbf{x}_n)} \tag{3.24}$$

so that

$$\delta\hat{L} = \frac{1}{\bar{V}} \sum_{l<n} \sum \sum_{\mathbf{K}} \Phi_K i\mathbf{K} e^{i\mathbf{K}\cdot(\mathbf{x}_l - \mathbf{x}_n)} \cdot \hat{\boldsymbol{\theta}}_{ln} \tag{3.25}$$

$$\hat{\boldsymbol{\theta}}_{ln} \equiv \frac{\partial}{\partial \mathbf{p}_l} - \frac{\partial}{\partial \mathbf{p}_n}$$

Inserting this form into the equation of motion (3.23) gives

$$\frac{\partial a_{(\mathbf{k})}}{\partial t} = \frac{1}{\bar{V}} \sum_{(\mathbf{k}')} \sum_{\mathbf{K}} \sum \sum_{l<n} e^{i\omega_{(\mathbf{k})}t} |\langle(\mathbf{k})|e^{i\mathbf{K}\cdot(\mathbf{x}_l - \mathbf{x}_n)}|(\mathbf{k}')\rangle \Phi_K i\mathbf{K} \cdot \hat{\boldsymbol{\theta}}_{ln}(a_{(\mathbf{k}')} e^{-i\omega_{(\mathbf{k}')}t}) \tag{3.26}$$

Let us evaluate the matrix element

$$\langle(\mathbf{k})|e^{i\mathbf{K}\cdot(\mathbf{x}_l - \mathbf{x}_n)}|(\mathbf{k}')\rangle = \frac{1}{V^N} \int d\mathbf{x}^N e^{-i\sum \mathbf{k}_s\cdot\mathbf{x}_s} e^{i\mathbf{K}\cdot(\mathbf{x}_l - \mathbf{x}_n)} e^{i\sum \mathbf{k}_s'\cdot\mathbf{x}_s} \tag{3.27}$$

For sufficiently large but finite volume, we introduce the representation

$$\int_{L^3} d\mathbf{x} e^{i\mathbf{k}\cdot\mathbf{x}} = V\delta(\mathbf{k}) \tag{3.28a}$$

Note that

$$\lim_{V\to\infty} \bar{V}\delta(\mathbf{k}) = (2\pi)^3 \delta(\mathbf{k}) \tag{3.28b}$$

We partition terms in the exponential of (3.27) as

$$\left[\sum_{j \neq l,n} \mathbf{x}_j \cdot (\mathbf{k}'_j - \mathbf{k}_j) \right] + \mathbf{x}_l \cdot (\mathbf{k}'_l - \mathbf{k}_l + \mathbf{K}) + \mathbf{x}_n \cdot (\mathbf{k}'_n - \mathbf{k}_n - \mathbf{K})$$

With the representation (3.28a), we obtain

$$\langle (\mathbf{k}) | e^{i\mathbf{K} \cdot (\mathbf{x}_l - \mathbf{x}_n)} | (\mathbf{k}') \rangle = \delta(\mathbf{K} - \mathbf{k}_l - \mathbf{k}'_l) \delta(\mathbf{k}'_n - \mathbf{k}_n - \mathbf{K}) \prod_{j \neq l,n}^{N} \delta(\mathbf{k}'_j - \mathbf{k}_j)$$

$$\equiv \Delta_{nl\mathbf{K}} \tag{3.29}$$

2.3.5 Selection Rules for Matrix Elements

The preceding finding implies a very important result: nonvanishing matrix elements occur only for $(\mathbf{k})$, $(\mathbf{k}')$ sequences that satisfy the relations

$$\begin{aligned} \mathbf{k}'_l &= \mathbf{k}_l - \mathbf{K} \\ \mathbf{k}'_n &= \mathbf{k}_n + \mathbf{K} \\ \mathbf{k}'_j &= \mathbf{k}_j, \qquad j \neq l, n \end{aligned} \tag{3.30}$$

We may conclude that in any nonvanishing matrix element the sequence $(\mathbf{k}')$ is the same as the sequence $(\mathbf{k})$ except for two $\mathbf{k}'$ vectors. Furthermore, these two vectors satisfy the conservation equation

$$\mathbf{k}'_l + \mathbf{k}'_n = \mathbf{k}_l + \mathbf{k}_n \tag{3.31}$$

Equation (3.26) now has the explicit form

$$\frac{\partial a_{(\mathbf{k})}}{\partial} = \frac{1}{V} \sum_{ln\mathbf{K}}' e^{i\omega_{(\mathbf{k})}t} \Phi_{\mathbf{K}} i\mathbf{K} \cdot \hat{\boldsymbol{\theta}}_{ln} (e^{i\omega_{(\mathbf{k}')}t} \Delta_{ln\mathbf{K}} a_{(\mathbf{k})}) \tag{3.32}$$

The summation $\sum_{ln\mathbf{K}}'$ is written for

$$\sum_{ln\mathbf{K}}' \equiv \sum_{l<n} \sum_{\mathbf{K}} \sum_{(\mathbf{k}')}$$

For the case of weak interactions, we set

$$\Phi_{ln} \rightarrow \varepsilon \Phi_{ln}$$

where $\varepsilon \ll 1$. Equation (3.32) then assumes the form

$$\frac{\partial a_{(\mathbf{k})}}{\partial t} = \varepsilon \hat{G} a_{(\mathbf{k})} = \varepsilon \sum_{ln\mathbf{K}}' \langle (\mathbf{k}) | \delta \hat{L}_{ln\mathbf{K}} | (\mathbf{k}') \rangle_t' a_{(\mathbf{k}')}(t) \tag{3.33}$$

where we have written

$$\langle (\mathbf{k}) | \delta L_{ln\mathbf{K}} | (\mathbf{k}') \rangle_t a_{(\mathbf{k}')} \equiv \frac{\Phi_{\kappa}}{\bar{V}} e^{i\omega_{(\mathbf{k})}t} i\mathbf{K} \cdot \hat{\boldsymbol{\theta}}_{ln} (e^{i\omega_{(\mathbf{k}')}t} \Delta_{ln\mathbf{K}} a_{(\mathbf{k}')}) \tag{3.34}$$

Note that this matrix element now carries an explicit time dependence. Substituting the perturbation expansion

$$a_{(\mathbf{k})} = a_{(\mathbf{k})}^{(0)} + \varepsilon a_{(\mathbf{k})}^{(1)} + \varepsilon^2 a_{(\mathbf{k})}^{(2)} \tag{3.35}$$

into (3.33) and equating coefficients of equal power in ε gives the following series of coupled equations.

$$\frac{\partial a_{(\mathbf{k})}^{(0)}}{\partial t} = 0, \qquad \frac{\partial a_{(\mathbf{k})}^{(1)}}{\partial t} = \hat{G} a_{(\mathbf{k})}^{(0)}, \qquad \frac{\partial a_{(\mathbf{k})}^{(2)}}{\partial t} = \hat{G} a_{(\mathbf{k})}^{(1)}, \ldots \tag{3.36}$$

These equations are simply integrated to obtain

$$a_{(\mathbf{k})}^{(1)} = \int_0^t dt' \hat{G}(t') a_{(\mathbf{k}')}^{(0)}$$

$$a_{(\mathbf{k})}^{(2)} = \int_0^t dt' \int_0^{t'} dt'' G(t') \otimes \hat{G}(t'') a_{(\mathbf{k}'')}^{(0)} \tag{3.37}$$

$$\vdots$$

Setting $t = 0$ in these expressions and in (28) indicates that $a_{(\mathbf{k})}^{(0)}$ is the initial value of $a_{(\mathbf{k})}$; that is, $a_{(\mathbf{k})}(0)$. The symbol $\otimes$ reminds us that, since $\hat{G}$ is a matrix, intermediate summation must be made in effecting the product $\hat{G}(t)\hat{G}(t'')$. Thus, the explicit form for $a_{(\mathbf{k})}^{(2)}$ is given by

$$a_{(\mathbf{k})}^{(2)} = \int_0^t \int_0^{t'} dt' \, dt'' \sum_{l n \mathbf{K}}{}'' \langle (\mathbf{k}) | \delta \hat{L}_{l n \mathbf{K}} | (\mathbf{k}') \rangle_{t'} \left(\sum_{l \bar{n} \bar{\mathbf{K}}}{}'' \langle (\mathbf{k}') | \delta \hat{L}_{l \bar{n} \bar{\mathbf{K}}} | (\mathbf{k}'') \rangle_{t''} a_{(\mathbf{k}'')}^{(0)} \right) \tag{3.38}$$

We may represent the sum

$$\sum_{l n \mathbf{K}}{}' \langle (\mathbf{k}) | \delta \hat{L}_{l n \mathbf{K}} | (\mathbf{k}') \rangle_t \tag{3.39}$$

by a diagram as described in the following section.

2.3.6 Properties of Diagrams

As noted above, we may represent the matrix element (3.39) by a diagram. Due to the section rules (3.30), such diagrams have one of the six forms shown in Fig. 2.6. These diagrams have the following four basic properties:

1. Each dot or vertex of a diagram represents an interaction in which two particles with given $\mathbf{k}$ values interact with each other and emerge with two different but conserved $\mathbf{k}$ values.
2. Two lines must always enter and emerge from any vertex.
3. If only three lines are connected to a dot as in the vertex ——●< , then one of the "emerging" $\mathbf{k}$ vectors is zero.

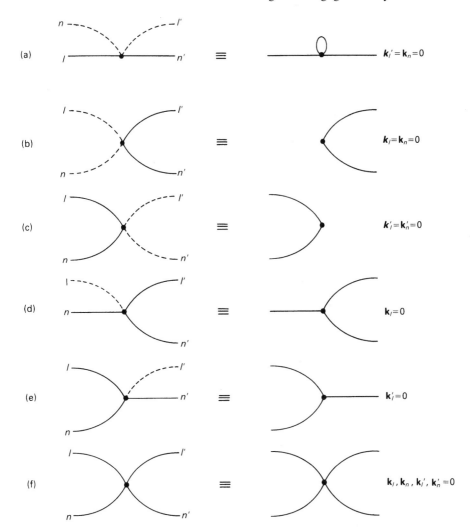

FIGURE 2.6. Six diagrammatic representations of the matrix element (3.39).

4. Since diagrams connect the initial $a_{(\mathbf{k}')}^{(0)}$ state to the present $a_{(\mathbf{k}')}^{(t)}$ state, we may associate the right *ket* side of any given diagram with an earlier time than the left *bra* side. Every vertex is a function of $\mathbf{k}_l$, $\mathbf{k}_n$, $\mathbf{K}$ and times.

As an example of the use of these diagrams, consider that at $t = 0$ the only $a_{(\mathbf{k})}$ coefficient that is different from zero is the a_1 coefficient. Then in equation (3.38), for the second-order coefficient $a_{(\mathbf{k})}^{(2)}$, the initial Fourier coefficient $a_{(\mathbf{k}')}^{(0)}$ becomes $a_1^{(0)}$ and its adjacent matrix element has only two possible diagrams, shown in Fig. 2.7. Due to the matrix product form $|(\mathbf{k}'')\rangle \langle (\mathbf{k}'')|$ in (3.38), these two diagrams can only couple to the five terms shown in Fig. 2.8. We thus obtain five possible contributions to $a^{(2)}(t)$. If we ask for $a_1^{(2)}(t)$, then only two

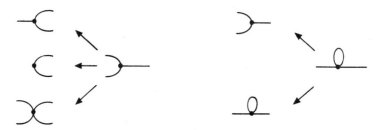

FIGURE 2.7. Diagrams coupling $a_{(\mathbf{k})}^{(2)}$ to $a_1(0)$.

FIGURE 2.8. Coupling diagrams of those in Figure 2.7.

diagrams contribute.

$$a_1^{(2)}(t) = \iint dt' \, dt'' \left\{ \quad \begin{matrix} \text{(diagram)} \end{matrix} \quad + \quad \begin{matrix} \text{(diagram)} \end{matrix} \quad \right\} \qquad (3.40)$$

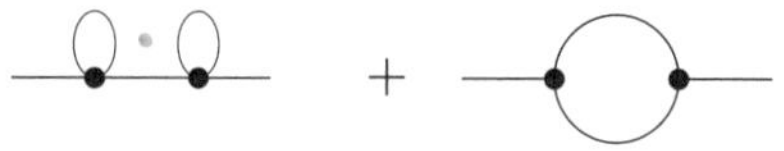

Summation over (ln), $(\bar{l}, \bar{n})$ is tacitly assumed. Each diagram in (3.40) represents a second-order interaction. For example, consider the diagram

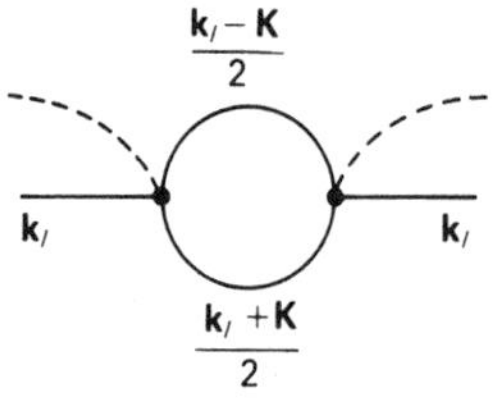

$$= \left\langle \mathbf{k}_l, 0 \left| \delta \hat{L} \right| \frac{\mathbf{k}_l - \mathbf{K}}{2}, \frac{\mathbf{k}_l + \mathbf{K}}{2} \right\rangle \left\langle \frac{\mathbf{k}_l - \mathbf{K}}{2}, \frac{\mathbf{k}_l + \mathbf{K}}{2} \left| \delta \hat{L} \right| \mathbf{k}_l, 0 \right\rangle_{t''}$$

Consider the example of a homogeneous gas for which $f_N(1 \cdots N)$ is independent of $\mathbf{x}^N$. We may write

$$f_N(\mathbf{p}^N, t) = a_{(0)}(\mathbf{p}^N, t)$$

which is seen to be consistent with (3.12) [setting $f_N(\mathbf{x}^N, \mathbf{p}^N = V^{-N} f(\mathbf{p}_N)]$. The equation of motion for a_0 or equivalently a_0 is given by (3.33). Let us assume that the only $a_{(\mathbf{k})}$ coefficients that are not zero initially are the a_0 coefficients. In the expressions (3.37), matrix elements serve to couple the state $\langle 0|$ to the state $|0\rangle$. Some *fourth-order* diagrams that come into play in this coupling are shown in Fig 2.9. These terms contribute to the $a_0^{(4)}(t)$ approximation. The meaning of the first diagram is shown in Fig. 2.10.

The meanings of the four segments of Fig. 2.10 are as follows:

$$D : \left\langle 0, 0, 0 \left| \delta \hat{L} \right| \left(\frac{\mathbf{K} + \mathbf{k}}{2} \right), -\left(\frac{\mathbf{K} + \mathbf{k}}{2} \right), 0 \right\rangle_{t_1}$$

FIGURE 2.9. Fourth-order diagrams.

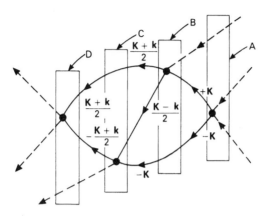

FIGURE 2.10. Interpretation of a diagram.

$$C : \left\langle \left(\frac{\mathbf{K}+\mathbf{k}}{2} \right), -\left(\frac{\mathbf{K}+\mathbf{k}}{2} \right), 0 \, \middle| \, \delta \hat{L} \, \middle| \, \left(\frac{\mathbf{K}+\mathbf{k}}{2} \right), -\mathbf{K} \left(\frac{\mathbf{K}-\mathbf{k}}{2} \right) \right\rangle_{t_2}$$

$$B : \left\langle \left(\frac{\mathbf{K}+\mathbf{k}}{2} \right), -\mathbf{K}, \left(\frac{\mathbf{K}-\mathbf{k}}{2} \right) \, \middle| \, \delta L \, \middle| \, \mathbf{K}, -\mathbf{K}, 0 \right\rangle_{t_3}$$

$$A : \langle \mathbf{K}, -\mathbf{K}, 0 | \delta L | 0, 0, 0 \rangle_{t_4}$$

The time integration of this term is

$$\int_0^t dt_1 \int_0^{t_1} dt_2 \int_0^{t_2} dt_3 \int_0^{t_3} dt_4$$

This term is of order ε^4. Note that each strip (A, B, C, D) involves the interplay of three particles. Thus, of the four fourth-order interaction diagrams shown in Fig. 2.9, only the term represents a fourth-order interaction that includes exclusively two-particle interactions.

2.3.7 Long-Time Diagrams: The Boltzmann Equation

If we restrict our discussion to rare gases, we expect that only two-body interactions are pertinent. Thus, in the homogenous, rare-gas limit, we consider diagrams like

Let us consider the second-order approximation to a_0. In the stated approximation, (3.38) becomes

$$a_0^{(2)}(t) = \int dt' \int dt'' \Leftrightarrow a_0(0)$$

$$= \int dt' \int dt'' \sum_{ln\mathbf{K}} \langle 00|\delta\hat{L}| - \mathbf{K}, \mathbf{K}\rangle_{t'} \langle -\mathbf{K}, \mathbf{K}|\delta\hat{L}|00\rangle_{t''} a_0(0)$$

$$= \int dt' \int dt'' \sum_{ln\mathbf{K}} \langle 00|\delta\hat{L}| - \mathbf{K}, \mathbf{K}\rangle_{t'} \langle 00|\delta\hat{L}| - \mathbf{K}, \mathbf{K}\rangle_{t''}^* a_0(0) \quad (3.41)$$

Here we have written a complex conjugate equality for matrix elements, which, we recall, are operators. Thus, for example,

$$\langle 00|\delta\hat{L}_{ln\mathbf{K}}| - \mathbf{K}, \mathbf{K}\rangle_{t'} = \frac{\mathbf{\Phi}_\kappa}{\bar{V}} i\mathbf{K} \cdot \hat{\boldsymbol{\theta}}_{ln} e^{i\mathbf{g}_{ln}\cdot\mathbf{K}t'}$$

$$\langle 00|\delta\hat{L}_{ln\mathbf{K}}| - \mathbf{K}, \mathbf{K}\rangle_{t'} = -\frac{e^{-i\mathbf{g}_{ln}\cdot\mathbf{K}t''}}{\bar{V}} \mathbf{\Phi}_{-\kappa} i\mathbf{K} \cdot \hat{\boldsymbol{\theta}}_{ln} \quad (3.42)$$

where, since only two $\mathbf{k}$ vectors enter these matrix elements,

$$\omega_{(\mathbf{k})} = \mathbf{K} \cdot \mathbf{v}_l - \mathbf{K} \cdot \mathbf{v}_n = \mathbf{K} \cdot \mathbf{g}_{ln}$$

Substituting these values into (3.41) gives

$$a_0^{(2)}(t) = \frac{1}{\bar{V}^2} \iint dt' \, dt'' \sum_{ln\mathbf{K}} |\mathbf{\Phi}_\kappa|^2 (\mathbf{K} \cdot \hat{\boldsymbol{\theta}}_{ln}) e^{i\mathbf{g}_{ln}\cdot\mathbf{K}(t'-t'')} \mathbf{K} \cdot \hat{\boldsymbol{\theta}}_{ln} a_0(0) \quad (3.43)$$

2.3.8 Reduction of Time Integrals

The time integrals in the preceding expressions may be reduced in the following manner. First note that the time integral in (3.43) is given by

$$I(t) = \int_0^t dt' \int_0^{t'} dt'' e^{i\alpha(t'-t'')} \quad (3.44)$$

The triangular domain of integration is shown in Fig. 2.11. To evaluate the integral, we change variables to (τ, T).

$$\tau = t' - t''$$
$$T = t' + t''$$

This transformation carries the Jacobian

$$J\left(\frac{t', t''}{\tau, T}\right) = \frac{1}{2}$$

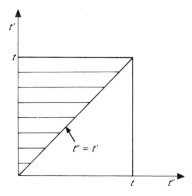

FIGURE 2.11. Domain of integration for (3.44).

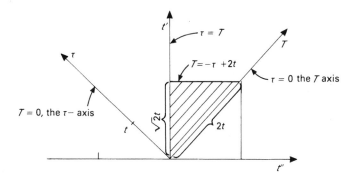

FIGURE 2.12. New domain of integration for (3.44).

The new domain of integration is shown in Fig 2.12. There results

$$I(t) = \frac{1}{2} \int_0^t d\tau \int_\tau^{2t-\tau} dT e^{i\tau\alpha} = \int_0^t d\tau e^{i\tau\alpha}(t - \tau)$$

$$I(t) = \left(t + i\frac{\partial}{\partial\alpha}\right) \int_0^t e^{i\tau\alpha} d\tau \qquad (3.45)$$

In the limit $t \to \infty$,

$$I(t) = \pi \left[t\delta(\alpha) + \frac{\partial}{\partial\alpha}\delta(\alpha)\right] = \pi t\delta(\alpha) \qquad (3.46)$$

As volume grows large, we may set the sum over $\mathbf{K}$ equal to an integral over $\mathbf{K}$ in accordance with (3.11a). Thus (3.43 becomes

$$a_0^{(2)}(t) = \frac{\pi t}{\bar{V}} \sum_{ln} \int d\mathbf{K}|\Phi_\kappa|^2(\mathbf{K} \cdot \hat{\boldsymbol{\theta}}_{ln})(\delta(\mathbf{g}_{ln} \cdot \mathbf{K})\mathbf{K} \cdot \hat{\boldsymbol{\theta}}_{ln}a_0(0)) \qquad (3.47)$$

Having obtained this explicit form, we conjecture on the form of $a_0(t)$ *to all orders in ε.* For weak concentrations, we expect the relaxation of the system to go as $\exp -(t/t_r)$, where $t_r \sim n^{-1}$, the inverse of the number density. So t_r

decreases as n increases. Thus

$$a_0(t) \sim e^{-tn} = \sum \frac{(-tn)^l}{l!}$$

This form suggests the following. In the sum (3.35), keep all terms that contain $(tn)^l$ amplitudes, irrespective of order in ε, but discard terms like $n^m(nt)^l$. From the expression (3.47) for $a_0^{(2)}(t)$, we find the nt dependence (note that this *is not a* dimensional statement):

$$a_0^{(2)}(t) \sim \frac{t}{L^3} \sum_{ln} \sim \frac{tN^2}{L^3} \sim N(tn)$$

Note in particular that the integral term cannot contribute any other V, t or N dependence. In like manner, we find that the density time dependence of the third-order term

also goes as Ntn. In this interaction, two particles start with momenta $\mathbf{k}_l = \mathbf{k}_n = 0$, then gain wave numbers $\mathbf{K}$ and $-\mathbf{K}$, respectively, then through a second interaction gain wave numbers $\mathbf{K}'$, $-\mathbf{K}$, and through a third interaction reassume the values $\mathbf{k}_l = \mathbf{k}_n = 0$. The diagram is therefore called a two-body diagram. On the other hand, the three-body diagram

has the density time dependence Ntn^2. The fourth-order product two-body diagram

$$\text{[diagram]} = \text{[diagram]}$$

has time dependence $N^2(tn)^2$. Continuing in this way, we find that only two-body diagrams must be retained in the low-density limit. It follows that in the expansion (3.35) of $a_0(t)$, terms giving density time dependence which goes as $(tn)^l$, may be grouped in the following manner (with summations over $\mathbf{K}$, l, n understood). Note also that *explicit time dependence has been extracted from diagrams*.

Long-time limit

$$a_0(t) = a_0(0) + t\left[\,\text{⬦}+\text{⬦⬦}+\text{⬦⬦⬦}+\cdots\right]a_0(0)$$
$$+ \frac{t^2}{2}\left[\,\text{⬦}+\text{⬦⬦}+\cdots\right]\left[\,\text{⬦}+\text{⬦⬦}+\cdots\right]a_0(0)$$
$$+\cdots$$

$$(3.48)$$

Let us call

$$\hat{S} \equiv \bullet\!\!-\!\!\bullet + \bullet\!\!-\!\!\bullet\!\!-\!\!\bullet + \bullet\!\!-\!\!\bullet\!\!-\!\!\bullet\!\!-\!\!\bullet + \cdots \tag{3.49}$$

Then the above terms may be collected and rewritten

$$a_0(t) = e^{t\hat{S}} a_0(0) \tag{3.50}$$

Differentiation gives

$$\frac{\partial a_0(t)}{\partial t} = \hat{S} a_0(t) \tag{3.51}$$

Keeping only the $0(\varepsilon)$ term $\bullet\!\!-\!\!\bullet$ gives

$$\frac{\partial a_0}{\partial t} = \bullet\!\!-\!\!\bullet \, a_0 \tag{3.52}$$

$$\frac{\partial a_0(t)}{\partial t} = \frac{\pi}{\bar{V}} \sum_{l<n} \sum \int d\mathbf{K} |\Phi_\kappa|^2 \mathbf{K} \cdot \hat{\boldsymbol{\theta}}_{ln} \delta(\mathbf{K} \cdot \mathbf{g}_{ln}) \mathbf{K} \cdot \hat{\boldsymbol{\theta}}_{ln} a_0$$

$$= \frac{\pi}{m^2 \bar{V}} \sum_{l<n} \sum \int d\mathbf{K} |\Phi_\kappa|^2 \mathbf{K} \cdot \left(\frac{\partial}{\partial \mathbf{v}_l} - \frac{\partial}{\partial \mathbf{v}_n} \right)$$

$$\times \, \delta[\mathbf{K} \cdot (\mathbf{v}_l - \mathbf{v}_n)] \mathbf{K} \cdot \left(\frac{\partial}{\partial \mathbf{v}_l} - \frac{\partial}{\partial \mathbf{v}_n} \right) a_0(t) \tag{3.53}$$

Note that these terms have the correct dimensions. With $V^{-1} = d\mathbf{K}$, the dimensions of the right side of (3.53) may be written

$$\left(\int d\mathbf{K} \mathbf{K} \cdot \hat{\boldsymbol{\theta}} \Phi_\kappa \right)^2 \delta(\mathbf{K} \cdot \mathbf{g}) = \left(\frac{1}{V} \frac{1}{L} \frac{1}{p} V E \right)^2 t = \frac{1}{t}$$

To obtain an equation for the one-particle distribution, we operate on (3.53) with

$$\int d\mathbf{v}_1 \ldots d\mathbf{v}_{r-1} \, d\mathbf{v}_{r+1} \ldots d\mathbf{v}_N$$

The contributions from the l, n summations may be listed in a triangle as shown in Fig. 2.13. All terms except those on the hatched lines are surface terms and give no contribution. The remaining terms give the summations

$$\sum_{l=1}^{r-1} \left(\frac{\partial}{\partial \mathbf{v}_l} - \frac{\partial}{\partial \mathbf{v}_r} \right) \cdots \left(\frac{\partial}{\partial \mathbf{v}_l} - \frac{\partial}{\partial \mathbf{v}_r} \right) + \sum_{n=r+1}^{N} \left(\frac{\partial}{\partial \mathbf{v}_r} - \frac{\partial}{\partial \mathbf{v}_n} \right) \cdots \left(\frac{\partial}{\partial \mathbf{v}_r} - \frac{\partial}{\partial \mathbf{v}_n} \right)$$

where the three dots represent between the explicit $\hat{\theta}$ operators. In the l sum, the lead $\partial/\partial \mathbf{v}_l$ derivatives give surface terms, as do the lead $\partial/\partial \mathbf{v}_n$ derivatives in the n sum. (Only the lead derivatives in these sums correspond to divergence.) So we are left with

$$\sum_{l=1}^{r-1} \left(-\frac{\partial}{\partial \mathbf{v}_r} \right) \cdots \left(\frac{\partial}{\partial \mathbf{v}_l} - \frac{\partial}{\partial \mathbf{v}_r} \right) + \sum_{l=r+1}^{N} \left(\frac{\partial}{\partial \mathbf{v}_r} \right) \cdots \left(\frac{\partial}{\partial \mathbf{v}_r} - \frac{\partial}{\partial \mathbf{v}_n} \right)$$

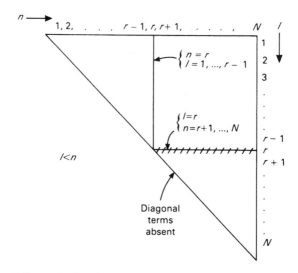

FIGURE 2.13. Triangle for enumerating terms in (3.53).

Changing the dummy variable n to l in the n sum gives

$$\sum_{l=1}^{r-1}\left(-\frac{\partial}{\partial\mathbf{v}_r}\right)\cdots\left(\frac{\partial}{\partial\mathbf{v}_l}-\frac{\partial}{\partial\mathbf{v}_r}\right)+\sum_{l=r+1}^{N}\left(\frac{\partial}{\partial\mathbf{v}_r}\right)\cdots\left(\frac{\partial}{\partial\mathbf{v}_r}-\frac{\partial}{\partial\mathbf{v}_l}\right)$$
$$=\sum_{l=1}^{N}\left(\frac{\partial}{\partial\mathbf{v}_r}\right)\cdots\left(\frac{\partial}{\partial\mathbf{v}_r}-\frac{\partial}{\partial\mathbf{v}_l}\right)$$

In this manner we obtain

$$\frac{\partial f_1(\mathbf{v}_r)}{\partial t}=\frac{\pi}{m^2\bar{V}}\sum_{l}\int d\mathbf{v}_l\int d\mathbf{K}|\Phi_K|^2\mathbf{K}\cdot\frac{\partial}{\partial\mathbf{v}_r}\Big(\delta(\mathbf{K}\cdot(\mathbf{v}_l-\mathbf{v}_n))$$
$$\times\mathbf{K}\cdot\left(\frac{\partial}{\partial\mathbf{v}_r}-\frac{\partial}{\partial\mathbf{v}_l}\right)f_2(\mathbf{v}_r,\mathbf{v}_l)\Big) \qquad (3.54)$$

where f_1 and f_2 are now normalized with respect to velocity. The presence of f_2 in the integrand stems from the fact that v_r and v_l have not been integrated out. The above equation may be rewritten

$$\frac{\partial f(\mathbf{v})}{\partial t}=\frac{\pi}{m^2\bar{V}}N\int d\mathbf{v}'\int d\mathbf{K}|\phi_K|^2\mathbf{K}\cdot\frac{\partial}{\partial\mathbf{v}}\Big(\delta(\mathbf{K}\cdot\mathbf{g})\mathbf{K}\cdot\left(\frac{\partial}{\partial\mathbf{v}}-\frac{\partial}{\partial\mathbf{v}'}\right)f_2(\mathbf{v},\mathbf{v}')\Big)$$
$$(3.55)$$

With molecular chaos, $f_2(\mathbf{v},\mathbf{v}')=f_1(\mathbf{v})f_1(\mathbf{v}')$, we obtain the Boltzmann equation. Corrections to this equation may be obtained incorporation third-order diagrams⚫⚫⚫ The Boltzmann equation is discussed at length in Chapter 3.

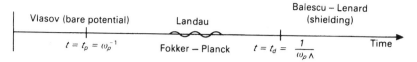

FIGURE 2.14. Time intervals and equations for a plasma.

2.3.9 Application of Diagrams to Plasmas

A plasma is a collection of ions and electrons maintained in separation by sufficiently high temperature. There are two fundamental times for a plasma: The plasma response time, t_p and the dielectric time, t_d.

If an extraneous electron is introduced into the plasma, the response time of the plasma to this disturbance is ω_p^{-1}, where (in cgs)

$$\omega_p^2 = \frac{4\pi n e^2}{m} \tag{3.56}$$

The parameter ω_p is called the *plasma frequency*. So we write

$$t_p = \frac{1}{\omega_p} \tag{3.57}$$

A shielding occurs about the foreign particle in what may be termed the *dielectric time*,

$$t_d = \frac{t_p}{\Lambda} = \frac{1}{\omega_p \Lambda} \tag{3.58}$$

Here Λ is the *plasma parameter*:

$$\Lambda = \frac{1}{4\pi n \lambda_d^3} \tag{3.59}$$

where λ_d is the *Debye distance*.[11] We note the relation

$$\lambda_d^2 \omega_p^2 = C^2 = \frac{k_B T}{m} \tag{3.60}$$

At high temperatures and low densities, $\Lambda \ll 1$, and the plasma is said to be *weakly coupled*. At low temperatures and high densities, $\Lambda \geq 1$, and the plasma is said to be *strongly coupled*.

The relevant kinetic equations that come into play in these intervals are listed in Fig. 2.14, appropriate to the weakly coupled domain.

So for early times, $t < t_p$, we look for decay behavior like

$$e^{-t/t_p} = e^{-\omega_p t} = \sum_l \frac{(-\omega_p t)^l}{l!} \tag{3.61}$$

---

[11]The Debye distance was encountered previously in Section 2.5.

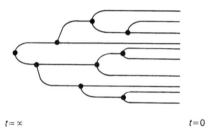

$t = \infty$ $t = 0$

FIGURE 2.15. Diagrams contributing to long-range interactions.

The terms correspond to diagrams with $(e^n nt)^l$ dependence.

For problems concerning the relaxation of an inhomogeneous plasma to a homogeneous plasma, the initial $a_1(t)$ coefficient that comes into play is $a_1(0)$, $l \neq 0$. Due to the long-range nature of the unshielded Coulomb potential, many particles interact and relevant diagrams are of the form shown in Fig 2.15. Summing such diagrams with proper $e^2 nt$ dependence gives the *Vlasov equation* encountered in Section 2.2.

For longer-time behavior, we look for decay like

$$e^{-t/t_d} = e^{-t\omega_p \Lambda} = \sum_l \frac{(t\omega_p \Lambda)^l}{l!}$$

$$\omega_p \Lambda = \frac{\omega_p}{n\lambda_d^3} = \frac{\omega_p^4}{n\lambda_d^3 \omega_p^3} = \frac{\omega_p^4}{nC^3} = \left(\frac{4\pi ne^2}{m}\right)^2 \frac{1}{nC^3} \sim e^4 n \quad (3.62)$$

Thus, dependence of relevant terms in this interval is $(e^4 nt)^l$. The kinetic equation for a homogeneous plasma in this time domain stems from diagrams that couple the initial interaction ⟩• to a final interaction •⟨ . Predominance of "grazing" collisions indicates that $\mathbf{k} = 0$ terms dominate the development of the plasma. Separation of terms with common $(e^4 nt)^l$ dependence gives the series:

$$\bigcirc \; + \; \oplus \; + \; \ominus \; + \; \oplus \; + \; \oplus \; + \; \cdots \; = \; \circledcirc$$

$$(3.63)$$

This is called the *ring approximation* and leads to the *Landau equation*.

Equations describing more general long-time behavior stem from summation of diagrams that carry dependence like $(e^2 n)^p (e^2 t)^q$. Summation of these terms leads to the *Balescu-Lenard equation*. This equation includes a shielding kernel in its interaction integral.[12]

---

[12]For more extensive discussion of the diagrammatic technique and its applications, see I. Prigogine, *Non Equilibrium Statistical Mechanics*, Wiley-Interscience, New York (1963); and R. Balescu, *Statistical Mechanics of Charged Particles*, Wiley-Interscience, New York (1963).

All these kinetic equations relevant to a plasma are discussed more fully in Chapter 4. The Landau equation is derived in Section 4.2.4. The Balescu-Lenard equation is derived in Section 4.2.5. The plasma parameter and the notions of strongly versus weakly coupled plasmas are more fully discussed in Chapter 4. Recall that the concepts of weakly and strongly coupled fluids were discussed earlier in the present chapter (Section 2.1).

2.4 Bogoliubov Hypothesis

2.4.1 Time and Length Intervals

In this section we turn to the Bogoliubov hypothesis for the equilibration of a nonequilibrium gas.[13][14] The hypothesis addresses an enclosed gas and defines three time intervals, which we label τ_1, τ_2, and τ_3. In the interval τ_1, two molecules are in each others interaction domain. The interval τ_2 is the mean-free-collision times, which is the mean times between collisions. The time τ_3 is the average time taken for a molecule to traverse the container in which the gas is confined. For a mole sample of a gas confined to a macroscopic container, we may write

$$\tau_1 \ll \tau_2 \ll \tau_3$$

We may relate three respective displacements to these three time intervals, which we label λ_1, λ_2, and λ_3. The displacement λ_1 is the *range of interaction*,[15] λ_2 is the *mean free path*,[16] and λ_3 is the edge length of the confining container. Typical values of these parameters are listed in Table 2.1.

2.4.2 The Three Temporal Stages

The Bogoliubov hypothesis addresses the functional dependence of the N-particle distribution function, $f_N(1, \ldots, N)$ relevant to three stages in the temporal development of the gas toward equilibrium. These stages are defined in Table 2.2.

The hypothesis relevant to the initial stage in well described by the following example. Consider that initially one of the N molecules in the gas is moving

---

[13]N. N. Bogoliubov in *Studies in Statistical Mechanics*, vol. 1, J. de Boer and G. E. Uhlenbeck, (eds.), Wiley, New York (1962), See also E. D. G. Cohen in *Fundamental Problems in Statistical Mechanics*, E. D. C. Cohen, (ed.), Wiley, New York (1962).

[14]An extension of the Bogoliubov hypothesis to dense fluids was made by R. L. Liboff, *Phys. Rev. A 31*, 1883 (1985); *A32*, 1909 (1985).

[15]The range of interaction was introduced previously in Section 2.1, where it was labeled r_0.

[16]The mean free path is discussed at length in the description of transport coefficients in Chapter 3, where it is labeled l.

TABLE 2.1. Bogoliubov length and time intervals for a gas with mean molecular speed of 300 m/s at standard conditions (container edge length is $\lambda_3 = 3$ cm)

	λ_1	λ_2	λ_3
cm	3×10^{-8}	3×10^{-5}	3
	τ_1	τ_2	τ_3
sec	10^{-12}	10^{-9}	10^{-4}

TABLE 2.2. Epochs in the Bogoliubov Hypothesis

$0 < t < \tau_1$	Initial stage
$\tau_1 < t \leq \tau_2$	Kinetic stage
$\tau_2 < t$	Hydrodynamic stage

in free flight (between collisions) at a speed greatly in excess of the mean speed of the remaining molecules. The N-body distribution function for this nonequilibrium configuration is given by (1.6.7).[17] In the initial interval there is no collisional exchange between molecules, and the initial nonequilibrium state experiences no equilibrating force. Thus it is hypothesized that in the *initial stage* no less than the full N-body distribution function is required to describe the state of the gas.

In the *kinetic stage*, molecules experience collisions and there is a tendency toward equilibration. It is hypothesized that in this interval all s-particle distributions are functionals of $f_1(1)$. That is

$$f_s = f_s(1, \ldots, s; f_1) \tag{4.1}$$

with explicit time dependence of f_s contained entirely in f_1. An example of the functional dependence of (4.1) is given by the case where molecules are statistically independent. For this case, we write

$$f_s = \prod_{i=1}^{s} f_1(i) \tag{4.2}$$

In the *hydrodynamic stage*, it is hypothesized that all distributions depend only on the hydrodynamic variables n, $\mathbf{u}$, and T, where n is number density, $\mathbf{u}$ is macroscopic fluid velocity, and T is temperature.[18]

Thus we have the following description. As the system passes from the initial nonequilibrium state to the finial equilibrium state, there is a diminishment in

---

[17]Note, however, that in (1.6.7) remaining molecules in the gas are at rest. A more correct distribution would include finite but small speeds of remaining molecules.

[18]Fluid dynamic variables are defined and more fully discussed in Chapter 3.

the level of description appropriate to the state of the fluid. In the initial stage, the full N-body distribution is required to describe the state. In equilibrium, the far less informative macroscopic variables n, $\mathbf{u}$, and T suffice.

2.4.3 Bogoliubov Distributions

Again we turn to the F_s distribution defined by (2.6):

$$F_s = V^s f_s \tag{4.3}$$

Note that, for a homogeneous fluid, F_s is dependent only on momentum. Rewriting BY_s (1.20) in terms of these distributions gives

$$\left(\frac{\partial}{\partial t} - \hat{L}_s\right) F_s - \frac{N-s}{V} \sum_{i=1}^{s} \int \hat{O}_{i,s+1} F_{s+1} \, d(s+1) = 0 \tag{4.4}$$

where $\hat{O}_{ij}$ is given by [see (1.7)]

$$\hat{O}_{ij} = -\mathbf{G}_{ij} \cdot \left(\frac{\partial}{\partial \mathbf{p}_i} - \frac{\partial}{\partial \mathbf{p}_j}\right) \tag{4.5}$$

and we have recalled the steps (1.15) and (1.16).

In the *thermodynamic limit*,

$$\lim_{\substack{N \to \infty \\ V \to \infty}} \left(\frac{N-s}{V}\right) = \lim \left(\frac{N}{V}\right) = \frac{1}{v} \tag{4.6}$$

where v, the *specific volume* (mean volume occupied per particle) is a finite parameter. In this limit, (4.4) becomes

$$\left(\frac{\partial}{\partial t} - \hat{L}_s\right) F_s - \frac{1}{v} \sum_{i=1}^{s} \int \hat{O}_{i,s+1} F_{s+1} \, d(s+1) = 0 \tag{4.7}$$

2.4.4 Density Expansion

In the kinetic stage, (4.1) is relevant, and we write, equivalently,

$$F_s = F_s(1, \ldots, s; F_1) \tag{4.8}$$

Since the explicit time dependence of F_s is contained entirely in F_1, we write

$$\frac{\partial F_s}{\partial t} = \left\langle \frac{\delta F_s}{\delta F_1}, \frac{\partial F_1}{\partial t} \right\rangle \tag{4.9}$$

The bracket notation denotes an inner product over all F_1 functions. As a simple example, again consider the case of statistical independence of molecules. In this event we write

$$F_s = \prod_{l=1}^{s} F_1(l) \tag{4.10a}$$

Substituting into (4.9), we obtain

$$\frac{\partial F_s}{\partial t} = \sum_k \frac{\partial F_1(k)}{\partial t} \prod_{l \neq k} F_1(l) \tag{4.10b}$$

As noted previously, the Bogoliubov analysis is relevant to a rare gas. With this motivation, we introduce the expansion

$$F_s(1, \ldots, s; F_1) = F_s^0 + \frac{1}{v} F_s^{(1)} + \frac{1}{v^2} F_s^{(2)} + \cdots \tag{4.11}$$

Substituting (4.11) into BY$_1$ [obtained from (4.7)], gives

$$\frac{\partial F_1}{\partial t} = -\frac{\mathbf{p}_1}{m} \cdot \frac{\partial F_1}{\partial \mathbf{x}_1} + \frac{1}{v} \hat{I}_{12} \left[F_2^{(0)} + \frac{1}{v} F_2^{(1)} + \cdots \right]$$

$$\equiv A^{(0)} + \frac{1}{v} A^{(1)} + \cdots \tag{4.12}$$

$$\hat{I}_{12} \equiv \int d2 \hat{O}_{12}$$

which serves to identify the $A^{(s)}$ coefficients. Substituting the preceding two expansions into the inner product (4.9), gives

$$\frac{\partial F_s}{\partial t} = \left\langle \frac{\delta F_s^{(0)}}{\delta F_1}, A^{(0)} \right\rangle + \frac{1}{v} \left\{ \left\langle \frac{\delta F_s^{(1)}}{\delta F_1}, A^{(0)} \right\rangle + \left\langle \frac{\delta F_s^{(0)}}{\delta F_1}, A^{(0)} \right\rangle \right\} + \cdots \tag{4.13}$$

This expansion may be written in the alternative operational form

$$\frac{\partial F_s}{\partial t} = \hat{D}^{(0)} F_s^{(0)} + \frac{1}{v} \left[\hat{D}^{(0)} F_s^{(1)} + \hat{D}^{(1)} F_s^{(0)} \right] + \cdots \tag{4.14}$$

where the $D^{(s)}$ operators are as implied.

Now we return to (4.7) and, inserting the expansion (4.11), obtain

$$\frac{\partial F_s}{\partial t} = \hat{L}_s \left[F_s^{(0)} + \frac{1}{v} F_s^{(1)} + \cdots \right] + \frac{1}{v} \sum_{i=1}^{s} \hat{I}_{i,s+1} \left[F_{s+1}^{(0)} + \frac{1}{v} F_{s+1}^{(1)} + \cdots \right] \tag{4.15}$$

with this equation at hand, we note that, together with (4.14) [which stems from (4.19) and BY$_1$], it comprises two equations for $\partial F_s / \partial t$. Setting

$$\frac{\partial F_s}{\partial t}(4.14) = \frac{\partial F_s}{\partial t}(4.15) \tag{4.16}$$

and equating terms of equal power in $1/v$ gives the sequence

$$\hat{D}^{(0)} F_s^{(0)} = \hat{L}_s F_s^{(0)} \tag{4.17a}$$

$$\hat{D}^{(0)} F_s^{(1)} + \hat{D}^{(1)} F_s^{(0)} = \hat{L}_s F_s^{(1)} + \sum_{i=1}^{s} \hat{I}_{i,s+1} F_{s+1}^{(0)} d(s+1) \tag{4.17b}$$

$$\vdots$$

$$\sum_{k=0}^{n} \hat{D}^{(k)} F_s^{(n-k)} = \hat{L}_s F_s^{(n)} + \sum_{i=1}^{s} \hat{I}_{i,s+1} F_{s+1}^{(n-1)} \qquad (4.17c)$$

$$\vdots$$

Equations (4.17) are self-contained functional equations that, in principle, determine the sequence $\{F_s^{(n)}\}$ in the expansion (4.11). A key element of this analysis, as will be found below, concerns the operator $\hat{L}_s$. We recall that this operator, as given by (1.11), pertains to a collection of s particles that interact with each other but are otherwise independent of the remaining aggregate.

We will not apply the sequence (4.17) to find a closed kinetic equation for F_1 in the low-density limit valid to order v^{-1}.

2.4.5   Construction of $F_2^{(0)}$

Our program for obtaining the said kinetic equation is as follows. To $O(\frac{1}{v})$, the relation (4.12) becomes

$$\frac{\partial F_1}{\partial t} + \frac{\mathbf{p}_1}{m} \cdot \frac{\partial F_1}{\partial \mathbf{x}_1} = \frac{1}{v} \int d2 \hat{O}_{12} F_2^{(0)} \qquad (4.18)$$

Thus, if we obtain $F_2^{(0)}(F_1)$, a closed kinetic equation for F_1 is established. To these ends, we recall (4.17a) and with $s = 2$ write

$$\hat{D}^{(0)} F_2^{(0)} = \hat{L}_2 F_2^{(0)} \qquad (4.19)$$

We must solve this equation for $F_2^{(0)}(F_1)$ subject to an appropriate boundary condition. We take this to be the condition that, when carried sufficiently far back in time, particles in a given aggregate are uncorrelated.

The Liouville operator, $\hat{L}_N$, was introduced in (3.1). For an isolated system of s we may write

$$\frac{\partial F_s}{\partial t} = [H_s, F_s] \equiv \hat{L}_s F_s \qquad (3.2a)$$

whose solution is given by

$$F_s(t) = e^{t\hat{L}_s} F_s(0) \equiv \hat{\Delta}_{-t}^{(s)} F_s(0) \qquad (4.20)$$

We have found previously in Section 1.5.3 that $\exp t\hat{L}_s$ propagates particle phase variables back in time[19] through the interval t to values determined by the Hamiltonian for the s-particle system. So we label the operator $\hat{\Delta}_{-t}^{(s)}$. Here are some properties of this operator:

$$\hat{\Delta}_0 = 1 \qquad (4.21a)$$

---

[19]With reference to Section 1.5.3, we note that $\hat{\Lambda}_s = i\hat{L}_s$. Recall also that $\hat{L}_s$ operates on the phase variables of F_s.

$$\hat{\Delta}_{-t_1}\hat{\Delta}_{-t_2} = \hat{\Delta}_{-(t_1+t_2)} \tag{4.21b}$$

$$\frac{\partial \hat{\Delta}_{-t}}{\partial t} = \hat{L}\hat{\Delta}_{-t} = \hat{\Delta}_{-t}\hat{L} \tag{4.21c}$$

Let us employ the $\hat{\Delta}_t$ operator to express the boundary condition on F_s described above:

$$\lim_{t\to\infty} \hat{\Delta}^{(s)}_{-t} F_s(1,\ldots,s;F_1) = \lim_{t\to\infty} \hat{\Delta}^{(s)}_{-t} \prod_{k=1}^{s} F_1(k) \tag{4.22}$$

Since this boundary condition holds for all $F_1(k)$, it also holds for $F_1' = \lim_{t\to\infty} \hat{\Delta}^{(1)}_t F_1(k)$. Substituting this value into (4.22) gives

$$\lim_{t\to\infty} \hat{\Delta}^{(s)}_{-t} F_s\left(1,\ldots,s;\hat{\Delta}^{(1)}_t F_1\right) = \lim_{t\to\infty} \hat{\Delta}^{(s)}_{-t} \prod_{k=1}^{s} \hat{\Delta}^{(1)}_t F_1(k) \tag{4.23}$$

As $\hat{\Delta}^{(1)}_t$ relates to a one-particle system, it operates on free particle phase variables. Thus

$$\text{RHS (4.23)} = \lim_{t\to\infty} \hat{\Delta}^{(s)}_{-t} \prod_{k=1}^{s} F_1\left(\mathbf{x}_k + \frac{\mathbf{p}_k}{m}t, \mathbf{p}_k\right)$$

$$= \lim_{t\to\infty} \prod_{k=1}^{s} F_1\left[\hat{\Delta}^{(s)}_{-t}\left(\mathbf{x}_k + \frac{\mathbf{p}_k}{m}t\right), \hat{\Delta}^{(s)}_{-t}\mathbf{p}_k\right]$$

$$= \prod_{k=1}^{s} F_1(\mathbf{x}_k^{(s)}, \mathbf{p}_k^{(s)}) \tag{4.24}$$

where $\mathbf{x}_k^{(s)}$ and $\mathbf{p}_k^{(s)}$ are phase values of the kth particle traced back in time to $t = -\infty$, under the interaction of s particles from respective starting values $\mathbf{x}_k + (\mathbf{p}_k/m)t$ and $\mathbf{p}_k$.

With these expressions at hand, we return to (4.19) and with the identification of (4.13), 4.14 write

$$\hat{D}^{(0)} F_s^{(0)}(1,\ldots,s;F_1) = \left\langle \frac{\delta F_s^{(0)}}{\delta F_1}, A^{(0)}(F_1) \right\rangle = \left\langle \frac{\delta F_s^{(0)}}{\delta F_1}, \left(\hat{L}_1 F_1\right) \right\rangle \tag{4.25}$$

A tractable representation of the functional derivative in the preceding equation is obtained as follows. Again we note that since (4.25) is true for all F_1 it is also true for $\hat{\Delta}^{(1)}_t F_1$. This permits us to write

$$\hat{D}^{(0)} F_s^{(0)}\left[1,\ldots,s;\hat{\Delta}^{(1)}_{-t}F_1\right] = \left\langle \frac{\delta F_s^{(0)}}{\delta\left[\hat{\Delta}^{(1)}_{-1}F_1\right]}, \left[\hat{L}_1\hat{\Delta}^{(1)}_{-1}F_1\right] \right\rangle \tag{4.26}$$

With the property (4.21c), the preceding may be rewritten

$$\hat{D}^{(0)} F_s^{(0)} = \left\langle \frac{\delta F_s^{(0)}}{\delta \left[\hat{\Delta}_{-t}^{(1)} F_1 \right]}, \frac{\partial \left[\hat{\Delta}_{-t}^{(1)} F_1 \right]}{\partial t} \right\rangle \tag{4.27}$$

which gives the desired representation

$$\hat{D}^{(0)} F_s^{(0)} \left[1, \dots, s; \hat{\Delta}_{-t}^{(1)} F_1 \right] = \frac{\partial}{\partial t} F_s^{(0)} \left[1, \dots, s; \hat{\Delta}_{-t}^{(1)} F_1 \right] \tag{4.28}$$

Substituting this relation into (4.17a) gives

$$\frac{\partial}{\partial t} F_s^{(0)} \left[1, \dots, s; \hat{\Delta}_{-t}^{(1)} F_1 \right] = \hat{L}_s F_s^{(0)} \left[1, \dots, s; \hat{\Delta}_{-1}^{(1)} F_1 \right] \tag{4.29}$$

With (4.20) it is evident that the solution to (4.29) is given by

$$F_s^{(0)} \left[1, \dots, s; \hat{\Delta}_{-t}^{(1)} F_1 \right] = \Delta_{-t}^{(s)} F_s^{(0)} [1, \dots, s; F_1] \tag{4.30}$$

Equivalently, we may write

$$F_s^{(0)} [1, \dots, s; F_1] = \Delta_{-1}^{(s)} F_s^{(0)} \left[1, \dots, s; \hat{\Delta}_t^{(1)} F_1 \right]$$

As this solution is valid for all t intervals, the limiting form

$$F_s^{(0)} [1, \dots, s; F_1] = \lim_{t \to \infty} \hat{\Delta}_{-1} F_s^{(0)} \left[1, \dots, s; \hat{\Delta}_t^{(1)} F_1 \right] \tag{4.31}$$

is likewise a solution. Incorporating our previously stated boundary-condition relations (4.23) and (4.24) gives the final desired solution:

$$F_s^{(0)}(1, \dots, s; F_1) = \prod_{k=1}^{s} F_1 \left[\mathbf{x}_k^{(s)}, \mathbf{p}_k^{(s)} \right] \tag{4.32}$$

Thus, the lowest-order solution $F_s^{(0)}$ in the Bogoliubov analysis is given by a product of one-particle distributions with phase variables $(\mathbf{x}^{(s)}, \mathbf{p}^{(s)})$ displaced back in time and coupled through the subspace s-particle Hamiltonian.

2.4.6 Derivation of the Boltzmann Equation

The final step in our program for obtaining a closed kinetic equation is to obtain a consistent representation for the $\hat{L}_2$ operator in (4.19).

To these ends, with (1.11) we write

$$\hat{L}_2 = -\hat{\kappa}_2 \hat{O}_{12} \tag{4.33}$$

where, we recall,

$$\hat{\kappa}_2 = \frac{\mathbf{p}_1}{m} \cdot \frac{\partial}{\partial \mathbf{x}_1} + \frac{\mathbf{p}_2}{m} \cdot \frac{\partial}{\partial \mathbf{x}_2}$$

and $\hat{O}_{12}$ is given by (4.5). With (4.33) we write

$$\hat{O}_{12} F_2^{(0)} = \hat{\kappa}_2 F_2^{(0)} + \hat{L}_2 F_2^{(0)} \tag{4.34}$$

We now find separate expressions for the two terms on the right of (4.34). With these expressions at hand, we return to (4.18) to obtain the desired kinetic equation. We start with $\hat{L}_2 F_2^{(0)}$.

Returning to (4.19) and employing (4.13) gives

$$\hat{L}_2 F_2^{(0)} = \hat{D}^{(0)} F_2^{(0)} = \left\langle \frac{\delta F_2^{(0)}}{\delta F_1}, A^{(0)} \right\rangle \tag{4.35}$$

where with (4.12) we write

$$A^{(0)} = -\frac{\mathbf{p}}{m} \cdot \frac{\partial F_1(\mathbf{x}, \mathbf{p})}{\partial \mathbf{x}} \tag{4.36}$$

Employing the solution for $F_2^{(0)}$ given by (4.32) in the inner product in (4.35) gives

$$\hat{L}_2 F_2^{(0)} = -\frac{\mathbf{p}_1^{(2)}}{m} F_1 \left[\mathbf{x}_2^{(2)}, \mathbf{p}_2^{(2)} \right] \cdot \frac{\partial}{\partial \mathbf{x}_1^{(2)}} F_1 \left[\mathbf{x}_1^{(2)}, \mathbf{p}_1^{(2)} \right]$$

$$- \frac{\mathbf{p}_2^{(2)}}{m} F_1 \left[\mathbf{x}_1^{(2)}, \mathbf{p}_1^{(2)} \right] \cdot \frac{\partial}{\partial \mathbf{x}_2^{(2)}} F_1 \left[\mathbf{x}_2^{(2)}, \mathbf{p}_2^{(2)} \right] \tag{4.37}$$

We wish now to find an alternative expression for the operator $\hat{\kappa}_2$. Introducing the transformation of variables $(\mathbf{x}_1, \mathbf{x}_2, \mathbf{p}_1, \mathbf{p}_2) \to (\mathbf{r}, \mathbf{x}_1, \mathbf{g}, \mathbf{p}_1)$, where

$$\mathbf{r} = \mathbf{x}_2 - \mathbf{x}_1$$
$$\mathbf{g} = \frac{\mathbf{p}_2 - \mathbf{p}_1}{m} \tag{4.38}$$

permits $\hat{\kappa}_2$ to be written

$$\hat{\kappa}_2 = \mathbf{g} \cdot \frac{\partial}{\partial \mathbf{r}} + \frac{\mathbf{p}_1}{m} \cdot \frac{\partial}{\partial \mathbf{x}_1} \equiv \hat{K}_B + \hat{K}_1 \tag{4.39}$$

The subscript B on the operator $\hat{K}_B = \mathbf{g} \cdot \partial/\partial \mathbf{r}$ denotes Boltzmann.

Inserting this expression into (4.34) and substituting the resulting form into (4.18) gives

$$\frac{\partial F_1}{\partial t} + \frac{\mathbf{p}_1}{m} \cdot \frac{\partial F_1}{\partial \mathbf{x}_1} = \frac{1}{v} \int d2 \hat{K}_B F_2^{(0)} + \frac{1}{v} \int \left(\hat{L}_2 - \hat{K}_1 \right) F_2^{(0)} d2 \tag{4.40}$$

If $F_2^{(0)}$ is homogeneous (that is, independent of $\mathbf{x}_1$ and $\mathbf{x}_2$) over the interaction domain, the second integral on the right side of (4.40) vanishes and the *collision integral* reduces to

$$\int d2 \hat{K}_B F_2^{(0)} = \int d\mathbf{x}_2 \, d\mathbf{p}_2 \mathbf{g} \cdot \frac{\partial}{\partial \mathbf{r}} \left[F_1 \left(\mathbf{p}_1^2 \right) F_1 \left(\mathbf{p}_2^{(2)} \right) \right] \tag{4.41}$$

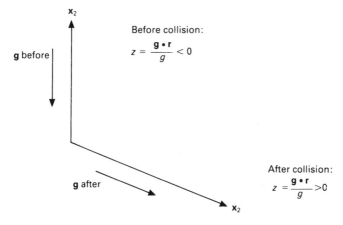

FIGURE 2.16. Asymptotic z values in the Bogoliubov derivation.

where $\mathbf{g}$ is the relative velocity defined in (4.38). The spatial integral in (4.41) may be evaluated as follows. We work in cylindrical coordinates with the cylinder axis parallel to $\mathbf{g}$. For the projection of $\mathbf{r}$ onto this axis, we have

$$z \equiv \frac{\mathbf{g} \cdot \mathbf{r}}{g} \qquad (4.42a)$$

The cylinder radius b and azimuthal angle ϕ complete the description, and the spatial volume in (4.41) becomes

$$d\mathbf{x}_2 = d\mathbf{r} = b\, d\phi\, db\, dz$$

$$ \qquad (4.42b)$$

$$\mathbf{g} \cdot \frac{\partial}{\partial \mathbf{r}} = g \frac{\partial}{\partial z}$$

Integrating the right side of (4.41) over z gives

$$\int_{-\infty}^{\infty} dz \frac{d}{dz} \left[F_1\left(\mathbf{p}_1^{(2)}\right) F_1\left(p_2^{(2)}\right) \right] = F_1\left(\mathbf{p}_1^{(2)}\right) F_1\left(p_2^{(2)}\right) \Big|_{-\infty}^{+\infty} \qquad (4.43)$$

The value $z = +\infty$ corresponds to the domain after interaction (see Fig 2.16). If $\mathbf{p}_1$ and $\mathbf{p}_2$ are traced back to the interval prior to collision, they assume values that give $\mathbf{p}_1$ and $\mathbf{p}_2$ after collision. Let us label these values of momenta $\mathbf{p}_1'$ and $\mathbf{p}_2'$. The value $z = -\infty$ corresponds to the domain prior to collision. What values do $\mathbf{p}_1$ and $\mathbf{p}_2$ attain if traced back to the infinite past from this domain? As particles are assumed not to be in each other's interaction domain prior to collision they move as free particles and the answer is $\mathbf{p}_1$, $\mathbf{p}_2$. Thus we write

$$F_1\left(\mathbf{p}_1^{(2)}\right) F_1\left(\mathbf{p}_2^{(2)}\right) \Big|_{-\infty}^{+\infty} = F_1\left(\mathbf{p}_1'\right) F_1\left(\mathbf{p}_2'\right) - F_1(\mathbf{p}_1) F_1(\mathbf{p}_2) \qquad (4.44)$$

Substituting these values into (4.41) and then into (4.39) (with the $\hat{L}_2 - \hat{K}_2'$ integral neglected and $d\mathbf{p}_2 = m\, d\mathbf{g}$) gives the standard form of the *Boltzmann equation*:

$$\frac{\partial F_1}{\partial t} + \frac{\mathbf{p}_1}{m} \cdot \frac{\partial F_1}{\partial \mathbf{x}_1} = \frac{2\pi}{v} \int d\mathbf{p}_2 \int b \, db \, g \left[F_1(\mathbf{p}_1')F_1(\mathbf{p}_2') - F_1(\mathbf{p}_1)F_1(\mathbf{p}_2) \right]$$

(4.45)

This equation is in a more canonical form than the previously obtained Fourier-structured equation (3.55). Again it is noted that a more direct derivation of the Boltzmann equation together with a full discussion of collision terms is given in Chapter 3.

As with the Prigogine technique, the Bogoliubov analysis also provides a recipe for generating corrections to the Boltzmann equation to higher density configurations.[20]

Bogoliubov's hypothesis relevant to the hydrodynamic stage finds strong corroboration in discussion of the Boltzmann $\mathcal{H}$-theorem in Section 2.5. Here it is found that the one-particle distribution of a fluid in local equilibrium is a functional of n, $\mathbf{u}$, and T, as stipulated by the Bogoliubov hypothesis. This important distribution is called the *local Maxwellian* and is discussed in detail in Section 3.3.10.

2.5   Klimontovich Picture

In this section we turn to an alternative description of kinetic theory formulated by Klimontovich.[21] In this formalism, phase densities are introduced, moments of which are found to be related to multiparticle distribution functions introduced earlier. Moments of deviations from the mean of these phase densities are similarly found to be related to previously encountered correlation functions. Applying these algorithms to an equation of motion for the first moment of the phase densities is found the first equation of the hierarchy (BY_1).

2.5.1   *Phase Densities*

The Klimontovich *phase density* for a system comprised of N particles is given by the following delta-function expression:

$$\mathcal{N}(\mathbf{z}, \{\mathbf{z}_i\}) = \sum_{i=1}^{N} \delta(\mathbf{z} - \mathbf{z}_i)$$

(5.1)

where

$$\mathbf{z}_i \equiv [\mathbf{x}_i(t), \mathbf{p}_i(t)]$$

(5.2a)

---

[20] See, for example, E. D. G. Cohen, *Acta Physica Austriaca, Suppl. X*, 157 (1973); J. R. Dorfman, *Physica 106A* , 77 (1981).

[21] Yu L. Klimontovich, *Statistical Theory of Non-Equilibrium Processes in a Plasma*, MIT Press, Cambridge, Mass. (1967).

represents the phase point of the ith particle and

$$\delta(\mathbf{z} - \mathbf{z}_i) \equiv \delta(\mathbf{x} - \mathbf{x}_i)\delta(\mathbf{p} - \mathbf{p}_i) \tag{5.2b}$$

The phase variable $\mathbf{z} = (\mathbf{x}, \mathbf{p})$ is a point in six-dimensional (for point particles) μ-space.

The normalization of $\mathcal{N}(\mathbf{z}, \{\mathbf{z}_i\})$ (dropping the $\{\mathbf{z}_i\}$ notation) is given by

$$\int \mathcal{N}(\mathbf{z})\, d\mathbf{z} = \sum_{i=1}^{N} \int \delta(\mathbf{z} - \mathbf{z}_i)\, d\mathbf{z} = N \tag{5.3}$$

Integrating over an infinitesimally small domain, $\Delta\mathbf{z}$, of μ-space and recalling the mean-value theorem gives

$$\mathcal{N}(\mathbf{z})\Delta\mathbf{z} = \Delta N \tag{5.4}$$

where ΔN is the number of particles in the infinitesimal phase volume $\Delta\mathbf{z}$.

2.5.2 Phase Density Averages

The average of $\mathcal{N}(z)$ is given by the rule (1.6.8), and we write[22]

$$\overline{\mathcal{N}(z)} = \int d\mathbf{z}_1 \cdots d\mathbf{z}_N\, f_N(\mathbf{z}_1, \ldots, \mathbf{z}_N) \sum_{i=1}^{N} \delta(\mathbf{z} - \mathbf{z}_i) \tag{5.5}$$

$$= \int d\mathbf{z}_2\, d\mathbf{z}_3 \cdots f_N(\mathbf{z}, \mathbf{z}_2, \ldots) + \int d\mathbf{z}_1\, d\mathbf{z}_3 \cdots f_N(\mathbf{z}_1, \mathbf{z}, \mathbf{z}_3, \ldots) + \cdots$$

Collecting terms gives

$$\overline{\mathcal{N}(\mathbf{z})} = N f_1(\mathbf{z}) \tag{5.6}$$

To obtain a parallel relation for the second moment,[23] $\overline{\mathcal{N}(\mathbf{z})\mathcal{N}(\mathbf{z}')}$, we first effect a division of double-sum delta functions as follows:

$$\sum_{i=1}^{N} \sum_{j=1}^{N} \delta(\mathbf{z} - \mathbf{z}_i)\delta(\mathbf{z}' - \mathbf{z}_j)$$

$$= \sum_{i} \sum_{\substack{j \\ i \neq j}} \delta(\mathbf{z} - \mathbf{z}_i)\delta(\mathbf{z}' - \mathbf{z}_j) + \delta_{i,j} \sum_{i} \delta(\mathbf{z} - \mathbf{z}_i)\delta(\mathbf{z}' - \mathbf{z}) \tag{5.7}$$

Again, with (1.6.8) we write

$$\overline{\mathcal{N}(\mathbf{z})\mathcal{N}(\mathbf{z}')} = \int d\mathbf{z}_1 \cdots d\mathbf{z}_N\, f_N(\mathbf{z}_1, \ldots, \mathbf{z}_N)$$

$$\times \left[\sum_{\substack{\cdot \\ i \neq j}} \sum \delta(\mathbf{z} - \mathbf{z}_i)\delta(\mathbf{z}' - \mathbf{z}_j) + \delta_{i,j} \sum \delta(\mathbf{z} - \mathbf{z}_i)\delta(\mathbf{z}' - \mathbf{z}) \right]$$

---

[22]Time dependence of distributions is tacitly assumed.

[23]That is, moments with respect to the N-particle distribution.

$$= N(N-1) \int d\mathbf{z}_i \, d\mathbf{z}_j \, f_2(\mathbf{z}_i, \mathbf{z}_j) \delta(\mathbf{z} - \mathbf{z}_i) \delta(\mathbf{z}' - \mathbf{z}_j)$$

$$+ N\delta(\mathbf{z} - \mathbf{z}') \int d\mathbf{z}_i \, f_1(\mathbf{z}_i) \delta(\mathbf{z} - \mathbf{z}_i) \tag{5.8}$$

In the last equality, $\mathbf{z}_i$ and $\mathbf{z}_j$ represent two arbitrary terms in the summations. Integrating gives the desired result:

$$\overline{\mathcal{N}(\mathbf{z})\mathcal{N}(\mathbf{z}')} = N(N-1)f_2(\mathbf{z}, \mathbf{z}') + N\delta(\mathbf{z} - \mathbf{z}')f_1(\mathbf{z}) \tag{5.9}$$

Continuing in this manner, for the third moment of $\mathcal{N}(\mathbf{z})$ we find

$$\overline{\mathcal{N}(\mathbf{z})\mathcal{N}(\mathbf{z}')\mathcal{N}(\mathbf{z}'')}$$

$$= 3!\binom{N}{3}f_3(\mathbf{z}, \mathbf{z}', \mathbf{z}'') + 2!\binom{N}{2}[\delta(\mathbf{z} - \mathbf{z}')f_2(\mathbf{z}, \mathbf{z}'') \tag{5.10}$$

$$+ \delta(\mathbf{z}-\mathbf{z}'')f_2(\mathbf{z}, \mathbf{z}') + \delta(\mathbf{z}'-\mathbf{z}'')f_2(\mathbf{z}, \mathbf{z}'')] + N\delta(\mathbf{z} - \mathbf{z}')\delta(\mathbf{z}' - \mathbf{z}'')f_1(\mathbf{z})$$

Equations (5.6), (5.9), and (5.10) may be concisely written in terms of the s-tuple distributions (1.6.14). There results

$$\overline{\mathcal{N}(\mathbf{z})} = F_1(\mathbf{z}) \tag{5.6a}$$

$$\overline{\mathcal{N}\mathbf{z}\mathcal{N}(\mathbf{z}')} = F_2(\mathbf{z}, \mathbf{z}') + \delta(\mathbf{z} - \mathbf{z}')F_1(\mathbf{z}) \tag{5.9a}$$

$$\overline{\mathcal{N}\mathbf{z}\mathcal{N}(\mathbf{z}')\mathcal{N}(\mathbf{z}'')} = F_2(\mathbf{z}, \mathbf{z}', \mathbf{z}'') + \delta(\mathbf{z} - \mathbf{z}')F_2(\mathbf{z}, \mathbf{z}'') + \delta(\mathbf{z} - \mathbf{z}'')F_2(\mathbf{z}, \mathbf{z}')$$

$$+ \delta(\mathbf{z}' - \mathbf{z}'')F_2(\mathbf{z}, \mathbf{z}'') + \delta(\mathbf{z} - \mathbf{z}')\delta(\mathbf{z}' - \mathbf{z}'')F_1(\mathbf{z}) \tag{5.10a}$$

Continuing in this way, we obtain a mapping between the moments of $\mathcal{N}(\mathbf{z})$ and the s-particle distribution functions.

2.5.3   Relation to Correlation Functions

Let us introduce the deviation of the phase density $\mathcal{N}(\mathbf{z})$ from its mean, which we label

$$\delta\mathcal{N}(\mathbf{z}) = \mathcal{N}(\mathbf{z}) - \overline{\mathcal{N}(\mathbf{z})}$$

With $\overline{\delta\mathcal{N}(\mathbf{z})} = 0$, we obtain

$$\overline{\mathcal{N}(\mathbf{z})\mathcal{N}(\mathbf{z}')} = \overline{\mathcal{N}(\mathbf{z})\mathcal{N}(\mathbf{z}')} + \overline{\delta\mathcal{N}(\mathbf{z})\delta\mathcal{N}(\mathbf{z}')} \tag{5.11}$$

In the limit $N \gg 1$, (5.6) and (5.9) give

$$\overline{\mathcal{N}(\mathbf{z})} = N f_1(\mathbf{z})$$

$$\overline{\mathcal{N}(\mathbf{z})\mathcal{N}(\mathbf{z}')} = N^2 f_2(\mathbf{z}, \mathbf{z}') + N\delta(\mathbf{z} - \mathbf{z}')f_1(\mathbf{z}) \tag{5.12}$$

Recalling the first equation in the sequence (2.19), which defines the correlation function $C_2(1, 2)$ (written in terms of the $\mathbf{z}, \mathbf{z}'$ variables), and comparing with (5.11) and (5.12), we obtain

$$\overline{\delta\mathcal{N}(\mathbf{z})\delta\mathcal{N}(\mathbf{z}')} = N^2 C_2(\mathbf{z}, \mathbf{z}') + N\delta(\mathbf{z} - \mathbf{z}')f_1(\mathbf{z}) \tag{5.13}$$

In similar manner, equations may be obtained connecting the higher-order correlation functions with higher-order mean products of deviation from the mean of the phase functions.

2.5.4   Equation of Motion

Finally, we turn to an equation of motion for the phase density (5.1). With arguments leading to the Liouville equation (1.4.7) taken to be appropriate to $\mathcal{N}(\mathbf{z})$ in μ-space, we obtain

$$\frac{\partial \mathcal{N}(\mathbf{z})}{\partial t} + \mathbf{v} \cdot \frac{\partial \mathcal{N}(\mathbf{z})}{\partial \mathbf{x}} + \dot{\mathbf{p}} \cdot \frac{\partial \mathcal{N}(\mathbf{z})}{\partial \mathbf{p}} = 0$$

(5.14)

$$\dot{\mathbf{p}} = \int \mathcal{N}(\mathbf{z}')\mathbf{G}(\mathbf{x}, \mathbf{x}') \, d\mathbf{z}'$$

The two-particle force $\mathbf{G}(\mathbf{x}, \mathbf{x}')$ is assumed to be of the symmetric form

$$\mathbf{G}(\mathbf{x}, \mathbf{x}') = \mathbf{G}(|\mathbf{x} - \mathbf{x}'|)$$

(5.14a)

Combining the latter two equations, we obtain

$$\frac{\partial \mathcal{N}(\mathbf{z})}{\partial t} + \mathbf{v} \cdot \frac{\partial \mathcal{N}(\mathbf{z})}{\partial \mathbf{x}} + \frac{\partial}{\partial \mathbf{p}} \cdot \int \mathbf{G}(\mathbf{x}, \mathbf{x}')\mathcal{N}(\mathbf{z})\mathcal{N}(\mathbf{z}') \, d\mathbf{z}' = 0$$

(5.15)

Taking the average of this equation gives

$$\frac{\partial \overline{\mathcal{N}(\mathbf{z})}}{\partial t} + \mathbf{v} \cdot \frac{\partial \overline{\mathcal{N}(\mathbf{z})}}{\partial \mathbf{x}} + \frac{\partial}{\partial \mathbf{p}} \cdot \int \mathbf{G}(\mathbf{x}, \mathbf{x}')\overline{\mathcal{N}(\mathbf{z})\mathcal{N}(\mathbf{z}')} \, d\mathbf{z}' = 0$$

(5.16)

With (5.6) and (5.9), (5.16) gives (in the limit $N \gg 1$)

$$\frac{\partial f_1(\mathbf{z})}{\partial t} + \mathbf{v} \cdot \frac{f_1(\mathbf{z})}{\partial \mathbf{x}} + N \frac{\partial}{\partial \mathbf{p}} \cdot \int \mathbf{G}(\mathbf{x}, \mathbf{x}') f_2(\mathbf{z}, \mathbf{z}') \, d\mathbf{z}' = 0$$

(5.17)

Here we have dropped the delta-function term in the integral owing to assumed symmetry (5.14a) of the interaction force (see Problem 2.8. Comparing (5.17) with the first of the hierarchy (1.24), BY_1, reveals them to be the same.

2.6   Grad's Analysis[24]

2.6.1   Liouville Equation Revisited

To this point we have encountered two derivations of the Boltzmann equation stemming from the BBKGY hierarchy. In the present section, Grad's derivation

---

[24]H. Grad, *Hand. d. Physik*, vol. 12, Springer, Berlin (1958).

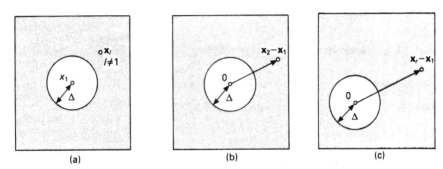

FIGURE 2.17. Configuration domain in Grad's derivation of the Boltzmann equation: (a) domain D; (b) domain D_2; (c) domain D_r.

is described. The starting point is the Liouville equation written in the form

$$\frac{\partial f_N}{\partial t} + \sum_{i}^{N} \frac{\partial}{\partial \mathbf{x}_i} \cdot (\mathbf{v}_i f_N) + \frac{1}{m} \sum_{i}^{N} \mathbf{G}_i \cdot \frac{\partial f_N}{\partial \mathbf{v}_i} = 0 \qquad (6.1)$$

The following formulas are needed in the derivation.

$$\int_{|\mathbf{y}-\mathbf{x}|>\Delta} \frac{\partial}{\partial \mathbf{y}} \cdot \mathbf{A}(\mathbf{x}, \mathbf{y}) \, d\mathbf{y} = \int_{|\mathbf{y}-\mathbf{x}|>\Delta} \text{div } \mathbf{A} \, d\mathbf{y} = - \oiint_{|\mathbf{y}-\mathbf{x}|=\Delta} \mathbf{A} \cdot d\mathbf{S} \qquad (6.2a)$$

$$\frac{\partial}{\partial \mathbf{x}} \cdot \int_{|\mathbf{y}-\mathbf{x}|>\Delta} \mathbf{A}(\mathbf{x}, \mathbf{y}) \, d\mathbf{y} = \int_{|\mathbf{y}-\mathbf{x}|>\Delta} \frac{\partial}{\partial \mathbf{x}} \cdot \mathbf{A} \, d\mathbf{y} - \oiint_{|\mathbf{y}-\mathbf{x}|=\Delta} \mathbf{A} \cdot d\mathbf{S} \qquad (6.2b)$$

2.6.2   Truncated Distributions

The derivation is further based on the introduction of the truncated distribution $f_1^{\Delta}(\mathbf{z}_1)$, where $\mathbf{z}_1 \equiv (\mathbf{x}_1, \mathbf{v}_1)$. This function is the probability of finding no molecules within a distance Δ of particle 1, with particle 1 in the state $\mathbf{z}$. To obtain the equation f_1^{Δ} satisfies, we integrate the Liouville equation over the domain D. This domain contains all states of particles $(2, \ldots, N)$ in which no particle is closer than Δ to particle 1 (see Fig. 2.17). There results

$$\int_D d2 \ldots dN \left\{ \frac{\partial f_N}{\partial t} + \sum_{i}^{N} \frac{\partial}{\partial \mathbf{x}_i} \cdot (\mathbf{v}_i f_N) + \frac{1}{m} \sum_{i}^{N} \mathbf{G}_i \cdot \frac{\partial f_N}{\partial \mathbf{v}_i} \right\} = 0 \quad (6.3)$$

$$f_1^{\Delta} = \int_D f_N \, d2 \ldots dN \qquad (6.3a)$$

$$D = \{|\mathbf{x}_1 - \mathbf{x}_2| > \Delta; \quad |\mathbf{x}_1 - \mathbf{x}_3| > \Delta, \ldots, |\mathbf{x}_1 - \mathbf{x}_N| > \Delta\}$$

If D_r denotes the domain $|\mathbf{x}_1 - \mathbf{x}_r| > \Delta$, then we may write

$$D = D_2 \times D_3 \times \cdots \times D_N \qquad (6.4)$$

From these equations we obtain

$$\frac{\partial f_1}{\partial t} + \sum_{}^{N} \int_D \left\{ \frac{\partial}{\partial \mathbf{x}_i} \cdot (\mathbf{v}_i f_N) + \frac{1}{m} \frac{\partial}{\partial \mathbf{v}_i} \cdot \mathbf{G}_i f_N \right\} d2 \cdots dN = 0 \qquad (6.5)$$

Let S_i be the sphere $|\mathbf{x}_i - \mathbf{x}_1| = \Delta$. Now consider the first term in the sum

$$\sum_{i=1}^{N} \frac{\partial}{\partial \mathbf{x}_i} \cdot \mathbf{v}_i f_N \qquad (6.6)$$

whose integral gives

$$\int_D \frac{\partial}{\partial \mathbf{x}_1} \cdot (\mathbf{v}_1 f_N) \, d2 \ldots dN = \frac{\partial}{\partial \mathbf{x}_1} \cdot (\xi_1 f_1^\Delta) + \sum_{i=2}^{N} \int \oiint_{S_i} f_2^\Delta(1, i) \mathbf{v}_i \cdot d\mathbf{S} \, d\mathbf{v}_i \qquad (6.7)$$

Here we have employed (6.2b), identifying the integral on the left of (6.7 with the first term on the right of (6.2b). The remaining terms in the sum (6.6 may be similarly integrated to obtain [with (6.2a)]

$$\sum_{i=2}^{N} \int_D \frac{\partial}{\partial \mathbf{x}_i} \cdot (\mathbf{v}_i f_N) \, d_2 \ldots dN = -\sum_{i=2}^{N} \int \oiint_{S_i} f_2^\Delta(1, i) \mathbf{v}_i \cdot d\mathbf{S} \, d\mathbf{v}_i \qquad (6.8)$$

In the force term in (6.5), all save the $\partial/\partial \mathbf{v}_1$ term reduce to surface integrals that vanish. The relevant factor in the remaining term is integrated to obtain

$$\int_D \mathbf{G}_1 f_N \, d_2 \ldots \, dN = \sum_{i=2}^{N} \int_D \mathbf{G}_{1i}(\mathbf{x}_1, \mathbf{x}_i) f_N \, d_2 \ldots \, dN$$

$$= (N - 1) \int_{D_2} \mathbf{G}_{12}(\mathbf{x}_1, \mathbf{x}_2) f_2(1, 2) \, d2 \qquad (6.9)$$

The domain D_2 contains all states of particles 1 and 2 in which they are not closer than the distance Δ (that is, $|\mathbf{x}_2 - \mathbf{x}_1| > \Delta$). The two-particle truncated distribution f_2^Δ is defined through the integral

$$f_2^\Delta(1, 2) = \int_{D'} f_N(1, \ldots, N) \, d3 \ldots dN$$
$$D' = D_3 \times D_4 \times \cdots \times D_N \qquad (6.10)$$

The domain D' contains all states of molecules $(3, \ldots, N)$ in which no particle is closer than Δ to particle 1.

2.6.3 Grad's First and Second Equations

Combining (6.5) and (6.9) gives *Grad's first equation*:

$$\frac{\partial f_1^\Delta}{\partial t} + \mathbf{v}_1 \cdot \frac{\partial f_1^\Delta}{\partial \mathbf{x}_1} + (N - 1) \int \oiint_{S_2} f_2^\Delta \, d\mathbf{S} \cdot (\mathbf{v}_1 - \mathbf{v}_2) \, d\mathbf{v}_2$$

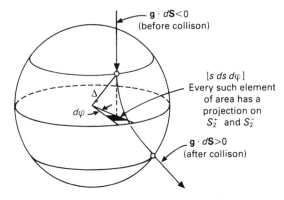

FIGURE 2.18. The S_2 sphere in Grad's derivation of the Boltzmann equation.

$$+\frac{N-1}{m}\frac{\partial}{\partial \mathbf{v}_1}\cdot\int_{D_2}\mathbf{G}_{12}f_2^{\Delta}\,d2=0 \qquad (6.11)$$

In obtaining this form, the surface integrals on the right side of (6.7) and (6.8) were recognized to be $N-1$ identical terms. This identity is made for much the same reasons as were employed in obtaining the $N-s$ identical terms in BBKGY sequence (1.20). The domain D_2 in (6.11) includes $|\mathbf{x}_2-\mathbf{x}_1|>\Delta$, while the surface integral in this same equation is over the surface $|\mathbf{x}_2-\mathbf{x}_1|=\Delta$. Suppose that $\mathbf{G}_{12}=0$ for $|\mathbf{x}_2-\mathbf{x}_1|>\Delta$ (such as for rigid spheres of diameter Δ). We then obtain *Grad's second* equation (see Problem 3.51).

$$\frac{\partial f_1^{\Delta}}{\partial_t}+\mathbf{v}_1\cdot\frac{\partial f_1^{\Delta}}{\partial \mathbf{x}_1}=-(N-1)\int\oiint_{S_2}f_2^{\Delta}\,d\mathbf{S}\cdot(\mathbf{v}_1-\mathbf{v}_2)\,d\mathbf{v}_2 \qquad (6.12)$$

The left side is the total time derivative of the function f_1^{Δ}. The surface integral indicates that the rate at which a set of particles in a given phase element changes is due only to the loss and gain of a pair of particles as they enter and leave each other's sphere of influence. Here we are interpreting Δ to be the range of interaction, previously labeled r_0 in Section 2.2.

2.6.4   The Boltzmann Equation

Grad's second equation is a precursor to the Boltzmann equation. To establish this property, we consider the sphere S_2 shown in Fig. 2.18. The points on S_2 are mapped onto the meridian disk by simple projection. Points on the disk have the radial coordinate (impact parameter) s and azimuth φ. To make the mapping one-to-one, the two hemispheres are distinguished according to

$$S_2^+ : \mathbf{g}\cdot d\mathbf{S}>0 \qquad \text{(after collision)}$$
$$S_2^- : \mathbf{g}\cdot d\mathbf{S}<0 \qquad \text{(before collision)}$$

The relative velocity $\mathbf{g}$ is given by

$$\mathbf{g} = \mathbf{v}_2 - \mathbf{v}_1 \tag{6.13}$$

For fixed $\mathbf{g}$,

$$S_2 = S_2^+ + S_2^-$$

The point on S_2^+ that projects onto (s, φ) is called $\mathbf{x}_2^+(s, \varphi)$, and the point on S_2^- that projects onto this point is called $\mathbf{x}_2^-(s, \varphi)$. There results (see Problem 2.16)

$$
\begin{aligned}
(\mathbf{v}_1 - \mathbf{v}_2) \cdot d\mathbf{S} &= -gs \, ds \, d\varphi \qquad \text{on } S_2^+ \\
(\mathbf{v}_1 - \mathbf{v}_2) \cdot d\mathbf{S} &= +gs \, ds \, d\varphi \qquad \text{on } S_2^-
\end{aligned}
\tag{6.14}
$$

If 2^+ denotes $(\mathbf{v}_2, \mathbf{x}_2^+)$ and 2^- denotes $(\mathbf{v}_2, \mathbf{x}_2^-)$, then, in the previously described coordinates, (6.12) appears as

$$
\begin{aligned}
\frac{\partial f_1^\Delta}{\partial t} + \mathbf{v}_1 \cdot \frac{\partial f_1^\Delta}{\partial \mathbf{x}_1} &= -(N-1) \left\{ \int_{S_2^+} + \int_{S_2^-} \right\} \\
&= (N-1) \int [(f_2^\Delta(1, 2^+) - f_2^\Delta(1, 2^-)] \, gs \, ds \, d\varphi \, d\mathbf{v}_2
\end{aligned}
\tag{6.15}
$$

At this point the following constraints are imposed:

1. Replace the arguments in $f_2^\Delta(1, 2^+ t)$ by the values they must be at the start of a binary collision (say $\bar{1}, \bar{2}, \bar{t}$) in order that they have the values $(1, 2^+)$ at the time t.

2. Set

$$
\begin{aligned}
f_2^\Delta(1, 2^-) &= f_1^\Delta(1) f_1^\Delta(2-) \\
f_2^\Delta(\bar{1}, \bar{2}) &= f_1^\Delta(\bar{1}) f_1^\Delta(\bar{2})
\end{aligned}
$$

which is the assumption of *molecular chaos*.

3. In the four functions $f_1^\Delta(1, t)$, $f_1^\Delta(2^-, t)$, $f_1^\Delta(\bar{1}, \bar{t})$, and $f_1^\Delta(\bar{2}, \bar{t})$, replace $\bar{\mathbf{x}}_2, \bar{\mathbf{x}}_1, \mathbf{x}_2^-$ by $\mathbf{x}_1$ and $\bar{t}$ by t.

Inserting these changes into (6.15) and neglecting one compared to N gives the Boltzmann equation (recall $F_1 = N f_1$):

$$\frac{\partial F_1}{\partial t} + \mathbf{v}_1 \cdot \frac{\partial F_1}{\partial \mathbf{x}_1} = \int [F_1(\bar{\mathbf{v}}_1) F_1(\bar{\mathbf{v}}_2) - F_1(\mathbf{v}_1) F_1(\mathbf{v}_2)] \, gs \, ds \, d\varphi \, d\mathbf{v}_2 \tag{6.16}$$

This completes the Grad derivation of the Boltzmann equation. As noted previously, the Boltzmann equation is discussed in greater detail in Chapter 3.

Note in particular that as (6.12) is a precursor to the Boltzmann equation, the neglected force integral in Grad's first equation (6.11) is a measure of the error associated with the Boltzmann equation. However, we may easily envision a perturbative technique of solution to Grad's first equation (6.11) in which Grad's second equation (6.12) [and resulting Boltzmann equation (6.16)] is the lead term. Thus, all three techniques as given by Prigogine, Bogoliubov, and Grad offer a means of obtaining higher-order corrections to the Boltzmann equation appropriate to denser fluids.

Problems

2.1. Establish the following *antinormalization* properties for the correlation functions C_s given by (2.21).[25]

$$\int C_s(1, \ldots, s)\, d1 = \int C_s(1, \ldots, s)\, d2 = \cdots = \int C_s(1, \ldots, s)\, ds = 0$$

2.2. (a) Write down BY_3. Define terms.
   (b) Integrate BY_3 to obtain BY_2.

2.3. Write down the explicit form of the Liouville equation for a two-component gas that contains $N/2$ molecules of mass m and $N/2$ molecules of mass M. The interparticle conservative forces are labeled $\mathbf{G}_m$, $\mathbf{G}_M$, and $\mathbf{G}_I$ relevant to $m - m$, $M - M$ and $m - M$ interactions, respectively.

2.4. A dumbbell molecule comprised of atoms with respective masses m_1 and m_2 that is in a uniform gravity field has the Lagrangian

$$L = \frac{\mu}{2}(a^2\dot{\theta}^2 + a^2\dot{\theta}^2 \sin^2\theta) + \frac{k}{2}\xi^2 + \frac{M}{2}(\dot{X}^2 + \dot{Y}^2 + \dot{Z}^2) - MgZ$$

The center of mass of the molecule is X, Y, Z, and ξ is the displacement from equilibrium separation of the two atoms.

   (a) Obtain expressions for the canonical momenta p_θ, p_ϕ, p_ξ, p_X, p_Y, and p_Z in terms of generalized velocities.
   (b) What is the Hamiltonian of the molecule?
   (c) A gas of N *noninteracting* identical molecules of this type occupy a cubical volume $V = L^3$. What is the one-particle Liouville equation for f_1, the one-particle *probability density* of the gas?
   (d) Let the molecules interact through the potential $V(R_{ij})$, where R_{ij} is the intermolecular center-of-mass separation. What is the explicit form of the *Liouville equation* for the gas? Leave triple sums intact.

2.5. Consider a system that contains N particles in equilibrium whose Hamiltonian is given by

$$H(\mathbf{x}^N, \mathbf{p}^N) = K(\mathbf{p}^N) + \Phi(\mathbf{x}^N)$$

Kinetic and potential energy terms have been written as K and Φ, respectively. Show that if the distribution function is of the product form

$$f_N(\mathbf{x}^N, \mathbf{p}^N) = X(\mathbf{x}^N)P(\mathbf{p}^N)$$

---

[25]For further discussion, see R. L. Liboff, *Phys. Fluids* 8, 1236 (1965); 9, 419 (1966). See also Problem 2.23.

FIGURE 2.19. Prigogine diagram from Problem 2.6.

where

$$P(\mathbf{p}^N) = \sum_{i=1}^{N} g_i(p_i^2)$$

then the Liouville equation implies the *canonical distribution*

$$f_c(\mathbf{x}^N, \mathbf{p}^N) = A \exp[-\beta H(\mathbf{x}^N, \mathbf{p}^N)]$$

where A and β are constants.[26]

2.6. Consider the diagram in the Prigogine analysis of the Liouville equation shown in Fig. 2.19.

    (a) What is the order of this diagram?
    (b) How many particles come into play in this diagram?
    (c) Calling the perturbation Liouville operator $\delta \hat{L}$, write down the product matrix element (in Dirac notation) that this diagram represents.
    (d) What is the integration in time that accompanies this diagram (list *only* differentials and limits of integration)?

2.7. Consider a system whose ensemble fluid occupies a *simply connected* domain R of Γ-space. A domain R is simply connected if any two points in R may be connected by a path contained entirely in R. Show that in the motion of the system R remains simply connected. That is, R does not separate into disconnected segments.

2.8. In the Klimontovich formulation, show that BY_1 as given by (5.17) follows from (5.16). Justify all steps.

2.9. A space is said to be a *metric space* if for any two points (x, y) in the space a "distance" function $d(x, y)$ exists that maps pairs of points in the space onto numbers and has the following properties:

    (1) $d(x, y) \geq 0$; $d(x, x) = 0$; $d(x, y) > 0$ if $x \neq y$ (positivity)
    (2) $d(x, y) = d(y, x)$ (symmetry)
    (3) $d(x, z) \leq d(x, y) + d(y, z)$ (triangle inequality)

Introduce a distance function for $2N$-dimensional Γ-space. Show that your choice satisfies the preceding properties, thereby establishing that Γ-space is a metric space. (*Hint*: Recall that Γ-space is a Euclidean space.)

---

[26]This theorem was previously discussed by R. L. Liboff and D. M. Heffernan, *Phys. Letts. 79A*, 29 (1980).

2.10. (a) What is the Bogoliubov boundary condition on the state of particles at $t = -\infty$? In what sense is this condition included in the Boltzmann equation?

(b) Establish the properties (4.21) relevant to the time-displacement operator $\hat{\Delta}_{-t}^{(s)}$.

2.11. What is the structure factor, $S(k)$, for an ideal gas?

2.12. Show that the equation of state of a fluid comprised of rigid sphere of radius a is given by

$$\frac{P}{nk_B T} = 1 + \frac{2\pi na^3}{3} g(a+)$$

where $g(a+)$ is the value of $g(r)$ at $r = a + \varepsilon, \varepsilon \to 0$. (*Hint:* Recall that the derivative of a step function is a delta function.)

2.13. For low-density fluids, the following form is found to be a reasonably good approximation for the radial distribution function.

$$g(r) \simeq \exp[-\Phi(r)/k_B T]$$

Consider that the interaction potential $\Phi(r)$ has a rigid repulsive core and a long-range "attractive tail." What are the corresponding values $g(0)$ and $g(\infty)$?

2.14. Show that the total correlation function $h(r)$ has at least one zero on the open r interval $(0+, \infty)$.

Answer

It follows from Problem 2.1 that

$$\int C_2(1, 2)\, d1\, d2 = 0$$

Inserting the defining relation (2.48) into this equation gives

$$\int f_0(p_1) f_0(p_2) h(r)\, d1\, d2 = 0$$

Introducing the transformation

$$\mathbf{r} = \mathbf{x}_2 - \mathbf{x}_1$$
$$2\mathbf{R} = \mathbf{x}_2 + \mathbf{x}_1$$

(which carries a unit Jacobian), we find

$$d\mathbf{x}_1\, d\mathbf{x}_2 = d\mathbf{r}_1\, d\mathbf{R}$$

and the preceding integral gives

$$\int_0^\infty 4\pi r^2\, dr h(r) = 0$$

Here we have assumed a sufficiently large volume permitting the upper limit of the integral to be set equal to infinity. As $h(0) = -1$, the above integral implies that $h(r)$ is zero at least once on the said open r interval.

2.15. What is the radial distribution function $g(r)$ corresponding to the structure factor

$$S(k) = e^{-ak}$$

where a is a constant length.

2.16. Establish the validity of (6.14).

Answer

The relation

$$\mathbf{g} \cdot d\mathbf{S} = g \Delta^2 d \cos \theta \, d\phi \cos \theta$$

together with

$$s^2 = \Delta^2 \sin^2 \theta$$

gives the desired result. Note that the angle θ is inscribed between Δ and the polar axis in Fig. 2.18.

2.17. A *Lie derivative* is defined as the form

$$\hat{\mathcal{L}} \equiv \mathbf{A}(\mathbf{x}) \cdot \frac{\partial}{\partial \mathbf{x}}$$

where $\mathbf{A}(\mathbf{x})$ is a vector function of $\mathbf{x}$. Show that Lie derivatives have the property

$$e^{t\hat{\mathcal{L}}} f(\mathbf{x}) = f(e^{t\hat{\mathcal{L}}}\mathbf{x}) = f[\mathbf{X}(\mathbf{x}, t)]$$

where

$$\frac{\partial \mathbf{X}(\mathbf{x}, t)}{\partial t} = \mathbf{A}[\mathbf{X}(\mathbf{x}, t)]$$
$$\mathbf{X}(\mathbf{x}, 0) = \mathbf{x}$$

2.18. **(a)** Show that the s-particle ($s \leq N$) distribution function satisfies the equation

$$\frac{\partial f_s}{\partial t} + \mathbf{\nabla}_s \cdot \mathbf{u}_s f_s = 0$$

where $\mathbf{\nabla}_s$ represents the gradient in s-dimensional Γ-space. In your derivation of the given kinetic equation, what does $\mathbf{u}_s$ represent?

(b) What is the physical interpretation of the given kinetic equation?

(c) Is the ensemble fluid for this subsystem incompressible? That is, does $\mathbf{\nabla}_s \cdot \mathbf{u}_s = 0$? If not, what is an appropriate expression for this term?

2.19. In the Bogoliubov analysis, we encounter the functional inner product (4.9). What is the explicit representation of the form

$$\left\langle \frac{\delta F_1(1) F_1(2)}{\delta F_1}, \frac{\mathbf{p}}{m} \cdot \frac{\partial}{\partial \mathbf{x}} F_1(\mathbf{x}, \mathbf{p}) \right\rangle$$

2.20. **(a)** Is symmetry of the N-body distribution function in its arguments assumed in the derivation of the Liouville equation? If so, where?

(b) Is this property assumed in the derivation of the BBKGY sequence? If so, where?

Answer

(a) This assumption was not made. Recall that it was found that *any arbitrary* function of the constants of motion of the system is a solution to the Liouville equation.
(b) This assumption was made. Thus, for example, if $f_3(1, 2, 3)$ is a symmetric distribution, then $\int f_3(1, 2, 3)\,d1\,d3$ and $\int f_3(1, 2, 3)\,d1\,d2$ give the same functional dependence on the remaining variable.

2.21. In BY$_s$ (1.20), discuss the difference between the force terms in the sum on O_{ij} and the force terms in the integral.

2.22. Obtain BY$_2$ within the Klimontovich formalism. *Hint*: Write the equation of motion for $\mathcal{N}(\mathbf{z})\mathcal{N}(\mathbf{z}')$ in twelve-dimensional $(\mathbf{z}, \mathbf{z}')$ phase space, take the ensemble average, assume that $(\mathbf{z} \neq \mathbf{z}')$ and introduce appropriate forms for terms containing $\dot{\mathbf{p}}$ and $\dot{\mathbf{p}}'$.

2.23. The correlation normalizations of Problem 2.1 as well as those implied by Problem 2.14 and equation (2.47) are valid for systems with vanishingly small fluctuations. More generally,[27] for the radial distribution function, one writes

$$\int [g(r) - 1]\,d\mathbf{r} + \frac{1}{n} = k_B T \kappa_T$$

where

$$\kappa_T \equiv -\frac{1}{V}\left(\frac{\partial V}{\partial P}\right)_T$$

is isothermal compressibility.

(a) What relation for κ_T is implied by the preceding integral equation for an ideal gas?
(b) In what limit is the normalization on the radial distribution implied by Problem 2.14 valid?

Answers

(a) For an ideal gas, $g(r) = 1$ and we obtain, $nk_B T \kappa_T = 1$.
(b) For a dense medium in which number density, n, is very large for which $\kappa_T = 0$.

[27]D. L. Goodstein, *State of Matter*, Prentice Hall, Englewood Cliffs, NJ (1975); F. Mohling, *Statistical Mechanics*, Wiley, New York (1982).

CHAPTER 3

The Boltzmann Equation, Fluid Dynamics, and Irreversibility

Introduction

This chapter begins with a review of scattering concepts important to derivation of the Boltzmann equation. Collisional invariants are then introduced that aid in obtaining fluid dynamic conservation equations and the $\mathcal{H}$ theorem. This theorem in turn leads to the well-known Maxwellian distribution. The central-limit theorem encountered previously in Chapter 1 is employed in the evaluation of the distribution of center-of-mass velocities of a finite stationary gas of molecules. A derivation of Poincaré's recurrence theorem stemming from the Liouville equation is given and is contrasted with the irreversibility of the $\mathcal{H}$ theorem of the Boltzmann equation.

Transport coefficients are introduced in the context of response to gradient perturbations of fluid variables. Maxwell's elementary mean-free-path estimate of transport coefficients is presented, as well as a brief account of their relation to autocorrelation functions. These notions serve as an introduction to the Chapman–Enskog expansion of the Boltzmann equation. Values obtained for transport coefficients through this method are shown to be in good agreement with experiment. The theory of the linear Boltzmann collision operator is discussed, and spectral properties of hard and soft potentials are noted. The chapter next turns to application of the Boltzmann equation to an electron gas immersed in an ionic fluid in the presence of a dc electric field. The resulting Druyvesteyn distribution is then employed in the derivation of an integral expression for electrical conductivity.

In the last section of the chapter the notion of irreversibility, previously discussed in reference to the $\mathcal{H}$ theorem, is reintroduced, and more recent contributions to this subject are noted. Ergodic and mixing flows are described and Birkhoff's contribution to the theory is reviewed. Action-angle variables are introduced together with Bruns's theorem on a limited number of constants of motion and the closely related concept of an integrable system. Numerical results relevant to a two-dimensional anharmonic oscillator lead to the concepts of resonant domains, chaotic motion, and the closely allied, often quoted, KAM theorem.

3.1   Scattering Concepts

3.1.1   Separation of the Hamiltonian

Basic notions of scattering theory are prerequisite to derivation of the Boltzmann equation. The theory of two-particle collisions is best formulated in the center-of-mass frame.

First we note that the Hamiltonian of two particles interacting under a central potential $V(r)$, in the "lab frame," is written

$$H(\mathbf{r}_1, \mathbf{r}_2, \mathbf{p}_1, \mathbf{p}_2) = \frac{p_1^2}{2m_1} + \frac{p_2^2}{2m_2} + V(r) \tag{1.1}$$

Particle masses are m_1 and m_2, respectively, and r is the relative displacement

$$r = |\mathbf{r}_2 - \mathbf{r}_2|$$

Transformation to the center of mass is effected through the mapping

$$\mathbf{p} = \frac{m_1\mathbf{p}_2 - m_2\mathbf{p}_1}{m_1 + m_2} = \mu\dot{\mathbf{r}}$$

$$\mathbf{r} = \mathbf{r}_2 - \mathbf{r}_1$$

$$\mathbf{P} = \mathbf{p}_1 + \mathbf{p}_2 \tag{1.2}$$

$$\mathbf{R} = \frac{m_1\mathbf{r}_1 + m_2\mathbf{r}_2}{m_1 + m_2}$$

Substituting (1.2) into (1.1) gives

$$H(\mathbf{R}, \mathbf{P}, \mathbf{r}, \mathbf{p}) = \frac{P^2}{2M} + \frac{p^2}{2\mu} + V(r) \tag{1.3}$$

where [as in (1.12)]

$$\mathbf{P} = M\dot{\mathbf{R}} \equiv (m_1 + m_2)\dot{\mathbf{R}}$$

is the momentum of the center mass,

$$\mu = \frac{m_1 m_2}{m_1 + m_2}$$

is the *reduced mass*, and

$$H_{\text{CM}} = \frac{P^2}{2M} \tag{1.4}$$

is the kinetic energy of the center of mass. The remaining part of the Hamiltonian,

$$H_{rel} = \frac{p^2}{2\mu} + V(r)$$

represents the Hamiltonian of a particle of mass μ in the potential field $V(r)$. H_{rel} is the Hamiltonian of the two-particle system relative to a frame moving with the center of mass. With this notation, (1.3) becomes

$$H = H_{CM} + H_{rel} \tag{1.5}$$

Since H is cyclic in $\mathbf{R}$, we may conclude that $\mathbf{P}$ is conserved, whence

$$H_{CM} = \text{constant}$$

This constant may be absorbed in $V(r)$ without affecting resulting equations of motion. Thus we may write

$$H = H_{rel} = \frac{p^2}{2\mu} + V(r) \tag{1.6}$$

So we find that the original two-body problem is reduced to a one-body problem of a particle of mass μ moving in the field of a central potential, $V(r)$. Since the force on this particle is radial, angular momentum is conserved, and we may conclude that the particle moves in a plane of constant orientation. See Fig.3.1. Thus the μ particle has two degrees of freedom, which we label r and ϕ. The angle ϕ is the inclination of $\mathbf{r}$ to an arbitrary axis in the plane. In terms of these coordinates, the Hamiltonian becomes

$$H = \frac{p_r^2}{2\mu} + \frac{p_\phi^2}{2\mu r^2} + V(r) \tag{1.7}$$

Since ϕ is a cyclic variable, p_ϕ is constant, and we set

$$p_\phi \equiv L$$

Furthermore, H is constant and may be identified with the energy of the system measured in the center-of-mass frame. Thus we write

$$E = \frac{p_r^2}{2\mu} + \frac{L^2}{2\mu r^2} + V(r) \tag{1.8}$$

With (1.7), Hamilton's equations (1.1.12) gives

$$\frac{p_r}{\mu} = \frac{dr}{dt} = \sqrt{\frac{2}{\mu}(E - V) - \frac{L^2}{\mu^2 r^2}}$$

$$\frac{d\phi}{dt} = \frac{L}{\mu r^2}, \qquad d\phi = \frac{L\,dt}{\mu r^2} \tag{1.9}$$

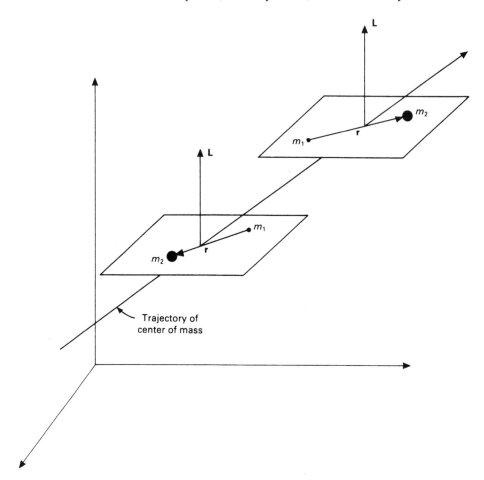

FIGURE 3.1. The radius vector **r** moves in a plane of constant orientation in the center-of-mass frame.

Substituting $d\phi$ for dt in the first expression gives

$$\phi = \int \frac{L\,dr/r^2}{\sqrt{2\mu[E - V(r)] - (L^2/r^2)}} + \text{constant} \qquad (1.10)$$

Note in particular that we are addressing scattering or unbound trajectories. With the convention that $V(\infty) = 0$, an unbound orbit is characterized by $E > 0$, where E is the energy as given by (1.8). A bound state corresponds to $E < 0$ (see Problem 3.1).

3.1.2 Scattering Angle

We wish to apply (1.9) to obtain the angle of scatter that two particles suffer in collision. A scattering event is divided into three epochs: before, during, and

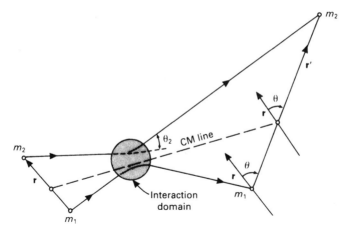

FIGURE 3.2. Scattering in the lab frame.

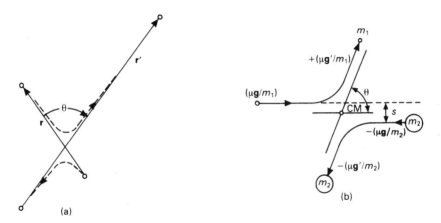

FIGURE 3.3. (a) Relative radius vector **r** before and after collision in the center-of-mass frame. (b) Scattering in the center-of-mass frame illustrating relative velocities of masses m_1 and m_2.

after the interaction. Both before and after collision, particles are free and do not interact with one another. A typical scattering in the lab frame is depicted in Fig. 3.2. The situation in the center-of-mass frame is depicted in Fig. 3.3.

The meaning of *before* and *after collision* is given in terms of the potential of interaction $V(r)$ or the range of interaction r_0. *Before collision* denotes the interval in which the particles are approaching one another and $r > r_0$, or, equivalently, $V(r) = 0$. *After collision* denotes the interval in which the particles are receding from one another and, again, $V(r) = 0$ and $r > r_0$.

It proves convenient in discussing scattering events to introduce the relative velocity vector before and after collision. Assuming that the interaction occurs

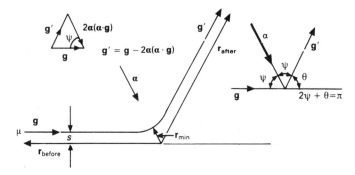

FIGURE 3.4. Scattering in the frame of the relative vector **r**. Note that we may write $\mathbf{g}' = \bar{\bar{S}}\mathbf{g}$, where $\bar{\bar{S}}$ is the scattering matrix $\bar{\bar{S}} = \bar{\bar{I}} - 2\overline{\mathbf{aa}}$.

in an interval about $t = 0$, we write

$$\mathbf{g} = \dot{\mathbf{r}}, \qquad t = \pm\infty \tag{1.11}$$

In that $V = 0$ before and after collision, with (1.8) we find

$$\frac{1}{2}\mu g^2 = \frac{1}{2}\mu g'^2 \tag{1.12}$$

and conclude that the magnitude of relative velocity is conserved over a collision.

A scattering event in the frame where the origin of the relative vector **r** is fixed (equivalent to the center-of-mass frame) is depicted in Fig. 3.4. The unit vector α bisects the angle between $-\mathbf{g}$ and $\mathbf{g}'$. The impact parameter s is such that the angular momentum of the system in this frame has magnitude $\mu g s$. The impact parameter s as well as the relative velocity vectors **g** and **g**′ are properties of the asymptotic collision states. The speed g and impact parameter s are both preserved in a given collision, and these two scalar quantities may be termed *properties of a collision*.

The scattering diagram (Fig. 3.4) is symmetric about the apse, $\mathbf{r}_{\min}$, so that, at $\mathbf{r}_{\min}$, $\dot{r} = 0$. This symmetry is a consequence of the conservation of angular momentum $L = \mu g s = \mu g' s'$ and the relative speed $g = g'$, so that $s = s'$. Inasmuch as $\dot{r}$ vanishes at the apse, it follows from (1.9) that at this point the denominator of the integral (1.10) vanishes. However, the integral remains bounded.

The angle ψ depicted in Fig. 3.4 is obtained by integrating (1.10) from $r_{\min}$ to ∞. There results

$$\phi(r_\infty) - \phi(r_{\min}) = \psi = \int_{r_{\min}}^{\infty} \frac{(L/r^2)\,dr}{\sqrt{2\mu(E - V) - (L^2/r^2)}} \tag{1.13}$$

The scattering angle θ is related to ψ as shown in Fig. 3.4.

$$\theta + 2\psi = \pi \tag{1.14}$$

A better representation of the integral (1.13) is given in terms of the inverse radius $u = r^{-1}$. Furthermore, the angular momentum $L = \mu g s$ is related to the energy $E = \frac{1}{2} \mu g^2$ through

$$E = \frac{L^2}{2 \mu s^2} \tag{1.15}$$

Substituting these relations in (1.13) gives

$$\psi = \int_0^{\bar{u}} \frac{s \, du}{\sqrt{1 - s^2 u^2 - (V/E)}} \tag{1.16}$$

$$1 - s^2 \bar{u}^2 - \frac{V(\bar{u})}{E} = 0$$

For potentials of the form

$$V(r) = K(r)^{-N} \equiv K u^N \tag{1.17}$$

(1.16) becomes

$$\psi = \int_0^{\bar{u}} \frac{s \, du}{\sqrt{1 - s^2 u^2 - (K u^N / E)}} \tag{1.18}$$

Substituting the nondimensional inverse radius

$$\beta \equiv s u \tag{1.19}$$

and the nondimensional impact parameter

$$b \equiv s \left(\frac{E}{K} \right)^{1/N} \tag{1.20}$$

in (1.18) gives

$$\psi(b) = \int_0^{\bar{\beta}} \frac{d\beta}{\sqrt{1 - \beta^2 - (\beta/b)^N}} \tag{1.21}$$

$$1 - \bar{\beta}^2 - \left(\frac{\bar{\beta}}{b} \right)^N = 0 \tag{1.21a}$$

This general result is employed below.

3.1.3   Cross Section

Differential scattering cross section is defined in the following manner. Imagine a uniform beam of particles of energy E and intensity I number/s-cm^2), which is incident on a scatterer located at the origin. The number of particles scattered into the element of solid angle $d\Omega$ about Ω is proportional to the incident intensity I and the element of solid angle $d\Omega$. The proportionality factor is σ, the differential scattering cross section (see Fig. 3.5). That is,

$$I\sigma(\Omega)d\Omega = \text{number deflected into } d\Omega \text{ about } \Omega \text{ per second} \qquad (1.22)$$

This number is the same number that passed through the differential of annulus, $s\,ds\,d\phi$, so that

$$I\sigma\,d\Omega = I\,d\phi s\,ds \qquad (1.23)$$

The azimuthal angle ϕ, which locates a section of the incident beam, is the same angle that appears in a spherical coordinate frame fixed with origin at the scatterer. That is, $d\Omega = d\cos\,d\phi$. Inserting this equality into (1.23) gives

$$\sigma(E,\theta) = s(E,\theta)\frac{ds(E,\theta)}{\sin\theta\,d\theta} \qquad (1.24)$$

This is the classical formula for the differential scattering cross section. The functional form $s(E,\theta)$ stems from the scattering integral (1.16).

Whereas σ pertains to the number of particles scattered into a specific direction, the *total scattering cross section*,

$$\sigma_T = \int_{4\pi} \sigma\,d\Omega = \pi r_0^2 \qquad (1.25)$$

relates to the total number scattered out of the incident beam. The range of the interaction is r_0, so particles with impact parameters $s > r_0$ are not scattered out of the beam. The total cross section σ_T represents the obstructional area that the scatterer presents to the incident beam. For a uniform beam of cross-sectional area $A \geq \sigma_T$, the quantity σ_T/A is the fraction of particles scattered (in all directions) out of the beam.

An important factor that enters the Boltzmann equation is $g\sigma d\cos\theta$. With (1.24), we write

$$g\sigma\,d\cos\theta = gs\,ds \qquad (1.26)$$

Recalling the nondimensional parameter b (1.20) permits (1.26) to be written

$$g\sigma\,d\cos\theta = \left(\frac{2K}{\mu}\right)^{2/N} g^{(N-4)/N} b\,db \qquad (1.27)$$

It is evident that the case $N = 4$ merits special attention. Molecules that interact under this potential are called *Maxwell molecules*.[1] For such molecules, the scattering weight $g\sigma\,d\cos\theta$ is independent of g. This property greatly facilitates calculations related to the *linearized Boltzmann equation*. These topics are fully discussed later in this chapter.

---

[1] For Maxwell's introduction of this potential, see *Scientific Papers of James Clark Maxwell*, W. D. Niven (ed.), Dover Publications, New York (1952).

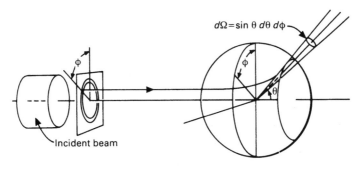

FIGURE 3.5. Coordinates for a scattering experiment .

The Coulomb cross section

Let us calculate σ for the Coulomb potential

$$V = \frac{K}{r}$$

Integrating (1.21), with $N = 1$, gives

$$b^2 = \frac{1}{4} \tan^2 \psi \tag{1.28}$$

In terms of $\theta (\theta + 2\psi = \pi)$, this relation appears as

$$b^2 = \frac{1}{4} ctn^2 \frac{\theta}{2} \tag{1.29}$$

Inserting this result into (1.27),

$$\sigma = \left(\frac{K}{E} \right)^2 \frac{b \, db}{d \cos \theta} \tag{1.30}$$

gives

$$\sigma = \left(\frac{K}{4E} \right)^2 \frac{1}{\sin^4(\theta/2)} \tag{1.31}$$

$$\sigma d \cos \theta = \frac{d \cos \psi \, K^2}{4 \cos^3 \psi \, E^2}$$

This is the well-known Rutherford cross section for Coulomb scattering. The divergence at $\theta = 0$ stems from the *long-range* nature of the Coulomb force.

Cross section for rigid spheres

The calculation of the scattering cross section for rigid spheres proceeds directly from geometry. If the two spheres have diameters σ_{01} and σ_{02}, respectively, then interaction does not occur for $r \geq (\sigma_{01} + \sigma_{02})/2$, where r is the displacement between sphere centers. A sketch of the scattering is depicted

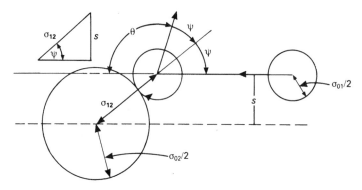

FIGURE 3.6. Scattering of rigid spheres.

in Fig. 3.6. The cross section is obtained directly from the triangular insert in Fig. 3.6, according to which

$$\sigma_{12} = \frac{\sigma_{01} + \sigma_{02}}{2}$$
$$s = \sigma_{12} \sin \psi \tag{1.32}$$
$$s \, ds = \sigma_{12}^2 \sin \psi \cos \psi \, d\psi = \frac{\sigma_{12}^2}{4} \sin \theta \, d\theta$$

With (1.24), we find

$$\sigma(\theta) = \frac{\sigma_{12}^2}{4} \tag{1.33}$$

So the scattering is isotropic in the center-of-mass frame. The total cross section σ_T is

$$\sigma_T = 2\pi \int_{-1}^{1} \sigma \, d \cos \theta = \pi \sigma_{12}^2 \tag{1.34}$$

the area of a disk of radius σ_{12}.

3.1.4 Kinematics

Two particles with respective momenta $\mathbf{p}_1$ and $\mathbf{p}_2$ collide. After collision the momenta are $\mathbf{p}_1'$ and $\mathbf{p}_2'$ collide. Conversation of momentum and energy equations for this event are

$$\mathbf{p}_1 + \mathbf{p}_2 = \mathbf{p}_1' + \mathbf{p}_2'$$
$$\frac{p_1^2}{2m_1} + \frac{p_2^2}{2m_2} = \frac{p_1'^2}{2m_1} + \frac{p_2'^2}{2m_2} \tag{1.35}$$

In the event that $m_1 = m_2$, these conservative equations become

$$\mathbf{v}_1 + \mathbf{v}_2 = \mathbf{v}_1' + \mathbf{v}_2'$$
$$v_1^2 + v_2^2 = v_1'^2 + v_2'^2 \tag{1.36}$$

Solving for $\mathbf{v}_1'$ and $\mathbf{v}_2'$ gives

$$\mathbf{v}_1' = \mathbf{v}_1 + \boldsymbol{\alpha}(\boldsymbol{\alpha} \cdot \mathbf{g})$$
$$\mathbf{v}_2' = \mathbf{v}_2 - \boldsymbol{\alpha}(\boldsymbol{\alpha} \cdot \mathbf{g}) \tag{1.37}$$

The unit apsidal vector $\boldsymbol{\alpha}$ is shown in Fig. 3.4. Subtracting these equations, we find

$$\mathbf{g}' = \mathbf{g} - 2\boldsymbol{\alpha}(\boldsymbol{\alpha} \cdot \mathbf{g}) \tag{1.38}$$

which, as noted previously in Fig. 3.4, may be written in the scattering matrix form

$$\mathbf{g}' = (\overline{\overline{I}} - 2\overline{\overline{\boldsymbol{\alpha}\boldsymbol{\alpha}}})\mathbf{g} \equiv \overline{\overline{S}}\mathbf{g} \tag{1.38a}$$

Inverse and reverse collision

Two symmetry properties of collisions are important in kinetic theory. In describing these symmetries, it proves convenient to speak of a collision in terms of two-particle momenta before $(\mathbf{p}_1, \mathbf{p}_2)$ and after $(\mathbf{p}_1', \mathbf{p}_2')$ collision. Thus we characterize a collision by the form

$$[(\mathbf{p}_1, \mathbf{p}_2) \to (\mathbf{p}_1', \mathbf{p}_2')]_s$$

The subscript s indicates that an impact parameter is a property of the collision. The inverse of a collision $[(\mathbf{p}_1, \mathbf{p}_2) \to (\mathbf{p}_1', \mathbf{p}_2')]_s$ is a collision containing the *final* state $(\mathbf{p}_1, \mathbf{p}_2)$. We see that if $[(\mathbf{p}_1, \mathbf{p}_2) \to (\mathbf{p}_1', \mathbf{p}_2')]_s$ satisfies the conservation equations (1.35) then $[(\mathbf{p}_1', \mathbf{p}_2') \to (\mathbf{p}_1, \mathbf{p}_2)]_s$ also satisfies them (read the conservation equations from right to left).

Symbolically, we may write

$$\text{inverse of } [(\mathbf{p}_1, \mathbf{p}_2) \to (\mathbf{p}_1', \mathbf{p}_2')]_s = [(\mathbf{p}_1', \mathbf{p}_2') \to (\mathbf{p}_1, \mathbf{p}_2)]_s \tag{1.39}$$

See Fig. 3.7.

The second type of collision that is important is the *reverse* collision. The reverse of a collision $[(\mathbf{p}_1, \mathbf{p}_2) \to (\mathbf{p}_1', \mathbf{p}_2')]_s$ is a collision containing the final state $(-\mathbf{p}_1, -\mathbf{p}_2)$. Again referring to the conservation equations, we conclude that $(-\mathbf{p}_1', -\mathbf{p}_2')_s$ are the desired initial momenta (see Fig. 3.8). That is, if $(-\mathbf{p}_1', \mathbf{p}_2')$ interact with the impact parameter s, they will yield $(-\mathbf{p}_1, -\mathbf{p}_2)_s$. Symbolically, we have

$$\text{reverse of } [(\mathbf{p}_1, \mathbf{p}_2) \to (\mathbf{p}_1', \mathbf{p}_2')]_s = [(-\mathbf{p}_1', \mathbf{p}_2') \to (-\mathbf{p}_1, \mathbf{p}_2)]_s \tag{1.40}$$

We recognize the reverse collision to be the time-reversed orbits of the original collision (see Section 1.1.5).

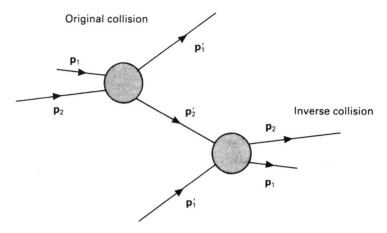

FIGURE 3.7. A collision and its inverse.

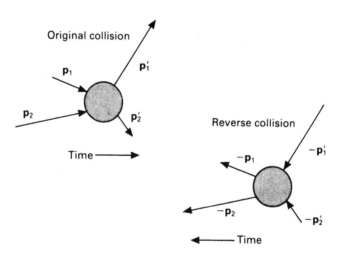

FIGURE 3.8. A collision and its reverse.

Due to their kinematic equivalence, we may conclude that the differential cross sections for the original, reverse, and inverse collisions are the same. These equalities relevant to the lab and center-of-mass frames are graphically depicted in Fig. 3.9.

The equality of these cross sections may be related to fundamental operations. Thus, equality of cross sections for the original and reverse collisions is due to invariance of physical laws under *time reversal*. Equality of cross sections for the original and inverse collisions is due to symmetric properties of the scattering matrix $\overline{\overline{S}}$ (1.38a). Thus, for example, $\overline{\overline{S}} = \overline{\overline{S}}^{-1}$, so that if $\mathbf{g}' = \overline{\overline{S}}\mathbf{g}$ then $\mathbf{g} = \overline{\overline{S}}\mathbf{g}'$ (see Problem 3.2).

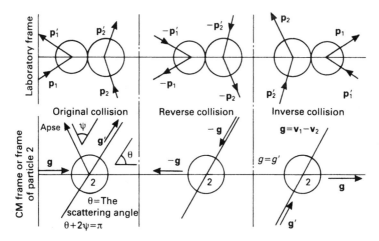

FIGURE 3.9. Scattering cross section is the same for all three collisions.

3.2 The Boltzmann Equation

The Boltzmann equation is an equation of motion for the one-particle distribution function and is appropriate to a rare gas. Various efforts have attempted
to derive this equation from first principles. Three of these, due respectively
to I. Prigogine, N. N. Bogoliubov, and H. Grad were described in Chapter 2.
An assortment of assumptions come into play in all such derivations, which
renders the analyses somewhat ad hoc.

In the present discourse, we revert to a more heuristic derivation, which
follows the spirit of Boltzmann's work, that is physically revealing and equally
ad hoc to more fundamental derivations.

3.2.1 Collisional Parameters and Derivation

In the limit of no interactions, particles are mutually independent and $F(\mathbf{x}, \mathbf{v}, t)$
(normalized to the total number of particles, N) satisfies the one-particle Liouville equation. This equation states that the net number of particles that enter
the phase element $\delta\mathbf{v}\delta\mathbf{x}$ following a particle's trajectory, in the time interval
δt, is zero. Let us denote this number by δR. Then we may write

$$\delta R = \delta\mathbf{x}\delta\mathbf{v}\delta t \left(\frac{\partial F}{\partial t} + \mathbf{v} \cdot \frac{\partial F}{\partial \mathbf{x}} + \frac{\mathbf{K}}{m} \cdot \frac{\partial F}{\partial \mathbf{v}} \right) = 0 \qquad (2.1)$$

where $\mathbf{K}$ is an externally supported force field. Consider now that particles
interact. Specifically, let r_0 denote the range of interaction. This parameter is
defined so that, for interparticle displacement $r > r_0$, interaction vanishes.
When particles enter the interaction domain ($r \lesssim r_0$), they experience a collision. Let the mean distance between collisions be l. This distance is also called
the *mean free path*. Our first criterion for the validity of the derivation of the

Boltzmann equation to follow is that

$$l \gg r_0 \qquad (2.2)$$

This constraint ensures rectilinear trajectories between collisions. If C is thermal speed, then we may introduce the times

$$\tau = \frac{l}{C}, \qquad \tau_0 = \frac{r_0}{C} \qquad (2.3)$$

An equivalent criterion to (2.2) may then be written

$$\tau \gg \tau_0 \qquad (2.4)$$

Collision terms

If interactions are turned on then the one-particle Liouville equation changes by virtue of collisions between molecules in the fluid. A measure of this effect is given by the interaction term in BY_1 (2.1.24). In the present model, this phenomenon is represented by the net rate at which collisions increase or decrease the number of molecules entering the phase volume $\delta x \delta v$. Thus in (2.1) we write

$$\delta R = \delta R_+ - \delta R_- \qquad (2.5)$$

The number of particles injected into $\delta v \delta x$ due to collisions in the interval δt is δR_+, while those ejected is δR_-.

First we consider δR_-. The velocity of all particles in the gas may be divided into two groups, the small band of velocities that fall into the interval δv about v and all other velocities denoted by the variable v_1. The number of particles that are removed from the phase element $\delta x \delta v$ in the time δt is simply the total number of collisions that the v particles have with all other particles (that is, the v_1 particles) in the time δt. It follows that to calculate δR_- we must account for all collisions between pairs of particles that eject one of them out of the interval δv about v, in other words, in every pair with the following properties:

1. One particle is in the phase element $\delta v \delta x$ about (v, x) and the other is in the phase element $\delta v_1 \delta x_1$ about (v_1, x_1).
2. The v_1 particles in δx_1 undergo a collision with the v particles in δx in the time δt.

By definition, the number of such particles is given by

$$\delta R_- = \int_1 F_2(z, z_1) \, \delta v_1 \, \delta x_1 \, \delta v \delta x \qquad (2.6)$$

To construct δx_1 so that it has property 2, we view a scattering event in the frame of the v particle (see Fig. 3.10). In (2.6) we have reintroduced the phase variable $z \equiv (x, v)$. Furthermore, recall that F_2 is normalized to $N(N-1)$, where N is the total number of particles.

With the aid of Fig. 3.10 we note that all v_1 particles in the cylinder shown of height $g \, \delta t$ and base area $s \, ds \, d\phi$ undergo a collision with the v particle in

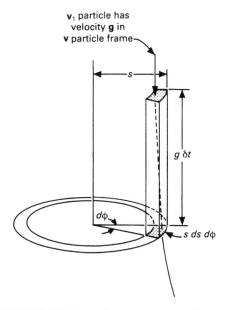

FIGURE 3.10. Scattering in the v-particle frame.

the time δt. Here we are writing

$$g = |\mathbf{v}_1 - \mathbf{v}|$$

It follows that

$$\delta \mathbf{x}_1 = g \delta t \, s \, ds \, d\phi \tag{2.7}$$

and (2.6) becomes

$$\delta R_- = \left(\int F_2 \, d\mathbf{v}_1 g \, s \, ds \, d\phi \right) \delta \mathbf{v} \, \delta \mathbf{x} \, \delta t \tag{2.8}$$

To calculate δR_+, we must account for all two-particle collisions that send one particle into the velocity interval $\delta \mathbf{v}$ about $\mathbf{v}$ in the time δt. But this is just the inverse of the original collision

$$(\mathbf{v}, \mathbf{v}_1) \to (\mathbf{v}', \mathbf{v}'_1)$$

That is,

$$(\mathbf{v}'', \mathbf{v}'_1) \to (\mathbf{v}, \mathbf{v}_1)$$

It follows that

$$\delta R_+ = \int_{1'} F_2(\mathbf{z}', \mathbf{z}'_1) \delta \mathbf{v}'_1 \delta \mathbf{x}'_1 \delta \mathbf{v}' \delta \mathbf{x}' \tag{2.9}$$

Here it is understood that primed variables refer to the inverse collision of unprimed variables in (2.8).

The next point in our derivation is to recall the integral invariants of Poincaré. These include the phase volume $\delta x \delta v \delta x_1 \delta v_1$ relevant to an N-particle system That is, we may write

$$\delta x \delta v \delta x_1 \delta v_1 = \delta x' \delta v' \delta x_1' \delta v_1' \tag{2.10}$$

Substituting these results into (2.5), (2.1) gives

$$\frac{DF}{DT} \equiv \frac{\partial F}{\partial t} + \mathbf{v} \cdot \frac{\partial F}{\partial x} + \frac{\mathbf{K}}{m} \cdot \frac{\partial F}{\partial \mathbf{v}} = \int [F_2(\mathbf{z}', \mathbf{z}_1') - F_2(\mathbf{z}, \mathbf{z}_1)] \, d\mathbf{v}_1 g s \, ds \, d\phi \tag{2.11}$$

We now assume that F_2 is homogeneous over the dimensions of the collision domain and write

$$F_2(\mathbf{z}, \mathbf{z}_1) = F_2(\mathbf{v}, \mathbf{v}_1)$$
$$F_2(\mathbf{z}', \mathbf{z}_1') = F_2(\mathbf{v}', \mathbf{v}_1') \tag{2.12}$$

Finally, we impose the assumption of *molecular chaos*, which states that particles in a rare gas are not correlated. That is,

$$F_2(\mathbf{v}, \mathbf{v}_1) = \frac{N-1}{N} F(\mathbf{v}) F(\mathbf{v}_1)$$
$$F_2(\mathbf{v}', \mathbf{v}_1') = \frac{N-1}{N} F(\mathbf{v}') F(\mathbf{v}_1') \tag{2.13}$$

Substituting (2.13) into (2.12) and neglecting 1 compared to N gives the *Boltzmann equation*:

$$\boxed{\frac{DF}{Dt} = \iint d\mathbf{v}_1 g \sigma \, d\Omega [F_1' F' - F_1 F]} \tag{2.14}$$

Here we have recalled that $s \, ds \, d\phi = \sigma \, d\Omega$ and have introduced the notation

$$F_1 \equiv F(\mathbf{v}_1), \qquad F' \equiv F(\mathbf{v}'), \qquad F_1' \equiv F(\mathbf{v}_1') \tag{2.14a}$$

Note also that, with the conservation solutions (1.37),

$$\mathbf{v}_1' = \mathbf{v}_1 + \boldsymbol{\alpha}(\boldsymbol{\alpha} \cdot \mathbf{g})$$
$$\mathbf{v}' = \mathbf{v} - \boldsymbol{\alpha}(\boldsymbol{\alpha} \cdot \mathbf{g}) \tag{2.14b}$$

These latter three equations, (2.14, 2.14a, 2.14b), comprise the *Boltzmann equation*.

Key assumptions brought into play in the derivation of this equation are as follows:

1. Range of interaction/mean free path $\ll 1$.
2. Particle trajectories are rectilinear before and after collision, as depicted in Fig. 3.10. This assumption is called the *stosszahlansatz*.
3. $F(\mathbf{x}, \mathbf{v})$ is homogeneous over the range of interaction.
4. Molecular chaos: $F_2(1, 2) = F(1)F(2)$.

3.2.2   Multicomponent Gas

Generalization of (2.14) to $\tilde{N}$ species is accomplished with aid of the following notation. First we rewrite (2.14) as

$$\frac{DF}{Dt} = \hat{J}(F \mid F) \tag{2.15}$$

where

$$\hat{J}(F \mid G) \equiv \iint d\mathbf{v}_1 g\sigma \, d\Omega (F'G'_1 - FG_1) \tag{2.16}$$

Then for a gas comprised of $\bar{N}$ species we have

$$\frac{DF_i}{Dt} = \sum_{j=1}^{\bar{N}} \hat{J}(F_i \mid F_j) \tag{2.17}$$

Note in particular that

$$\hat{J}(F_i \mid F_j) = \iint d\mathbf{v}_j g_{ij}\sigma_{ij} \, d\Omega_{ij} (F'_i F'_j - F_i F_j) \tag{2.18}$$

The set of $\bar{N}$ coupled equations (2.17) comprises the generalization of the Boltzmann equation to $\tilde{N}$ species. These relations are returned to in Section 3.7 in the derivation of the Druyvesteyn distribution relevant to electron transport in an ionic medium immersed in an electric field.

3.2.3   Representation of Collision Integral for Rigid Spheres

Consider a gas comprised of rigid spheres of diameter σ_{01}. With (1.33), we recall

$$\sigma(\theta) = \frac{\sigma_{01}^2}{4} \tag{2.19}$$

We wish to work with the angle ψ in place of the scattering angle θ (see Fig. 3.4).

$$\theta + 2\psi = \pi$$

so that

$$\cos\theta = 1 - 2\cos^2\psi$$

and

$$d\cos\theta = -4\cos\psi \, d\cos\psi$$

and

$$d\Omega = -d\phi \, d\cos\theta = 4\cos\psi \, d\cos\psi \, d\phi \tag{2.20}$$

The collision integral is a function of $\mathbf{v}$ and angles may be defined with respect to this vector. Thus, with $\mathbf{v}$ as the polar axis, $\mathbf{v}_1$ is given the polar angle λ and azimuth κ. Combining results, we write

$$J(F \mid F) = \sigma_{01}^2 \int g v_1^2 dv_1 \, d\cos\lambda \, d\kappa \int \cos\psi \, d\cos\psi \, d\phi [F_1' F' - F_1 F] \tag{2.21}$$

Recalling the definition of g,

$$\mathbf{g} = \mathbf{v}_1 - \mathbf{v} \tag{2.22}$$

we obtain

$$g^2 = v^2 + v_1^2 - 2vv_1 \cos\lambda \tag{2.23}$$

Referring to Fig. 3.4, we see that

$$\boldsymbol{\alpha} \cdot \mathbf{g} = g\cos\psi \tag{2.24}$$

This permits (2.14b) to be rewritten

$$\mathbf{v}_1' = \mathbf{v}_1 + \boldsymbol{\alpha} \cos\psi$$
$$\mathbf{v}' = \mathbf{v} - \boldsymbol{\alpha} g \cos\psi$$

so that

$$F_1' = F[\mathbf{v}_1 + \boldsymbol{\alpha} g \cos\psi]$$
$$F' = F[\mathbf{v} - \boldsymbol{\alpha} g \cos\psi] \tag{2.25}$$

To measure $\boldsymbol{\alpha}$, we choose the polar axis $\mathbf{g}$. The azimuth ϕ is measured with respect to the plane specified by the vectors $\mathbf{v}$, $\mathbf{v}_1$ (see Fig. 3.11). The Cartesian components of $\boldsymbol{\alpha}$ in this frame are

$$\boldsymbol{\alpha} = (\sin\psi \cos\phi, \sin\psi \sin\phi, \cos\psi) \tag{2.26}$$

In the spirit of this representation, we integrate over ψ, holding $\mathbf{v}_1$ and $\mathbf{g}$ fixed prior to integrating over $\mathbf{v}_1$.

Explicit representation of F' and F_1' as given by (2.25) requires knowledge of $\boldsymbol{\alpha} \cdot \mathbf{v}$ and $\boldsymbol{\alpha} \cdot \mathbf{v}_1$. With the aid of Fig. 3.11 and the addition law of cosines, we find

$$\boldsymbol{\alpha} \cdot \mathbf{v} = v[\cos\psi \cos\lambda + \sin\psi \sin\lambda \cos\phi]$$
$$\boldsymbol{\alpha} \cdot \mathbf{v}_1 = v_1[\cos\psi \cos(\lambda + \gamma) + \sin\psi \sin(\lambda + \gamma)\gamma \cos\phi]$$

where

$$\cos\gamma = \frac{v_1^2 - v^2 + g^2}{2v_1 g}$$

This completes our representations of $\hat{J}(F \mid F)$ for the scattering of rigid spheres. Note that, in the more general case, in place of σ given by (2.19) we have $\sigma = \sigma(\theta, g)$, in which case σ remains in the integrand of the $\cos\psi$ integration.

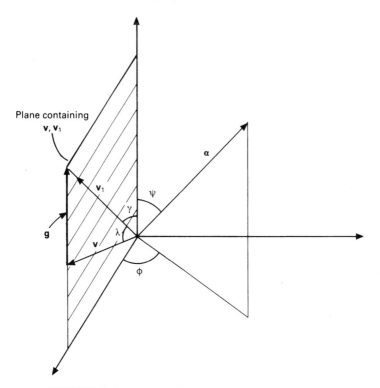

FIGURE 3.11. The coordinates of the apsidal vector α.

It should be noted that the preceding derivation addresses smooth spheres. Such types of spheres exchange no angular momentum in collision (see Problems 3.53 and 3.54). We note also that Grad's second equation (2.6.12) is better suited to the study of interacting rigid spheres than is the Boltzmann equation.[2]

3.3 Fluid Dynamic Equations and the Boltzmann $\mathcal{H}$ Theorem

3.3.1 Collisional Invariants

We wish to derive the fundamental fluid dynamic equations from the Boltzmann equation (2.14). Such equations express the basic conservation principles

---

[2]As has been previously noted, the assumption of Maxwell molecules greatly simplifies the collision integral [see (1.27)]. With the additional assumption that $g\sigma = $ constant, the Boltzmann equation has been solved exactly for a spatially uniform fluid: A. V. Bobylev, *Doc. Acad. Nauk. SSSR 225*, 1296 (1975); M. Krook and T. S. Wu, *Phys. Fluids 20*, 1589 (1977). See also R. M. Ziff, *Phys. Rev. A 24*, 509 (1981).

of physics: mass (or number) of particles, momentum, and energy. Thus a necessary requirement for a kinetic equation to be a valid physical relation is that it imply these conservation equations. To show that this requirement is satisfied by the Boltzmann equation, we introduce the notion of collisional (or summational) invariants.

Toward these ends, we recall the collisional integral of (2.14):

$$\hat{J}(F) \equiv \iint \sigma \, d\Omega \, d\mathbf{v}_1 g(F_1' F' - F_1 F) \tag{3.1}$$

Let us identify the linear operator:

$$\hat{I}[\phi(\mathbf{v})] \equiv \iint \hat{J}(F)\phi(\mathbf{v}) \, d\mathbf{v}$$

$$= \iiint \sigma \, d\Omega \, d\mathbf{v}_1 \, d\mathbf{v} g(F_1' F' - F_1 F)\phi(\mathbf{v}) \tag{3.2}$$

Changing variables,

$$(\mathbf{v}, \mathbf{v}_1) \rightarrow (\mathbf{v}_1, \mathbf{v})$$

gives the equality

$$\hat{I}(\phi) = I(\phi_1) \tag{3.3}$$

Consider next the integral

$$\hat{I}(\phi') = \iiint \sigma \, d\Omega \, d\mathbf{v}_1 \, d\mathbf{v} g \phi(\mathbf{v}')(F_1' F' - F_1 F) \tag{3.4}$$

The change of variables

$$(\mathbf{v}, \mathbf{v}_1) \rightarrow (\mathbf{v}', \mathbf{v}_1')$$

carries a unit Jacobian due to Liouville's theorem for a two-particle system. Therefore,

$$d\mathbf{v}_1 \, d\mathbf{v} = d\mathbf{v}_1' \, d\mathbf{v}'$$

The measure $g\sigma \, d\Omega$ is also invariant under this change in variables and (3.4) becomes

$$\hat{I}(\phi') = \iiint \sigma' \, d\Omega' \, d\mathbf{v}_1' \, d\mathbf{v}' g' \phi'(F_1 F - F_1' F')$$

$$= -\hat{I}(\phi) \tag{3.5}$$

Finally, exchanging variables $(\mathbf{v}_1', \mathbf{v}') \rightarrow (\mathbf{v}', \mathbf{v}_1')$ gives

$$\hat{I}(\phi') = \hat{I}(\phi_1') \tag{3.6}$$

Combining (3.3), (3.5), and (3.6) gives

$$4\hat{I}(\phi) = \hat{I}(\phi) + \hat{I}(\phi_1) - \hat{I}(\phi') - \hat{I}(\phi_1')$$

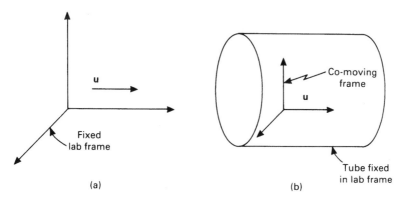

FIGURE 3.12. (a) The absolute frame is fixed in the lab frame. (b) The relative frame moves with the local mean velocity of the fluid.

Due to the linearity of the $\hat{I}$ operator, this latter result may be rewritten

$$\hat{I}(\phi) = \frac{1}{4}\hat{I}(\phi + \phi_1 - \phi' - \phi_1')$$ (3.7)

A function $\psi(\mathbf{v})$ is a *summation (or collisional) invariant* if

$$\psi_1 + \psi = \psi_1' + \psi'$$ (3.8)

That is, $\psi(\mathbf{v})$ is a property of a molecule that is preserved in a collision, such as mass, momentum, and energy.

If $\psi(\mathbf{v})$ is a collisional invariant, then with (3.7) and (3.8) we see that

$$\hat{I}(\psi) = 0, \qquad \psi = \text{collisional invariant}$$ (3.9)

The three fundamental collisional invariants described above correspond to $\psi = 1$, $\psi = \mathbf{v}$, and $\psi = v^2$, and we write

$$\hat{I}(1) = \hat{I}(\mathbf{v}) = \hat{I}(v^2) = 0$$ (3.10)

3.3.2   Macroscopic Variables and Conservative Equations

Prior to learning the manner in which collisional invariants come into play in obtaining the conservation equations, we turn to the definition of fluid-dynamic variables and the equations they satisfy. Fluid dynamic (or macroscopic) variables are commonly defined in one of two frames: the *absolute* and *relative* (or *co-moving*) frames. In the absolute frame the fluid moves with respect to a coordinate frame fixed in the lab frame. In the relative frame, fluid variables are measured with respect to a coordinate frame that moves with the local velocity of the fluid (see Fig. 3.12).

Conservation equations typically appear in the form

$$\frac{\partial A}{\partial t} + \nabla \cdot \mathbf{\Gamma}_A = 0$$ (3.11)

where $\mathbf{\Gamma}_A$ is a flux vector corresponding to the quantity A. Integrating this equation over a small volume V and employing Gauss' theorem gives

$$\frac{\partial}{\partial t} \int_N A \, d\mathbf{x} = - \oiint_S \mathbf{\Gamma}_A \cdot d\mathbf{S} \tag{3.11a}$$

This equation says that the quantity $\int_V A \, d\mathbf{x}$ can change only by virtue of the net flow out of the volume as measured by the flux integral over the surface S.

Consider the number density $n(\mathbf{x}, t)$ [encountered previously in (1.6.12)] and macroscopic mean velocity $\mathbf{u}(\mathbf{x}, t)$. These variables are defined by (with $F = Nf$)

$$n(\mathbf{x}, t) = \int F \, d\mathbf{v} \tag{3.12}$$

$$n\mathbf{u}(\mathbf{x}, t) = \int F\mathbf{v} \, d\mathbf{v} \tag{3.13}$$

and obey the *continuity* equation

$$\frac{\partial n}{\partial t} + \nabla \cdot n\mathbf{u} = 0 \tag{3.14}$$

This equation is an expression of the conservation of matter. It says that the mass in a given volume can change only by virtue of a net flux of mass out of the volume.

Note that with (3.13) we may write [recall (1.6.8)]

$$\mathbf{u} = \frac{\int F\mathbf{v} \, d\mathbf{v}}{\int F \, d\mathbf{v}} = \langle \mathbf{v} \rangle \tag{3.13a}$$

Thus $\mathbf{u}$ is appropriately termed *mean* macroscopic velocity.

Next we consider the *pressure tensor* $\bar{\bar{p}}$, *heat flow vector* $\mathbf{q}$, and *kinetic energy density* e_K, all measured in the absolute frame. These are given by the integrals[3]

$$\bar{\bar{p}} = \int Fm\overline{\overline{\mathbf{v}\mathbf{v}}} \, d\mathbf{v} = n\langle m\overline{\overline{\mathbf{v}\mathbf{v}}}\rangle \tag{3.15}$$

$$\mathbf{q} = \int F\frac{1}{2}mv^2\mathbf{v} \, d\mathbf{v} = n\left\langle\frac{1}{2}mv^2\mathbf{v}\right\rangle \tag{3.16}$$

$$e_K = \int F\frac{1}{2}mv^2 \, d\mathbf{v} = n\langle\frac{1}{2}mv^2\rangle \tag{3.17}$$

The pressure component p_{xy}, say, represents the transport of x momentum per second across a y surface (the unit vector $\hat{\mathbf{y}}$ is normal to a y surface). Evidently, $\mathbf{q}$ represents a flux of kinetic energy.

---

[3] The energy density e_K was previously introduced in (1.6.16) where it was labeled E_K.

Conservation of momentum gives

$$\frac{\partial}{\partial t}\rho\mathbf{u} + \nabla \cdot \bar{\bar{p}} = \rho\mathbf{K} \tag{3.18}$$

Here we have written $m\mathbf{K}$ for an externally supported force field that permeates the fluid, where m is molecular mass and

$$\rho(\mathbf{x}, t) = mn(\mathbf{x}, t) \tag{3.18a}$$

is *mass density*.

The conservation equation for energy is given by

$$\frac{\partial e_K}{\partial t} + \nabla \cdot \mathbf{q} = \rho\mathbf{K} \cdot \mathbf{u} \tag{3.19}$$

which, together with (3.14) and (3.18), constitutes the three fundamental conservation equations of fluid dynamics (in the absolute frame).

3.3.3 Conservation Equations and the Boltzmann Equation

To obtain the conservation equations (3.14), (3.18), and (3.19) from the Boltzmann equation (2.14), first we rewrite the Boltzmann equations as

$$\frac{DF}{Dt} = \hat{J}(F) \tag{3.20}$$

$$\frac{DF}{Dt} \equiv \frac{\partial F}{\partial t} + \mathbf{v} \cdot \frac{\partial F}{\partial \mathbf{x}} \tag{3.20a}$$

First note that the left side of (3.20) gives

$$\int d\mathbf{v} \frac{DF}{Dt} \begin{pmatrix} 1 \\ m\mathbf{v} \\ \dfrac{mv^2}{2} \end{pmatrix} = \begin{pmatrix} \dfrac{\partial n}{\partial t} + \nabla \cdot n\mathbf{u} \\ \dfrac{\partial \rho\mathbf{u}}{\partial t} + \nabla \cdot \bar{\bar{p}} \\ \dfrac{\partial e_K}{\partial t} + \nabla \cdot \mathbf{q} \end{pmatrix} \tag{3.21}$$

The right side of (3.20) gives

$$\int d\mathbf{v}\hat{J}(F) \begin{pmatrix} 1 \\ m\mathbf{v} \\ \dfrac{m}{2}v^2 \end{pmatrix} = \begin{bmatrix} \hat{I}(1) \\ m\hat{I}(\mathbf{v}) \\ \dfrac{m}{2}\hat{I}(v^2) \end{bmatrix} = \begin{pmatrix} 0 \\ 0 \\ 0 \end{pmatrix} \tag{3.22}$$

Equating (3.21) to (3.22) reproduces the three conservation equations (3.14), (3.18), and (3.19). We may conclude that the Boltzmann equation satisfies the required property of implying the conservation equations.

Relative macroscopic variables

We turn next to rewriting these conservation equations in terms of more physically pertinent *relative* macroscopic variables defined in terms of the deviation-from–the-mean microscopic velocity **c** given by

$$\mathbf{c} = \mathbf{v} - \mathbf{u} = \mathbf{v} - \langle \mathbf{v} \rangle \tag{3.23}$$

The velocity **c** is that of a fluid element measured in a frame moving with the mean velocity of the fluid, **u**. The relative pressure tensor $\bar{\bar{P}}$, heat flow vector **Q**, and (kinetic) energy density E_K are given by the parallel equations to (3.15), (3.16), and (3.17), with **v** replaced by **c**. There results

$$\bar{\bar{P}} = \int F F m \overline{\overline{\mathbf{cc}}} \, d\mathbf{c} = \rho \langle \overline{\overline{\mathbf{cc}}} \rangle \tag{3.24}$$

$$\mathbf{Q} = \int F \frac{1}{2} m c^2 \mathbf{c} \, d\mathbf{c} = \rho \left\langle \frac{1}{2} c^2 \mathbf{c} \right\rangle \tag{3.25}$$

$$E_K = \int F \frac{1}{2} m c^2 \, d\mathbf{c} = \rho \left\langle \frac{1}{2} c^2 \right\rangle \tag{3.26}$$

Substituting **c** as given by (3.23) into these expressions gives

$$\bar{\bar{p}} = \bar{\bar{P}} + \rho \overline{\overline{\mathbf{uu}}} \tag{3.27}$$

$$\mathbf{q} = \mathbf{Q} + \bar{\bar{p}} \cdot \mathbf{u} + \mathbf{u}(e_K - \rho u^2) = \mathbf{Q} + \bar{\bar{P}} \cdot \mathbf{u} + \mathbf{u} E_K + \mathbf{u} \frac{\rho u^2}{2} \tag{3.28}$$

$$e_K = E_K + \frac{\rho u^2}{2} \tag{3.29}$$

The relation (3.27) says that absolute pressure is greater than relative pressure by the flow energy $\rho \overline{\overline{\mathbf{uu}}}$. Consider water flowing in a pipe with $\mathbf{u} = (u_x, 0, 0)$. Thus $\rho \overline{\overline{\mathbf{uu}}} = \rho u^2$. If $\bar{\bar{p}}$ is measured, the *fixed* pressure meter measures this additional momentum flux due to the macroscopic flow of the water in the pipe. This is not what we mean by pressure. We measure pressure in a frame moving with the fluid. The meter that measures $\bar{\bar{P}}$ does not see the macroscopic flow velocity **v**.

Inserting the relation (3.27), (3.28), and (3.29) into the macroscopic equations (3.14), (3.18), and (3.19) gives the normal form of the conservation equations:

$$\rho \left(\frac{\partial}{\partial t} + \mathbf{u} \cdot \nabla \right) \mathbf{u} + \nabla \cdot \bar{\bar{P}} = p\mathbf{K} \tag{3.30}$$

$$\frac{\partial}{\partial t} E_K + \nabla \cdot (\mathbf{u} E_K) + \bar{\bar{P}} : \nabla \mathbf{u} + \nabla \cdot \mathbf{Q} = 0 \tag{3.31}$$

(see Problem 3.3). The tensor notation of the double inner product $\overline{\overline{P}} : \overline{\overline{\nabla \mathbf{u}}}$ is as follows:

$$\overline{\overline{P}} : \overline{\overline{\nabla \mathbf{u}}} = \mathrm{Tr}(\overline{\overline{P}}\,\overline{\overline{\nabla \mathbf{u}}}) = P_{ij}\frac{\partial}{\partial x_j}u_i$$

The right side of the preceding equations is written in the *Einstein convention* in which repeated indexes are summed. [4]

Equations (3.30) and (3.31) will be returned to in Section 3.5 in our discussion of transport coefficients.

3.3.4 Temperance: Variance of the Velocity Distribution

The temperature T of a fluid comprised of molecules with no internal degrees of freedom is given by

$$E_K = \int F(c)\frac{m^2}{2}\,d\mathbf{c} = \frac{3}{2}nk_BT \tag{3.32}$$

where

$$k_B = 1.381 \times 10^{-16}\,\mathrm{erg/K}$$

represents *Boltzmann's constant*.

The significance of defining T as given by (3.32) in the relative frame is as follows. Consider a distribution of molecules that are all at rest. Now let the fluid move as a rigid body with a speed $\mathbf{u}$. With (3.32), we see that $T \propto E_K = 0$, but that e_K as given by (3.29) is greater than zero. So had we defined T with respect to absolute variables we would obtain the erroneous result that the temperature of a gas of stationary molecules is greater than zero, by virtue of the macroscopic motion of the whole body of molecules.

Note that, (3.23), (3.32) may be rewritten

$$\frac{3k_BT}{m} = \langle(\mathbf{v} - \langle\mathbf{v}\rangle)^2\rangle \tag{3.32a}$$

which, recalling (1.8.2b), indicates that $3k_BT/m$ is a measure of the variance of the velocity probability density. Thus, in general, T is a measure of the spread of this density about the mean. Thus a wide distribution (many velocities contribute to F) corresponds to a high temperature. If the distribution of speeds is sharply peaked about the mean $\langle\mathbf{v}\rangle$, then temperature is small (see Fig. 3.13).

Having defined the fundamental fluid-dynamic variables ($n, \mathbf{u}, T, E_K, \mathbf{Q}$, and $\overline{\overline{P}}$) as velocity moments of $F(\mathbf{x}, \mathbf{c}, t)$ it is evident that such relations may be generalized to define additional macroscopic variables. Thus consider the

---

[4]Tensor notation is reviewed in Appendix A.

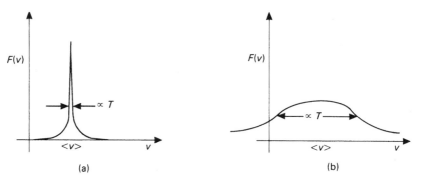

FIGURE 3.13. In (a) most molecules move with mean speed $\langle v \rangle$ corresponding to low temperature. In (b) the variance of $F(c)$ is large, as is the temperature. Measurement of v in this instance finds large fluctuations about the mean, $\langle v \rangle$.

tensor form

$$n \Lambda_{i_1, i_2, \ldots, i_n} = \int d\mathbf{c} F(\mathbf{c}, \mathbf{x}, t) c_{i_1} c_{i_2} \ldots c_{i_n} \tag{3.33}$$

The fluid variable $\Lambda(\mathbf{x}, t)$ is an nth-rank tensor in three dimensions.[5,6] The indexes i_k run from 1 to 3, corresponding to the Cartesian components of $\mathbf{c}$.

The preeminence of the lower-order variables (n, $\mathbf{u}$, T, E_K, $\mathbf{Q}$, and $\bar{\bar{P}}$) is due to their presence in the three conservation equations.

3.3.5   Irreversibility

Preparatory to our discussion of the $\mathcal{H}$ theorem, it is appropriate at this point to discuss the concept of irreversibility. Consider that 1 mole of gas is comprised of identical inert molecules save for the fact that half of the sample molecules are labeled color A and the remaining half color B. The entire gas is confined to an isolated enclosure with A molecules and B molecules separated by a partition located at the midplane of the enclosure. The entire gas is in equilibrium at some given temperature. The partition is removed (ideally, without incurring any other perturbation on the system). After sufficient time, the gases mix and the A and B molecules become uniformly distributed over the entire enclosure (see Fig. 3.14).

It is universally observed that the starting state of this process does not reoccur. Why is this observation peculiar to the laws of nature? In Section 1.1.5, we found that if a system's Hamiltonian has a trajectory $[q(t)p(t)]$ then it also has a trajectory $[q(-t), -p(-t)]$. It is evident that we may view the starting

---

[5]Thus we have the quip, "A distribution function is worth a thousand macroscopic variables."

[6]Grad's method of moments takes advantage of the tensor quality of (3.33). See Section 5.10.

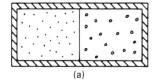 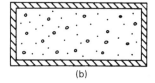

(a)   (b)

FIGURE 3.14. (a) Confined gases. (b) Gases expand to fill the whole interior region. For mole concentrations, state (a) is not observed to reoccur.

state (a) in Fig. 3.7 as the time-reversed state of (b). Thus, for example, say that in the state (b), at a given instant, the momentum of all particles in the gas is reversed. Then, as depicted in Fig. 1.5, the motion should reverse itself, thereby restoring the initial coordinate configuration. However, such reversal does not occur and we conclude the motion is irreversible.

More generally, an irreversible law is characterized by an equation that has a solution peculiar to a particular direction of time. This is a quality of irreversibility, because if the motion corresponding to this solution is observed, we may conclude that time is flowing, say, in the forward direction. A reversible law on the other hand implies no such distinction.

3.3.6 Poincaré Recurrence Theorem

The evident irreversibility of the process described in Fig. 3.14 also plays havoc with a theorem due to Poincaré. This theorem states the following: The system trajectory of a bounded isolated system of finite energy will, after sufficient time, return arbitrarily close to its initial location in Γ-space.

To establish this theorem, we consider that the initial state $z_0 \equiv (q_0, p_0)$ is contained in the set of phase points Ω_0. That is, $z \in \Omega_{10}$. Due to the preceding description, the system point moves on a surface of finite measure in Γ-space. It follows that after sufficient time the path swept out by the set Ω_0 must intersect itself. Let us show this. Let $\hat{T}$ denote an operator that displaces Ω_0 in unit time. Then, due to Liouville's theorem

$$\Omega_0, \hat{T}\Omega_0, \hat{T}^2\Omega_0, \ldots$$

all have the same measure. If these sets did not intersect, the surface on which they move would have to be of infinite measure. This contradicts our assumption.

Therefore, we may write

$$\hat{T}^k\Omega_0 \cap \hat{T}^n\Omega_0 = \bar{\Omega} \neq \varnothing \tag{3.34}$$

for some value of the integers k and n. The empty set is $\varnothing$. Due to uniqueness of trajectories, as previously described in Chapter 1, $\hat{T}$ is a one-to-one mapping. This property permits us to write

$$\hat{T}(A \cap B) = \hat{T}(A) \cap \hat{T}(B) \tag{3.35}$$

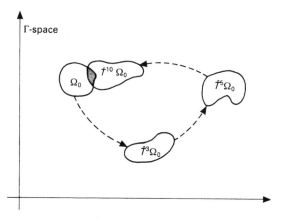

FIGURE 3.15. Shaded domain represents the intersection $\hat{T}^{10}\Omega_0 \cap \Omega_0$.

Operating on (3.34) with $\hat{T}^{-n}$ gives

$$\hat{T}^{-n}(\hat{T}^{k}\Omega_0 \cap \hat{T}^{n}\Omega_0) = \hat{T}^{-n}\bar{\Omega} \neq \varnothing$$

With (3.35), the preceding equation becomes

$$\hat{T}^{k-n}\Omega_0 \cap \Omega_0 \neq \varnothing \qquad (3.36)$$

Thus, after $k - n$ units of time, the set Ω_0 has a finite intersection with itself. Taking the measure of Ω_0 to be arbitrarily small establishes the theorem. The process described by (3.36) is depicted in Fig. 3.15. (The group property of the $\hat{T}$ operators is discussed in Problem 3.4). Applying this result to the process described in Fig. 3.14 indicates that, after sufficient time, the system point returns arbitrarily close to configuration (a).

Another approach to the question of irreversibility is offered by the Boltzmann $\mathcal{H}$ theorem, which is described in the following section.

3.3.7 Boltzmann and Gibbs Entropies

The Gibbs entropy, $\mathcal{H}_N$, is given by

$$\mathcal{H}_N = \int f_N \ln f_N d1 \ldots dN \qquad (3.37)$$

To discover how this function changes in time, we first recall the Liouville equation

$$\frac{df_N}{dt} = \frac{\partial f_N}{\partial t} + [f_N, H] = 0 \qquad (3.38)$$

Operating on this equation with

$$\int d1 \cdots dN(1 + \ln f_N)$$

reveals that[7]

$$\frac{d\mathcal{H}_N}{dt} = \int (1 + \ln f_N)\frac{df_N}{dt}d1\cdots dN = 0 \tag{3.39}$$

Thus $\mathcal{H}_N$ obeys a reversible equation [that is, if $\mathcal{H}_N(t)$ is a solution to (3.39), so is $\mathcal{H}_N(-t)$] and is constant.

The Gibbs entropy is related to thermodynamic entropy through the formula

$$S = -k_B\mathcal{H}_N \tag{3.40}$$

This is the kinetic prescription for the entropy of an isolated N-body system. The second law of thermodynamics stipulates that, for any process of an isolated system, $\Delta S \geq 0$, the equality holding for reversible processes. Since the Liouville equation is reversible, the fact that it implies $S = $ constant is seen to be consistent with the second law of thermodynamics.

For vanishing correlations,

$$f_N = \prod_{i=1}^{N} f_1(i)$$

and

$$\mathcal{H}_N = \sum_{i=1}^{N} \int \prod_{j=1}^{N} f_1(j)\ln f_1(i)d1\cdots dN$$

$$= \sum_{i=1}^{N} \mathcal{H}(i) = N\mathcal{H} \tag{3.41}$$

where

$$\mathcal{H} = \int f_1 \ln f_1 \, dx \, dp \tag{3.42}$$

denotes Boltzmann entropy and with (3.40) gives the Boltzmann formula

$$S = -Nk_B\mathcal{H} \tag{3.43}$$

We will find later that, away from equilibrium, Boltzmann's equation implies that $d\mathcal{H}/dt \equiv \dot{\mathcal{H}} < 0$. This is an irreversible equation because, as noted above, it implies one distinct motion with the positive flow of time, that is, a decaying solution. On the other hand, a reversible equation that implies a decaying solution must also imply a growing solution. That the laws of nature are reversible indicates then that there is no experiment to discern if time is flowing in one direction. On the other hand, the law $\dot{\mathcal{H}} < 0$ does imply a preferred direction in time, since with time flowing in the forward direction,

---

[7]Note that this same result occurs for any functional $\psi(f_N)$ with a bounded derivative $|d\psi/df_N| < \infty$. See Problem 3.58

$\mathcal{H}(t)$ decreases. If we were to observe $\mathcal{H}(t)$ increasing, we could conclude by virtue of the statement $\dot{\mathcal{H}} < 0$ that time was flowing backward.

With this background, we turn to derivation of the $\mathcal{H}$ theorem (put forth by Boltzmann near the turn of the century).

3.3.8   Boltzmann's $\mathcal{H}$ Theorem

This proof begins with the Boltzmann equation (2.14)

$$\frac{\partial f}{\partial t} + \mathbf{v} \cdot \frac{\partial f}{\partial \mathbf{x}} + \frac{\mathbf{K}}{m} \cdot \frac{\partial f}{\partial \mathbf{v}} = \hat{J}(f) \tag{3.44}$$

where $\hat{J}(f)$ has been written for the collision integral on the right side of (2.14). (Note also that we are working in velocity space.) Operating on (3.44) with

$$\int d\mathbf{x} \, d\mathbf{v} (1 + \ln f)$$

gives

$$\frac{\partial}{\partial t} \int d\mathbf{x} \, d\mathbf{v} f \ln f + \int d\mathbf{x} \, d\mathbf{v} \frac{\partial}{\partial \mathbf{x}} \cdot \mathbf{v} \ln f + \frac{\mathbf{K}}{m} \cdot \int d\mathbf{x} \, d\mathbf{v} \frac{\partial}{\partial \mathbf{v}} f \ln f$$

$$= \int d\mathbf{x} \, d\mathbf{v} \hat{J}(f)(1 + \ln f)$$

$$= \int d\mathbf{x} \hat{I}(1 + \ln f) \tag{3.45}$$

where $\hat{I}$ is defined by (3.2). Passing to large volume and dropping surface terms in (3.45) gives

$$\frac{1}{m} \frac{d}{dt} \mathcal{H} = \int d\mathbf{x} \hat{I}(1 + \ln f) \tag{3.46}$$

With the property (3.7), we may write

$$4\hat{I}(1 + \ln f) = \hat{I}(1 + \ln f) + \hat{I}(1 + \ln f_1) - \hat{I}(1 + \ln f') - \hat{I}(1 + \ln f_1')$$

$$= \hat{I} \left(\ln \frac{f_1 f}{f_1' f'} \right) = -\hat{I} \left(\ln \frac{f_1' f'}{f_1 f} \right) \tag{3.47}$$

Thus we may write

$$4\hat{I}(1 + \ln f) = -\int d\mathbf{v} \int d\mathbf{v}_1 g\sigma \, d\Omega (f_1' f' - f_1 f) \ln \left(\frac{f_1' f'}{f_1 f} \right)$$

Setting $X \equiv f_1' f'$ and $Y \equiv f_1 f$, this equation may be rewritten

$$4\hat{I} = -\int d\mathbf{v} \int d\mathbf{v}_1 g\sigma \, d\Omega (X - Y) \ln \left(\frac{X}{Y} \right)$$

$$= -\int d\mathbf{v} \int d\mathbf{v}_1 g\sigma \, d\Omega L(X, Y) \tag{3.48}$$

Here we have defined

$$L(X, Y) \equiv (X - Y) \ln \left(\frac{X}{Y} \right) \tag{3.49}$$

With X and Y positive, we examine $X = Y$, $X > Y$, and $X < Y$. For all three cases, $L \geq 0$. It follows that

$$\hat{I}(1 + \ln f) \leq 0 \tag{3.50}$$

since $d\mathbf{v}\, d\mathbf{v}_1 g\sigma\, d\Omega$ is a positive measure. The equality in (3.50) corresponds to $X = Y$; that is, $f_1' f' = f_1 f$. Inserting (3.50) into (3.46) gives

$$\frac{1}{m} \frac{d}{dt} \mathcal{H} = \int d\mathbf{x} \hat{I}(1 + \ln f) \leq 0$$

or, more simply,

$$\frac{d\mathcal{H}}{dt} \leq 0 \tag{3.51}$$

which is Boltzmann's $\mathcal{H}$ theorem.[8] The relation (3.51) states that, for arbitrary initial $f(\mathbf{x}, \mathbf{v}, t)$, $\mathcal{H}$ decreases until

$$f_1' f' = f_1 f \tag{3.52}$$

whereafter it remains constant.

Source of irreversibility

Equation (3.51) is evidently an irreversible statement. What is the source of this irreversibility? To answer this question, we recall that (3.51) stems from the Boltzmann equation (3.44), which is irreversible. To see this, consider the transformation, the left side of (3.44), which is irreversible. To see this, consider the transformation $\mathbf{x} \rightarrow \mathbf{x}' = \mathbf{x}$, $\mathbf{v} \rightarrow \mathbf{v}' = -\mathbf{v}$, and $t \rightarrow t = -t$. Under the transformation, the left side of (3.44) changes sign, whereas the collision integral on the right side maintains its sign. This is so because the integration measure of $\hat{J}(f)$ is positive and $f \geq 0$.

3.3.9 Statistical Balance

The statement (3.52) is sometimes called *statistical balance*. At this value, $\delta R_+ = \delta R_-$ in (2.5). That is, this rate of gain of particles into the volume element $\delta \mathbf{x} \delta \mathbf{v}$ is equal to the rate of loss of particles from this volume element.

Note in particular that with

$$\frac{1}{m} \frac{d\mathcal{H}}{dt} = \frac{\partial}{\partial t} \int d\mathbf{x}\, d\mathbf{v} f \ln f = \int d\mathbf{x}\, d\mathbf{v} \frac{\partial f}{\partial t} (1 + \ln f) \tag{3.53}$$

---

[8] An $\mathcal{H}$ theorem relevant to quantum systems is derived in Section 5.4.3.

the condition of equilibrium, $\partial f / \partial t = 0$, implies that $d\mathcal{H}/dt = 0$. We have found above that this latter condition is satisfied only when

$$f_1' f' = f_1 f \tag{3.54}$$

Thus the conditions (3.54) is necessary for equilibrium. For a force-free spatially homogeneous fluid, this condition is evidently also sufficient for equilibrium, and we may conclude that statistical balance is necessary and sufficient for the equilibrium of such a fluid.

3.3.10   The Maxwellian

To find the equilibrium distribution that is a solution to (3.54), we first take the ln of both sides. Labeling the solution f_0 gives

$$\ln f_{01}' + \ln f_0' = \ln f_{01} + \ln f_0 \tag{3.55}$$

This equation asserts that $\ln f_0$ is a collisional invariant [see (3.8)]. Thus we may write

$$\ln f_0(\mathbf{v}) = -A(\mathbf{v} - \mathbf{v}_0)^2 + \ln B \tag{3.55a}$$

or, equivalently,

$$f_0(\mathbf{v}) = B e^{-A(\mathbf{v} - \mathbf{v}_0)^2} \tag{3.56}$$

The constants A and B are determined through the relations (3.12), (3.13), and (3.26).

$$n = N \int B e^{-A(\mathbf{v} - \mathbf{v}_0)^2} \, d\mathbf{v} \tag{3.57a}$$

$$n\mathbf{u} = N \int B \mathbf{v} e^{-A(\mathbf{v} - \mathbf{v}_0)^2} \, d\mathbf{v} \tag{3.57b}$$

$$\frac{3}{2} n K_B T = N \int B \frac{mc^2}{2} e^{-A(\mathbf{v} - \mathbf{v}_0)^2} \, d\mathbf{c}$$

With these equations, we obtain

$$F_0(\mathbf{v}) = N f_0(\mathbf{v}) = \frac{n}{(2\pi RT)^{3/2}} \exp \left(-\frac{(\mathbf{v} - \mathbf{u})^2}{2RT} \right) \tag{3.58}$$

where[9]

$$R \equiv \frac{k_B}{m}$$

The distribution (3.58) is quite important in kinetic theory. It goes by either of the names the *Maxwellian*, the *Boltzmann distribution*, or the *Maxwell–Boltzmann distribution*.

---

[9]In kinetic theory, the symbol R often denotes the *gas constant* $k_B N_0$, where N_0 is Avogadro's number.

Although the Maxwellian (3.58) is a solution to the statistical balance equation, (3.54), there is another more general and physically relevant solution. Thus we observe that the distribution (3.58) remains is a solution to the statistical balance equation, (3.54), there is another more general and physically relevant solution. Thus we observe that the distribution (3.58) remains a solution to (3.54) with the constant values of n, $\mathbf{u}$, and T replaced by corresponding functions of $\mathbf{x}$ and t. The resulting equilibrium distribution is called the *local Maxwellian* and is given by

$$F^0(\mathbf{x}, \mathbf{v}, t) = \frac{n(\mathbf{x}, t)}{[2\pi RT(\mathbf{x}, t)]^{3/2}} \exp\left(-\frac{(\mathbf{v} - \mathbf{u}(\mathbf{x}, t))^2}{2RT(\mathbf{x}, t)}\right) \qquad (3.59)$$

With the normalization conditions (3.57), we find

$$n(\mathbf{x}, t) = \int F^0(\mathbf{x}, \mathbf{v}, t)\, d\mathbf{v} \qquad (3.60a)$$

$$n(\mathbf{x}, t)\mathbf{u}(\mathbf{x}, t) = \int F^0(\mathbf{x}, \mathbf{v}, t)\mathbf{v}\, d\mathbf{v} \qquad (3.60b)$$

$$\frac{3}{2}n(\mathbf{x}, t)k_B T(\mathbf{x}, t) = \int F^0(\mathbf{x}, \mathbf{c}, t)\frac{mc^2}{2}\, d\mathbf{c} \qquad (3.60c)$$

In the Maxwellian given by (3.58), the parameters n, $\mathbf{u}$, and T are constant in $\mathbf{x}$ and t. To distinguish this distribution from (3.59), it is called the *absolute Maxwellian*. To recapitulate,

$$F_0(\mathbf{v}) : \text{absolute Maxwellian}, \ n, \mathbf{u}, T = \text{constants}$$
$$F^0(\mathbf{x}, \mathbf{v}, t) : \text{local Maxwellian}, \ n, \mathbf{u}, T = \text{functions of } \mathbf{x}, t$$

We recall that $\dot{\mathcal{H}}(F^0) = \dot{\mathcal{H}}(F_0) = 0$. Thus both these distributions may be termed equilibrium distributions. However, thermodynamic equilibrium for a force-free fluid implies constant values of all macroscopic variables. This equilibrium is better described by $F_0(\mathbf{v})$. Prior to this state, the fluid is in a local equilibrium described by $F^0(\mathbf{x}, \mathbf{v}, t)$. This sequence of states in the approach to equilibrium of a fluid is well described in a sketch of $\mathcal{H}(t)$ (see Fig. 3.16).

We may speculate on the forms of the relaxation times τ_L and τ_A shown in Fig. 3.16. Let the fluid be confined to the volume $V = L^3$, and let l denote the mean free path of molecules between collisions. For $n \gg 1$, it is evident that $l \ll L$. With the thermal speed given by (3.32a),

$$\langle c^2 \rangle = \frac{3k_B T}{m} \equiv 3C^2 \qquad (3.61a)$$

we write

$$\tau_L \simeq \frac{l}{C} \quad \tau_A \simeq \frac{L}{C} \qquad (3.61b)$$

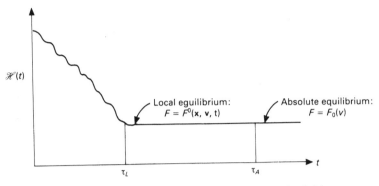

FIGURE 3.16. The approach to equilibrium of a fluid.

Having found that the local Maxwellian (3.59) is an equilibrium solution to the Boltzmann equation, we momentarily return to the Bogoliubov hypothesis (Section 2.4). We recall that this hypothesis states that, in the final hydrodynamic stage of a fluid's approach to equilibrium, distributions become functionals of the fluid variables n, $\mathbf{u}$, and T. It is evident that this conjecture finds corroboration in the structure of the local Maxwellian given by (3.59), which is seen to be a functional of the said fluid dynamic variables.

Local equilibrium and macroscopic variables

We have found that the local equilibrium (3.59) satisfies statistical balance (3.55) and therefore renders the collision integral of the Boltzmann equation zero. However, when substituted into the Boltzmann equation (2.14), it is evident that, in general, the left side of this equation does not vanish. As space dependence of F^0 is contained entirely in the macroscopic variables n, $\mathbf{u}$, and T, it suffices to find their dependence on space to obtain a complete equilibrium solution. For non-force-free fluids, this may accomplished directly from the Boltzmann equation as demonstrated in the following section.

As we will find in Section 3.5, the local Maxwellian also comes into play in the Chapman–Enskog expansion of the Boltzmann equation. In this procedure, F^0 emerges as the lowest-order solution with n, $\mathbf{u}$, and T occurring as functions of $(\mathbf{x}, t)$. Accompanying lowest-order conservation equations serve to determine these macroscopic variables, thereby establishing $F^0(\mathbf{x}, \mathbf{v}, t)$ as the corresponding lowest-order solution to the Boltzmann equation in the nonequilibrum limit.

3.3.11 The Barometer Formula

Consider that an externally supported conservative force field $\mathbf{K}$ permeates a fluid. We may write

$$\mathbf{K} = -\frac{\partial \Phi}{\partial \mathbf{x}}$$

Let us call the equilibrium distribution for this configuration $\bar{F}_0$, so that $\partial \bar{F}_0/\partial t = 0$. The remaining terms in the left side of the Boltzmann equation (2.14) give zero, providing

$$m\mathbf{v} \cdot \frac{\partial \bar{F}_0}{\partial \mathbf{x}} = \frac{\partial \Phi}{\partial \mathbf{x}} \cdot \frac{\partial \bar{F}_0}{\partial \mathbf{v}} \tag{3.62a}$$

A more general solution to the statistical balance equation (3.55) is given by

$$\ln \bar{F}_0 = \frac{-A(\mathbf{v} - \mathbf{v}_0)^2 + \ln B - 2A\psi(\mathbf{x})}{m} \tag{3.62b}$$

Substituting this form into (3.62a) indicates that $\psi(x) = \Phi(x)$, providing $\mathbf{v}_0 \cdot \nabla\Phi = 0$. With this constraint imposed, (3.62b) is the desired equilibrium solution.

Fitting $\bar{F}_0$ to the constraints (3.57) gives

$$\bar{F}_0(\mathbf{x}, \mathbf{v}) = \frac{n_0}{(2\pi R T_0)^{3/2}} \exp\left\{ \frac{-[m(\mathbf{v} - \mathbf{u}_0)^2/2 + \Phi(\mathbf{x})]}{k_B T_0} \right\} \tag{3.63}$$

where n_0, $\mathbf{u}_0$, and T_0 are constants and $\mathbf{u}_0$ is normal to $\nabla\Phi(\mathbf{x})$.

The equilibrium number density that follows from (3.63) is given by

$$n(\mathbf{x}) = \int \bar{F}_0(\mathbf{x}, \mathbf{v}) \, d\mathbf{v} = n_0 \exp\left[-\frac{\Phi(\mathbf{x})}{k_B T} \right] \tag{3.64}$$

so that n_0 is the value of $n(\mathbf{x})$ where $\Phi(\mathbf{x}) = 0$. The equilibrium temperature is given by

$$3n(\mathbf{x})RT = \int \bar{F}_0(\mathbf{x}, \mathbf{c})c^2 \, d\mathbf{c} = 3n_0 R T_0 \exp\left[-\frac{\Phi(\mathbf{x})}{k_B T} \right] \tag{3.65}$$

so that $T = T_0 = $ constant. Thus we may write

$$\bar{F}_0(\mathbf{x}, \mathbf{v}) = \frac{n(\mathbf{x})}{(2\pi R T)^{3/2}} \exp\left[-\frac{(\mathbf{v} - \mathbf{u}_0)^2}{2RT} \right] \tag{3.66}$$

where $T = T_0$ and $n(\mathbf{x})$ is given by (3.64).

For a stationary column of gas in a gravity field, we have

$$\Phi(z) = mg(z - z_0)$$

where g is acceleration due to gravity and z is vertical displacement. Inserting this potential into (3.64) gives

$$n(z) = n_0 \exp\left[-\frac{mg(z - z_0)}{k_B T} \right] \tag{3.67}$$

This exponential decay in number density is commonly called the *barometer formula* (also, *the law of atmospheres*).

3.3.12   Central-Limit Theorem Revisited

We wish to apply the central-limit theorem (Section 1.8 to the problem of finding the probability distribution of the total momentum of a fluid of N molecules
in equilibrium at temperature T. As P denotes probability, to avoid confusion in terminology, we will call total momentum $\mathbf{K}$ and individual molecular
momentum $\mathbf{k}_i$. Thus

$$\mathbf{K} = \sum_{i=1}^{N} \mathbf{k}_i \tag{3.68}$$

If we assume that individual momenta are uncorrelated, then $\mathbf{K}$ as given by
(3.68) may be viewed as a sum of statistically independent random variables,
and the central-limit theorem is appropriate.

The characteristic function for the random variable $\mathbf{k}_i$ is (deleting the
subscript i)

$$\phi(\mathbf{a}) = \int d\mathbf{k} e^{i\mathbf{k}\cdot\mathbf{a}} P(\mathbf{k}) \tag{3.69}$$

where $P(\mathbf{k})$ is single-particle probability density. Expanding $\phi(\mathbf{a})$ about $\mathbf{a} = 0$
gives

$$\phi(\mathbf{a}) = 1 + i(\mathbf{k}) \cdot \mathbf{a} - \frac{1}{2}\langle\overline{\overline{\mathbf{kk}}}\rangle : \overline{\overline{\mathbf{aa}}} + \cdots \tag{3.70}$$

from which we obtain[10]

$$\ln \phi(\mathbf{a}) = i\langle\mathbf{k}\rangle \cdot \mathbf{a} - \frac{1}{2}\overline{\overline{\langle\mathbf{kk} - \langle\mathbf{k}\rangle\langle\mathbf{k}\rangle\rangle}} : \overline{\overline{\mathbf{aa}}} + \cdots \tag{3.71}$$

If the fluid is at rest, then $\langle\mathbf{k}\rangle = 0$, whereas if the fluid is isotropic, we may
set

$$\langle\overline{\overline{\mathbf{kk}}}\rangle = \frac{1}{3}\langle k^2\rangle\overline{\overline{I}} \tag{3.72}$$

which gives the consistent statement

$$\mathrm{Tr}\langle\overline{\overline{\mathbf{kk}}}\rangle = \mathrm{Tr}\,\frac{1}{3}\langle k^2\rangle\overline{\overline{I}} = \langle k^2\rangle$$

Substituting these results into (3.71) gives

$$\ln \phi(\mathbf{a}) = -\frac{1}{6}\langle k^2\rangle a^2 \tag{3.73}$$

Comparing this finding with (1.8.7) gives

$$\mathcal{E}(\mathbf{k}) = 0$$
$$D(\mathbf{k}) = \frac{1}{3}\langle k^2\rangle \tag{3.74}$$

---

[10]Let $\phi = 1 + \psi$. Then, with $\psi < 1$, $\ln \phi = \ln(1 + \psi) = \psi - (\psi^2/2) + \cdots$.

With these results at hand, we turn to the central-limit theorem. The generalization of (1.8.29b) to three dimensions for the present problem is given by

$$P(\mathbf{K}, N) = \frac{1}{[2\pi N D(\mathbf{k})]^{3/2}} \exp\left\{\frac{-[\mathbf{K} - N\mathcal{E}(\mathbf{k})]^2}{2N D(\mathbf{k})}\right\} \tag{3.75}$$

Substitution of our findings (3.74) into the preceding gives

$$P(\mathbf{K}, N) = \frac{1}{[(2\pi/3)N\langle k\rangle]^{3.2}} \exp\left[-\left(\frac{K^2}{\frac{2}{3}N\langle k^2\rangle}\right)\right] \tag{3.76}$$

With the equipartition theorem, we write

$$\langle k^2\rangle = 3mk_B T \tag{3.77}$$

With $M = Nm$ representing total mass of the fluid, substitution of (3.77) into (3.76) gives the desired result:

$$P(\mathbf{K}) = \frac{1}{(2\pi M k_B T)^{3/2}} \exp\left(-\frac{K^2}{2M k_B T}\right) \tag{3.78}$$

This finding may be written in a more concise form in terms of the variable

$$\kappa^2 \equiv M k_B T$$

which permits (3.78) to be written

$$P(K) = \frac{1}{(2\pi\kappa^2)^{3/2}} e^{-K^2/2\kappa^2} \tag{3.78a}$$

With reference to (1.8.29), we see that the latter distribution is Gaussian (in three dimensions) with normalization

$$\int P(K)\,d\mathbf{K} = \frac{1}{(2\pi\kappa^2)^{3/2}} \int_0^\infty 4\pi K^2\,dK e^{-K^2/2\kappa^2} = 1$$

and variance

$$\sigma^2 = \kappa^2 \tag{3.79}$$

(which has dimensions of squared momentum).

Consider that the fluid has a mass of 1 g and is at room temperature (300 K). We find

$$\sigma \simeq 2 \times 10^{-7} \text{ g-cm/s}$$

Referring to (1.8.38), the preceding result indicates that there is a 70% probability that the center-of-mass speed (K/M) of the fluid is less than 2×10^{-7} cm/s. Furthermore, with the asymptotic relation (1.8.39) we find the probability that this period is in excess of 1 cm/s is given by

$$P(K/M > 1\text{cm/s}) \simeq \sqrt{\frac{2}{\pi}} \frac{e^{-\lambda^2/2}}{\lambda}\left(1 - \frac{1}{\lambda^2} + \cdots\right) \ll 1 \tag{3.80}$$

where $\lambda = 0.5 \times 10^7$. Thus there is an immeasurably small probability that the fluid surges in a given direction with a macroscopic speed.

Statistical and dynamical pictures

The preceding example indicates that there is a finite probability that measurement finds $\mathbf{K} \neq 0$. This is an evident contradiction of the law of conservation of momentum as the system was assumed to be at rest. This situation arises as a result of the probabilistic nature of the analysis. Molecular momenta are viewed as random variables with given expectations and variances. This is the statistical approach. In the dynamical picture, we calculate momenta as a function of time from dynamical equations. In this representation, there is no uncertainty in the outcome of measurement, and momentum is conserved.

3.4   Transport Coefficients

3.4.1   Response to Gradient Perturbations

Consider that a fluid is in equilibrium (or steady state) at given uniform values of n, $\mathbf{u}$, and T. Gradients of these variables develop under perturbation. The fluid responds to these gradients in a manner to restore equilibrium (Le Chatelier's principle). Thus, with $\mathcal{R}$ written for "response" we may write

$$\mathcal{R}[\nabla n] = n\mathbf{u} \tag{4.1a}$$

$$\mathcal{R}[\overline{\nabla \mathbf{u}}] = S \tag{4.1b}$$

$$\mathcal{R}[\nabla T] = \mathbf{Q} \tag{4.1c}$$

$$\mathcal{R}[\nabla \Phi] = \mathbf{J} \tag{4.1d}$$

That is, fluid motion will develop in response to gradients in density; components of the strain tensor will develop in response to gradients in fluid velocity; heat flow will develop in response to gradients in temperature, and currents will develop in response to a gradient in electric potential.

The coefficients that relate gradients of the perturbation to response motions are called transport coefficients, which are defined as follows.

Diffusion coefficient

This coefficient stems from the response relation (4.1a). It reads

$$n\mathbf{u} = -D\nabla n \tag{4.2}$$

The specific configuration that this equation describes is as follows. Consider a two-component gas. Let us call atoms of one component, 1-particles and atoms of the other component, 2-particles. These components have respective densities n_1 and n_2. In equilibrium, species 1 and 2 are homogeneously intersperse. Let the gas suffer a perturbation in density subject to the constraint

$n_1 + n_2 = $ constant. Then velocities will develop in the gas to restore the homogeneous state in accord with (4.2).

In a one-component gas, one may conceptually label a certain subset of particles with 1 and the complement, 2. The resulting motion of the conceptually labeled subgroup of particles is called *self-diffusion*. Diffusion in a two-component system is called *mutual diffusion*.

Thermal conductivity

This coefficient stems from (4.1c) and we write

$$\mathbf{Q} = -\kappa \nabla T \tag{4.3}$$

In the event that internal energy is proportional to T, (4.3) indicates that heat flow will emerge in response to gradients in internal energy.

Coefficient of viscosity

It is conventional to partition the pressure tensor in the following manner:

$$\overline{\overline{P}} = \overline{\overline{I}} p - \overline{\overline{S}} \tag{4.4}$$

In this expression, p denotes scalar pressure and $\overline{\overline{S}}$ represents the component of $\overline{\overline{P}}$ that emerges in response to gradients in velocity. The following two properties are assumed for $\overline{\overline{S}}$:

1. $\overline{\overline{S}}$ contains no terms other than $\overline{\overline{\nabla \mathbf{u}}}$ terms since $\overline{\overline{S}} = 0$ if $\overline{\overline{\nabla \mathbf{u}}} = 0$.
2. $\overline{\overline{S}} = 0$ if the fluid is in a state of uniform rotation. Rigid-body rotation of a fluid is characterized by a constant rotational frequency vector, $\mathbf{\Omega}$, such that the macroscopic motion of an element of fluid at $\mathbf{r}$ is

$$\mathbf{u} = \mathbf{\Omega} \times \mathbf{r}$$

Thus, property 2 states that $S = 0$ if $\mathbf{u} = \mathbf{\Omega} \times \mathbf{r}$. This property is satisfied by the tensor

$$a \left(\frac{\partial u_i}{\partial x_k} + \frac{\partial u_k}{\partial x_i} \right) + b \delta_{ik} \nabla \cdot \mathbf{u}$$

where a and b are arbitrary constants. With this form at hand, we write

$$S_{ik} = \eta \left(\frac{\partial u_i}{\partial x_k} + \frac{\partial u_k}{\partial x_i} - \frac{2}{3} \delta_{ik} \frac{\partial u_l}{\partial x_l} \right) + \zeta \delta_{ik} \frac{\partial u_l}{\partial x_l} \tag{4.5}$$

In this expression, η is the *coefficient of shear viscosity* and ζ is the *coefficient of bulk viscosity*. Recall that a fluid is incompressible, providing $\nabla \cdot \mathbf{u} = 0$.

The expression (4.5) may be written in terms of the *symmetric strain tensor*:[11]

$$\Lambda_{ik} = \frac{1}{2}\left(\frac{\partial u_i}{\partial x_k} + \frac{\partial u_k}{\partial x_i}\right) \tag{4.6a}$$

$$\mathrm{Tr}\,\overline{\overline{\Lambda}} = \nabla \cdot \mathbf{u} \tag{4.6b}$$

This permits (4.4) to be written in vector form:

$$\overline{\overline{P}} = \overline{\overline{I}}p - 2\eta(\overline{\overline{\Lambda}} - \tfrac{1}{3}\overline{\overline{I}}\nabla \cdot \mathbf{u}) - \zeta\overline{\overline{I}}\nabla \cdot \mathbf{u} \tag{4.7}$$

Note that this form has the property

$$\mathrm{Tr}\,\overline{\overline{P}} = 3p - 2\eta[(\mathrm{Tr}\,\overline{\overline{\Lambda}}) - \nabla \cdot \mathbf{u}] - 3\zeta\nabla \cdot \mathbf{u} = 3p - 3\zeta\nabla \cdot \mathbf{u}$$

For an incompressible fluid,

$$\mathrm{Tr}\,\overline{\overline{P}} = 3p$$

The term in (4.7) containing the coefficient of shear viscosity η should, by arguments surrounding (4.1), enter as a response force that tends to diminish gradients in fluids velocity. Let us consider a simple example that illustrates this property.

The force on a fluid element stems from the pressure tensor $\overline{\overline{P}}$. For an incompressible fluid, (4.7) reduces to

$$\overline{\overline{P}} = \overline{\overline{I}}p - 2\eta\overline{\overline{\Lambda}}$$

Consider further that the fluid is in a state of shear given by the fluid velocity

$$\mathbf{u} = [u_x(z), 0, 0]$$

The force on an infinitesimally thin slab of fluid of area $\Delta x\,\Delta y$ that lies normal to the z axis at a specific value of z is given by

$$F_x = P_{xz}\Delta x\,\Delta y = -\eta\frac{\partial u_x}{\partial z}\Delta x\,\Delta y$$

We see that the force F_x is opposite to the gradient of u_x. It slows down the slab, thereby diminishing the gradient in u_x. A molecular mechanism for this viscous, frictional force is described in Section 4.2.

Momentum equation

An important consequence of transport coefficients is that their defining relations serve as additional equations that contribute to closing the conservation

---

[11]Multiplying $\overline{\overline{\Lambda}}$ by Δt gives the symmetric strain tensor relevant to solid-state physics. For further discussion, see R. P. Feynman, R. B. Leighton, and M. Sands, *Feynmann Lectures on Physics*, Vol. II, Addison–Wesley, Reading, Mass. (1964), Chapters 38 and 39. In general, strain is the deformation of a medium that occurs in response to stress.

equations. Thus, for example, with (4.7) inserted into the momentum equation (3.30), we find

$$\rho \left[\frac{\partial \mathbf{u}}{\partial t} + (\mathbf{u} \cdot \nabla)\mathbf{u} \right] + \nabla p - \eta \nabla^2 \mathbf{u} - \left(\zeta + \frac{\eta}{3} \right) \nabla(\nabla \cdot \mathbf{u}) = 0 \qquad (4.8)$$

This vector equation represents three equations in five unknowns. The continuity equation that relates ρ and $\mathbf{u}$ together with a remaining scalar equation in these variables closes the system. With $\zeta = 0$, (4.8) is part of a system of fluid dynamic equations called the Navier–Stokes equations, which are more fully described later in this chapter.

Electrical conductivity and mobility

We recall Ohm's law,

$$\mathbf{J} = \sigma_c \mathcal{E} \qquad (4.9)$$

where σ_c is conductivity. The mobility coefficient μ is defined through the equation

$$\mathbf{u} = \langle \mathbf{v} \rangle = \mu \mathcal{E} \qquad (4.10)$$

Drude model

The Drude model permits a simple relation between σ and μ to be found. We adopt the following Langevin equation:

$$m \langle \dot{\mathbf{v}} \rangle = e\mathcal{E} - vm \langle \mathbf{v} \rangle \qquad (4.11)$$

where the second term on the right side represents a friction term. In equilibrium, $\langle \dot{\mathbf{v}} \rangle = 0$ and (4.11) reduces to[12]

$$\mathbf{u} = \frac{e}{vm} \mathcal{E}$$
$$\mu = \frac{e}{vm} \qquad (4.12)$$

and

$$\mathbf{J} = en\mathbf{u} = \frac{e^2 n}{vm} \mathcal{E} \qquad (4.13)$$

$$\sigma_2 = \frac{e^2 n}{vm} = en\mu \qquad (4.14)$$

Values of these parameters for some typical metals are listed in Table 5.1 in Chapter 5.

---

[12]This topic is returned to in Section 3.7 of the present chapter and in Chapter 5.

Electrochemical potential

Suppose a gradient in density is present in an aggregate of charge, in addition to an electric field. In this event, two driving forces act on the charges, one due to electric force and the other due to diffusion. To describe this situation, the Langevin equation (4.11) is generalized to read

$$m\langle \dot{\mathbf{v}} \rangle = e\varepsilon - vm\langle \mathbf{v} \rangle + \langle \mathbf{F}_D \rangle$$

An expression for the diffusion force $\langle \mathbf{F}_D \rangle$ is obtained from (4.2). Thus

$$\langle \mathbf{F}_D \rangle = -vm\langle \mathbf{x} \rangle_D = vmD\frac{\nabla n}{n}$$

Combining the latter two expressions, together with (4.13), in steady state we find[13]

$$\mathbf{J} = \frac{e^2 n}{vm}\varepsilon + eD\nabla n \tag{4.15}$$

Introducing the electric potential

$$\varepsilon = -\nabla\Phi$$

and recalling (4.14), we obtain

$$\mathbf{J} = \frac{\sigma_c}{e}\nabla(\psi_D e\Phi) \tag{4.15a}$$

where

$$\psi_D \equiv \frac{e^2 D\eta}{\sigma}$$

and $\psi_D - e\Phi$ is called the *electrochemical potential*. Note, in particular, that if $\psi_0 = e\Phi+$ constant, electric force is balanced by diffusion force and there is no current flow. The notion of electrochemical potential plays an important role in the theory of the diffusion of ions through semipermeable membranes.[14]

3.4.2 Elementary Mean-Free-Path Estimates

Basic notions

This formulation of transport coefficients rests on two key concepts. First, consider a small volume of a fluid in equilibrium with number density of

---

[13]The relation (4.15) is often employed in the study of charge carriers in a semiconductor where it is termed the drift-diffusion equation. For further discussion, see S. Selberherr, *Analysis and Simulation of Semiconductor Devices*, Springer Verlag, New York (1984).

[14]See A. Katchalsky and P. Curran, *Non-Equilibrium Thermodynamics in Biophysics*, Harvard University Press, Cambridge, Mass. (1965).

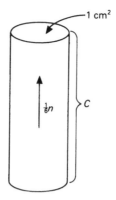

FIGURE 3.17. A flux of $nC/6$ particles passes through the upper face per second.

particles, n. We introduce the mean speed C,[15]

$$\langle v^2 \rangle = C^2$$

For point particles, equipartition of energy gives the relation[16]

$$mC^2 = 3k_B T$$

If particles have the mean speed C, what is the mean flux of particles, Γ (number/cm²-s) that move in any of six Cartesian directions at any instant of time? To answer this question consider a cylinder of height C and unit cross-sectional area (see Fig. 3.17). Since particles have the mean speed C, on the average, one-sixth of the particles in the cylinder will pass through the upper surface per second. If we call this direction the z direction, we have

$$\Gamma_z = \frac{1}{6} nC \qquad (4.16)$$

The second concept that comes into play in this analysis is the mean free path, l. It is assumed that the only means for transport of information in the fluid is via collisions. Thus, for example, equilibration of a gradient in n will be made through collisions that carry particles from domains of large n to domains of small n.

Another relation important to transport phenomena relates total scattering cross section[17] σ to particle number density and mean free path. It appears as

$$n\sigma l \simeq 1 \qquad (4.17)$$

---

[15]More accurately, C here represents the rms molecular speed. In other parts of the text, for purposes of writing the Maxwellian in concise form, we set $mC^2 \equiv k_B T$ [see (3.61a) and Problem 3.57].

[16]Each degree of freedom of a fluid in equilibrium has $k_B T/2$ units of energy.

[17]Previously labeled σ_T. See (1.25). Recall also that σ_c denotes electrical conductivity.

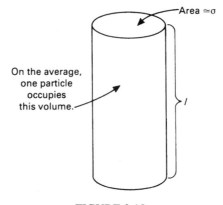

FIGURE 3.18.

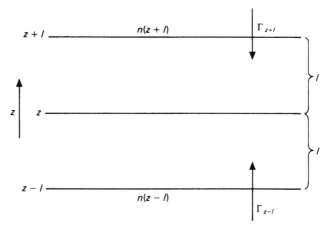

FIGURE 3.19. A particle that has a collision at $z - l$ has no further collisions until it reaches z.

The argument that supports this relation is that, on the average, one particle will be found in the volume $l\sigma$ (see Fig. 3.18).

Self-diffusion

Here we are speaking of the self-diffusion of a conceptually labeled subgroup of particles of a single species. Let the density of such particles be n, and let there be a gradient of n of these particles in the z direction (see Fig. 3.19). The flux of particles in the $+z$ direction that crosses the z-plane is equal to the number that reach the z plane from $(z - l)$, minus those that reach it from $(z + l)$. That is,

$$\Gamma_z = \Gamma_{z-1} - \Gamma_{z+l}$$

With (4.16), we have

$$\Gamma_2 = \frac{1}{6}C[n(z-l) - n(z+l)]$$

$$\Gamma_z = \frac{2lC}{6}\left[\frac{n(z-l) - n(z+l)}{2l}\right]$$

$$\Gamma_z = nu_z \simeq -\frac{1}{3}lC\frac{\partial n}{\partial z}$$

Comparing with (4.2), we may conclude

$$D + \frac{1}{3}lC \qquad (4.18)$$

Mutual diffusion

We consider two distinct gases with respective densities n_1 and n_2. The gas occupies a cylindrical volume with unit cross section. Labeling the midplane of the cylinder $z = 0$, the system is initially constrained so that particles 1 occupy the domain $z > 0$ and particles 2 the domain $z < 0$. This constraint is removed, and the two species diffuse across the midplane subject to the previously mentioned conditions:

$$n_1 + n_2 = n = \text{constant} \qquad (4.19)$$

Thus we require that the net flux of particles 1 across the midplane be balanced by the flux of particles 2 across the midplane.

$$\Gamma_1 = n_1 u_1, \qquad \Gamma_2 = n_2 u_2$$

$$\Gamma_1 = -\Gamma_2 \qquad (4.20)$$

Signs in this equation denote direction of flow. With this constraint and the defining equalties

$$\Gamma_1 = -D_{12}\frac{dn_1}{dz}, \qquad \Gamma_2 = -D_{21}\frac{dn_2}{dz} \qquad (4.21a)$$

we find that the mutual diffusion coefficients are equal:

$$D_{12} = D_{21} \qquad (4.21b)$$

Furthermore, in the limit that the two species become identical, D_{12} must reduce to the coefficient of self-diffusion (4.18).

Let us consider the force exerted on gas 1 due to transfer of momentum between the two components.[18] First, let M_{12} denote average z momentum transferred from gas 1 to gas 2 per unit volume per unit time due to collisions.

---

[18]This derivation follows that of J. C. Maxwell, *Phil. Trans. Roy. Soc. 157*, 49 (1867). An account of this approach may also be found in R. D. Present, *Kinetic Theory of Gases*, McGraw–Hill, New York (1958).

With the law of partial pressures for ideal gases (that is, $p_1 = n_1 k_B T$), we may write this force as

$$dp_1 = k_B T \, dn_1 = -M_{12} \, dz \qquad (4.22)$$

We will calculate M_{12} for the case of rigid spheres. As was shown in Section 3.1, the scattering of rigid spheres in the center-of-mass frame is isotropic. So the average velocity of a molecule in the lab frame after collision is equal to the velocity of the center of mass. We may conclude that the average decrement in momentum suffered by a type 1 molecule per collision is

$$\Delta \mathbf{p}_1 = m_1(u_1 - u_c) = \mu(u_1 - u_2) \qquad (4.23)$$

where μ is reduced mass and u_c is the speed of the center of mass.[19]

Let σ represent the total scattering cross section for the interaction of these two types of particles. Then the related collision frequency per unit volume is given by

$$v_{12} = n_1 n_2 \sigma C_{12} \qquad (4.24)$$

where

$$C_{12}^2 = \frac{3 k_B T}{\mu} \qquad (4.25)$$

is an effective two-particle thermal speed. Thus the average rate of momentum transfer per unit volume is

$$M_{12} = v_{12} \Delta \mathbf{p}_1$$
$$M_{12} = n_1 n_2 (u_1 - u_2) \kappa_{12} \qquad (4.26)$$

where

$$\kappa_{12} \equiv \mu C_{12} \sigma$$

Recalling the equality (4.20) permits (4.26) to be rewritten as

$$M_{12} = n \Gamma_1 \kappa_{12} \qquad (4.27)$$

where $n = n_1 + n_2$ is total particle density. Inserting (4.22) into the latter equation gives

$$k_B T \frac{dn_1}{dz} = -M_{12} = -n \Gamma_1 \kappa_{12}$$

whence

$$\Gamma_1 = -\frac{k_B T}{n \kappa_{12}} \frac{dn_1}{dz} = -D_{12} \frac{dn_1}{dz} \qquad (4.28)$$

---

[19]Recall (1.2); $\dot{R} \equiv u_c = (m_1 u_1 + m_2 u_2)/(m_1 + m_2)$.

This gives the desired symmetric expression

$$D_{12} = \frac{(k_B T/3\mu)^{1/2}}{n\sigma} \tag{4.29}$$

The cross section for scattering of rigid spheres is given by (1.34):

$$\sigma = \pi \left(\frac{\sigma_{01} + \sigma_{02}}{2} \right)^2 \equiv \pi \sigma_{12}^2 \tag{4.30}$$

where σ_{01} and σ_{02} are respective particle diameters. The result (4.29) then becomes

$$D_{12} = \frac{(k_B T/3\mu)^{1/2}}{n\pi \sigma_{12}^2} \tag{4.31}$$

which is seen to be independent of the proportions of the mixture. In the limit that the two gases become identical, $\sigma_{12} \to \sigma_0$, the diameter of a sphere, n, becomes the particle density of a single species (with $n_1 = n_2 = n/2$), and

$$\frac{k_B T}{3\mu} \to \frac{2k_B T}{3m} = \frac{2}{9} C^2$$

Thus, with (4.18) we may write

$$D_{12} \simeq D$$

in the said limit. The approximate form of the equality reflects the rough estimates made in obtaining these coefficients.

As will be found later in the text, a more accurate form of the mutual diffusion coefficient compared to the form (4.31) is given by

$$D_{12} = \frac{3}{8} \frac{(\pi k_B T/2\mu)^{1/2}}{n\pi \sigma_{12}^2} \tag{4.31a}$$

Setting $\mu = m/2$ and $\sigma_{12} = \sigma_0$ gives the related self-diffusion coefficient.

Viscosity

We consider a spatially uniform fluid in a state of shear such that

$$\mathbf{u} = [u_x(z), 0, 0]$$

The fluid velocity that is only in the x direction varies only in the z direction (see Fig. 3.20). Note first that for this configuration (4.7) (for an incompressible fluid) reduces to

$$P_{xz} = -S_{xz} = -\eta \frac{\partial u_x}{\partial z} \tag{4.32}$$

Each particle at $z - l$ that suffers a collision and moves in the $+z$ direction, on the average, carries an x component of momentum from the region at $z - l$, that is, $mu_x(z - l)$. The flux of particles carrying this momentum, again, is

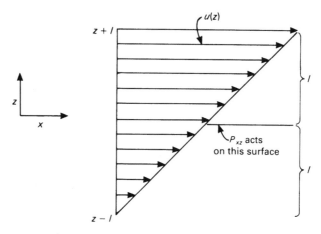

FIGURE 3.20. The state of shear in a fluid.

given by (4.16). Thus, the mean x component of momentum flux across the z plane due to transport of particles in the z direction is the difference between the gain term

$$\Gamma_p^+ = \frac{1}{6} n C m u_x (z - l)$$

and the loss term

$$\Gamma_p^- = \frac{1}{6} n C m u_x (z + l)$$

The net force per area on the z plane (in the x direction) is[20]

$$P_{xz} = \Gamma_p^+ - \Gamma_p^-$$
$$P_{xz} = \frac{1}{6} \rho C 2l \left[\frac{u_x(z - l) - u_x(z + l)}{2l} \right]$$
$$P_{xz} = -\frac{1}{3} \rho C l \frac{\partial u_x}{\partial z}$$

Comparing with (4.32) gives Maxwell's classic estimate:

$$\eta = \frac{1}{3} \rho C l \qquad (4.33)$$

With (4.17), we see that our expression for η is independent of fluid density.

---

[20]This force is akin to the retardation of a moving train due to loading of heavy parcels from rest onto the train.

Thermal conductivity

Continuing in this manner, and assuming a gradient in energy per particle $e_K(z)$, we find (with $\Gamma_Q = \frac{1}{6} nue_K$)

$$Q_z = -\frac{1}{3} nCl \frac{\partial e_K}{\partial z}$$

With

$$c_V = \frac{\partial e_K}{\partial T}$$

written for the specific heat per particle, the preceding equation becomes

$$Q_z = -\frac{1}{3} nClc_V \frac{\partial T}{\partial z}$$

With (4.3), we may then write

$$\kappa = \frac{1}{3} nClc_V \qquad (4.34)$$

Setting $\rho = mn$ and comparing with (4.33) gives the relation

$$\frac{\kappa}{\eta} = \frac{c_V}{m} \equiv \bar{c}_V \qquad (4.35)$$

A compilation of the preceding expressions is given in Table 3.1.

Kinetic-theory parameters

The preceding relations suggest that certain ratios of transport coefficients are constants. Thus we find

$$\frac{\eta \bar{c}_V}{\kappa} \simeq 1 \qquad (4.36)$$

A closely related parameter, called the *Prandtl number*, is given by

$$\mathcal{P} \equiv \frac{\eta \bar{c}_p}{\kappa} = \frac{\eta \bar{c}_V}{\kappa} \frac{\bar{c}_p}{\bar{c}_V} \simeq \gamma \qquad (4.37a)$$

Here γ is written for the ratio of specific heats, $\bar{c}_p/\bar{c}_V$, which for monatonic ideal gases has the value $\frac{5}{3}$.

Another closely allied parameter in kinetic theory is the *Schmidt number*,

$$\mathcal{L} \equiv \frac{\eta}{\rho D} \simeq 1 \qquad (4.37b)$$

Values of η, κ, and the ratio $\kappa/\eta \bar{c}_V$ for some common gases are listed in Table 3.2. Values of $\mathcal{P}$ and $\mathcal{L}$ are listed in Table 3.3. In Table 3.2 we see that the ratio $\kappa/\eta \bar{c}_V$ is closer to 2.5 than 1 for monatomic gases. As we will find in the following section, a more detailed kinetic analysis gives the value $\frac{5}{2}$

TABLE 3.1. Elementary Mean-Free-Path Expressions of Transport Coefficients

Coefficient	Expressions
Self-diffusion	$D = \frac{1}{3}lC$
Mutual-diffusion	$D_{12} = \dfrac{(k_B T/3\mu)^{1/2}}{n\sigma}$
Viscosity	$\eta = \frac{1}{3}\rho Cl$
Heat conductivity	$\kappa = \frac{1}{3}nClc_V$
Ratio	$\dfrac{\kappa}{\eta} = \dfrac{c_V}{m} = \bar{c}_V$
Electrical conductivity	$\sigma_c = \dfrac{e^2 n}{vm}$
Mobility	$\mu = \dfrac{e}{vm}$
Ratio	$\dfrac{\mu}{D} = \dfrac{e}{k_B T}$ (Einstein relation)

TABLE 3.2. Observed Values of Viscosity and Thermal Conductivity

	$\eta \times 10^7$ g/cm-s	$\kappa \times 10^3$, cal/cm-s-K	$\kappa/\eta\bar{c}_V$	$\dfrac{9\gamma - 5}{4}$
He	1875	0.344	2.44	2.49
H_2	840	0.416	2.06	1.92
CO_2	1377	0.034	1.64	1.70
CH_4	1027	0.072	1.75	1.70

for this ratio.[21] The last column of Table 3.2 suggests that a uniformly good approximation for this ratio is the value $(9\gamma - 5)/4$.[22] In Table 3.3 we see that $\mathcal{P}$ and L lie close to unity for all materials listed. A more detailed kinetic study gives the values $\mathcal{P} = \frac{2}{3} = 0.67$ and $\mathcal{L} = \frac{5}{6} = 0.83$.[22]

3.4.3 Diffusion and Random Walk

In Section 1.7.5, we found that, relevant to the displacement l in n steps of a random walk, $\langle l^2 \rangle = n$. Equivalently, with displacement written x and n replaced by time t, $\langle x^2 \rangle = t$. We wish now to show the relation between this result and diffusion.

---

[21] See Table 3.4.

[22] For further discussion see A. Isihara, *Statistical Physics*, Academic Press, New York (1971).

TABLE 3.3. Observed Values of the Schmidt and Prandtl Numbers

	$\mathcal{L}$	$\mathcal{P}$
Ne	0.73	0.66
Ar	0.75	0.67
N_2	0.74	0.71
CH_4	0.70	0.74
O_2	0.74	0.72
CO_2	0.71	0.75
H_2	0.73	0.71

Combining (4.2) with the continuity equation (3.14) gives the diffusion equation

$$\frac{\partial n}{\partial t} = D\nabla^2 n \tag{4.38}$$

Consider that, at time $t = 0$,

$$n(\mathbf{r}, 0) = N_0 \delta(\mathbf{r}) \tag{4.39}$$

The solution to (4.38) corresponding to this intial value is the Gaussian distribution [recall (1.8.34)]

$$n(r, t) = \frac{N_0}{(4\pi Dt)^{3/2}} \exp\left[-\left(\frac{r^2}{4Dt}\right)\right] \tag{4.40}$$

Note that this distribution maintains the normalization

$$\int_0^\infty dr 4\pi r^2 n(r, t) = N_0 \tag{4.41}$$

We may identify

$$\frac{4\pi r^2 n(r, t)\, dr}{N_0} = P(r, t)\, dr \tag{4.42}$$

as the probability of finding a particle in the shell dr about r at the time t. This probability density permits calculation of the mean square displacement

$$\langle r^2 \rangle = \int_0^\infty P(r, t) r^2\, dr = \int_0^\infty \frac{4\pi r^4 e^{-r^2/4Dt}}{(4\pi Dt)^{3/2}}\, dr = 6Dt \tag{4.43}$$

$$\langle r^2 \rangle = 6Dt$$

which is relevant to a three-dimensional random walk[23] and is seen to maintain the fundamental relation $\langle r^2 \rangle \propto t$, with $6D$ playing the role of a proportionality constant. This result was found previously (1.8.31) employing the central-limit theorem.

---

[23]Random walk in various dimensions is discussed by R. L. Liboff, *Phys. Rev.* *141*, 222 (1966).

3.4.4 Autocorrelation Functions, Transport Coefficients, and Kubo Formula

Self-diffusion

In this section we present a brief description of an alternative approach to the evaluation of transport coefficients. We begin with the diffusion coefficient.[24] Our starting equation is (4.38).

With the Fourier transform

$$\bar{n}(\mathbf{k}, t) = \int d\mathbf{r} e^{i\mathbf{k}\cdot\mathbf{r}} n(\mathbf{r}, t) \qquad (4.44)$$

we find that the transform of (4.38) is given by

$$\frac{\partial \bar{n}}{\partial t} = -k^2 D\bar{n} \qquad (4.45)$$

which has the solution

$$\bar{n}(\mathbf{k}, t) = \exp(-k^2 D t) \qquad (4.46)$$

We wish to recapture the result (4.43). From (4.46), we find

$$\frac{\partial^2 \bar{n}}{\partial k^2}\Big|_{k=0} = -2Dt \qquad (4.47)$$

whereas differentiation of (4.44) gives [with $n(\mathbf{r}, t)$ isotropic]

$$\frac{\partial^2 \bar{n}}{\partial k^2}\Big|_{k=0} = -\frac{1}{3}\int_0^\infty 4\pi r^4 n(r, t)\, dr \qquad (4.48)$$

Assuming that

$$\int_0^\infty 4\pi r^2 n(r, t)\, dr = 1$$

again permits us to interpret $n(\mathbf{r}, t)$ as a probability density. Thus (4.48) gives

$$\frac{\partial^2 \bar{n}}{\partial k^2}\Big|_{k=0} = -\frac{1}{3}\langle r^2(t)\rangle \qquad (4.49)$$

Equating this result to (4.47) recaptures (4.43):

$$\langle r^2(t)\rangle = 6Dt$$

If the particle begins its random walk at $\mathbf{r}_0$ instead of at the origin we obtain

$$\langle |\mathbf{r}(t) - \mathbf{r}_0|^2\rangle = 6Dt \qquad (4.50)$$

We wish to express this result in terms of a time correlation function.

---

[24] An extensive review of these techniques is given by R. Zwanzig, *Ann. Rev. Phys. Chem 16*, (1965). Our presentation follows that of E. Helfand, *Phys. Fluids 4*, 681 (1961).

Suppose that the velocity $\mathbf{v}(t)$ of the particle in question is known. Then we may write

$$\mathbf{r}(t) - \mathbf{r}_0 = \int_0^t \mathbf{v}(t')\, dt'$$

and it follows that

$$[\mathbf{r}(t) - \mathbf{r}_0]^2 = \int_0^t dt' \int_0^t dt' \int_0^t dt'' \mathbf{v}(t'') \cdot \mathbf{v}(t')$$

Taking the ensemble average of both sides of this equation gives

$$\langle [\mathbf{r}(t) - \mathbf{r}_0]^2 \rangle = \int_0^t dt' \int_0^t dt'' \langle \mathbf{v}(t'') \cdot \mathbf{v}(t') \rangle \qquad (4.51)$$

Assuming the process to be homogeneous in time, we write

$$\langle \mathbf{v}(t') \cdot \mathbf{v}(t'') \rangle = \langle \mathbf{v}(t' - t'') \cdot \mathbf{v}(0) \rangle \qquad (4.52)$$

Introducing the variable $\tilde{t} = t' - t''$ and $T = t' + t''$ (see Fig. 2.12) gives the integral equality

$$\int_0^t dt' \int_0^t dt'' g(t' - t'') = 2t \int_0^1 \left(1 - \frac{\tilde{t}}{t}\right) g(\tilde{t})\, d\tilde{t}$$

This relation permits (4.51) to be written as

$$\langle [\mathbf{r}(t) - \mathbf{r}_0]^2 \rangle = 2t \int_0^t \left(1 - \frac{\tilde{t}}{t}\right) \langle \mathbf{v}(0) \cdot \mathbf{v}(\tilde{t}) \rangle\, d\tilde{t}$$

whence, with (4.50),

$$6D = 2 \int_0^t \left(1 - \frac{\tilde{t}}{t}\right) \langle \mathbf{v}(0) \cdot \mathbf{v}(\tilde{t}) \rangle\, d\tilde{t}$$

Assuming rapid decay of the time correlation $\langle \mathbf{v}(0) \cdot \mathbf{v}(\tilde{t}) \rangle$ permits the preceeding to be written (with $\tilde{t}$ replaced by t) as

$$D = \frac{1}{3} \int_0^\infty \langle \mathbf{v}(0) \cdot \mathbf{v}(t) \rangle\, dt \qquad (4.53)$$

which is the desired result, expressing D in terms of the velocity autocorrelation function (see Section 1.8.8 and Problem 3.63 addressing "long-time tails").

Recalling the Einstein relation in Table 3.1, which relates mobility and diffusion coefficients, the preceding equation permits us to write[25]

$$\mu = \frac{e}{3k_B T} \int_0^\infty \langle \mathbf{v}(0) \cdot \mathbf{v}(t) \rangle\, dt \qquad (4.53a)$$

---

[25] As noted in Section 1.8.4, fluctuation in the number of particles is termed *shot noise*. Fluctuation in velocities, such as included in (4.53), is termed *Johnson noise* (named for J. B. Johnson).

As a simple application of (4.53), consider that

$$\mathbf{v}(t) = \mathbf{v}(0)e^{-t/\tau}$$

where τ is a relaxation time. Substituting this result into (4.53) gives

$$D = \frac{1}{3} \int_0^\infty \langle v^2(0)e^{-t/\tau} \rangle \, dt$$

Taking the integral inside the ensemble average gives

$$D = \frac{1}{3} \langle \tau v^2(0) \rangle$$

Assuming that the relaxation time, τ, is independent of the starting velocity, we find

$$D = \frac{1}{3} \tau \langle v^2(0) \rangle = \frac{1}{3} \tau C^2$$

$$D = \frac{1}{3} lC$$

in agreement with Maxwell's result (4.18). Proceeding in a manner similar to that in obtaining (4.53), time-correlation expressions for other transport coefficients, such as viscosity and thermal conductivity, may also be so expressed.[26]

Linear response theory

The preceding development may be generalized to arbitrary Hamiltonian systems in the following manner. Let $A(q, p)$ be a dynamical function that is not explicitly dependent on time. With (1.1.25), we write

$$\frac{dA}{dt} = -[H, A] = -\hat{L}A \tag{4.54}$$

whose solution may be written

$$A(t) = e^{-t\hat{L}} A(0) \tag{4.55}$$

The Liouville operator, $\hat{L}$, was introduced previously in the Liouville equation (2.1.3), whose solution in turn may be written

$$f(t) = e^{t\hat{L}} f(0) \tag{4.56}$$

The expectation of A at the time t is given by

$$\langle A \rangle = \int A(0) f(t) \, dq \, dp \tag{4.57}$$

---

[26]See, for example, D. A. McQuarrie, *Statistical Mechanics*, Harper & Row, New York (1973).

Here we have introduced the notation

$$A(0) = A(q, p) \tag{4.57a}$$

With (4.56), (4.57) may be written

$$\langle A \rangle = \int A(0)(e^{t\hat{L}} f(0)) \, dq \, dp \tag{4.58}$$

Consider that the Hamiltonian, H_0, of a given system is perturbed so that

$$H = H_0 - \varepsilon \bar{H} F(t) \tag{4.59}$$

where ε is a parameter of smallness, $\bar{H}(q, p)$ is time independent, and $F(t)$ is independent of q and p. The distribution is likewise perturbed and we write

$$f = f^{(0)} + \varepsilon \Delta f \tag{4.59a}$$

Substituting the latter two relations into the Liouville equation gives the following $0(\varepsilon)$ equation:

$$\frac{\partial \Delta f}{\partial t} - \hat{L}_0 \Delta f - [f^{(0)}, \bar{H}] F(t) = 0 \tag{4.60}$$

where $\hat{L}_0$ is the Liouville operator corresponding to H_0. Operating on (4.60) with $\exp -t\hat{L}_0$ gives

$$\frac{\partial}{\partial t} [e^{-t\hat{L}_0} \Delta f] = e^{-t\hat{L}_0} [f^{(0)}, \bar{H}] F(t)$$

Integrating, we find

$$\Delta f = \int_0^t e^{(t-t')\hat{L}_0} [f^{(0)}, \bar{H}] F(t') \, dt' \tag{4.61}$$

The average of a response B to the perturbation is then given by

$$\langle B \rangle = \int B \Delta f(t) \, dq \, dp$$

where we have assumed that $\langle B \rangle = 0$ in equilibrium and have dropped the bookkeeping parameter ε. Substituting (4.61) into this relation gives [in the notation of (4.57)]

$$\langle B \rangle = \int B(0) \int_0^t e^{(t-t')\hat{L}_0} [f^{(0)}, \bar{H}] F(t') \, dt' \, dq \, dp \tag{4.62}$$

or, equivalently,

$$\langle B(t) \rangle = \int_0^t \phi(t - t') F(t') \, dt' \tag{4.63}$$

where the function ϕ is as implied. As (4.63) is a convolution integral, the Fourier transform gives the following widely employed result:

$$\langle B(\omega) \rangle = \phi(\omega) F(\omega) \tag{4.64}$$

The term $F(\omega)$ is the transform of the perturbing force, $B(\omega)$ is the transform of the response to the perturbation, and $\phi(\omega)$ plays the role of the transform of the related transport coefficient.

A concise form of $\phi(t - t') \equiv \phi(\tilde{t})$ is obtained as follows. With (4.62) and (4.63), we write (with $\tilde{t}$ replaced by t)

$$\phi(t) = \int B(0) e^{t\hat{L}_0} [f^{(0)}, \bar{H}] \, dq \, dp$$

A single parts integration gives, with (4.55),

$$\phi(t) = \int B(t) [f^{(0)}, \bar{H}] \, dq \, dp \tag{4.65}$$

Again integrating by parts, we find

$$\phi(t) = \int [\bar{H}, B(t)] f^{(0)} \, dq \, dp$$

or, equivalently,

$$\phi(t) = \langle [\bar{H}, B(t)] \rangle_0 \tag{4.66}$$

which is the desired result. The zero subscript reminds us that the average is calculated with the equilibirum distribution, $f^{(0)}$.

The quantum analog of (4.66) is given by

$$\phi(t) = \frac{1}{i\hbar} \text{Tr} \{ \hat{\rho}^{(0)} [\bar{H}, \hat{B}(t)] \} \tag{4.67}$$

where $\hat{\rho}$ is the density matrix and Tr denotes the trace.[27] The preceding relation is often referred to as *Kubo's formula*. [28]

Electrical mobility

Let us apply (4.64) in the calculation of the Fourier transform of electrical mobility. We identify the electric field, $e\mathcal{E}$, as the perturbing force, and particle velocity $\mathbf{v}$ is the response. We will make our calculation with respect to the Cartesian coordinate x_i. In the notation of (4.59), we then identify

$$\bar{H} = x_i, \qquad F_i = \varepsilon \mathcal{E}_i$$

and the response $B = v_j$. With these relations, (4.64) gives

$$\langle v_j(\omega) \rangle = \mu_{ji}(\omega) \mathcal{E}_i(\omega) \tag{4.68}$$

where μ_{ji} is the mobility tensor:

$$\mu_{ji}(\omega) = e\phi_{ji}(\omega) \tag{4.69}$$

---

[27]These quantum concepts are fully developed in Section 5.2.

[28]R. Kubo, *J. Phys. Soc. Japan* 12, 570 (1957). For further discussion, see R. Kubo, *Statistical Mechanics*, Wiley, New York (1965), Chapter 6.

The Fourier transform $\phi(\omega)$ is obtained from (4.66). That is, with our identifications,

$$\phi_{ji}(t) = \langle [x_i, v_j(t)] \rangle_0 \qquad (4.70)$$

Again assuming a simple relaxation model, we write

$$\mathbf{v}(t) = \mathbf{v}(0)e^{-t/\tau}$$

where τ is a constant relaxation time. Thus (4.70) becomes

$$\phi_{ji}(t) = \langle [x_i, v_j]e^{-t/\tau} \rangle_0$$

where, in the spirit of (4.57a), we have set $v(0) = v$. The preceding equation gives

$$\phi_{ji}(t) = \frac{\delta_{ij}}{m}e^{-t/\tau} \qquad (4.71)$$

Introducing the Fourier transform

$$\phi_{ji}(\omega) = \int_0^\infty dt\,\phi_{ji}(t)e^{i\omega t}$$

and inserting (4.71), we obtain

$$\phi_{ji}(\omega) = \frac{\delta_{ij}}{m}\left[\frac{1}{(1/\tau) - i\omega}\right] \qquad (4.72)$$

which is the desired result. In the DC limit, we set $\omega = 0$ to obtain

$$\phi_{ji}(0) = \frac{\delta_{ij}\tau}{m}$$

Inserting this result into (4.69), we find

$$\mu_{ji}(0) = \delta_{ij}\frac{e\tau}{m}$$

which with (4.68) gives

$$\langle \mathbf{v} \rangle = \frac{e\tau}{m}\varepsilon \qquad (4.73)$$

These latter two relations agree with our previous mean-free-path estimates listed in Table 3.1.

With these elementary notions of transport coefficients behind us, we turn to one of the more powerful techniques of solving the Boltzmann equation. As will be demonstrated, an integral component of this method of solution addresses the calculation of transport coefficients.

3.5   The Chapman–Enskog Expansion

3.5.1   Collision Frequency

The collision integral of the Boltzmann equation (2.14) may be written

$$\hat{J}(F \mid F) \equiv -F \iint \sigma \, d\Omega g \, d\mathbf{v}_1 F(\mathbf{v}_1) + \iint \sigma \, d\Omega g \, d\mathbf{v}_1 F' F_1' \qquad (5.1)$$

It is evident from Fig. 3.10 that the quantity

$$\nu(\mathbf{v}) = \iint \sigma \, d\Omega g \, d\mathbf{v}_1 F(\mathbf{v}_1) \qquad (5.2)$$

represents a collision frequency.[29]
Let us rewrite the Boltzmann equation as

$$\frac{DF}{Dt} = \iint \sigma \, d\Omega g \, d\mathbf{v}_1 (F' F_1' - F - F F_1) \equiv \hat{I} F \qquad (5.3)$$

which serves to define the collision–integral operator $\hat{I}$. With (5.2), we see that $\hat{I}$ has dimensions of frequency s^{-1}. Thus we may write

$$\hat{I} = \nu_0 \hat{\hat{I}} \qquad (5.4)$$

where $\hat{\hat{I}}$ is nondimensional and ν_0 has been written for a constant with dimensions s^{-1}. In this notation, (5.3) appears as

$$\frac{DF}{Dt} = \nu_0 \hat{\hat{I}} F \qquad (5.5)$$

The Chapman–Enskog expansion is relevant to the domain of large collision frequency. With C taken as thermal speed, we write

$$C \simeq l\nu \qquad (5.6a)$$

or, equivalently, with (4.17),

$$\nu \simeq n\sigma C \qquad (5.6b)$$

The relation (5.6a) indicates that large collision frequency is equivalent to small mean free path, whereas (5.6b) indicates that such extremes are attained in the limit that the product $n\sigma C$ grows large.

---

[29]Discussion of this parameter is returned to in Section 4.2.1 in derivation of the Krook–Bhatnager–Gross equation and in Section 6.4 relevant to the distinction between hard soft potentials.

3.5.2   The Expansion[30]

The *first step* in the Chapman–Enskog expansion is to write the Boltzmann equation as

$$\left(\frac{\partial}{\partial t} + \hat{D}\right) F = \frac{1}{\varepsilon}\hat{I}F \tag{5.7}$$

where the dimensionless parameter $\varepsilon \ll 1$. This parameter may be thought of as a bookkeeping parameter in that it is eventually set equal to 1. With (5.5), we see that the form (5.7) is equivalent to stipulating that the gas is dominated by large collision frequency. The parameter $\hat{D}$ in (5.7) is given by

$$\hat{D} \equiv \mathbf{v} \cdot \frac{\partial}{\partial \mathbf{x}} + \mathbf{K} \cdot \frac{\partial}{\partial \mathbf{v}} \tag{5.7a}$$

Step 2 in the Chapman–Enskog procedure is to introduce the expansion

$$F = F^{(0)} + \varepsilon F^{(1)} + \varepsilon^2 F^{(2)} + \cdots \tag{5.8}$$

The normalization of F follows (1.6.14a). Furthermore, F satisfies the moment relations (3.12), (3.13) and (3.32):

$$n = \int F\, d\mathbf{v}, \qquad n\mathbf{u} = \int F\mathbf{v}\, d\mathbf{v}, \qquad \frac{3}{2}nk_B T = \int \frac{mc^2}{2} F\, d\mathbf{v} \tag{5.9}$$

Step 3 in the Chapman–Erskog expansion stipulates that the variables $(n, \mathbf{u}, T)$ are all $0(1)$ quantities and stem from $F^{(0)}$, whereas terms in the series (5.8) corresponding to $F^{(i)}$, $i > 0$, contribute to the higher moments $\mathbf{Q}$ and $\bar{\bar{P}}$. Thus we write

$$\int F^{(0)} \begin{pmatrix} 1 \\ \mathbf{v} \\ c^2 \end{pmatrix} d\mathbf{v} = \begin{pmatrix} n \\ n\mathbf{u} \\ 3nk_B T/m \end{pmatrix} \tag{5.10a}$$

$$\int F^{(i)} \begin{pmatrix} 1 \\ \mathbf{v} \\ c^2 \end{pmatrix} d\mathbf{v} = \begin{pmatrix} 0 \\ 0 \\ 0 \end{pmatrix}, \qquad i > 0 \tag{5.10b}$$

$$\mathbf{Q} = \sum_l \varepsilon^l \mathbf{Q}^{(l)} = \frac{1}{2} \sum_l \varepsilon^l \int F^{(l)} \mathbf{c} mc^2\, d\mathbf{c} \tag{5.10c}$$

$$\bar{\bar{P}} = \sum_l \varepsilon^l \bar{\bar{P}}^{(l)} = \sum_l \varepsilon^l \int F^{(l)} m\overline{\overline{\mathbf{cc}}}\, d\mathbf{c} \tag{5.10d}$$

---

[30]This technique was developed in a series of independent papers by S. Chapman and D. Enskog over the second decade of this century. See references. This method of solution is also described in S. Chapman and T. G. Cowling, *The Mathematical Theory of Non-Uniform Gases*, 3rd ed., Cambridge University Press, New York (1970).

Expansion of $\hat{\mathcal{D}}$ and $\hat{J}$

Substitution of the series (5.8) into $\hat{\mathcal{D}}F$ gives

$$\hat{\mathcal{D}}F = \hat{\mathcal{D}}F^{(0)} + \varepsilon\hat{\mathcal{D}}F^{(1)} + \cdots \qquad (5.11)$$

For the collision integral, we obtain

$$\hat{J}(F \mid F) = \hat{J}\left(\sum_{l=0}^{\infty}\varepsilon^l F^{(l)} \;\middle|\; \sum_{n=0}^{\infty}\varepsilon^n F^{(n)}\right) = \sum_{l=0}^{\infty}\sum_{n=0}^{\infty}\varepsilon^{l+n}\hat{J}(F^{(l)} \mid F^{(n)}) \qquad (5.12)$$

It proves convenient to introduce the ordered operator

$$\hat{J}^{(s)}(F^{(0)}, F^{(1)}, \ldots, F^{(s)}) = \sum_{\substack{n \quad l \\ (n+l=s)}} \hat{J}(F^{(l)} \mid F^{(n)}) \qquad (5.13)$$

The expansion (5.12) may then be written

$$\hat{J}(F \mid F) = \hat{J}^{(0)}(F^{(0)}) + \varepsilon\hat{J}^{(1)}(F^{(0)}, F^{(1)})$$
$$+ \varepsilon^2\hat{J}^{(2)}(F^{(0)}, F^{(1)}, F^{(2)}) + \cdots \qquad (5.14)$$

Thus, for example,

$$J^{(1)}(F^{(0)}, F^{(1)}) = \hat{J}(F^{(0)} \mid F^{(1)}) + \hat{J}(F^{(1)} \mid F^{(0)}) \qquad (5.14a)$$

Expansion of the time derivative

The *fourth step* in the Chapman–Enskog expansion addresses the time derivative in (5.7). It stipulates that the time dependence of F is solely dependent on the hydrodynamic variables n, $\mathbf{u}$, T so that

$$\frac{\partial F}{\partial t} = \frac{\partial F}{\partial n}\frac{\partial n}{\partial n} + \frac{\partial F}{\partial \mathbf{u}}\cdot\frac{\partial \mathbf{u}}{\partial t} + \frac{\partial F}{\partial T}\frac{\partial T}{\partial t} \qquad (5.15)$$

The time derivatice is expanded as

$$\frac{\partial}{\partial t} = \frac{\partial_0}{\partial t} + \varepsilon\frac{\partial_1}{\partial t} + \varepsilon^2\frac{\partial_2}{\partial t} + \cdots \qquad (5.16)$$

The physical meaning of this expansion is that lowest-order terms vary most rapidly, whereas higher-order terms are more slowly varying. Explicit expressions for the time derivations in (5.15) follow from ε-ordering of the conservation equations (3.14), (3.30), and (3.31). We obtain [recall (3.31) for notation]

$$\frac{\partial_0 n}{\partial t} = -\nabla \cdot n\mathbf{u},$$

$$\frac{\partial_r n}{\partial t} = 0, \qquad r > 0$$

$$\frac{\partial_0 \mathbf{u}}{\partial t} = -(\mathbf{u} \cdot \nabla)\mathbf{u} + \tilde{\mathbf{K}} - \frac{1}{\rho}\nabla \cdot \overset{=}{P}{}^{(0)}$$

$$\frac{\partial_r \mathbf{u}}{\partial t} = \frac{-1}{\rho}\nabla \cdot \overset{=}{P}{}^{(r)}, \qquad r > 0 \tag{5.17}$$

$$\frac{\partial_0 T}{\partial t} = -\mathbf{u} \cdot \nabla T - \frac{2}{3}\frac{1}{nk_B}[\overset{=}{P}{}^{(0)} : \overline{\nabla\mathbf{u}} + \nabla \cdot \mathbf{Q}^{(0)}]$$

$$\frac{\partial_r T}{\partial t} = -\frac{2}{3nk_B}[\overset{=}{P}{}^{(r)} : \overline{\nabla\mathbf{u}} + \nabla \cdot \mathbf{Q}^{(r)}], \qquad r > 0$$

When written in terms of the symmetric strain tensor $\overset{=}{\Lambda}$ (4.6a), the last two equalities in (5.17) appear as

$$\frac{\partial_0 T}{\partial t} = -\mathbf{u} \cdot \nabla T - \frac{2}{3}\frac{1}{nk_B}[\nabla \cdot \mathbf{Q}^{(0)} + \overset{=}{\Lambda} : \overset{=}{P}{}^{(0)}]$$

$$\frac{\partial_r T}{\partial t} = -\frac{2}{3nk_B}[\nabla \cdot \mathbf{Q}^{(r)} + \overset{=}{\Lambda} : \overset{=}{P}{}^{(r)}], \qquad r > 0 \tag{5.17a}$$

Substituting the preceding expansions for F, $\partial F/\partial t$, $\partial/\partial t$, $\hat{\mathcal{D}}F$, and $\hat{J}(F \mid F)$ into the Boltzmann equation (5.7) gives

$$\varepsilon\left[\left(\frac{\partial_0}{\partial t} + \varepsilon\frac{\partial_1}{\partial t}\cdots\right)(F^{(0)} + \varepsilon F^{(1)} + \cdots) + (\hat{\mathcal{D}}F^{(0)} + \varepsilon\hat{\mathcal{D}}F^{(1)} + \cdots)\right]$$

$$+ [\hat{J}^{(0)}(F^{(0)}) + \varepsilon\hat{J}^{(1)}(F^{(0)}, F^{(1)}) + \cdots] \tag{5.18}$$

where, with (5.16), we write

$$\frac{\partial_l F^{(r)}}{\partial t} = \frac{\partial F^{(r)}}{\partial n}\frac{\partial_1 n}{\partial t} + \frac{\partial F^{(r)}}{\partial \mathbf{u}} \cdot \frac{\partial_l \mathbf{u}}{\partial t} + \frac{\partial F^{(r)}}{\partial T}\frac{\partial_1 T}{\partial t} \tag{5.18a}$$

Equating coefficients of equal powers of ε gives the desired series of coupled integral equatins for $F^{(r)}$:

$$0 = \hat{J}^{(0)}(F^{(0)}) \tag{5.19a}$$

$$\left(\frac{\partial_0}{\partial t} + \hat{\mathcal{D}}\right)F^{(0)} = \hat{J}^{(1)}(F^{(0)}, F^{(1)}) \tag{5.19b}$$

$$\left(\frac{\partial_0}{\partial t} + \hat{\mathcal{D}}\right)F^{(1)} + \frac{\partial_1}{\partial t}F^{(0)} = \hat{J}^{(2)}(F^{(0)}, F^{(1)}, F^{(2)}) \tag{5.19c}$$

The first-order solution is obtained by solving (5.19a):

$$\hat{J}(F^{(0)} \mid F^{(0)}) = 0 \qquad (5.20)$$

First-order solution

Recalling our discussion of the Boltzmann $\mathcal{H}$ theorem (Section 3.4), the solution to (5.20) is the local Maxwellian F^0 (3.58):

$$F^{(0)} = F^0 = \frac{n}{(2\pi RT)^{3/2}} \exp\left(\frac{-c^2}{2RT}\right) \qquad (5.21)$$

In this manner we find that the first-order solution in the small mean-free-path approximation is the local Maxwellian F^0. This solution may be used to obtain a first-order set of hydrodynamical equations for the purpose of evaluating n, $\mathbf{u}$, and T. To construct these, we first calculate the heat conductivity $\mathbf{Q}$ and the stress $\bar{\bar{P}}$. From previous definitions (3.24) and (3.25), we write

$$\mathbf{Q} = \frac{m}{2} \int c^2 \mathbf{c} F \, d\mathbf{c}$$

$$\bar{\bar{P}} = m \int \overline{\mathbf{c}\mathbf{c}} F \, d\mathbf{c}$$

Substituting F^0 into these two formulas gives

$$\mathbf{Q}^{(0)} = 0 \qquad (5.22a)$$

$$\bar{\bar{P}}^{(0)} = \bar{\bar{I}}_p = \bar{\bar{I}} n k_B T \qquad (5.22b)$$

To lowest order, there is no heat flow and the pressure tensor is diagonal. Substituting these values into the conservation equatins (3.14), (3.30), and (3.31) gives the first-order approximation to these equations, which are called the *Euler* equations:

$$\frac{\partial n}{\partial t} + \nabla \cdot n\mathbf{u} = 0 \qquad (5.23a)$$

$$\rho\left(\frac{\partial}{\partial t} + \mathbf{u} \cdot \nabla\right)\mathbf{u} + \nabla p = \rho\mathbf{K} \qquad (5.23b)$$

$$\left(\frac{\partial}{\partial t} + \mathbf{u} \cdot \nabla\right)\left(\frac{p}{n^{5/3}}\right) = 0 \qquad (5.23c)$$

The adiabatic quality of flow (5.23c) in this approximation is seen to be consistent with the property that there is no heat flow (5.22a). Equations (5.21) to (5.23) constitute the first-order solutions to the Boltzmann equation in the Chapman–Enskog procedure. Solving the Euler equations (5.23) gives $n = n(\mathbf{x}, t)$, $\mathbf{u} = \mathbf{u}(\mathbf{x}, t)$, and $T = T(\mathbf{x}, t)$, which when substituted in (5.21) completely determines $F^0(\mathbf{x}, \mathbf{v}, t)$.

Each successive iterate in the Chapman–Enskog expansion yields a more detailed set of hydrodynamic equations, better suited to higher-order spatial fluctuations in the fluid. As noted above, the first iterate gives the Euler equations (5.23). Equations stemming from higher-order approxiamtions are named as follows. The second iterate gives the *Navier–Stokes* equations [of which (4.8) is the momentum equation]. The third iterate gives the *Burnett equations*.

3.5.3 Second-Order Solution

The second-order solution is obtained from (5.19b):

$$\left(\frac{\partial_0}{\partial t} + \hat{D}\right) F^{(0)} = \hat{J}^{(1)}(F^{(0)}, F^{(1)}) \tag{5.24}$$

We introduce the function Φ,

$$F^{(1)} = F^0 \Phi \tag{5.25a}$$

and [with (5.14a)] the $\hat{\Box}$ operator

$$\hat{\Box}\Phi \equiv \frac{1}{F^0} \hat{J}^{(1)}(F^0, F^0\Phi) = \frac{1}{F^0}[\hat{J}(F^0 \mid F^0\Phi) + \hat{J}(F^0\Phi \mid F^0)]$$

$$\iint \sigma \, d\Omega g \, d\mathbf{v}_1 F^0(v_1)[\Phi'_1 + \Phi' - \Phi_1 - \Phi] \tag{5.25b}$$

The starting equation (5.24) may then be written

$$\frac{1}{F^0}\left(\frac{\partial_0}{\partial t} + \hat{D}\right) F^0 = \hat{\Box}\Phi \tag{5.26}$$

Note in particular that $\hat{\Box}$ is a *linear operator*. Terms on the left of (5.26) give

$$\frac{1}{F^0}\frac{\partial_0 F^0}{\partial t} = \left[\frac{1}{n}\frac{\partial_0 n}{\partial t} + 2\boldsymbol{\xi} \cdot \frac{\partial_0 \mathbf{u}}{\partial t} + \left(\xi^2 - \frac{3}{2}\right)\frac{1}{T}\frac{\partial_0 T}{\partial t}\right]$$

$$\frac{1}{F^0}\mathbf{v} \cdot \frac{\partial F^0}{\partial \mathbf{x}} = \mathbf{v} \cdot \left[\frac{1}{n}\frac{\partial n}{\partial \mathbf{x}} + 2\boldsymbol{\xi} \cdot \frac{\partial \mathbf{u}}{\partial \mathbf{x}} + \left(\xi^2 - \frac{3}{2}\right)\frac{1}{t}\frac{\partial T}{\partial \mathbf{x}}\right] \tag{5.27}$$

where[31]

$$\xi^2 \equiv \frac{c^2}{2RT} \equiv \frac{c^2}{\bar{C}^2} \tag{5.28}$$

and $\boldsymbol{\xi} \cdot \partial \mathbf{u}/\partial \mathbf{x}$ is summed over parallel components of $\boldsymbol{\xi}$ and $\mathbf{u}$. Replacing time derivatives in (5.27) by related expressions in (5.17) gives (with $\mathbf{K} = 0$)

$$\boxed{\sqrt{2RT}\left(\xi^2 - \frac{5}{2}\right)\boldsymbol{\xi} \cdot \nabla \ln T + 2\left(\overline{\overline{\boldsymbol{\xi}\boldsymbol{\xi}}} - \frac{1}{3}\xi^2\overline{\overline{\mathbf{I}}}\right) : \overline{\overline{\nabla\mathbf{u}}} = \hat{\Box}\Phi} \tag{5.29}$$

---

[31] In most of the text, $mC^2 \equiv k_B T$ (save for Section 4.2, where C was used to denote rms molecular speed). To avoid confusion here, we set $m\bar{C}^2 = 2k_B T$.

which is seen to be a *linear inhomogeneous integral equation* for the distribution Φ.

If this equation is solved for Φ, then F is known to second order:

$$F = F^0[1 + \Phi] \tag{5.30}$$

The general solution to (5.29) is a linear combination of homogeneous, Φ_h, and inhomogeneous, Φ_i, solutions, where

$$\hat{\Box}\Phi_h = 0 \tag{5.31}$$

and Φ_i is a *particular* solution of (5.29).

From the structure of $\hat{\Box}$, it is evident that Φ_h is any linear combination of the three summational invariants.

$$\Phi_h = \alpha + \boldsymbol{\beta} \cdot m\mathbf{c} + \frac{1}{2}\gamma mc^2 \tag{5.32}$$

where α, $\boldsymbol{\beta}$, γ are arbitrary constants. To arrive at the particular solution to (5.29), we note that the left side is in the form

$$\text{LHS (5.29)} = \mathbf{X}(\boldsymbol{\xi}) \cdot (2RT)^{1/2} \nabla \ln T + \overline{\overline{\mathbf{Y}}}(\boldsymbol{\xi}) : \overline{\overline{\nabla \mathbf{u}}} \tag{5.33}$$

Since $\tilde{\Box}$ is a linear operator and Φ is a scalar, (5.33) suggests that we take the particular solution to (5.29) to have the form[32]

$$\Phi_i = \mathbf{A}(\boldsymbol{\xi}) \cdot (2RT)^{1/2} \nabla \ln T + 2\overline{\overline{B}}(\boldsymbol{\xi}) : \overline{\overline{\nabla \mathbf{u}}} \tag{5.34}$$

Thus, to find the inhomogeneous solution Φ_i, we must obtain the vector function $\mathbf{A}$ and the tensor function $\overline{\overline{B}}$. Inserting this form into (5.29) and equating coefficients of the different components of $\nabla \ln T$ and $\overline{\overline{\nabla \mathbf{u}}}$ gives the following equations for $\mathbf{A}$ and $\overline{\overline{B}}$:

$$\hat{\Box}\mathbf{A} = \boldsymbol{\xi}\left(\xi^2 - \frac{5}{2}\right) \tag{5.35a}$$

$$\hat{\Box}\overline{\overline{B}} = \left(\overline{\overline{\boldsymbol{\xi}\boldsymbol{\xi}}} - \frac{1}{3}\xi^2\overline{\overline{I}}\right) \equiv \overset{\circ}{\overline{\overline{\boldsymbol{\xi}\boldsymbol{\xi}}}} \tag{5.35b}$$

Note that the only variables in $\mathbf{A}$ are $\boldsymbol{\xi}$, n, and T. The only vector that can be formed from these elements is $\boldsymbol{\xi}$ itself. Thus we write

$$\mathbf{A} =' bs\, A(\xi^2)\boldsymbol{\xi} \tag{5.36}$$

where A is some scalar function.

The linearity of (5.35b) and the form of its inhomogeneous term imply that $\overline{\overline{B}}$ is a symmetric, traceless tensor. Again $\overline{\overline{B}}$ depends only on $\boldsymbol{\xi}$, n, and T. The

---

[32]Note that the dimension of $\hat{\Box}$ is frequency and that of $\mathbf{A}$ and $\overline{\overline{B}}$ is time.

only symmetric traceless tensor that can be constructed from these variables is $\overline{\overline{\boldsymbol{\xi}\boldsymbol{\xi}}} - \frac{1}{3}\xi^2\overline{\overline{I}}$, so

$$\overline{\overline{B}} = \mathcal{B}(\xi^2)\left(\overline{\overline{\boldsymbol{\xi}\boldsymbol{\xi}}} - \frac{1}{3}\xi^2\overline{\overline{I}}\right) \tag{5.36a}$$

The scalar functions $\mathcal{A}$ and $\mathcal{B}$ satisfy the integral equations

$$\hat{\Box}(\boldsymbol{\xi}\mathcal{A}) = \boldsymbol{\xi}\left(\xi^2 - \frac{5}{2}\right) \tag{5.37a}$$

$$\hat{\Box}\left[\left(\overline{\overline{\boldsymbol{\xi}\boldsymbol{\xi}}} - \frac{1}{3}\xi^2\overline{\overline{I}}\right)\mathcal{B}\right] = \left(\overline{\overline{\boldsymbol{\xi}\boldsymbol{\xi}}} - \frac{1}{3}\xi^2\overline{\overline{I}}\right) \tag{5.37b}$$

Returning to the homogeneous solution (5.30), we note that the constants α, β, and γ contained in Φ_h are determined by the constraint conditions (5.10b). Inserting

$$F_1 = F^0[\Phi_h + \Phi_i] \tag{5.38}$$

into these constraint equations yields the three integral conditions (see Appendix B, Section B.1):

$$\int F^0\left(\alpha + \gamma\frac{1}{2}mc^2\right)d\mathbf{c} = 0 \tag{5.39a}$$

$$\int F^0[\mathcal{A}(\xi^2)\nabla\ln T + m\boldsymbol{\beta}]mc^2\,d\mathbf{c} = 0 \tag{5.39b}$$

$$\int F^0\left(\alpha + \frac{1}{2}mc^2\gamma\right)\frac{1}{2}mc^2\,d\mathbf{c} = 0 \tag{5.39c}$$

Equations (5.39a) and (5.39c) imply that

$$\alpha = \gamma = 0 \tag{5.40}$$

whereas (5.39b) implies that $\boldsymbol{\beta}$ is in the direction of $\nabla\ln T$, so it may be absorbed into the $\nabla\ln T$ term in Φ_i.

The total solution of the Boltzmann equation to terms of second order then appears as

$$F = F^0[1 + (2RT)^{1/2}\mathbf{A}\cdot\nabla\ln T + 2\overline{\overline{B}}:\overline{\overline{\nabla\mathbf{u}}}]$$

$$= F^0\left[1 + (2RT)^{1/2}\mathcal{A}(\xi)\boldsymbol{\xi}\cdot\nabla\ln T + \mathcal{B}(\xi)\left(\overline{\overline{\boldsymbol{\xi}\boldsymbol{\xi}}} - \frac{1}{3}\xi^2\overline{\overline{I}}\right):\overline{\overline{\nabla\mathbf{u}}}\right] \tag{5.41}$$

where $\mathcal{A}$ and $s\mathcal{B}$ are particular solutions to integral equations (5.37).

3.5.4 *Thermal Conductivity and Stress Tensor*

With the form of the second-order solution (5.41) at hand, it is possible to obtain expressions for corresponding nonvanishing second-order contributions to $\mathbf{Q}$ and $\overline{\overline{P}}$. These are obtained by inserting (5.41) into the defining equations (4.3) and (4.7) (with $\zeta = 0$).

$$\mathbf{Q} = \frac{1}{2} m (2RT)^{3/2} \int \xi^2 \boldsymbol{\xi} F \, d\mathbf{c} = 0 + \mathbf{Q}^{(1)} \tag{5.42a}$$

$$\overline{\overline{P}} = m2RT \int \overline{\overline{\boldsymbol{\xi}\boldsymbol{\xi}}} F \, d\mathbf{c} = \overline{\overline{I}} p + \overline{\overline{P}}^{(1)} \tag{5.42b}$$

Coefficient of thermal conductivity

Substituting (5.41) into (5.42), we find

$$\mathbf{Q} = \frac{m\bar{C}^4}{2} \nabla \ln T \cdot F^0 \xi^2 \overline{\overline{\boldsymbol{\xi}\boldsymbol{\xi}}} \mathcal{A}(\xi^2) \, d\mathbf{c} \tag{5.43}$$

(Recall that $\int F^0 \, d\mathbf{c}$ has dimensions of number density.) With reference to (B.B1.3), the preceding reduces to

$$\mathbf{Q} = \frac{2}{3} \frac{k_B^2 T}{m} \nabla T \int F^0 \xi^4 \mathcal{A}(\xi^2) \, d\mathbf{c} \tag{5.44}$$

With $\int F^0 \mathcal{A}(\xi^2) \xi^2 \, d\mathbf{c} = 0$ [which corresponds to setting $\beta = 0$ in (5.39b)], the preceding equation may be rewritten

$$\mathbf{Q} = \frac{2}{3} \frac{k_B^2 T}{m} \nabla T \int \boldsymbol{\xi} \mathcal{A}(\xi^2) \cdot F^0 \left(\xi^2 - \frac{5}{2} \right) \boldsymbol{\xi} \, d\mathbf{c} \tag{5.45}$$

With (5.35a), we find

$$\mathbf{Q} = \left(\frac{2}{3} \frac{k_B^2 T}{m} \int F^0 \mathbf{A} \cdot \hat{\Box} \mathbf{A} \, d\mathbf{c} \right) \nabla T \tag{5.46}$$

or, equivalently,

$$\mathbf{Q} = \frac{2}{3} \frac{k_B^2 T}{m} \langle \mathbf{A} \mid \hat{\Box} \mathbf{A} \rangle \nabla T \tag{5.47}$$

Note, in particular, that the bracket symbol $\langle | \rangle$ in (5.47) includes the local Maxwellian and an inner product. The latter expression implies the following form for the coefficient of thermal conductivity:

$$\kappa = -\frac{2}{3} \frac{k_B^2 T}{m} \langle \mathbf{A} \mid \hat{\Box} \mathbf{A} \rangle \tag{5.48}$$

Viscosity

To obtain the related expression for viscosity, we recall (5.42b) and write

$$\overline{\overline{P}}^{(1)} = \int Fm\overline{\overline{cc}}\,d\mathbf{c} = m \int F^0 \Phi \overline{\overline{cc}}\,d\mathbf{c} \tag{5.49}$$

With (5.41), the latter equation may be written (deleting the superscript 1)

$$\overline{\overline{P}} = 4k_B T \int F^0 B(\xi^2)(\overset{\circ}{\overline{\overline{\xi\xi}}} : \overline{\overline{\nabla \mathbf{u}}})\overline{\overline{\xi\xi}}\,d\mathbf{c} \tag{5.50}$$

Employing the results of Problem 3.6 permits (5.50) to be written

$$\overline{\overline{P}} = \frac{4k_B T}{5}\overset{\circ}{\overline{\overline{\nabla \mathbf{u}}}} \int F^0 B(\xi^2)(\overset{\circ}{\overline{\overline{\xi\xi}}} : \overset{\circ}{\overline{\overline{\xi\xi}}})\,d\mathbf{c} \tag{5.51}$$

$$= \frac{4k_B T}{5}\overset{\circ}{\overline{\overline{\nabla \mathbf{u}}}} \int F^0 \overset{\circ}{\overline{\overline{\xi\xi}}} : \overline{\overline{B}}\,d\mathbf{c}$$

With (5.35b), we write

$$\overline{\overline{P}} = \frac{4k_B T}{5}\overset{\circ}{\overline{\overline{\nabla \mathbf{u}}}} \int F^0 \overline{\overline{B}} : \hat{\square}\overline{\overline{B}}\,d\mathbf{c}$$

$$\overline{\overline{P}} = \frac{4k_B T}{5}\langle \overline{\overline{B}} \mid \hat{\square}\overline{\overline{B}}\rangle \overset{\circ}{\overline{\overline{\nabla \mathbf{u}}}} \tag{5.52}$$

Note that the inner product $\langle \mid \rangle$ now includes the trace operation, which renders it a scalar. Note that we have written [recall (4.6a)]

$$\overset{\circ}{\overline{\overline{\nabla \mathbf{u}}}} = \overline{\overline{\Lambda}} - \frac{1}{3}\overline{\overline{I}}\,\mathrm{Tr}\,\overline{\overline{\nabla \mathbf{u}}}$$

The shear stress component of (4.7) may then be written

$$\overline{\overline{P}} = -2\eta \overset{\circ}{\overline{\overline{\nabla \mathbf{u}}}}$$

which with (5.52) gives

$$\eta = -\frac{2}{5}k_B T \langle \overline{\overline{B}} \mid \hat{\square}\overline{\overline{B}}\rangle \tag{5.53}$$

3.5.5 Sonine Polynomials

In this manner we find that transport coefficients depend on the matrix elements of the interaction operator $\hat{\square}$. To evaluate these expressions, we work with

Sonine polynomials,[33] defined as follows (see Problems 3.7):

$$S_m^{(n)}(x) = \sum_{p=0}^{n} \frac{(m+n)!(-x)^p}{(m+p)!(n-p)!p!} \qquad (5.54a)$$

Some leading values are

$$S_m^{(0)}(x) = 1, \quad S_m^{(1)}(x) = m + 1 - x$$

$$S_m^{(2)}(x) = \frac{(m+1)(m+2)}{2} - x(m+2) + \frac{x^2}{2}$$

The orthogonality of these polynomials is given by

$$\int_0^\infty x^m e^{-x} S_m^{(n)}(x) S_m^{(q)}(x)\, dx = 0, \qquad\qquad n \neq q$$

$$= \frac{(m+n)!}{n!}, \qquad n = q \qquad (5.54b)$$

Expansions of $\mathbf{A}$ and $\overline{\overline{B}}^{(r)}$ are given by

$$\mathbf{A} = \sum_{r=1}^{\infty} a_r S_{3/2}^{(r)}(\xi^2)\boldsymbol{\xi} \equiv \sum a_r \mathbf{A}^{(r)} \qquad (5.55a)$$

$$\overline{\overline{B}} = \sum_{r=1}^{\infty} b_r S_{5/2}^{(r-1)}(\xi^2)\overline{\overline{\boldsymbol{\xi}\boldsymbol{\xi}}} \equiv \sum b_r \overline{\overline{B}}^{(r)} \qquad (5.55b)$$

Note that $\mathbf{A}^{(r)}$ and $\overline{\overline{B}}^{(r)}$ are dimensionless.

First approximation for A

To find the a_r coefficients, we multiply (5.35a) by $F^0 \mathbf{A}^{(l)}$ and integrate:

$$\langle \mathbf{A}^{(l)} \mid \hat{\Box}\mathbf{A} \rangle = \int F^0 \left(\xi^2 - \frac{5}{2} \right) \boldsymbol{\xi} \cdot \mathbf{A}^{(l)}\, d\mathbf{c} \equiv n\alpha_l \qquad (5.56)$$

Note that α_i, as well as $\mathbf{A}^{(l)}$, is dimensionless and n denotes number density. Inserting (5.55a) into (5.56) gives

$$n\alpha_l = \sum_r a_r \langle \mathbf{A}^{(l)} \mid \hat{\Box}\mathbf{A}^{(r)} \rangle \equiv \sum_{r=1}^{\infty} A_{lr} a_r \qquad (5.57)$$

---

[33]N.J. Sonine, *Math. Ann. 16*, 41 (1880). First introduced in the present context by D. Burnett, *Proc. London Math. Soc.*. Ser. 2, *39*, 385 (1935). Subsequently shown by C. S. Wang Chang and G. E. Uhlenbeck to be eigenfunctions of the $\hat{\Box}$ operator for Maxwell molecules. *Univ. Michigan Engr. Rept.* CM-681 (1952). This topic is discussed in the present work in Section 6.5. We note further that, apart from a multiplicative constant, Sonine and Laguerre polynomials are identical. Laguerre polynomials come into play in the solution of the Schroedinger equation for the hydrogen atom. See R. L. Liboff, *Introductory Quantum Mechanics*, 4th ed., Addison-Wesley, San Francisco, CA. (2002).

The α_l are known parameters (5.56) as are the $\mathbf{A}^{(l)}$ functions (5.55a). Thus (5.57) comprises an infinite-dimensional matrix equation for the coefficients a_r.

Varying orders of solution (within the second-order Chapman–Enskog approximation) are obtained by cutting off the A_{lr} matrix. Thus, for example, keeping only the leading terms in (5.57) gives

$$a_1 = \frac{n\alpha_1}{A_{11}}$$

Substituting into (5.55a), we find

$$\mathbf{A} = \frac{n\alpha_1}{A_{11}}\mathbf{A}^{(1)}$$

whence

$$\left\langle \mathbf{A} \mid \hat{\Box}\mathbf{A} \right\rangle = a_1^2 A_{11} = \frac{n^2\alpha_1^2}{A_{11}} \tag{5.58}$$

Evaluating α_1 from (5.56), we find (see Problem 3.8)

$$\left\langle \mathbf{A} \mid \hat{\Box}\mathbf{A} \right\rangle = \left(\frac{15^2}{4}\right)\frac{n^2}{A_{11}} \tag{5.59}$$

Substituting into (5.48) gives the following generic form for the lowest-order approximation to the coefficient of thermal conductivity within the second Chapman–Enskog approximation:

$$\kappa = -\frac{25}{4}\frac{\bar{c}_v k_B T n^2}{A_{11}} \tag{5.60}$$

Here we have written $\bar{c}_V$ for c_V/mass, which for particles with no internal structure has the value $3k_B/2m$. The A_{11} element contains specifics of the particle interaction. However, a very explicit form of this term may be written in terms of integration over scattering parameters. We obtain

$$A_{11} = -4n^2\Omega^{(2,2)} \tag{5.61}$$

where, in general (for study of one-component gases)

$$\Omega^{(l,q)} \equiv \sqrt{\frac{4\pi k_B T}{m}} \int_0^\infty \int_0^\infty e^{-y^2} y^{2q+3}(1 - \cos^l \theta)s\, ds\, dy \tag{5.62a}$$

where

$$y^2 \equiv \frac{q^2}{2\bar{C}^2} = \frac{mg^2}{4k_B T}$$

With (1.16) and following, we recall that the scattering angle $\theta = \theta(s, g)$. Thus (5.62a) may be rewritten

$$\Omega^{(l,q)} = \sqrt{\frac{4\pi k_B T}{m}} \int_0^\infty e^{-y^2} y^{2q+3} Q^{(l)}\, dy \tag{5.62b}$$

where $Q^{(l)}$ is the weighted cross section.[34]

$$Q^{(l)} = \int_0^\infty (1 - \cos^l \theta) s \, ds \tag{5.62c}$$

Note that the dimensions of $\Omega^{(l,q)}$ are the same as v/n.

Substituting (5.62a) into (5.60) gives

$$\kappa = \frac{25}{16} \frac{\bar{c}_V k_B T}{\Omega^{(2,2)}} \tag{5.63}$$

First approximation for $\overline{\overline{B}}$

To find the b_r coefficients in (5.55b), we multiply (5.35b) by $F^0 \overline{\overline{B}}^{(l)}$ and integrate.

$$\langle \overline{\overline{B}}^{(l)} \mid \hat{\Box} \overline{\overline{B}}^{(l)} \rangle = \int F^0 \overset{\circ}{\overline{\overline{\boldsymbol{\xi}\boldsymbol{\xi}}}} : \overline{\overline{B}}^{(l)} \, d\mathbf{c} \equiv n\beta_l \tag{5.64}$$

Introducing (5.55b) gives

$$\sum_r b_r \langle \overline{\overline{B}}^{(l)} \mid \hat{\Box} \overline{\overline{B}}^{(r)} \rangle = n\beta_l$$

$$n\beta_l = \sum_r B_{lr} b_r \tag{5.65}$$

Again we obtain a matrix equation of infinite dimension for the coefficients b_r. Keeping the leading term in (5.65) gives

$$b_1 = \frac{n\beta_1}{B_{11}} \tag{5.66}$$

With (5.55b), we find

$$\langle \overline{\overline{B}} \mid \hat{\Box} \overline{\overline{B}} \rangle = b_1^2 B_{11} = \frac{n^2 \beta_1^2}{B_{11}}$$

Evaluating β_1^2, we find (see Problem 3.9)

$$\langle \overline{\overline{B}} \mid \hat{\Box} \overline{\overline{B}} \rangle = \frac{25}{4} \frac{n^2}{B_{11}} \tag{5.67}$$

Substituting this expression into (5.53) gives the following lowest-order expression for the coefficient of viscosity, within the second approximation of

---

[34]Comparison of notation for $Q^{(l)}$ as given by (5.62c) [that is, $Q^{(l)}$ here] with that found in Chapman and Cowling (CC) and Hirshfelder, Curtis, and Bird (HCB) is as follows: $Q^{(l)}$ (here) $= (2\pi)^{-1} Q^{(l)}$(HCB) $= g^{-1} Q^{(l)}$(CC). Note also that: $\langle \mid \rangle$ (here) $= -n^2[\,,\,]_1$(CC).

the Chapman–Enskog expansion.

$$\eta = -\frac{5}{2}\frac{k_B T n^2}{B_{11}} \tag{5.68}$$

Further reduction of B_{11} gives

$$B_{11} = -4n^2\Omega^{(2,2)} \tag{5.69}$$

and we find

$$\eta = \frac{5}{8}\frac{k_B T}{\Omega^{(2,2)}} \tag{5.70}$$

Comparison with (5.63) gives

$$\frac{\kappa}{\eta} = \frac{5}{2}\bar{c}_V$$

Note that our earlier elementary mean-free-path calculation (Table 3.2) is in good agreement with this more detailed finding.

3.5.6   Application to Rigid Spheres

We wish to evaluate the integrals (5.62) for the case of a gas of rigid spheres of diameter σ_0. Recalling (1.32), we write

$$s\,ds = \frac{\sigma_0^2}{4}\,d\cos\theta$$

Substituting into (5.62b) gives

$$\begin{aligned}
Q^{(l)} &= \int (1 - \cos^l\theta)s\,ds \\
&= \frac{\sigma_0^2}{4}\int_{-1}^{1}(1 - \mu^l)\,d\mu \\
&= \frac{\sigma_0^2}{4}\left[2 - \frac{1}{l+1}(-1)^l)\right] \\
&\equiv \frac{\sigma_0^2}{4}H(l) \tag{5.71a}
\end{aligned}$$

Note that we set $\mu \equiv \cos\theta$. For $\Omega^{(l,q)}$, we obtain

$$\Omega^{(l,q)} = \frac{\sigma_0^2}{4}\left(\frac{\pi k_B T}{m}\right)^{1/2}H(l)(q+1)! \tag{5.71b}$$

Thus for interactions of rigid spheres we find $|H(2) = \frac{4}{3}]$

$$\Omega^{(2,2)} = 2\sigma_0^2\left(\frac{\pi k_B T}{m}\right)^{1/2} \tag{5.72}$$

Substituting into (5.63) gives

$$\kappa = \frac{25}{32} \frac{\bar{c}_V}{\sigma_0^2} \left(\frac{m k_B T}{\pi} \right)^{1/2}$$

$$\kappa = \frac{75}{64} \frac{1}{\sigma_0^2} \left(\frac{k_B^3 T}{m \pi} \right)^{1/2}$$

(5.73)

With (5.70) we find

$$\eta = \frac{5}{16} \frac{1}{\sigma_0^2} \left(\frac{m k_B T}{\pi} \right)^{1/2}$$

(5.74)

3.5.7 Diffusion and Electrical Conductivity

As discussed previously, mutual diffusion is a property of a two-component medium. In this case, coupled Boltzmann equations come into play as described in Section 3.2.2. Following a procedure similar to that described above, we find that the interaction integrals (5.62) again emerge with the modification

$$y^2 = \left(\frac{\mu}{2 k_B T} \right) g^2$$

(5.75)

$$\mathbf{g} = \mathbf{v}_2 - \mathbf{v}_1$$

where, we recall, μ is reduced mass. The relation (5.62b) becomes

$$\Omega_{12}^{(l,q)} = \frac{\sigma_{12}^2}{4} \left(\frac{\pi k_B T}{2\mu} \right)^{1/2} H(l)(q+1)1$$

(5.76)

For interacting rigid spheres, (5.71b) becomes

$$\Omega_{12}^{(l,q)} = \frac{\sigma_{12}^2}{4} \left(\frac{\pi k_B T}{2\mu} \right)^{1/2} H(l)(q+1)!$$

(5.76a)

Specifically,

$$\Omega_{12}^{(1,1)} = \sigma_{12}^2 \left(\frac{\pi k_B T}{2} \right)^{1/2}$$

(5.77)

where, we recall,

$$\sigma_{12} \equiv \frac{1}{2}(\sigma_{01} + \sigma_{02})$$

The lowest-order estimate within the second Chapman–Enskog approximation for the coefficient of mutual diffusion is then given by

$$D_{12} = \frac{3 k_B T}{16 n \mu \Omega_{12}^{(1,1)}}$$

(5.78)

For spherical molecules, there results

$$D_{12} = \frac{3}{16n\sigma_{12}^2} \left(\frac{2k_B T}{\pi \mu} \right)^{1/2}$$
$$n = n_1 + n_2$$

(5.79)

This finding was previously cited in (4.31a). In the limit that $n_1 = n_2 = n/2$ and $\sigma_{01} = \sigma_{02} \equiv \sigma_0$, (5.79) gives the coefficient of self-diffusion:

$$D = \frac{3}{8n\sigma_0^2} \left(\frac{k_B T}{\pi m} \right)^{1/2}$$

(5.79a)

Electrical conductivity

Again stemming from the study of a two-component gas comprised of ions of charge Ze and electrons of charge e, we find that conductivity is directly related to mutual diffusion as

$$\sigma_c = \frac{n_i n_e n}{\rho^2 k_B T} (Zem - eM)^2 D_{ie}$$

(5.80)

where ρ is mass density and M is ion mass. In the limit that $m/M \ll 1$, this relation reduces to

$$\sigma_c = \frac{n_e n_e^2}{k_B T n_i} D_{ie}$$

(5.80a)

For singly charged ions, $n_e = n_i$. Further setting $\mu \simeq m$ in (5.77), the preceding relation reduces to

$$\sigma_c = \frac{3}{8} \frac{e^2}{m \Omega^{(1,1)}}$$

(5.81)

Comparison with the Drude model result (4.15) indicates that $\Omega^{(1,1)}$ takes the place of collision frequency times volume per particle. These results relevant to rigid spheres are listed in Table 3.4

3.5.8 Expressions of $\Omega^{(l,q)}$ for Inverse Power Interaction Forces

Consider the interaction potential between molecules to be given by (1.17),

$$V = Kr^{-N}$$

(5.82)

corresponding to the radial force of magnitude

$$F = KNr^{-(N+1)}$$

TABLE 3.4. Transport Coefficients in the Chapman–Enskog Expansion for Rigid-Sphere Interaction[a]

Coefficient	Value
κ, thermal conductivity	$\dfrac{75}{64} \dfrac{1}{\sigma_0^2} \left(\dfrac{k_B^3 T}{m\pi} \right)^{1/2}$
η, viscosity	$\dfrac{5}{16} \dfrac{1}{\sigma_0^2} \left(\dfrac{m k_B T}{\pi} \right)^{1/2}$
D_{12}, mutual diffusion	$\dfrac{3}{16 n \sigma_{12}^2} \left(\dfrac{2 k_B T}{\pi \mu} \right)^{1/2}$
D, self-diffusion	$\dfrac{3}{8 n \sigma_0^3} \left(\dfrac{k_B T}{\pi m} \right)^{1/2}$
σ_c, electrical conductivity	$\dfrac{n_i n_e n}{\rho^2 k_B T} (Z e M - e M)^2 D_{ie}$
$\sigma_c \quad (m/M \ll 1)$	$\dfrac{n_e n e^2}{k_B T n_i} d_{ie}$

[a]With $\bar{c}_V = 3 k_B / m$, these results give $\kappa / n \bar{c}_V = \frac{5}{2} = 2.5$, $\mathcal{P} = \frac{2}{3} = 0.67$ and $\mathcal{L} = \frac{5}{6} = 0.83$, which are seen to be in reasonable agreement with measured values given in Table 3.3.

Then, for $Q^{(l)}$, with (1.20) and following, we find

$$
\begin{aligned}
Q^{(l)} &= \int_0^\infty (1 - \cos^l \theta) s \, ds \\
&= \left(\frac{2K}{\mu} \right)^{2/N} g^{-4/N} \int_0^\infty (1 - \cos^l \theta) b \, db \\
Q^{(l)} &\equiv \left(\frac{2K}{\mu} \right)^{2/N} g^{-4/N} A_l(N) \\
A_l(N) &\equiv \int_0^\infty (1 - \cos^l \theta) b \, db
\end{aligned}
\tag{5.83}
$$

Inserting this result into (5.76) gives

$$
\begin{aligned}
\Omega^{(l,q)} &= \frac{\sqrt{\pi} A_l(N) (2K)^{2/N} (2 k_B T)^{(N-4)/2N}}{\mu^{1/2}} \int_0^\infty e^{-y^2} y^{2q+3-(4/N)} \, dy \\
\Omega^{(l,q)} &= \frac{\sqrt{\pi} A_l(N) (2K)^{2/N} (2 k_B T)^{(N-4)/2N}}{2 \mu^{1/2}} \Gamma\left(q + 2 - \frac{2}{N} \right)
\end{aligned}
\tag{5.84}
$$

where $\Gamma(x)$ is the gamma function. Note in particular the simplifying property for Maxwell molecules, $N = 4$.

Nature of approximation

Having come to this point in the Chapman–Enskog expansion, it is evident that the technique of solution involves nested approximations within each iterate (5.19) of the expansion. As we have seen, the second approximation yields the integral equations (5.35). To solve these equations, an additional approximation comes into play: casting these equations in matrix form permits an approximation corresponding to the order of the truncation of the related infinite matrices. In the results presented above, related approximation within the second Chapman–Enskog iterate involved the lowest approximation, 1×1 matrices.

3.5.9 Interaction Models and Experimental Values

Combining preceding results with various models of interactions between molecules gives the following list for viscosities to lowest-order results within the second Chapman–Enskog approximation.

1. *Rigid elastic spheres of diameter σ_0.*

$$\eta = \frac{5}{16\sigma_0^2} \left(\frac{k_B m T}{\pi} \right)^{1/2} \tag{5.85}$$

2. *Repulsive potential, $V = K r^{-N}$.*

$$\eta = \frac{5}{8} \left(\frac{m k_B T}{\pi} \right)^{1/2} \left(\frac{k_B T}{K} \right)^{2/N} \bigg/ A_2(N) \Gamma \left(4 - \frac{2}{N} \right) \tag{5.86}$$

3. *Sutherland model.*[35] Attractive spheres of diameter σ_0 and interaction potential $V = -K/r^N$.

$$\eta = \frac{5}{16\sigma_0^2} \left(\frac{m k_B T}{\pi} \right)^{1/2} \bigg/ \left(1 + \frac{S}{k_B T} \right) \tag{5.87}$$

In (5.87), S is the temperature-independent Sutherland constant[36]

$$S = \frac{I(N)K}{\sigma_0^N} \tag{5.88}$$

Note that S is proportional to the potential energy of the molecules in contact. The function $I(N)$ is given by

$$I(N) = 4 \int_0^1 \bar{\beta}^{4-N} (2\bar{\beta}^2 - 1)(1 - \bar{\beta}^2)^{1/2} \int_0^{\bar{\beta}} \beta^N (1 - \beta^2)^{-3/4} \, d\beta \, d\bar{\beta}$$

---

[35] W. Sutherland, *Phil. Mag. 36*, 507 (1893); *17*, 320 (1909).
[36] S. Chapman and T. G. Cowling, ibid., Sec. 10.41. In this work, the Sutherland constant is written as (5.88) divided by k_B. In the present work, S carries dimensions of energy.

TABLE 3.5. Viscosity of CO_2 and N_2 in the Sutherland Model

	CO_2	
T (°C)	$\eta \times 10^7$ (observed)	$\eta \times 10^7$ (evaluated)
−20.7	1294	1284
15.0	1457	1462
99.1	1861	1857
182.4	2221	2216
302.0	2681	2686

	N_2	
T (°C)	$\eta \times 10^7$ (observed)	$\eta \times 10^7$ (evaluated)
−76.3	1275	1269
−37.9	1465	1469
16.1	1728	1728
51.6	1880	1884
100.2	2084	2086
200.0	2461	2461
250.1	2629	2633

Some leading values of $I(N)$ are as follows:

N	$I(N)$
2	$\frac{1}{8}(\pi^2 - 8) = 0.2337$
3	$\frac{8}{3}(3\ln 3 - 2) = 0.2118$
4	$\frac{3}{2}(10 - \pi^2) = 0.1956$
6	$\frac{5}{24}(9\pi^2 - 88) = 0.1722$
8	$\frac{7}{45} = 0.1556$

In applying (5.87) to experimental observation, it proves convenient to remove the dependence on numerator coefficients of this expression by rewriting it in the form

$$\eta(T) = \eta(T') \left(\frac{T}{T'}\right)^{3/2} \frac{S + k_B T'}{S + k_B T} \tag{5.89}$$

With values,

$$\eta(T = 273\text{K}, \ CO_2) = 1388 \times 10^{-7} \ \text{g cm}^{-1}\text{s}^{-1}$$
$$\eta(T = 273\text{K}, \ N_2) = 1654 \times 10^{-7} \ \text{g cm}^{-1}\text{s}^{-1}$$

and

$$S(CO_2) = 239.7k_B, \qquad S(N_2) = 104.7k_B$$

the expression (5.89) gives the values shown in Table 3.5, in very good agreement with experiment.

3.5.10   The Method of Moments

In this concluding portion of the present section, we present a brief description of an alternative solution to the Boltzmann equation due to Grad,[37] which has come to be known as the method of moments.

We begin by defining another nondimensional microscopic velocity, $\bar{\mathbf{c}}$, closely related to ξ introduced above: $\bar{\mathbf{c}} \equiv \mathbf{c}/C = \sqrt{2}\xi$. The nondimensional distribution function $\bar{f}(\mathbf{x}, \bar{\mathbf{c}}, t)$ is given by

$$F(\mathbf{x}, \bar{\mathbf{c}}, t) = \frac{n}{C^3} \bar{f}(\mathbf{x}, \bar{\mathbf{c}}, t) \tag{5.90}$$

The Maxwellian (5.2) may be written

$$F^0 = \frac{n}{C^3} \bar{f}^0(\bar{c}) \equiv \frac{n}{C^3} \omega(\bar{c}) \tag{5.91a}$$

$$\omega(\bar{c}) \equiv \frac{1}{(2\pi)^{3/2}} e^{-\bar{c}^2/2} \tag{5.91b}$$

Tensor Hermite polynomials

In the method of moments, the distribution function $\bar{f}(\mathbf{x}, \bar{\mathbf{c}}, t)$ is expanded in a series of tensor Hermite polynomials, $H_{\mathbf{i}}^{(n)}(\bar{\mathbf{c}})$. The double index of $H_{\mathbf{i}}^{(n)}(\bar{\mathbf{c}})$ reflects the fact that it is a tensor polynomial of rank n in three-dimensional space. The subscript $\mathbf{i}$ denotes a sequence of n indexes $(i_1, \ldots, i_n)$, where i_k may have any of the values $(1, 2, 3)$ corresponding to the three Cartesian directions.

Here are some fundamental properties of these polynomials. They are defined with respect to $\omega(c)$ given by (5.91b). Thus

$$H_{i_1, i_2, \ldots, i_n}^{(n)}(\mathbf{c}) = (-)^n \frac{1}{\omega(c)} \left(\frac{\partial}{\partial c_{i_1}} \frac{\partial}{\partial c_{i_1}} \cdots \frac{\partial}{\partial c_{1_n}} \right) \omega(c) = \frac{(-)^n}{\omega} \nabla^n \omega \tag{5.92}$$

These polynomials also follow from the generating function:

$$\exp\left[-\frac{1}{2}(\sigma^2 - c^2) \right] = \sum_{n=0}^{\infty} H_{i_1, \ldots, i_n}^{(n)}(\mathbf{c}) c_{i_1}' c_{i_2}' \cdots c_{i_n}' \tag{5.93}$$

$$\sigma = \mathbf{c}' - \mathbf{c}$$

The orthogonality of the tensor Hermite polynomials is given by

$$\int H_{i_1 \cdots i_n}^{(n)} H_{j_1 \cdots j_n}^{(n)} \omega(c) \, d\mathbf{c} = n! \Delta_{j_1, j_2, \ldots, j_n}^{i_1, i_2, \ldots, i_n} \tag{5.94}$$

where

$$\Delta_{j_1, \ldots, j_n}^{i_1, \ldots, i_n} = 1$$

[37]H. Grad, *Comm. Pure and App. Math.* 2, 331 (1949).

for $\{i_k\}$ a permutation of $\{j_k\}$ and zero otherwise.

The first four polynomials of this set are

$$H^{(0)}(\mathbf{c}) = 1 \tag{5.95a}$$
$$H_i^{(1)}(\mathbf{c}) = c_i \tag{5.95b}$$
$$H_{ij}^{(2)}(\mathbf{c}) = c_i c_j - \delta_{ij} \tag{5.95c}$$
$$H_{ijk}^{(3)}(\mathbf{c}) = c_i c_j c_k - (c_i \delta_{jk} + c_j \delta_{ki} + c_k \delta_{ij}) \tag{5.95d}$$

Here are some recurrence relations among these polynomials [deleting the subscript on $H_i^{(n)}(\mathbf{c})$].

$$\frac{\partial}{\partial c_1} H^{(n)} = I_i H^{(n-1)} \tag{5.96a}$$
$$c_i H^{(n)} = H^{(n+1)} + I_i H^{(n-1)} \tag{5.96b}$$
$$c_i \frac{\partial}{\partial c_i} H^{(n)} = n H^{(n)} + 2 I H^{(n-2)} \tag{5.96c}$$

Here we have written $\mathbf{I}$ for the identity matrix. It is such that the expression $I_i A$ represents the sum of all products in which i is attached to $\mathbf{I}$. This operation is demonstrated in application of (5.96b) to the case $n = 2$:

$$c_i H_{jk}^{(2)} = H_{ijk}^{(3)} + \left(\sum I_i H^{(1)} \right)_{ijk}$$
$$H_{ijk}^{(3)} = c_i H_{jk}^{(2)} - (\delta_{ij} H_k^{(1)} + \delta_{ik} H_j^{(1)})$$
$$= c_i (c_j c_k - \delta_{jk}) - \delta_{ij} c_k - \delta_{ik} c_j$$

which agrees with (5.95d).

The expansion

The distribution $\bar{f}(\mathbf{x}, \bar{\mathbf{c}}, t)$ is expanded in Hermite polynomials as follows:

$$\bar{f}(\mathbf{x}, \bar{\mathbf{c}}, t) = \bar{f}^0(\bar{c}) \sum_{n=0}^{\infty} \frac{1}{n!} a_{\mathbf{i}}^{(n)}(\mathbf{x}, t) H_{\mathbf{i}}^{(n)}(\bar{\mathbf{c}}) \tag{5.97}$$

which in explicit form appears as

$$\bar{f} = \bar{f}^0 \left[a^{(0)} H^{(0)} + a_i^{(1)} H_i^{(1)} + \frac{1}{2!} a_{ij}^{(2)} H_{ij}^{(2)} + \frac{1}{3!} a_{ijk}^{(3)} H_{ijk}^{(3)} + \cdots \right] \tag{5.97a}$$

Inverting the series (5.98) gives

$$a_{\mathbf{i}}^{(n)}(\mathbf{x}, t) = \int \bar{f}(\mathbf{x}, \bar{\mathbf{c}}, t) H_i^{(n)}(\bar{\mathbf{c}}) \, d\mathbf{c} \tag{5.98}$$

Since $H_i^{(n)}(\bar{\mathbf{c}})$ are polynomials in the velocity $\bar{\mathbf{c}}$, the coefficients $a_i^{(n)}(\mathbf{x}, t)$ are moments of the distribution $\bar{f}$. Each such moment corresponds to a macroscopic fluid dynamic variable [recall (3.33)]. Thus, for example, we may make the following identifications:

$$a^{(0)} = 1$$
$$a_i^{(1)} = 0$$
$$a_{ij}^{(2)} = \frac{P_{ij} - p\delta_{ij}}{p}, \qquad \mathrm{Tr}\,\overline{\overline{a}}^{(2)} = 0 \qquad (5.99)$$
$$a_{ijj}^{(3)} = \frac{2Q_i}{p\sqrt{RT}}$$

(a repeated index gives the trace).

Macroscopic equations

With (5.97), we see that construction of the distribution $\bar{f}(\mathbf{x}, \bar{\mathbf{c}}, t)$ depends on knowledge of the tensor coefficients $a_i^{(n)}$. These coefficients are obtained as follows. The series (5.97) is substituted into the Boltzmann equation (2.14). Following this substitution, we operate on the equation with $\int d\bar{\mathbf{c}}\, H_i^{(n)}(\bar{\mathbf{c}})$ to obtain (with, we recall, $C = \sqrt{RT}$)

$$\frac{Da^{(n)}}{Dt} + \left(\frac{\partial \mathbf{u}}{\partial x_i}\right) a_i^{(n)} + \frac{n}{2C^2}\left(\frac{DC^2}{Dt}\right) a^{(n)} + C\frac{\partial a_i^{(n+1)}}{\partial x_i}$$
$$+ Ca_i^{(n+1)}\frac{\partial}{\partial x_i}\ln(\rho C^{n+1}) = C\sum a^{(n-1)}$$
$$+ \left[\frac{1}{C}\frac{D\mathbf{u}}{Dt} + C\sum \ln(\rho C^{n+1})\right] a^{(n-1)} + \frac{1}{C}\left(\frac{\partial C^2}{\partial x_1}\right)\mathbf{I}a_i^{(n+1)}$$
$$+ \left(\frac{1}{C^2}\frac{DC^2}{Dt}\mathbf{I} + \sum \mathbf{u}\right)a^{(n-2)} + \frac{1}{C}\left(\sum C^2\right)\delta a^{(n-3)} = J^{(n)} \qquad (5.100)$$

Note that all terms in this equation have dimensions of t^{-1} and the nth-order tensor polynomials. [The coefficient n in the third term of (5.100) is tensor order, not number density.] Subscripts on tensor terms are tacitly assumed save for repeated indexes, in which case products are summed. Indexes of the **I** term are unequal to remaining indexes of a term. Thus, for example, consider the term

$$\frac{\partial}{\partial x_i}C^2\mathbf{I}a_i^{(n+1)} = \frac{\partial}{\partial x_i}C^2\delta_{lk}a_{imro\cdots}^{(n-1)} = T_{klmro\cdots}^{(n)}$$

where $T^{(n)}$ is written for an nth-order tensor.

The convective derivative is D/Dt, while the symbol $\sum$ represents the symmetric sum of gradients. For instance,

$$\sum a^{(2)} \equiv \frac{\partial a_{ij}^{(2)}}{\partial x_k} + \frac{\partial a_{kl}^{(2)}}{\partial x_j} + \frac{\partial a_{jk}^{(2)}}{\partial x_i} \qquad (5.101)$$

The collision term $J^{(n)}$ is written for

$$J^{(n)} \equiv \frac{n}{2} \int \bar{f}(\bar{\mathbf{c}}) \bar{f}(\bar{\mathbf{c}}_1) I^{(n)}(\bar{\mathbf{c}}, \bar{\mathbf{c}}_1) \, d\bar{\mathbf{c}} \, d\bar{\mathbf{c}}_1 \qquad (5.102)$$

where

$$I^{(n)} \equiv \int g\sigma \, d\Omega [H^{(n)}] \qquad (5.102a)$$

$$[H^{(n)}] \equiv H^{(n)}(\bar{\mathbf{c}}') + H^{(n)}(\bar{\mathbf{c}}_1') - H^{(n)}(\bar{\mathbf{c}}) - H^{(n)}(\bar{\mathbf{c}}_1)$$

Nature of approximation

The technique of solution is to approximate the series (5.97) with a finite amount of terms. Inasmuch as the leading term in the expansion is the local Maxwellian, the closer the has is to the equilibrium state, the less terms suffice to describe the state.

The simplest approximation is the "second-order" one, where we set

$$\bar{f} = \bar{f}^0 \left[1 + \frac{1}{2} a_{ij}^{(2)} H_{ij}^{(2)} \right] \qquad (5.103)$$

Dropping all terms containing $a^{(3)}$ in (5.100) gives

$$\frac{\partial a_{ij}^{(2)}}{\partial t} + u_r \frac{\partial}{\partial x_r} a_{ij}^{(2)} + a_{ir}^{(2)} \frac{\partial u_j}{\partial x_r} + a_{ri}^{(2)} \frac{\partial u_j}{\partial x_r}$$

$$+ \frac{\partial u_i}{\partial x_j} + \frac{\partial u_j}{\partial x_i} + [a_{ij}^{(2)} + \delta_{ij}] \frac{1}{C^2} \frac{DC^2}{Dt} = J_{ij}^{(2)} \qquad (5.104)$$

If all terms beyond $a^{(2)}$ are neglected in the calculation of $J^{(2)}$, then this last equation involves $\{\rho, \mathbf{u}, T, a_{ij}^{(2)}\}$ or, equivalently, $\{\rho, \mathbf{u}, T, P_{ij}\}$ Inasmuch as $a_{ij}^{(2)}$ is a symmetric, traceless tensor, there are only six relevant components of $\overline{\overline{a}}^{(2)}$. These, together with the five variables $\{\rho, \mathbf{u}, T\}$, constitute eleven scalar quantities. These eleven variables define the state of the system in the second-order approximation.

Equation (5.104) comprises six scalar equations. These equations together with the continuity equation

$$\frac{\partial \rho}{\partial t} + \nabla \cdot \rho \mathbf{u} = 0$$

the momentum equation

$$\rho \left(\frac{\partial \mathbf{u}}{\partial t} + \mathbf{u} \cdot \nabla \mathbf{u} \right) + \nabla \cdot \overline{\overline{P}} = 0$$

and the condition $\mathrm{Tr} \, \overline{\overline{a}}^{(2)} = 0$ are eleven scalar equations, which serve to close the system.

3.6   The Linear Boltzmann Collision Operator

We return to the linear collision operator (5.25) introduced in the Chapman-Ernskog solution of the Boltzmann equation.

$$\hat{\Box}\Phi \equiv \iint \sigma\, d\Omega g\, d\mathbf{v}_1 F^0(v_1)[\Phi_1' + \Phi' - \Phi_1 - \Phi] \tag{6.1}$$

As evidenced by the expressions (5.53) and (5.56), this operator plays a key role in evaluation of transport coefficients. In this section a number of basic properties of the $\hat{\Box}$ operator are derived.

3.6.1   Symmetry of the Kernel

Let us label

$$d\mu \equiv F^0(v)\, d\mathbf{v} \tag{6.2}$$

Then (6.1) may be written

$$\hat{\Box}\Phi = -\int d\mu_1 \hat{K}(\mathbf{v}, \mathbf{v}_1)\Phi(\mathbf{v}_1) \tag{6.3}$$

where the kernel $\hat{K}$ is as implied. We wish to show that $\hat{K}$ is symmetric; that is,

$$\hat{K}(\mathbf{v}, \mathbf{v}_1) = \hat{K}(\mathbf{v}_1, \mathbf{v}) \tag{6.4}$$

Let ϕ and ψ be any two elements of $\mathcal{L}_2$ space. The matrix element of the operator $\hat{\Box}$ with respect to these two functions is given by

$$\begin{aligned}
\Box_{\phi\psi} &= \int d\mu\phi\hat{\Box}\psi \\
&= -\iint d\mu\, d\mu_1 \phi(\mathbf{v})\hat{K}(\mathbf{v}, \mathbf{v}_1)\psi(\mathbf{v}_1) \\
&= -\iiint d\mu\, d\mu_1 \sigma\, d\Omega g\phi[\psi_1 + \psi - \psi_1' - \psi']
\end{aligned} \tag{6.5}$$

We first wish to establish that $\Box_{\phi\psi}$ is a symmetric matrix; that is,

$$\Box_{\phi\psi} = \Box_{\psi\phi} \tag{6.6}$$

Repeating arguments described in Section 3.1 on collisional invariants, we obtain

$$\Box_{\phi\psi} = -\frac{1}{4}\iiint d\mu\, d\mu_1 \sigma\, d\Omega g(\phi_1 + \phi - \phi_1' - \phi')(\psi_1 + \psi - \psi_1' - \psi') \tag{6.7a}$$

This may be rewritten

$$\Box_{\phi\psi} = -\iiint d\mu\, d\mu_1 \sigma\, d\Omega g\psi(\phi_1 + \phi - \phi_1' - \phi') = \Box_{\psi\phi} \tag{6.7b}$$

which establishes the validity of (6.6).

With (6.7a), we may conclude that the diagonal elements $\Box_{\phi\phi}$ are negative except when ϕ is a linear combination of summational invariants. That is,

$$\Box_{\phi\phi} \leq 0 \tag{6.8a}$$

$$\Box_{\phi\phi} = 0 \qquad \text{only if } \phi = \alpha + \boldsymbol{\beta} \cdot \mathbf{v} + \gamma v^2 \tag{6.8b}$$

Proceeding with our derivation, we consider the difference

$$0 = \Box_{\psi\phi} - \Box_{\phi\psi} = \iint d\mu_1\, d\mu(\phi \hat{K}\psi_1 - \psi \hat{K}\phi_1) \tag{6.9}$$

Interchanging $\mathbf{v}$ and $\mathbf{v}_1$ in the second integral permits (6.9) to be written

$$0 = \iint d\mu_1\, d\mu[\hat{K}(\mathbf{v}, \mathbf{v}_1) - \hat{K}(\mathbf{v}_1, \mathbf{v})]\phi(\mathbf{v})\phi(\mathbf{v}_1) \tag{6.10}$$

Here it is understood that $\hat{K}(\mathbf{x}, \mathbf{y})$ operates only on functions of $\mathbf{y}$.

Introducing the operator

$$\hat{T}(\mathbf{v}, \mathbf{v}_1) \equiv \hat{K}(\mathbf{v}, \mathbf{v}_1) - \hat{K}(\mathbf{v}_1, \mathbf{v}) \tag{6.11}$$

permits (6.10) to be written

$$\iint d\mu\, d\mu_1 \hat{T}(\mathbf{v}, \mathbf{v}_1)G(\mathbf{v}, \mathbf{v}_1) = 0$$

$$G(\mathbf{v}, \mathbf{v}_1) \equiv \phi(\mathbf{v})\psi(\mathbf{v}_1) \tag{6.12}$$

Suppose $\hat{T}$ is not identically zero. This means that a function G exists such that $\hat{T}G \neq 0$ over a domain of $\mathbf{v}, \mathbf{v}_1$ phase space. It follows that there is a subdomain D of this domain in which $\hat{T}G$ does not change sign. Let $\tilde{G} \equiv G$ in D and $\tilde{G} \equiv 0$ elsewhere. Then for this choice of function $\int TG \neq 0$, which violates (6.12). We may conclude that the only way to ensure that $\int \hat{T}\psi\phi$ vanishes for all ψ and ϕ in L_2 is for $\hat{T}$ to vanish identically. Thus $\hat{K}$ is a symmetric operator.

3.6.2 Negative Eigenvalues

Let us suppose that the collision operator $\hat{\Box}$ has a discrete spectrum. Then we may write

$$\hat{\Box}\psi_n = \nu_n \psi_n \tag{6.13}$$

Due to the symmetry of $\hat{\Box}$ (6.6), the eigenfunctions ψ_n comprise an orthogonal sequence. Thus we obtain

$$\Box_{nm} = \int d\mu\, \psi_n \hat{\Box}\psi_m = \nu_m \int d\mu\, \psi_n \psi_m$$

$$= \nu_m \delta_{nm} \|\psi_m\|^2 \tag{6.14}$$

where $\|\psi_m\|$ represents the norm of ψ_m. The diagonal elements of $\hat{\Box}$ obey the properties (6.8), and we may write

$$0 \geq \Box_{nn} = \nu_n \int d\mu \psi_n^2 \equiv \nu_n \|\psi_n\|^2$$

so that

$$\nu_n \leq 0 \qquad\qquad (6.15)$$

The equality in this relation occurs if ψ_n is any of the five independent scalar summational invariants. Thus the $\hat{\Box}$ operator has a fivefold degenerate eigenvalue as the origin on the real ν line. All remaining eigenvalues are negative. We conclude that the eigenvalues of $\hat{\Box}$ lie on the negative real axis with a fivefold degeneracy as the origin.

The significance of the negative quality of these eigenvalues is seen from the following discussion. Let a spatially homogeneous gas suffer a small displacement from equilibrium in velocity space. We recall the perturbation (5.30)

$$F = F^{(0)}(1 + \Phi) \qquad\qquad (6.16)$$

Substituting this form into the Boltzmann equation (2.14) and keeping terms linear in Φ, we obtain

$$\frac{\partial \Phi}{\partial t} = \hat{\Box}\Phi \qquad\qquad (6.17)$$

With this equation at hand, we may conclude that the nonpositive spectrum of $\hat{\Box}$ implies that Φ decays in time and that consequently the system returns to the Maxwellian state. This conclusion, stemming from the linearized Boltzmann equation (6.17), is seen to be consistent with our previous finding concerning the decay of Boltzmann $\mathcal{H}$ and accompanying reduction to the Maxwellian state, stemming from the nonlinear Boltzmann equation (Section 3.8).

3.6.3 Comparison of Boltzmann and Liouville Operators

It is interesting at this point to offer a brief comparison of properties of the Boltzmann collision operator (both linear $\hat{\Box}$ and nonlinear $\hat{J}$) with those of the Liouville operator $\hat{L}_N$ (2.1.3). The Liouville operator contains the full N-particle Hamiltonian, and the related Liouville equation is reversible. That is, if $F_N(\mathbf{x}^N, \mathbf{p}^N, t)$ is a solution, then so is $F_N(\mathbf{x}^N, -\mathbf{p}^N, -t)$ [recall Section 1.1.5]. We also found that eigenvalues of $\hat{L}_N$ are pure imaginary (see Section 2.3.1), which evidently permits oscillatory behavior of the N-particle distribution function. Furthermore, by virtue of the Poincaré recurrence theorem [Section 3.6], we may conclude that for a bounded system, solutions to the Liouville equation exist that, after sufficient time, return arbitrarily close to their initial values.

The Boltzmann collision operator, on the other hand, accounts only for two-particle interactions. Furthermore, owing to the assumption of molecular chaos $[F_2(1, 2) = F_1(1)F_1(2)]$, we may say that particles in collision lose memory of this collision after interaction. This property evidently contributes to irreversibility.

Distinctions between $\hat{J}(f)$ and $\hat{L}_N(f_N)$ were also evident in our preceding discussion of the $\mathcal{H}$ theorem (Sections 3.7 and 3.8). There we found that the Gibbs entropy, as implied by the Liouville equation, is constant, whereas the Boltzmann entropy, as implied by the Boltzmann equation, increases with time (recall that $S \propto -\mathcal{H}$).

3.6.4   Hard and Soft Potentials

We may distinguish between hard and soft interaction potentials in the following manner. Consider the set of potentials (1.17)

$$V(r) = Kr^{-N} \tag{6.18}$$

where in general the constant K is dependent on the number N. We have found previously (1.27) that for Maxwell molecules ($N = 4$) the integration measure $g\sigma\, d\cos\theta$ is independent of the relative speed g. Let us adopt the convention that potentials with $N > 4$ be labeled *hard* and those with $N < 4$ be labeled *soft*.

The separation potential with $N = 4$ corresponding to Maxwell molecules permits exact evaluation of the spectrum of the $\hat{\Box}$ operator. Results of this calculation are prescribed in the following section.

3.6.5   Maxwell Molecule Spectrum

As noted above, for Maxwell molecules the integration measure $g\sigma\, d\cos\theta$ is independent of relative speed g. For this case, the eigenvalue spectrum of the $\hat{\Box}$ operator is discrete. Eigenvalues and eigenfunctions for this case were first discovered by S. C. Wang Chang and G. E. Uhlenbeck.[38]

Working in spherical coordinates (ξ, θ, ϕ), eigenfunctions of the $\hat{\Box}$ collision operator are given by

$$\psi_{nlm}(\xi, \theta, \phi) = N_{nlm} S_{l+(1/2)}^{(n)}(\xi^2)\xi^l Y_l^m(\theta, \phi) \tag{6.19}$$

---

[38]C. S. Wang Chang and G. E. Uhlenbeck (1952), ibid. These reports are reprinted in J. deBoer and G. E. Uhlenbeck, *Studies in Statistical Mechanics*, North-Holland, Amsterdam (1970). A concise description of these findings may be found in G. E. Uhlenbeck and G. W. Ford, *Lectures in Statistical Mechanics*, American Math. Soc., Providence, R. I. (1963). See also L. Waldman, *Hand. der Physik XIII*, 295 Springer-Verlag, Berlin (1958).

where $Y^m(\theta, \phi)$ are the spherical harmonics,[39] N_{nlm} are normalization constants, $S_{l+(1/2)}^{(n)}(x)$ are Sonine polynomials, encountered previously in Section 5.5 and defined by (3.54). With the integration measure $d\mu$ (6.2), the sequence (6.19) is an orthogonal set and is a basis of $\mathcal{L}_2$ space.

Eigenvalues corresponding to the functions (6.19) are given by

$$\langle \psi_{nlm} | \hat{\Box} | \psi_{nlm} \rangle = v_{nl} = 2\pi \int_0^\pi d\theta \sin\theta \, F(\theta) \left[\cos^{2n+l} \frac{\theta}{2} P_l \left(\cos\frac{\theta}{2} \right) \right.$$
$$\left. + \sin^{2n+l} \frac{\theta}{2} P_l \left(\sin\frac{\theta}{2} \right) - 1 - \delta_{n0}\delta_{l0} \right] \tag{6.20}$$

where $P_l(x)$ are Legendre polynomials (see Table 3.6) and $F(\theta)$ is given by [recall (1.27)]

$$g\sigma(g, \theta) = \sqrt{\frac{2K}{\mu}} F(\theta), \qquad F(\theta) = \frac{b \, db}{d \cos\theta} \tag{6.21}$$

Since the azimuthal m number is absent from v_{nl}, we see that each eigenvalue is $2l + 1$ degenerate.[40]

As is evident from (1.23), all classical total cross sections corresponding to potentials with unbounded range are infinite. This singular behavior resides in the eigenvalues (6.20) through their dependence on $F(\theta)$, whose properties are now discussed.

The function $F(\theta)$ was first evaluated by Maxwell.[41] It is given by

$$F(\theta) = \frac{1}{4} \frac{(\cos 2\bar{\theta})^{1/2}}{\sin\theta \sin 2\bar{\theta} [\cos^2 \bar{\theta} \, \bar{K}(\sin\bar{\theta}) - \cos 2\bar{\theta} \, \bar{E}(\sin\bar{\theta})]} \tag{6.22}$$

where

$$\psi = \frac{\pi - \theta}{2} = (\cos 2\bar{\theta})^{1/2} \hat{K}(\sin\bar{\theta}) \tag{6.22a}$$

and $\bar{K}(x)$, $\bar{E}(x)$ are, respectively, complete elliptic integrals of the first and second kind.[42]

$$\bar{E}(k) = \int_0^{\pi/2} (1 - k^2 \sin^2\phi)^{1/2} \, d\phi, \qquad \bar{K}(k) = \int_0^{\pi/2} \frac{d\phi}{(1 - k^2 \sin^2\phi)^{1/2}}$$

---

[39] A concise list of properties of these functions may be found in R. L. Liboff, *Introductory Quantum Mechanics* (2002) ibid., Chapter 9.

[40] This degeneracy corresponds to the rotational symmetry of the $\hat{\Box}$ operator with respect to ϕ. Such degeneracy is often encountered in quantum mechanics; see R. L. Liboff, ibid.

[41] *Scientific Papers of James Clerk Maxwell*, W. D. Niven, (ed.), Dover, New York (1952). See Chapter XXVIII.

[42] E. T. Whittaker and G. N. Watson, *A course of Modern Analysis*, Cambridge University Pres, New York (1952), Chapter 22. See also M. Abramoxitz and I. A. Stegun, *Handbook of Mathematical Functions*, Dover, New York (1965), Chapter 17.

Evaluation of $F(\theta)$ as given by (6.22) indicates that it is a monotonic decreasing function of θ. For small θ, we obtain

$$F(\theta) \simeq \frac{(3\pi)^{1/2}}{16}\theta^{-5/2}\left(1 + \frac{35}{24\pi}\theta + \cdots\right) \qquad (6.23)$$

At $\theta = \pi$,

$$F(\pi) = \frac{1}{4\bar{K}^2(\pi/4)} = 0.0727$$

The result (6.23) indicates that $F(\theta)$ diverges at small θ, which, as noted above, is related to the unbounded range of $V(r)$. With the property (6.23), we see that the eigenvalue (6.20) likewise diverges at $\theta \simeq 0$. Resolution of this singular behavior is discussed at the close of this section.

3.6.6 Further Spectral Properties

We return to the notion of hard and soft potentials introduced above. Our discussion involves the collision frequency (5.2) now written relative to a Maxwellian state.

$$\nu(v) = \iint \sigma \, d\Omega g \, d\mathbf{v}_1 F^0(v_1) \qquad (6.24)$$

We first evaluate $\nu(v)$ for a gas of rigid spherical molecules of diameter σ. Recalling (1.33), we write

$$\sigma(\theta) = \frac{\sigma^2}{4} \qquad (6.25)$$

where σ is the diameter of a molecule. Inserting this value into (6.24) and performing the integration, we find (see Problem 3.12)

$$\nu(\xi) = \nu_0\left[e^{-\xi^2} + \left(2\xi + \frac{1}{\xi}\right)\frac{\sqrt{\pi}}{2}\,\mathrm{erf}(\xi)\right] \qquad (6.26)$$

where $\mathrm{erf}(\xi)$ represents the error function (see Appendix B):

$$\mathrm{erf}(\xi) \equiv \frac{2}{\sqrt{\pi}}\int_0^\xi e^{-x^2}\,dx$$

with ξ given by (5.28) and

$$\nu_0 \equiv n\sigma^2\sqrt{2\pi RT}$$

At large ξ, (6.26) gives

$$\nu(\xi) \sim \sqrt{\pi}\,\nu_0\xi, \qquad \xi \gg 1 \qquad (6.27)$$

whereas at $\xi \simeq 0$

$$\mathrm{erf}(\xi) = \frac{2}{\sqrt{\pi}}\left(\xi - \frac{\xi^3}{3} + \cdots\right)$$

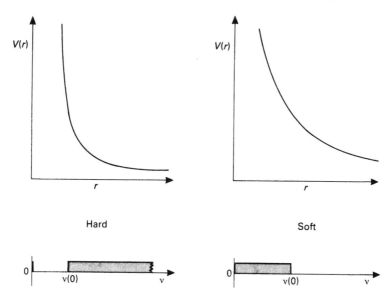

FIGURE 3.21. Sketch illustrating the eigenvalue spectra of $-\hat{\square}$ for hard and soft potentials, respectively. For hard potentials, the fivefold degenerate null eigenvalue is seen to be isolated from the continuum.

which gives

$$\nu(0) = 2\nu_0 \tag{6.28}$$

We may attribute this persistence of collision frequency for a stationary particle to the fact that the particle is situated in a gaseous medium at finite temperature. Note in particular that, as $T \rightarrow 0$, $\nu(0) \rightarrow 0$.

The preceding results demonstrate that for a gas of rigid sphere molecules $\nu(\xi)$ ranges from $\nu(0) = 2\nu_0 > 0$ to infinity.

A similar property for hard potentials, in general, has been established by Grad.[43] These properties address the spectrum of the linear collision operator, $\hat{\square}$. They are as follows:

1. For hard potentials the null eigenvalue, $\nu = 0$ is isolated from the rest of the spectrum of $\hat{\square}$. In addition to $\nu = 0$, the spectrum of $-\hat{\square}$ includes a continuum, which is the values of the collision frequency $\nu(\xi)$ and, consistent with our finding above, ranges from a minimum $\nu(0)$ to infinity. For soft potentials, the eigenvalue continuum ranges from a maximum $\nu(0)$ to zero. These properties are sketched in Fig. 3.21.

---

[43]H. Grad, *Phys. Fluids* 6, 147 (1963), "Asymptotic Theory of the Boltzmann Equation, II," in *Rarefied Gas Dynamics Symposium*, vol. I, J. Laurmann (ed.), Academic Press, New York (1963).

We have noted previously that the eigenvalue v_n contributes a term like $\exp(-v_n t)$ to the solution of an initial-value problem. The same is true for a continuous spectrum, where the exponential term contributes to an integral in place of a sum. Thus, for hard potentials, where v is bounded away from zero, the slowest decay of the distribution goes as $\exp[-v(0)t]$. For soft potentials (such as for the Coulomb or Newtonian interaction), v is not bounded away from zero, and we may expect long-time nonexponential decay for such cases.

2. Write $\hat{\Box}$ in the form

$$\hat{\Box} = \hat{K} - \mathbf{v}(\xi) \tag{6.29}$$

where $v(\xi)$ is given by (6.24). Then

(a) The operator $\hat{K}$ is bounded. That is, for any ψ in $\mathcal{L}_2$ a constant M exists such that $\|\hat{K}\psi\| \leq M\|\psi\|$.

(b) The operator $\hat{K}$ is completely continuous. A bounded linear operator $\hat{K}$ is completely continuous if, for any bounded sequence $\{\psi_n\}$ of elements of $\mathcal{L}_2$ (that is, $\|\psi_n\| \leq R$ for some R and all n), the sequence $\{\hat{K}\psi_n\}$ contains at least one convergent subsequence. As a consequence of this property and the *spectrum theorem*[44] $\hat{K}$ has a discrete point spectrum with the origin as the only accumulation point.

3. In Section 6.5, we found that the unbounded range of the Maxwell molecule force law gave rise to singular eigenvalues. This singular behavior is circumvented if the force law is cut off at some finite range. The resulting eigenfunctions are tensor Hermite polynomials, $H_{\mathbf{i}}^{(n)}$, encountered earlier in Section 5.10. Eigenvalues of the collision operator $\hat{\Box}$ may then be written

$$v_{\mathbf{i}}^{(n)} = \langle H_{\mathbf{i}}^{(n)} | \hat{\Box} | H_{\mathbf{i}}^{(n)} \rangle \tag{6.30}$$

Tensor Hermite polynomials are relevant to expansion of the velocity component of the distribution function in Cartesian coordinates. Sonine polynomials [as seen in (6.19)] are relevant to expansion in spherical coordinates. Thus, for example, working with the nondimensional distribution, (5.90), these respective expansions appear as

$$\bar{f}(\mathbf{x}, \bar{\mathbf{c}}, t) = \sum_{n,l,m} \frac{n!}{\Gamma(n+l+(3/2))} a_{nlm}(\mathbf{x}, t) e^{-\bar{c}} S_{l+(1/2)}^{(n)}(\bar{c}^2) \bar{c}^l Y_l^m(\theta, \phi) \tag{6.31a}$$

$$\bar{f}(\mathbf{x}, \bar{\mathbf{c}}, t) = \sum_{n=0}^{\infty} \frac{1}{n!} a_{\mathbf{i}}^{(n)}(\mathbf{x}, t) \omega(\bar{c}) H_{\mathbf{i}}^{(n)}(\bar{\mathbf{c}}) \tag{6.31b}$$

---

[44]This theorem and related topics are discussed by H. H. Stone, *Linear Transformations in Hilbert Space and Their Application to Analysis*, American Mathematical Society, New York (1932); I. Stakgold, *Boundary Value Problems of Mathematical Physics*, vol. 1, Macmillan, New York (1967), Section 2.10.

In either representation, coefficients of expansion, a, are related to fluid dynamic variables. Factorial factors and weight functions in (6.31) are related to orthogonality properties [recall (5.54b) and (5.94)].

3.7   The Druyvesteyn Distribution[45]

3.7.1   Basic Parameters and Starting Equations

To this point in our discourse, we have considered only the kinetic theory of one-component fluids. As an application of the kinetic theory of a two-component fluid, we turn to an important problem related to the theory of conduction of electrons through a medium consisting of heavier ions or neutral particles.[46]

Our two basic assumptions are (1) electron–electron collisions may be neglected compared to electron–ion collisions, and (2) the background in distribution is given by the Maxwellian

$$F(v_1) = n \left(\frac{M}{2\pi k_B T} \right)^{3/2} \exp \left(-\frac{M v_i^2}{2k_B T} \right) \tag{7.1}$$

$$\int F(v_1)\, d\mathbf{v}_1 = n \tag{7.1a}$$

Here we have written M and $\mathbf{v}_1$ for ion mass and velocity, respectively, and n is ion number density.

Our starting equation is the Boltzmann equation (2.17) for a two-component fluid comprised of ions and electrons in the presence of an electric field $\mathcal{E}$. Neglecting electron–electron collisions, only one collision integral remains in (2.17) and, passing to the equilibrium limit, we obtain

$$\frac{e\mathcal{E}}{m} \cdot \frac{\partial f}{\partial \mathbf{v}} = \iint (f' F_1' - f F_1)\sigma g\, d\Omega\, d\mathbf{v}_1 \tag{7.2}$$

where $f(\mathbf{v})$ is the electron distribution function and e/m is electron charge-to-mass ratio. In this discussion we take $f(v)$, as $F(v)$ [see (7.1a)], to be normalized to n.

---

[45]Named for M. J. Druyvesteyn, *Physica 10*, (1930). For further discussion of this topic, see S. Chapman and T. G. Cowling, *The Mathematical Theory of Non-Uniform Gases*, (1974) ibid.; L. B. Loeb, *Fundamental Processes of Electrical Discharges in Gases*, Wiley, New York (1947); W. P. Allis, *Hand. d. Physik* vol. XXI, Springer-Verlag, Berlin (1956); M. J. Druyvesteyn and F. M. Penning, *Rev. Mod. Phys. 12*, 87 (1940).

[46]We will in general call these particles "ions."

Writing

$$a \equiv \frac{m}{m + M}$$

$$a_1 = \frac{M}{m + M} \tag{7.3}$$

so that

$$a + a_1 = 1$$

and

$$\mathbf{V} \equiv a\mathbf{v} + a_1\mathbf{v}_1 = \frac{m\mathbf{v} + M\mathbf{v}_1}{m + M} \tag{7.4}$$

is the velocity of the center of mass of a colliding electron–ion pair of particles. The kinematic relations connecting these velocities may be written

$$\begin{pmatrix} \mathbf{g} \\ \mathbf{V} \end{pmatrix} = \begin{pmatrix} 1 & -1 \\ a & a_1 \end{pmatrix} \begin{pmatrix} \mathbf{v} \\ \mathbf{v}_1 \end{pmatrix} \tag{7.5a}$$

$$\begin{pmatrix} \mathbf{g}' \\ \mathbf{V}' \end{pmatrix} = \begin{pmatrix} 1 & -1 \\ a & a_1 \end{pmatrix} \begin{pmatrix} \mathbf{v}' \\ \mathbf{v}_1' \end{pmatrix} \tag{7.5b}$$

$$\begin{pmatrix} \mathbf{v} \\ \mathbf{v}_1 \end{pmatrix} = \begin{pmatrix} a_1 & 1 \\ -a & 1 \end{pmatrix} \begin{pmatrix} \mathbf{g} \\ \mathbf{V} \end{pmatrix} \tag{7.5c}$$

$$\begin{pmatrix} \mathbf{v}' \\ \mathbf{v}_1' \end{pmatrix} = \begin{pmatrix} a_1 & 1 \\ -a & 1 \end{pmatrix} \begin{pmatrix} \mathbf{g}' \\ \mathbf{G} \end{pmatrix} \tag{7.5d}$$

Energy conservation gives

$$g = g' \tag{7.6a}$$

whereas conservation of momentum gives

$$\mathbf{V} = \mathbf{V}' \tag{7.6b}$$

The Jacobian of these transformations (7) are all seen to be unity. Thus, for example, we may write

$$d\mathbf{v}\, d\mathbf{v}_1 = d\mathbf{g}\, d\mathbf{V} \tag{7.7}$$

From the defining relations

$$\mathbf{g} = \mathbf{v} - v_1 \tag{7.8a}$$
$$\mathbf{g}' = \mathbf{v}' - \mathbf{v}_1' \tag{7.8b}$$
$$\mathbf{V} = a\mathbf{v}\, a_1\mathbf{v}_1 \tag{7.8c}$$
$$\mathbf{V}' = a\mathbf{v}' + a_1\mathbf{v}_1' \tag{7.8d}$$

we obtain

$$\mathbf{g} - \mathbf{g}' = \mathbf{v}' - \mathbf{v}_1' - \mathbf{v} - \mathbf{v}_1 \tag{7.9a}$$
$$0 = a\mathbf{v}' + a_1\mathbf{v}_1' - a\mathbf{v} - a_1\mathbf{v}_1' \tag{7.9b}$$

Multiplying (7a) by a and $-a_1$, respectively, and adding to (7b) gives

$$\mathbf{v}' - \mathbf{v} = a_1(\mathbf{g}' - \mathbf{g}) \tag{7.10a}$$
$$\mathbf{v}_1' - \mathbf{v}_1 = -a(\mathbf{g}' - \mathbf{g}) \tag{7.10b}$$

From the relations

$$\mathbf{v}' = \mathbf{V} + a_1\mathbf{g}' \tag{7.11a}$$
$$\mathbf{v} = \mathbf{V} + a_1\mathbf{g} \tag{7.11b}$$

we find

$$v'^2 = V^2 + 2a_1\mathbf{V} \cdot \mathbf{g}' + a_1^2 g'^2 \tag{7.12a}$$
$$v^2 = V^2 + 2a_1\mathbf{V} \cdot \mathbf{g} + a_1^2 g^2 \tag{7.12b}$$

Subtracting these equations and recalling the conservation equation (7.6) gives

$$v'^2 - v^2 = 2a_1\mathbf{V} \cdot (\mathbf{g}' - \mathbf{g}) \tag{7.13}$$

Our third assumption concerns the electron-to-ion mass ratio m/M. We set

$$\frac{m}{M} \simeq 1 - a_1 = \varepsilon^2 \tag{7.14}$$

where ε is a parameter of smallness. Finally, we assume thermal equilibrium between electrons and set

$$\frac{3}{2}k_B T = \left\langle \frac{1}{2}Mv_1^2 \right\rangle = \left\langle \frac{1}{2}mv^2 \right\rangle \tag{7.15}$$

With the preceding relations we find

$$v_1 \simeq \varepsilon v \tag{7.16}$$
$$a = \varepsilon^2$$
$$a_1 = 1 - \varepsilon^2 \tag{7.17}$$

And with (7.10) and (7.14) we obtain the order-of-magnitude relations

$$\mathbf{g} = \mathbf{v} + 0(\varepsilon) \tag{7.18a}$$
$$\mathbf{g}' = \mathbf{v}' + 0(\varepsilon) \tag{7.18b}$$
$$\mathbf{V} = \mathbf{v}_1 = \mathbf{v}_1' = 0(\varepsilon) \tag{7.18c}$$

See Fig. 3.22. Having obtained these kinematic relations, we turn next to the Legendre expansion of $f(\mathbf{v})$.

3.7.2 Legendre Polynomial Expansion

Let $\hat{\mathcal{E}}$ and $\hat{\mathbf{v}}$ denote unit vectors. Then we define

$$\mu = \hat{\mathcal{E}} \cdot \hat{\mathbf{v}} \tag{7.19}$$

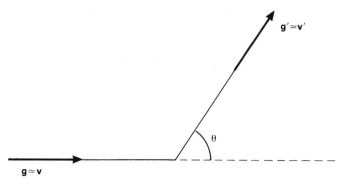

FIGURE 3.22. Approximate vector equalities in the scattering diagram for $m/M \ll$ 1. Note that the scattering element of the solid angle, $d\Omega$, lies in the direction of $\mathbf{g}'$.

With μ so defined, toward the end of solving (7.2), we introduce the following expansion of $f(\mathbf{v})$ in Legendre polynomials $P_l(\mu)$ (see Table 3.6).[47,48]

$$f(\mathbf{v}) = \sum_{l=0}^{\infty} P_l(\mu) f_l(v) \tag{7.20}$$

In the present study, $\mathcal{E}$ is taken to be a moderate field, in which case we expect small departure from isotropy of $f(\mathbf{v})$. Accordingly, we keep terms only to $l = 1$ in (7.20) and obtain

$$f(\mathbf{v}) = f_0(v) + \mu f_1(v) \tag{7.21}$$

Substituting this expansion into the left side of (7.2), we find (see Problem 3.13)

$$\frac{e\mathcal{E}}{m} \cdot \frac{\partial f}{\partial \mathbf{v}} = \frac{e\mathcal{E}}{m} \left[\mu \frac{df_0}{dv} + \frac{1}{3} \left(\frac{df_1}{dv} + \frac{2f_1}{v} \right) \right] \tag{7.22}$$

For the collision terms, we obtain

$$\begin{aligned}
\hat{J}(f, F) &= \iint (f' F_1' - f F_1) \sigma g \, d\Omega \, d\mathbf{v}_1 \\
&= \iint (f_0' F_1' - f_0 F_1) \sigma g \, d\Omega \, d\mathbf{v}_1 \\
&\quad + \iint (\mu' f_1' F_1' - \mu f_1 F_1) \sigma g \, d\Omega \, d\mathbf{v}_1
\end{aligned} \tag{7.23}$$

---

[47] This expansion for an anisotropic distribution function was first introduced by H. A. Lorentz, *The Theory of Electrons*, B. G. Teubner, Leipzig (1909). See Appendix D.

[48] The convergence of such expansions is discussed by W. P. Allis in *Electrical Breakdown and Discharges in Gases*, E. E. Kunhardt and L. H. Leussen (eds.), Plenum, New York (1983).

TABLE 3.6. Properties of the Legendre Polynomials

Generating formulas:

$$(1 - 2\mu s + s^2)^{-1/2} = \sum_{l=0}^{\infty} P_l(\mu)s^l$$

$$P_l(\mu) = \frac{1}{2^l l!} \frac{d^l}{d\mu^l}(\mu^2 - 1)^l \begin{cases} -1 \leq \mu \leq 1 \\ l = 0, 1, 2, 3, \ldots \end{cases}$$

Legendre's equation:

$$(1 - \mu^2)\frac{d^2 P_l(\mu)}{d\mu^2} - 2\mu\frac{d P_l(\mu)}{d\mu} + l(l+1)P_l(\mu) = 0$$

Recurrence relations:

$$(l+1)P_{l+1}(\mu) = (2l+1)\mu P_l(\mu) - l P_{l-1}(\mu)$$

$$(1 - \mu^2)\frac{d}{d\mu}P_l(\mu) = l\mu P_l(\mu) + l P_{l-1}(\mu)$$

Normalization and orthogonality:

$$\int_{-1}^{1} P_l(\mu)P_m(\mu)\,d\mu = \frac{2}{2l+1} \qquad (l = m)$$

$$= 0 \qquad (l \neq m)$$

First few polynomials:

$P_0 = 1$	$P_2 = \frac{1}{2}(3\mu^2 - 1)$	$P_4 = \frac{1}{8}(35\mu^4 - 30\mu^2 + 3)$
$P_1 = \mu$	$P_3 = \frac{1}{2}(5\mu^3 - 3\mu)$	$P_5 = \frac{1}{8}(63\mu^5 - 70\mu^3 + 15\mu)$

Special Values:

$$P_l(\mu)(-1)^l P_l(-\mu) \qquad P_l(1) = 1$$

where

$$\mu' = \hat{\mathcal{E}} \cdot \hat{\mathbf{v}}'$$

A note of caution is in order here: we should keep in mind that f_1 in (7.23) denotes the first-order correction to f as given in (7.21) so that (as opposed to Boltzmann notation) $f_1' \equiv f_1(v')$. However, in keeping with Boltzmann notation, $F_1 \equiv F(\mathbf{v}_1)$ and $F' = F(\mathbf{v}_1')$. In parallel form to (7.23), we set

$$\hat{J}(f \mid F) = \hat{J}_0(f_0 \mid F) + \hat{J}_1(f_1 \mid F) \tag{7.24}$$

where

$$\hat{J}_0(f_0 \mid F) = \iint (f_0' F_1' - f_0 F_1)\sigma g \, d\Omega \, d\mathbf{v}_1 \tag{7.25a}$$

$$\hat{J}_1(f_1 \mid F) = \iint (\mu' f_1' F_1' - \mu f_1 F_1)\sigma g \, d\mathbf{v}_1 \tag{7.25b}$$

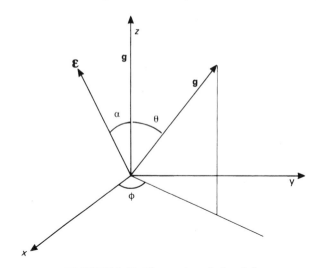

FIGURE 3.23. The angles of $\mathbf{g}'$ and $\mathcal{E}$.

Combining (7.2), (7.22), and (7.24) gives the expanded form

$$\frac{e\mathcal{E}}{m}\left[\mu\frac{df_0}{dv} + \frac{1}{3}\left(\frac{df_1}{dv} + \frac{2f_1}{v}\right)\right] = \hat{J}_0(f_0 \mid F) + \hat{J}_1(f_1 \mid F) \qquad (7.26)$$

We turn next to reduction of the $\hat{J}_1(f_1 \mid F)$ collision integral. With (7.6a) and (7.18a, b), we may write

$$|\mathbf{v} + 0(\varepsilon)| = |\mathbf{v}' + 0(\varepsilon)|$$

whence

$$v = v' + 0(\varepsilon)$$

and

$$f_1(v') = f_1(v) + 0(\varepsilon) \qquad (7.27a)$$

With (7.18c), we write

$$F_1' = F_1 + 0(\varepsilon) \qquad (7.27b)$$

Thus (7.25a) becomes

$$\hat{J}_1(f_1 \mid F) = nf_1 \int (\mu' - \mu)\sigma g \, d\Omega + 0(\varepsilon) \qquad (7.28a)$$

To further reduce this integral, we go into a representation where $\mathbf{g}$ is taken as the polar axis (see Fig. 3.23). We note [recall (7.19)]

$$\mu = \hat{\mathbf{v}} \cdot \hat{\mathcal{E}} = \hat{\mathbf{g}} \cdot \hat{\mathcal{E}} + 0(\varepsilon) = \cos\alpha + 0(\varepsilon) \qquad (7.29a)$$
$$\mu' = \hat{\mathbf{v}}' \cdot \hat{\mathcal{E}} = \hat{\mathbf{g}}' \cdot \hat{\mathcal{E}} + 0(\varepsilon) \qquad (7.29b)$$

From Fig. 3.23 we note the Cartesian components of $\hat{\mathcal{E}}$ and $\hat{\mathbf{g}}'$,

$$\hat{\mathcal{E}} = (\sin \alpha, 0, \cos \alpha)$$
$$\hat{\mathbf{g}}' = (\sin \theta \cos \phi, \sin \theta \sin \phi, \cos \theta)$$

where

$$\hat{\mathbf{g}}' \cdot \varepsilon = \cos \alpha \cos \theta + \sin \alpha \sin \theta \cos \phi \tag{7.30}$$

Substituting this value into (7.29b) permits the Ω integral of the μ' component of (7.28) to be written

$$\int \mu' \, d\Omega = \iiint \sin \theta \, d\theta \, d\phi (\cos \alpha \cos \theta + \sin \alpha \sin \theta \cos \phi) + 0(\varepsilon)$$

which reduces to

$$\int \mu' \, d\Omega = \int \mu \cos \theta \, d\Omega + 0(\varepsilon) \tag{7.31}$$

Substituting this result into (7.28) gives

$$\hat{J}_1(f_1 \mid F) = -n f_1 \mu \int (1 - \cos \theta) \sigma g \, d\Omega + 0(\varepsilon) \tag{7.32}$$

This finding may be written in the more serious form:

$$J_1 = -n f_1 \mu v Q + 0(\varepsilon) \tag{7.33}$$

where Q is the weighted cross section,

$$Q \equiv \int (1 - \cos \theta) \sigma \, d\Omega \tag{7.34}$$

[compare with (5.83)].

Further introducing the relaxation time τ and mean free path l [recall (2.2) and following],

$$\tau \equiv \frac{1}{nvQ}, \qquad l \equiv \frac{1}{nQ} = v\tau \tag{7.35}$$

allows (7.33) to be rewritten

$$\hat{J}_1(f_1 \mid F) = -\frac{\mu f_1}{\tau} + 0(\varepsilon) \tag{7.36}$$

With this result at hand, we return to the starting equation (7.26). Equating terms of equal Legendre polynomial order and keeping terms of $0(1)$, we find

$$\frac{e\mathcal{E}}{3m} \left(\frac{df_1}{dv} + \frac{2f_1}{v} \right) = \hat{J}_0(f_0 \mid F) \tag{7.37a}$$

$$\frac{e\mathcal{E}}{m} \frac{df_0}{dv} = -\frac{f_1}{\tau} \tag{7.37b}$$

3.7.3   Reduction of $\hat{J}_0(f_0 \mid F)$

It is evident from the preceding two equations that elimination of f_1 would lead to a second-order integrodifferential equation for f_0. However, an alternative path leads to a simpler relation. We note that (7.37a) may be rewritten

$$\frac{e\mathcal{E}}{3m} \frac{1}{v^2} \frac{d}{dv}(v^2 f_1) = \hat{J}_0(f_0 \mid F) \tag{7.38}$$

We wish to integrate this equation from $v = 0$ to a finite value, which we label $\tilde{v}$. Introducing the notation

$$\int_{v < \tilde{v}} d\mathbf{v} \equiv \int_0^{\tilde{v}} \int dv v^2 \, d\Omega$$

there results

$$I \equiv \frac{e\mathcal{E}}{3m} \int_{v < \tilde{v}} \frac{1}{v^2} \frac{d}{dv}[v^2 f_1(v)] \, d\mathbf{v} = \iiint_{v < \tilde{v}} (f_0' F_1' - f_0 F_1)\sigma g \, d\Omega \, d\mathbf{v}_1 \, d\mathbf{v} \tag{7.39}$$

Switching primed and unprimed velocities in (7.39) gives

$$\iiint_{v < \tilde{v}} f_0' F_1' \sigma g \, d\Omega \, d\mathbf{v}_1 \, d\mathbf{v} = \iiint_{v' < \tilde{v}} f_0 F_1 \sigma g \, d\Omega \, d\mathbf{v}_1 \, d\mathbf{v} \tag{7.40}$$

Here we have noted the invariance of the measure $\sigma g \, d\Omega \, d\mathbf{v}_1 \, d\mathbf{v}$ under the said transformation. Now note that with

$$v' = v - \Delta v$$

the condition $v' < \tilde{v}$ becomes

$$v < \tilde{v} + \Delta v \tag{7.40a}$$

where Δv is to be determined. Writing

$$d\mathbf{v} = v^2 \, dv \, d\Omega_v$$

together with the preceding equation, we find

$$I = \iiint \left[\int_{\tilde{v}}^{\tilde{v} + \Delta v} f_0 F_1 \sigma g v^2 \, dv \right] d\Omega \, d\Omega_v \, d\mathbf{v}_1 \tag{7.41}$$

To evaluate Δv, we recall (7.13)

$$v'^2 - v^2 = 2a_1 \mathbf{V} \cdot (\mathbf{g}' - \mathbf{g})$$

and (7.18)

$$\mathbf{V} = \mathbf{v}_1 = 0(\varepsilon)$$

To keep this order of magnitude in mind, we write

$$\mathbf{v}_1 \to \varepsilon \mathbf{v}_1$$

(and eventually set $\varepsilon = 1$). To this same order, we find

$$v'^2 - v^2 = 2a_1\varepsilon \mathbf{v}_1 \cdot (\mathbf{v}' - \mathbf{v})$$
$$= 2(1 - \varepsilon^2)\varepsilon \mathbf{v}_1 \cdot (\mathbf{v}' - \mathbf{v})$$
$$v'^2 = v^2 + 2\varepsilon \mathbf{v}_1 \cdot (\mathbf{v}' - \mathbf{v})$$
$$v' = v + \varepsilon \mathbf{v}_1 \cdot (\hat{\mathbf{v}}' - \hat{\mathbf{v}}) + 0(\varepsilon^2)$$

Thus the condition $v' < \tilde{v}$ is equivalent to

$$v + \varepsilon \mathbf{v}_1 \cdot (\hat{\mathbf{v}}' - \hat{\mathbf{v}}) < \tilde{v}$$

or

$$v < \tilde{v} + \varepsilon \mathbf{v}_1 \cdot (\hat{\mathbf{v}} - \hat{\mathbf{v}}')$$

which, with (7.40a), implies that

$$\Delta v = \varepsilon \mathbf{v}_1 \cdot (\hat{\mathbf{v}} - \hat{\mathbf{v}}')$$

That is, Δv is an $0(\varepsilon)$ quantity. Recalling our previous rule, we set $\varepsilon = 1$ to obtain

$$\Delta v = \mathbf{v}_1 \cdot (\hat{\mathbf{v}} - \hat{\mathbf{v}}') \tag{7.42}$$

3.7.4 Evaluation of I_0 and I_1 Integrals

Having obtained this expression for Δv, we return to (7.41) and expand $f_0(v)$ about $v = \tilde{v}$. As the integral is evaluated over a small interval about $\tilde{v}$, we write

$$f_0(v) = f_0(\tilde{v}) + (v - \tilde{v})\frac{df_0}{dv}\bigg|_{\tilde{v}} + \cdots$$

Substituting this expansion into (7.41) gives corresponding I_0 and I_1 integrals. We first examine I_0.

$$I_0 = f_0(\tilde{v}) \iiint \left[\int_{\tilde{v}}^{\tilde{v}+\Delta v} F_1 \sigma g v^2 \, dv \right] d\Omega \, d\Omega_v \, d\mathbf{v}_1 \tag{7.43}$$

$$I_0 \equiv f_0(\tilde{v})\psi(\tilde{v})$$

where $\psi(\tilde{v})$, as implied, is independent of electric field $\mathcal{E}$.

Next we turn to evaluation of I_1, given by

$$I_1 = \iiint \left[\int_{\tilde{v}}^{\tilde{v}+\Delta v} (v - \tilde{v})\frac{df_0}{d\tilde{v}} F_1 \sigma g v^2 \, dv \right] d\Omega \, d\Omega_v \, d\mathbf{v}_1 \tag{7.44}$$

where we have set

$$\frac{df_0}{dv}\bigg|_{\tilde{v}} \equiv \frac{df_0}{d\tilde{v}}$$

Note that with (7.18)

$$gv^2 = v^3 + 0(\varepsilon)$$

Furthermore, we write

$$\sigma(\mathbf{g}') = \sigma(v, \hat{\mathbf{g}}') + 0(\varepsilon)$$

and expand the v-dependent part of σ about $v = \tilde{v}$.

$$\sigma(v) = \sigma(\tilde{v}) + \frac{d\sigma}{d\tilde{v}}(v - \tilde{v}) + \cdots$$

There results

$$\int_{\tilde{v}}^{\tilde{v}+\Delta v} (v - \tilde{v})v^3 \sigma(\mathbf{g}')\, dv = \sigma(\tilde{v}) \int_{\tilde{v}}^{\tilde{v}+\Delta v} (v - \tilde{v})v^3 \, dv$$

$$+ \frac{d\sigma}{d\tilde{v}} \int_{\tilde{v}}^{\tilde{v}+\Delta v} (v - \tilde{v})v^3 (v - \tilde{v})\, dv$$

However, note that over the interval of integration

$$v - \tilde{v} \le \Delta v$$

Thus

$$\int_{\tilde{v}}^{\tilde{v}+\Delta v} (v - \tilde{v})v^3 (v - \tilde{v})\, dv \le \Delta v \int_{\tilde{v}}^{\tilde{v}+\Delta v} (v - \tilde{v})\, dv$$

And we may write

$$\int_{\tilde{v}}^{\tilde{v}+\Delta v} (v - \tilde{v})v^3 \sigma(\mathbf{g}')\, dv = \sigma(\tilde{v}) \int_{\tilde{v}}^{\tilde{v}+\Delta v} (v - \tilde{v})v^3 \, dv + 0(\varepsilon)$$

$$\simeq \sigma(\tilde{v})\frac{\tilde{v}^3}{2}(\Delta v)^2 \tag{7.45}$$

providing[49]

$$\left\| \frac{d\sigma}{d\tilde{v}} \right\| \le \| \sigma \|$$

Substituting the result (7.45) into (7.44) yields

$$I_1 = \frac{\tilde{v}^3}{2}\frac{df_0}{d\tilde{v}} \iiint F_1 \sigma(\tilde{v}, \hat{\mathbf{g}}')[\mathbf{v}_1 \cdot (\hat{\mathbf{v}} - \hat{\mathbf{v}}')]^2 \, d\Omega \, d\Omega_v \, dv_1 \tag{7.46}$$

Now note that we may write

$$[\mathbf{v}_1 \cdot (\hat{\mathbf{v}} - \hat{\mathbf{v}}')]^2 = v_1^2(\hat{\mathbf{v}} - \hat{\mathbf{v}}') \cdot \hat{\mathbf{v}}_1 \hat{\mathbf{v}}_1 \cdot (\hat{\mathbf{v}} - \hat{\mathbf{v}}')$$

---

[49]Note that this condition fails for the Coulomb cross section (1.31) at low energies.

With the relation (see Problem 3.15)

$$\int \hat{\mathbf{v}}_1 \hat{\mathbf{v}}_1 \, d\mathbf{v}_1 = \frac{\overline{\overline{I}}}{3} \int d\mathbf{v}_1$$

(7.46) may be rewritten

$$\begin{aligned}
I_1 &= \tilde{v}^3 \frac{df_0}{d\tilde{v}} \iiint F_1 \sigma \frac{v_1^2}{6} (\hat{\mathbf{v}} - \hat{\mathbf{v}}') \cdot (\hat{\mathbf{v}} - \hat{\mathbf{v}}') \, d\Omega \, d\Omega_v \, d\mathbf{v}_1 \\
&= \tilde{v}^3 \frac{df_0}{d\tilde{v}} \iiint F_1 \sigma \frac{v_1^2}{3} (1 - \hat{\mathbf{v}} \cdot \hat{\mathbf{v}}') \, d\Omega \, d\Omega \, d\mathbf{v}_1
\end{aligned} \tag{7.47}$$

Noting

$$\hat{\mathbf{v}} \cdot \hat{\mathbf{v}}' = \hat{\mathbf{g}} \cdot \hat{\mathbf{g}}' + 0(\varepsilon) = \cos\theta + 0(\varepsilon)$$

and writing

$$\int d\Omega_v \int F_1(v_1) \frac{M}{2} v_1^2 \, d\mathbf{v}_1 = 4\pi \left(\frac{3}{2}\right) nk_B T$$

gives, finally,

$$I_1 = \tilde{v}^3 \frac{df_0}{d\tilde{v}} 4\pi \frac{nk_B T}{M} \int \sigma(1 - \cos\theta) \, d\Omega$$

which, with (7.34), may be written

$$I_1 = \tilde{v}^3 \frac{df_0}{d\tilde{v}} \frac{nk_B T}{M} 4\pi Q$$

Recalling (7.35), $Q = 1/nl$, the preceding relation becomes

$$I_1 = 4\pi \frac{k_B T}{M} \frac{\tilde{v}^3}{l} \frac{df_0}{d\tilde{v}} \tag{7.48}$$

With (7.39) we write

$$\frac{4\pi}{3} \frac{e\mathcal{E}}{m} \tilde{v}^2 f_1(\tilde{v}) = I_0 + I_1 = f_0(\tilde{v})\psi(\tilde{v}) + 4\pi \frac{k_B T}{M} \frac{\tilde{v}^3}{l} \frac{df_0}{d\tilde{v}} \tag{7.49}$$

Let us find an expression for $\psi(\tilde{v})$. In that this function is independent of $\mathcal{E}$, we may evaluate it at $\mathcal{E} = 0$. At this value

$$f_0(\tilde{v}) = n \left(\frac{m}{2\pi k_B T}\right)^{3/2} \exp\left(-\frac{m\tilde{v}^2}{2k_B T}\right)$$

so that

$$\frac{df_0}{d\tilde{v}} = -\frac{m\tilde{v}}{k_B T} f_0$$

With this value substituted into (7.49), we find

$$\psi(\tilde{v}) = 4\pi \frac{m}{M} \frac{\tilde{v}^4}{l} \tag{7.50}$$

Inserting this value into (7.49) with $\mathcal{E} \neq 0$ gives the desired equation (replacing $\tilde{v}$ with v):

$$\frac{1}{3} \frac{e\varepsilon}{m} f_1 = \frac{k_B T}{M} \frac{v}{l} \frac{df_0}{dv} + \frac{m}{M} \frac{v^2}{l} f_0 \tag{7.51}$$

which is equivalent to the integration of (7.37a).

3.7.5 Druyvesteyn Equation

Having reduced (7.37a) to (7.51), we are now prepared to obtain a single equation for f_0. Substituting (7.51) into (7.37b) gives

$$\left[\frac{1}{3} \frac{l}{v} \left(\frac{e\mathcal{E}}{m} \right)^2 + \frac{k_B T}{M} \frac{v}{l} \right] \frac{df_0}{dv} + \frac{m}{M} \frac{v^2}{l} f_0 = 0 \tag{7.52}$$

where we have written, with (7.35),

$$v = \frac{l}{\tau}$$

Introducing the speeds

$$C_1^2 \equiv \frac{k_B T}{m} \qquad C_2^4 = \frac{1}{3} \frac{M}{m} \left(\frac{e\mathcal{E}l}{m} \right)^2$$

permits (7.52) to be rewritten

$$(C_2^4 + C_1^2 v^2) \frac{df_0}{dv} = -v^3 f_0$$

With the mean free path l given by (7.35), we may not, in general, assume that it is independent of v. Integrating (7.53) gives the *Druyvesteyn distribution*

$$\boxed{f_0(v) = K \exp\left(-\int \frac{v^3 \, dv}{C_2^4 + v^2 C_1^2} \right)} \tag{7.53}$$

where K is a normalization constant.

The Druyvesteyn equation (7.52) may be cast in still simpler form when written in terms of the nondimensional electric field

$$s \equiv \frac{M}{6m} \left(\frac{e\mathcal{E}l}{k_B T} \right)^2 \tag{7.53a}$$

and nondimensional energy

$$x \equiv \frac{mv^2}{2k_B T} \tag{7.53b}$$

With these parameters at hand, (7.52) becomes

$$(s + x) \frac{df_0}{dx} + x f_0 = 0 \tag{7.54}$$

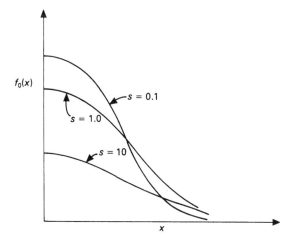

FIGURE 3.24. Sketch of the variation of the Druyvesteyn distribution with varying electric field (s).

which, assuming l to be independent of velocity, is simply integrated to yield that which we term the *absolute Druyvesteyn distribution*:

$$f_0(x) = K(x+s)^s e^{-x} \qquad (7.55)$$

A sketch of this distribution is given in Fig. 3.24 at various values s, illustrating the manner in which the spread of the energy distribution increases with electric field. This behavior evidently stems from acceleration of electrons due to the electric field.

Review of assumptions

A brief recapitulation of the assumptions that led to the Druyvesteyn distribution (7.53) is in order at this point. These are as follows:

1. Neglect electron–electron collisions.
2. Background ions are in a Maxwellian distribution (7.1).
3. $m/M \ll 1$ (7.14).
4. Moderate electric field. Small anisotropy in $f(\mathbf{v})$ (7.21).
5. Introduction of phenomenological coefficients τ and l (7.35), with $v \simeq l/\tau$.
6. Electrons scatter elastically from ions.

3.7.6 Normalization, Velocity Shift, and Electrical Conductivity

The normalization of $f_0(x)$ is obtained as follows. Returning to (7.21) and with (7.37b), we write

$$f(v) = f_0(v) - \mu\tau \frac{e\mathcal{E}}{m} \frac{df_0}{dv} \qquad (7.56)$$

Thus

$$n = \iint \left(f_0(v) - \mu\tau \frac{e\mathcal{E}}{m} \frac{df_0}{dv} \right) v^2 \, dv \, d\Omega$$

The second term vanishes and we obtain

$$n = 4\pi \int_0^\infty f_0(v) v^2 \, dv \tag{7.57}$$

Converting to x notation, (7.53b), we find

$$\frac{3}{2} x^{1/2} \, dx + \left(\frac{m}{2k_B T} \right)^{3/2} 3v^2 \, dv$$

so that

$$n = \frac{4\pi}{2} \left(\frac{2k_B T}{m} \right)^{3/2} \int f_0(x) \sqrt{x} \, dx \tag{7.58}$$

With (7.55), we write

$$n = 2\pi \left(\frac{2k_B T}{m} \right)^{3/2} K \int_0^\infty (x + s) e^{-x} \sqrt{x} \, dx \tag{7.59}$$

which determines the normalization constant K. At $s = 0$ (no electric field), (7.59) gives

$$n = 2\pi \left(\frac{2k_B T}{m} \right)^{3/2} K \frac{\sqrt{\pi}}{2}$$

$$n = \left(\frac{2\pi k_B T}{m} \right)^{3/2} K$$

$$K = \frac{N}{(2\pi k_B T/m)^{3/2}}$$

which is the correct normalization for the Maxwellian [see (7.1)].

Finally, we turn to the shift in electron drift velocity and closely allied electrical conductivity due to the presence of the electric field. With (3.13), we write

$$n\mathbf{u} = \int f(\mathbf{v})\mathbf{v} \, d\mathbf{v} \tag{7.60}$$

Substituting the form for $f(\mathbf{v})$ as given by (7.56) into the preceding gives

$$n\mathbf{u} = \int f_0(v)\mathbf{v} \, d\mathbf{v} - \frac{e\mathcal{E}}{m} \int \tau\mu \frac{df_0}{dv} \mathbf{v} \, d\mathbf{v}$$

As f_0 is rotationally symmetric, the first term in the preceding equation vanishes, and we obtain

$$n\mathbf{u} = -\frac{e\mathcal{E}}{m} \int_0^\infty \tau \, dv v^3 \frac{df_0}{dv} \int_{-1}^1 d\cos\theta \cos\theta \int_0^{2\pi} d\phi(\hat{\mathbf{v}}_x, \hat{\mathbf{v}}_y, \hat{\mathbf{v}}_z)$$

Here we have taken $\mathcal{E}$ to lie in the z direction. [Recall (7.19).] All but the z integral vanish, and we obtain

$$nu = -\frac{2\pi e \mathcal{E}}{m} \int_0^\infty \tau \, dv v^3 \frac{df_0}{dv} \int_{-1}^1 d\mu \mu^2$$

So we find

$$\mathbf{u} = \hat{\mathcal{E}}\mu \qquad nu = -\frac{4\pi}{3}\frac{e\mathcal{E}}{m} \int \tau \, dv v^3 \frac{df_0}{dv} \qquad (7.61)$$

Integrating by parts and dropping surface terms gives

$$nu = \frac{4\pi e \mathcal{E}}{m} \int_0^\infty \tau f_0 v^2 \, dv \qquad (7.62)$$

Substituting the expression for τ given by (7.35), we obtain

$$nu = \frac{4\pi e \mathcal{E} l}{m} \int_0^\infty f_0 v \, dv \qquad (7.63)$$

Thus we may conclude that a nonvanishing mean speed of electrons is incurred due to the presence of the electric field.

In the Drude model of conductivity [see (4.11) and following], we write

$$J = enu = \sigma_c \mathcal{E}$$

where σ_c is electrical conductivity. With (7.63), we obtain

$$\sigma_c = 4\pi \frac{e^2 \mathcal{E} l n}{mC} \int_0^\infty \bar{f}_0 \bar{v} \, d\bar{v}$$

where   (7.64)

$$f_0 = \frac{n}{C^3} \bar{f}_0, \qquad v = C\bar{v}$$

With

$$\nu = \frac{C}{l}$$

(7.64) becomes

$$\sigma_c = \alpha \sigma_c^{(0)} \qquad (7.65)$$

where α is the Druyvesteyn correction factor,

$$\alpha \equiv 4\pi \int_0^\infty \bar{f}_0 \bar{v} \, d\bar{v} \qquad (7.65a)$$

and

$$\sigma_c^{(0)} = \frac{e^2 n}{m\nu} \qquad (7.65b)$$

is the Drude conductivity as given previously in (4.15).

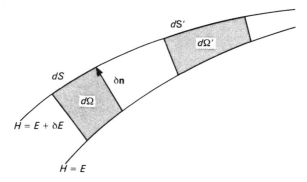

FIGURE 3.25. The normal displacement $\delta\mathbf{n}$ between constant energy surfaces varies as $\delta H/|\nabla H|$.

3.8   Further Remarks on Irreversibility

In this final section of the present chapter we return to the problem of irreversibility encountered earlier in Sections 3.5 and 3.8. First, a brief review of ergodic and mixing flows is presented. We then pass to a description of action-angle variables in preparation for a discussion of more recent developments in the theory of irreversibility.

3.8.1   Ergodic Flow

Prior to discussion of ergodic theory, it is necessary to introduce the following geometrical notion. As noted in Chapter 1, volumes in phase space are invariant under a canonical transformation or the natural motion of the system. In keeping with this property, an *energy shell* was introduced in Γ-space to describe the motion of an ensemble relevant to an isolated system. However, the language of ergodic theory addresses an *energy surface*. Accordingly, a measure-preserving element of surface must be introduced. Let the energy shell be defined by the surfaces $H = E$ and $H = E + \delta E$. The displacement between constant energy surfaces varies as $\delta H/|\nabla H|$. Thus, if dS represents differential surface on the energy shell, invariant differential volume of the energy shell $d\Omega$, varies as $dE\,dS/|\nabla H|$ (see Fig. 3.25). This description permits identification of

$$d\Sigma \equiv \frac{dS}{|\nabla H|} \tag{8.1}$$

as the invariant measure of the element of surface dS.[50] Thus we have two differential elements in Γ-space that are invariant under the natural motion of

---

[50] A quantum version of this construction is given in Section 5.3.

the system:

$$d\Omega \rightarrow d\Omega' = d\Omega$$
$$d\Sigma \rightarrow d\Sigma' = d\Sigma \qquad (8.2)$$

The invariant measure of a set, A, of points on the energy surface is given by

$$\mu(A) = \int_{A \in E} \frac{dS}{|\nabla H|} \qquad (8.3a)$$

The measure of the set of points E comprising the whole energy surface is

$$\mu(E) = \int_E \frac{dS}{|\nabla H|} \qquad (8.3b)$$

Let the natural motion of the set A be governed by the operator $\hat{T}$ [introduced in Section 3.6], which we now write as $\hat{T}(t)$. Then, if in the interval t, $A \rightarrow A'$, we may write

$$A \rightarrow A' = \hat{T}(t)A \qquad (8.4)$$

By virtue of the Liouville theorem, it follows that

$$\mu(A') = \mu(A) \qquad (8.5)$$

We are now prepared to discuss the ergodic hypothesis.

Ergodic hypothesis

This hypothesis states that *almost all* orbits (that is, except for a set of measure zero) on the energy surface pass through every domain of positive measure and remain in these regions for an average time equal to the ratio of their measure to that of E.

An equivalent statement may be made in terms of the average of a dynamical variable $g(\mathbf{z})$. It is evident that the ergodic hypothesis implies that all domains on the energy surface of equal measure are equally probable. The corresponding distribution function is[51]

$$f(\mathbf{z}) = \frac{1}{\mu(E)}, \qquad \text{for } \mathbf{z} \text{ on } H = E \qquad (8.6)$$

which has the normalization

$$\int_{H=E} f(\mathbf{z}) \, d\Sigma = 1 \qquad (8.6a)$$

With (8.3), the phase average of $g(\mathbf{z})$ is given by

$$\langle g \rangle_\Gamma = \frac{1}{\mu(E)} \int_{H=E} g(\mathbf{z}) \, d\Sigma \qquad (8.7)$$

---

[51]In equilibrium statistical mechanics, (8.6) is called the microcanonical distribution. This distribution is further discussed in Section 5.2.1.

In equilibrium, we may also define a time average of $g(\mathbf{z})$ as

$$\langle g \rangle_\tau = \lim_{\tau \to \infty} \frac{1}{\tau} \int_{t_0}^{t_0 + \tau} g[\mathbf{z}(t)]\, dt \tag{8.8}$$

The ergodic theorem seeks to establish the equality of $\langle g \rangle_\Gamma$ and $\langle g \rangle_\tau$. For this equality to be valid, it is evident that $\langle g \rangle_\tau$ must be independent of the initial time t_0 and its phase point $\mathbf{z}(0)$.

Birkhoff's theorem

Birkhoff's theorem addresses these questions.[52] The theorem rests on the assumption that the displacement transformations $\hat{T}(t)$ for the system at hand are *metrically transitive* (or, equivalently, that the energy surface is *metrically indecomposable*). The transformation $\hat{T}(t)$ on the energy surface is metrically transitive if the only sets left invariant under $\hat{T}(t)$ are the whole set or sets of measure zero.[53] Alternatively, we may say that the energy surface is metrically indecomposable if it cannot be divided into two invariant parts each of positive measure. An invariant part of phase space means that all points in this part remain in that part during the motion of the system. With this assumption, Birkhoff's theorem states that the limit $\langle g \rangle_\tau$ exists everywhere on the energy surface, except for a set of measure zero, and that $\langle g \rangle_\tau$ is independent of the initial time t_0 as well as the initial phase point $\mathbf{z}_0$.

Furthermore, the theorem implies that

$$\langle \langle g \rangle_\tau \rangle_\Gamma = \langle g \rangle_\Gamma \tag{8.9}$$

Since $\langle g \rangle_\tau$ is independent of $\mathbf{z}_0$, it is constant on the energy surface:

$$\langle g \rangle_\tau = A = \text{constant}$$

Thus

$$\langle \langle g \rangle_\tau \rangle_\Gamma = A$$

and we obtain

$$\langle \langle g \rangle_\tau \rangle_\Gamma = \langle g \rangle_\tau \tag{8.9a}$$

---

[52]G. D. Birkhoff, *Proc. Nat. Acad. Sci. U.S.A.* **17**, 656 (1931). For a more extensive discussion of these topics, see I. E. Farquhar, *Ergodic Theory in Statistical Mechanics*, Wiley-Interscience, New York (1964).

[53]I. Oxtoby and S. Ulam suggest that almost every group of continuous transformations is metrically transitive: *Ann. of Maths.* **42**, 874 (1941). M. Kac indicates that it is virtually impossible to decide if a given Hamiltonian generates metrically transitive transformations: *Probability and Related Topics in Physical Theory*, Wiley-Interscience, New York (1959). G. Sinai has argued that this property is satisfied by an aggregate of colliding hard spheres: *Statistical Mechanics*, T. Bak, (ed.), Benjamin, Menlo Park, Calif. (1967).

With (8.9), the latter relation gives the desired result.[54]

$$\langle g \rangle_\tau = \langle g \rangle_\Gamma \tag{8.10}$$

The significance of this relation to irreversibility is as follows. First note that (8.6) is an equilibrium distribution.[55] It follows that in writing (8.10), we assert that, independent of initial conditions, the time average goes to the equilibrium average (8.7). This result identifies (8.6) as a preferred state, which is an alien notion to reversible dynamics.

The following property is important to subsequent discussion. A bounded system with no integral of the motion other than the energy is ergodic.[56] The reason for this is as follows. Let there be an additional constant of the motion, B, other than the energy for the given system. Then the system point moves on the hypersurface which is the intersection of the surfaces $B = $ constant and $H = E$. This intersection is a subdomain of the energy surface that the system point is constrained to move on, and the motion is nonergodic. Ergodic motion results if this constraint is removed.

We may conclude from these statements that, for an ergodic system, points move about the energy surface visiting all domains of equal measure with equal frequency with no preference given to any single domain. Whereas this description suggests that a given distribution will spread with time to cover the whole energy surface, counterexamples[57] exhibit distributions that, although ergodic, retain their initial geometrical shape and do not approach an equilibrium distribution.

3.8.2 Mixing Flow and Coarse Graining

Mixing flow causes any initial distribution to spread throughout the energy surface. Note in particular that mixing flow is ergodic but ergodic flow is not always mixing. A system is mixing if for any two functions $h(\mathbf{z})$ and $f(\mathbf{z})$ in $\mathcal{L}_2$, and defined on $H = E$,

$$\lim_{t \to \pm\infty} \frac{1}{\mu(E)} \int_{H=E} h(\mathbf{z}) f[\mathbf{z}(t)] \, d\Sigma = \frac{\int_{H=E} h(\mathbf{z}) \, d\Sigma \int_{H=E} f(\mathbf{z}) \, d\Sigma}{[\mu(E)]^2} \tag{8.11}$$

An immediate consequence of this definition is as follows. In the above equality, let $f(\mathbf{z})$ be a nonequilibrium distribution with normalization (8.6a). Then

---

[54]Birkhoff's theorem does not imply that the system point passes through every point in the energy surface. Thus, whereas $\langle g \rangle_\tau$ involves a single trajectory, it is known that a curve that cannot intersect itself cannot fill a surface in two or more dimensions.

[55]In equilibrium statistical mechanics (8.6) incorporates the principle of equi-a priori probabilities; that is, all states on the energy surface are equally probable. The corresponding distribution is called the microcanonical distribution.

[56]For further discussion, see N. G. Van Kampen, *Physica 53*, 98 (1971).

[57]See, for example, L. Reichl, *A Modern Course in Statistical Physics*, Chapter 8, University of Texas Press, Austin (1980).

with (8.11) we may write

$$\langle h \rangle = \int_{H=E} h(\mathbf{z}) f[\mathbf{z}(t)] d\Sigma \xrightarrow[t \to \pm\infty]{} \frac{1}{\mu(E)} \int_{H=E} h(\mathbf{z}) d\Sigma \qquad (8.12)$$

Thus, for mixing flow, $\langle h \rangle$ approaches an average with respect to the equilibrium distribution (8.6): $f = [\mu(E)]^{-1}$.

However, the following should be borne in mind. For either ergodic or mixing flow, the ensemble density does not change in the neighborhood of a moving phase point. This is a consequence of the Liouville equation, which, in addition to incorporating the incompressibility of the ensemble fluid, further implies an unchanging Gibbs entropy (3.39). The mixing flow of an initial ensemble distribution in phase space with time is often compared to the mixing of water and oil. After sufficient stirring,[58] the fluids appear to be homogeneous. However, more precise observation reveals that the oil and water have remained separate. Furthermore, we may envision a careful stirring that reestablishes the original configuration.

Irreversibility in classical mechanics my be attained under *coarse graining*. Coarse graining begins with dividing the energy shell into subdomains $\{E_n\}$ such that the nth subdomain has volume Ω_n. If $D(\mathbf{z}, t)$ is the ensemble density at the time t, then the coarse-grained density is given by

$$\Pi(\mathbf{z}, t) \equiv \frac{1}{\Omega_n} \int_{\mathbf{z} \in E_n} D(\mathbf{z}', t) d\mathbf{z}' \qquad (8.13)$$

We note that $\Pi(\mathbf{z}, t)$ is constant over each cell and has the same normalization as $D(\mathbf{z}, t)$.

The Gibbs's coarse-grained entropy is defined by

$$\eta(t) = \sum_n \Omega_n \Pi_n \ln \Pi_n \qquad (8.14)$$

which may alternatively be written

$$\eta(t) = \int \Pi(\mathbf{z}, t) \ln \Pi(\mathbf{z}, t) d\mathbf{z}$$

$$= \int D(\mathbf{z}, t) \ln \Pi(\mathbf{z}, t) d\mathbf{z} \qquad (8.14a)$$

Consider the special case that $D(\mathbf{z}, t)$ is initially constant over each cell.[59] Then $D(\mathbf{z}, t) = \Pi(\mathbf{z}, 0)$ and we may write

$$\eta(t) - \eta(0) = \int \Pi(t) \ln \Pi(t) d\mathbf{z} - \int D(0) \ln D(0) d\mathbf{z}$$

---

[58]For consistency of this model, the oil must not separate into disconnected segments at any time during the stirring. See Problem 2.7.

[59]In this event, D is a *simple set function*. For further discussion, see R. L. Liboff, *J. Stat. Phys. 11*, 343 (1974).

$$= \int [D(t) \ln \Pi(t) \, d\mathbf{z} - D(t) \ln D(t)] \, d\mathbf{z} \qquad (8.15)$$

where we have employed (8.14a). Adding

$$\int D \, d\mathbf{z} - \int \Pi \, d\mathbf{z} = 0$$

to the right side of (8.15) gives

$$\eta(t) - \eta(0) = \int D \left[\ln \left(\frac{\pi}{D} \right) + 1 - \left(\frac{\Pi}{D} \right) \right] d\mathbf{z} \qquad (8.16)$$

Since $1 + \ln y \geq y$ and $D \geq 0$, we may write

$$\eta(t) - \eta(0) \leq 0 \qquad (8.17)$$

We may conclude that, with D initially constant over individual cells, if η changes, it decreases. Furthermore, it ceases to increase when $\Pi(\mathbf{z})$ is uniform (see Problem 3.16).

It is evident that the irreversible statement (8.17) is a consequence of the coarse-graining definition (8.12). In going into this representation, the level of information on the state of the system is reduced. It is further evident that $\Pi(\mathbf{z}, t)$ is less informative than $D(\mathbf{z}, t)$ is about the state of the system. It appears to be a general rule that irreversibility occurs when working in a representation that is less informative than the Liouville picture.

3.8.3 Action-Angle Variables

In this section, we consider finite systems that occupy a bounded region of phase space and that undergo periodic or nearly periodic motion. A system of coordinates and momenta particularly well suited to periodic motion are action-angle variables. These variables are closely related to a formalism called *Hamilton–Jacobi theory*. In this formalism, we seek to discover a canonical transformation that renders all new coordinates (q') cyclic. As all new momenta are then constants of the motion, Hamilton's equation (1.1.12) is simply integrated to yield the orbits in the new coordinate frame, that is, $q' = q'(t)$. The dynamical problem is then reduced to algebraically transforming the $q' = q'(t)$ motion to the original $q = q(t)$ frame.

Such a canonical transformation may be obtained from an appropriate generating function. Consider specifically the function $G_2(q, p')$ (introduced in Section 1.2.1) relevant to a system with N degrees of freedom. For uniformity with commonly adopted notation, G_2 is relabeled S. Recalling (1.2.14), we write

$$p_i = \frac{\partial S_i(q^N, p'^N)}{\partial q_i} \qquad (8.18)$$

$$q_i' = \frac{\partial S_i(q^N, p'^N)}{\partial p_i'} \qquad (8.19)$$

We assume that this generating function renders all momenta constant. Let us call these constant momenta α^N, in which case $S = S(q^N, \alpha^N)$. Assuming the motion to be *separable*, we write

$$S(q^N, \alpha^N) = \sum_i S_i(q_i, \alpha^N) \tag{8.20}$$

The relation (8.18) gives

$$p_i = \frac{d S_i(q_i, \alpha^i)}{d q_i} \tag{8.21}$$

or, equivalently,

$$S_i = \int p_i \, dq_i \tag{8.22}$$

The motion is periodic if $p_i(q_i, \alpha^N)$ is a periodic function of q_i (rotation) or is a closed orbit in the (q_i, p_i) plane (vibration). In either event, the integrals (8.22) about cyclic orbits are evidently constant and are taken as the new momenta. These are assigned the symbol J_i and are called *action* variables.

$$\Delta S_i \equiv J_i = \oint p_i \, dq_i \tag{8.23}$$

With the new Hamiltonian cyclic in new coordinates, we write

$$H' = H'(J_1, \ldots, J_N) \tag{8.24}$$

Coordinates conjugate to action variables, J_i, are called *angle variables*, labeled Θ_i. With (8.19), we write

$$\Theta_i = \frac{\partial S(q^N, J^N)}{\partial J_i}, \qquad \left[p_i = \frac{\partial S(q^N, J^N)}{\partial q_i} \right] \tag{8.25}$$

Hamilton's equations for the new coordinates are given by

$$\dot{\Theta}_i = \frac{\partial H'}{\partial J_i} = v_i(J_1, \ldots, J_N) \tag{8.26}$$

which are simply integrated to yield

$$\Theta_i = v_i T + \Theta_i(0) \tag{8.27}$$

To discover the physical meaning of Θ_i, we evaluate $\delta_i \Theta_j$, the net change in Θ_j due to carrying q_i through a complete cycle.

$$\delta_i \Theta_j = \oint \frac{\partial \Theta_j}{\partial q_i} \, dq_i = \oint\!\!\!\oint \frac{\partial^2 S}{\partial q_i \partial J_j} \, dq_i$$

which, assuming separability (8.20), reduces to

$$\delta_i \Theta_j = \frac{\partial}{\partial J_j} \oint p_i \, dq_i = \frac{\partial}{\partial J_j} J_i = \delta_{ji} \tag{8.28}$$

So Θ_j changes by unity if q_i goes through a complete cycle, but remains unchanged otherwise.

Let τ_i be the period of oscillation of q_i. Then with (8.27) and (8.28) we have

$$\Theta_i(\tau_i) - \Theta_i(0) = v_i \tau_i = 1 \qquad (8.29)$$

Thus we may identify v_i as the frequency of oscillation of q_i. Angular frequency is given by $\omega_i = 2\pi v_i$ so that, with (8.27), $\theta_i = 2\pi \Theta_i$ may be identified with angular displacement of the periodic motion. When written in terms of these angular variables, (8.26) through (8.29) become

$$\dot{\theta}_i = \frac{\partial H'}{\partial J_i} = \omega_i(J_1, \ldots, J_N) \qquad (8.26a)$$

$$\theta_i = \omega_i t + \theta_i(0) \qquad (8.27a)$$

$$\delta_i \theta_j = 2\pi \delta_{ij} \qquad (8.28a)$$

$$\theta_I(\tau_i) - \theta_i(0) = \omega_i \tau_i = 2\pi \qquad (8.29a)$$

In (8.26a), H' was written for $2\pi \times H'$ of (8.26).

A note of caution is in order at this point. In the canonical mapping $(q, p) \rightarrow (J, \theta)$, the variable θ is a many-valued function of old variables so that the transformation $(q, p) \rightarrow (J, \theta)$ is not one to one and θ is not a well-defined dynamical variable. This situation is easily remedied for periodic motion. Consider that the θ interval corresponding to (q, p) periodicity is 2π. The mapping is rendered one to one either by introducing the function $\theta \bmod 2\pi$ or by working with any trigonometric function of θ. The function $\theta \bmod 2\pi$ of, say, $\theta = \theta_1 + n2\pi$, where n is an integer and $\theta_1 < 2\pi$, is θ_1.

It is evident that a well-defined canonical transformation must be a one to one mapping between old and new variables. Thus, for example, if the mapping $(q, p) \leftrightarrow (\theta, J)$ is not one to one, then knowledge of a constant in the (θ, J) frame does not determine an image constant in the (q, p) frame.

As described in Section 3.1, the two-body problem is reduceable to the motion in a plane of an effective particle with reduced mass μ. It might be thought that action-angle variables are relevant only to interactions that give a closed orbit in the plane of motion. The fact is that of all attractive interactions of the form $V = Kr^n$ only $n = 2$ (harmonic oscillator) and $n = -1$ (Kepler problem) give closed orbits.[60] However, for nonclosed orbits, $r(t)$ and $\theta(t)$ may still be periodic but with incommensurate frequencies, in which case action-angle variables remain appropriate. (See Problem 3.17.) We now turn briefly to the technique of obtaining the generating function S, which renders all new coordinates cyclic.

---

[60] V.I. Arnold, *Mathematical Methods of Classical Mechanics*, Springer-Verlag, New York (1978).

3.8.4 Hamilton–Jacobi Equation

For all generating functions previously discussed, old and new Hamiltonians, H and H', respectively, are related through $H' = H + \partial S/\partial t$. With S not explicitly dependent on time, we obtain $H = H'$. We may identify the constant H' with, say, the first of the ordered constant momenta of the system. Calling this constant E, we obtain

$$H\left(q_1, \ldots, q_N, \frac{\partial S}{\partial q_1}, \ldots, \frac{\partial S}{\partial q_N}\right) = E \qquad (8.30)$$

This equation is called the *Hamilton–Jacobi equation* for *Hamilton's characteristic function S*. The solution to this first-order partial differential equation in N independent variables is a function of N constants. Thus we may write

$$S = S(q_1, \ldots, q_N; \alpha_1, \ldots, \alpha_N) \qquad (8.31)$$

This is the desired structure of S with $\{\alpha_i\}$ identified as the new constant momenta and new coordinates given by $q_i' = \partial S/\partial \alpha_i$. As previously assigned, the new Hamiltonian $H' = \alpha_1$, which, as sought, is cyclic in all q_i'. Note in particular that

$$\dot{p}_i' = -\frac{\partial H'}{\partial q_i'} = -\frac{\partial \alpha_1}{\partial q_1'} = 0$$

$$\dot{q}_i' = \frac{\partial H'}{\partial p_i'} = \frac{\partial \alpha_1}{\partial \alpha_i} = \delta_{i1} \qquad (8.32)$$

which give

$$q_1' = \frac{\partial S}{\partial \alpha_1} = t + \beta_1$$

$$q_1' = \frac{\partial S}{\partial \alpha_i} = \beta_i \qquad (i \neq 1) \qquad (8.32a)$$

Harmonic oscillator

Let us apply the Hamilton–Jacobi equation (8.30) to the problem of finding the orbit $q = q(t)$ of the harmonic oscillator with the Hamiltonian

$$H = \frac{p^2}{2m} + \frac{kq^2}{2} \qquad (8.33)$$

where m is mass and k is the spring constant. The relation (8.30) becomes

$$\frac{1}{2m}\left(\frac{\partial S}{\partial q}\right)^2 + \frac{kq^2}{2} = E$$

which gives

$$S(q, E) = \sqrt{mk} \int dq \sqrt{\frac{2E}{k} - q^2} + A \qquad (8.34)$$

where A is a constant of integration. Differentiating, we find, with (8.32a),

$$\frac{\partial S}{\partial E} = t + \beta = \frac{1}{\omega} \int \frac{dq}{\sqrt{(2E/k)^2 - q^2}}$$

$$= \frac{1}{\omega} \sin^{-1} q \sqrt{\frac{K}{2E}} \qquad (8.35)$$

where we have set

$$\omega^2 \equiv \frac{k}{m} \qquad (8.35a)$$

Inverting (8.35) gives the desired solution:

$$q = \sqrt{\frac{2E}{k}} \sin \omega(t + \beta) \qquad (8.36)$$

In action-angle formalism,

$$J = \oint \frac{\partial S}{\partial q} dq = \sqrt{mk} \oint dq \sqrt{\frac{2E}{k} - q^2}$$

$$J = \frac{2\pi E}{\omega} = \frac{E}{\nu} \qquad (8.37)$$

The new Hamiltonian is

$$H'(J) = \nu J \qquad (8.37a)$$

which on differentiation returns (8.27) for angle variable $\Theta(t)$.

3.8.5   Conditionally Periodic Motion and Classical Degeneracy

We return to the discussion of Section 8.1 relevant to systems that are separable (8.20) and whose independent coordinates execute periodic motion.

Consider a dynamical function of the state of this system, $F(q^N, p^N)$. When expressed in terms of action-angle variables, this function is periodic in the angle variables with the period of each θ_i variable equal to 2π. Thus the function $F(q^N, p^N)$, when expressed in terms of action-angle variables, may be expanded as a multiple Fourier series:

$$F(J^N, \theta^N) = \sum_{n_1=-\infty}^{\infty} \cdots \sum_{n_N=-\infty}^{\infty} A_{n_1 \cdots n_N} \exp[i(n_1\theta_1 + \cdots + n_N\theta_N)] \qquad (8.38)$$

where $\{n_i\}$ are integers. The time dependence of this function is obtained from (8.26) and (8.27). Again, setting $H' = E$ gives

$$\theta_i = \frac{\partial E}{\partial J_i} t + a_i \qquad (8.39)$$

where a_i is a constant. Substituting these value into (8.38) and absorbing $\{a_i\}$ into the Fourier coefficients $A_{\{n\}}$ gives the time-dependent form

$$F(t) = \sum_{n_1} \cdots \sum_{n_N} A_{n_1 \cdots n_N} \exp[it(n_1\omega_1 + \cdots + n_N\omega_N)] \qquad (8.40)$$

where, with (8.26a), we have labeled

$$\omega_i = \frac{\partial E}{\partial J_i} \qquad (8.40a)$$

Thus we see that each term in the sum (8.39) is a periodic function with frequency

$$\omega_{n_1 \cdots n_N} = n_1\omega_1 + \cdots + n_N\omega_N \qquad (8.41)$$

Since these frequencies are in general not commensurate, the sum (8.40) is not in general a periodic function. However, with Poincaré's recurrence theorem (Section 3.6), we may conclude that after a sufficient time the system, in developing from a given initial state, will return arbitrarily close to the initial state. Thus the dynamical function $F(t)$ is said to be *conditionally periodic*.

Degeneracy

In the event that two or more frequencies are commensurate (for example, $\bar{n}_1\omega_1 = \bar{n}_2\omega_2$, where $\bar{n}_1$ and $\bar{n}_2$ are integers), the system is said to be *degenerate*. For *complete degeneracy*, all frequencies are commensurate, and we may write

$$\omega_{n_1 \cdots n_N} = \omega_1(n_1 + n_2\bar{k}_2 + n_3\bar{k}_3 + \cdots + n_K\bar{k}_N) \equiv s\omega_1 \qquad (8.42)$$

where $\bar{k}_i$ and s are rational numbers and ω_1 represents the minimum of the values $\{\omega_1\}$. In this event, F, as given by (8.40) is periodic. Let us calculate the frequency of this periodic oscillation. Toward these ends, we set $s = \bar{n}_a/\bar{n}_b$, where $\bar{n}_a$ and $\bar{n}_b$ are integers, to obtain

$$\omega_{n_1 \cdots n_2} = \bar{n}_a\left(\frac{\omega_1}{\bar{n}_b}\right) \qquad (8.42a)$$

So, for complete degeneracy, we may associate a frequency with each term in the sum (8.40), which is equal to either the *fundamental*, $\omega_1/\bar{n}_b$, or one of its harmonics, $\bar{n}_a(\omega_1/\bar{n}_b)$. Thus the angular frequency of the function (8.40) for complete degeneracy is $\omega_1/\bar{n}_b \leq \omega_1$.

Consider the specific example of twofold degeneracy with $\omega_1 = k\omega_2$, where k is a rational number. Again we set $k = \bar{n}_2/\bar{n}_1$. The frequency sum (8.41) becomes

$$\omega = \frac{\omega_1}{\bar{n}_2}n + n_3\omega_3 + \cdots + n_N\omega_N \qquad (8.43)$$

where n is an integer that replaces the $\bar{n}_1$ and $\bar{n}_2$ integers in the multiple Fourier series (8.40). Thus degeneracy leads to reduction in the number of independent frequencies for the system. As we will find immediately below, a similar property pertains to the number of independent constants of motion on which the energy depends.

Again considering the case that only two frequencies are commensurate [see (8.40)], we write

$$\bar{n}_1 \frac{\partial E}{\partial J_1} = \bar{n}_2 \frac{\partial E}{\partial J_2}$$

This relations yields $E = E(\bar{n}_2 J_1 + \bar{n}_1 J_2)$. Thus

$$E = E(\bar{n}_2 J_1 + \bar{n}_1 J_2, J_3, \ldots, J_N) \tag{8.44}$$

and we may conclude that, when degeneracy occurs, there is a reduction of the number of independent constants on which the energy depends.[61]

In addition to the preceding properties concerning degeneracy, we note the following. With (8.39), we see that the form

$$A_{12} = \theta_1 \frac{\partial E}{\partial J_2} - \theta_2 \frac{\partial E}{\partial J_1} \tag{8.45}$$

is a constant of motion. With degeneracy, $\bar{n}_1 \omega_1 = \bar{n}_2 \omega_2$, the preceding gives

$$B_{12} \equiv \frac{\bar{n}_2 A_{12}}{\omega_1} = \bar{n}_1 \theta_1 - \bar{n}_2 \theta_2 \tag{8.46}$$

If either θ_1 or θ_2 changes by 2π, B_{12} changes by an integer $\times 2\pi$. So replacing the right side of (8.46) with its value mod 2π gives a single-valued integral of the motion. On the other hand, for nondegeneracy, it is evident that (8.45) does not yield a single-valued integral of the motion.

An example of an additional constant for degenerate systems occurs for the Kepler problem. In this case the additional constant is the Laplace–Runge–Lenz vector[62]

$$\mathbf{A} = \mathbf{p} \times \mathbf{L} - \mu \mathbf{r} V(r) \tag{8.47}$$

where $\mathbf{L} = \mathbf{r} \times \mathbf{p}$ is angular momentum, μ is reduced mass, and the potential $V(r) = -K/r$. With the Hamiltonian of the system given by

$$H = \frac{p^2}{2\mu} + V(r)$$

---

[61] A similar situation occurs in quantum mechanics. The Hamiltonian of a particle moving in a central potential is invariant under rotation about the origin. This invariance is a twofold symmetry. Consequently, eigenenergies are dependent only on the principal quantum number and are independent of orbital or azimuthal quantum numbers.

[62] Properties of this vector are discussed by R. J. Finkelstein, *Nonrelativistic Mechanics*, Benjamin, Menlo Park, Calif (1973). See also H. Golstein, *Classical Mechanics*, 2nd. ed., Addison-Wesley, Reading, Mass. (1980). An additional example of this degeneracy property is given in Problem 3.17.

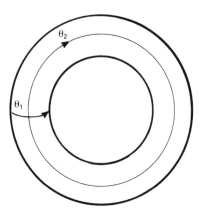

FIGURE 3.26. Angular displacements lie on the surface of a torus, which is a two-dimensional hypersurface in four-dimensional phase space.

the equation of motion (1.1.25) returns $\dot{\mathbf{A}} = 0$. Incorporating relations among the constants $\mathbf{A}$ and $\mathbf{L}$ and the energy E indicates that only one scalar property of $\mathbf{A}$ is independent of $\mathbf{L}$ and E as, for example,

$$A^2 = \mu^2 K^2 + 2mEL^2 \tag{8.48}$$

This observation permits the following conclusion. The constant A^2 is an example of a dynamical function that commutes with the Hamiltonian but that, nevertheless, is not a function of the Hamiltonian. When coupled with (1.1.26g), the preceding remarks indicate that whereas $G = G(H) \Rightarrow [G, H] = 0$ the converse is not in general a valid statement.[63]

Invariant tori

Consider a system with two degrees of freedom with coordinates q_1, q_2, which individually execute periodic motion. The related angular displacements θ_1, θ_2 (mod 2π) lie on a constant toroidal surface, which may be identified with the constant actions J_1 and J_2 (see Fig. 3.26). If the motion is degenerate, then $\bar{n}_1 \omega_1 = \bar{n}_2 \omega_2$. It follows that in the time interval $2\pi \bar{n}_2/\omega_1$, over which θ_1 makes $\bar{n}_2$ revolutions, θ_2 makes $\bar{n}_1$ revolutions, resulting in a closed orbit. As the $(\theta, J) \leftrightarrow (q, p)$ mapping is one to one, a closed orbit likewise results in (q, p) space, with which we may associate a new constant J value. Again, degeneracy leads to an additional constant of motion.

---

[63]The same rule applies in quantum mechanics. For a particle moving in a central potential, $[\hat{L}_z, \hat{H}] = 0$, which gives $d\langle L_z \rangle/dt = 0$. The degeneracy of the eigenstates of $\hat{H}$ stems from rotational symmetry, which infers that $\hat{H}$ is invariant under operation by $\hat{L}_z$. This is the reason for the null commutation property, and not that $\hat{L}_z = \hat{L}_z(\hat{H})$.

3.8.6   Bruns's Theorem

The three-body problem refers to the following configuration: three bodies of given masses interacting under attractive potentials and executing bounded motion. This system has 9 degrees of freedom and, in general, trajectories are given in terms of 18 constants of motion. Ten such constants are given by

$$\mathbf{P}, \frac{\mathbf{P}t}{M} - \mathbf{R}, \quad \mathbf{L}, \quad E \qquad (8.49)$$

where $\mathbf{R}$, $\mathbf{P}$, and M are center-of-mass displacement, momentum, and mass, respectively, $\mathbf{L}$ is total angular momentum, and E is energy.

In 1887, Heinrich Bruns showed that the only algebraic integrals for the three-body problem are the ten known integrals (8.49).[64] We may conclude that, in general, the three-body problem is nonintegrable so that the N-body problem is nonintegrable for $N \geq 3$. [See Problem 3.61.]

Integrable motion may be defined as follows. If a canonical transformation to a representation exists in which all coordinates are cyclic, then the motion of the related system is said to be *integrable*. For bounded integrable systems, we may identify new momenta with action variables so that the transformed Hamiltonian appears as

$$H = H(J_1, \ldots, J_N)$$

Therefore, in general, a bounded system is integrable providing a canonical transformation exists in which the new Hamiltonian is given as above. As discussed in the previous section, when such constants are expressed in terms of angle variables, they are referred to an invariant tori (see Fig. 3.26).

3.8.7   Anharmonic Oscillator

We may gain deeper insight into the preceding discussion through study of certain anharmonic oscillator systems. Thus, for example, consider a system with two degrees of freedom with the Hamiltonian[65]

$$H = \frac{1}{2}(p_1^2 + p_2^2 + q_1^2 + q_2^2) + q_1 q_2^2 - \frac{1}{3}q_1^3 \qquad (8.50)$$

---

[64]H. Bruns, *Berichte der Kgl. Sächs. Ges. der Wiss.* (1887). See also H. Poincaré, *Les Method Nouvelles de la Mechanique Céleste*, Gauthier-Villars et fils, Paris (1892, 1893, 1899). Reprinted by Dover Publications, New York (1957), Vol. 1, Chapter V; E. T. Whitaker, *A Treatise on the Analytical Dynamics of Particles and Rigid Bodies*, 4th ed., Dover, New York (1944). *Note*: Special solutions to the three-body problem were shown to exist for restricted motion on a plane by Lagrange (1772). This configuration was cast in a symmetric more revealing form by R. Broucke and H. Lass, *Celestial Mech. 8*, 5 (1973). For further discussion, see D. Hestenes, *New Foundations for Classical Mechanics*, D. Reidel, New York (1986).
[65]M. Hénon and C. Heiles, *Astron. J. 69*, 73 (1964). A closely allied problem was examined by E. N. Lorentz, *J. Atmos. Science 20*, 130 (1963).

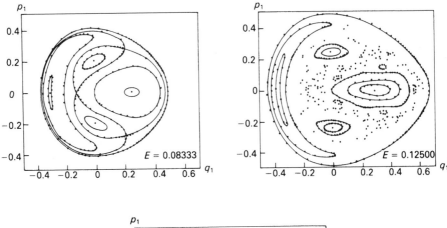

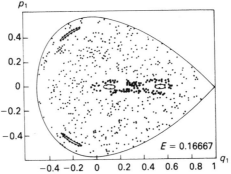

FIGURE 3.27. Development of chaotic motion corresponding to the Hénon–Heiles Hamiltonian (8.50). From M. Hénon and C. Heiles, *Astron. J. 69*, 73 (1964). Used by permission of Carl E. Heiles.

Trajectories lie on the surface $H = E =$ constant in four-dimensional Γ-space. We may obtain a two-dimensional image (that is, a *Poincaré map*) of the trajectory in the following manner. Observe the trajectory each time it passes through the $q_2 = 0$ surface with $p_2 > 0$ in the q_1, p_1 plane. (We will refer to this plane as the $\overline{q_1 p_1}$ plane.) If the energy is the only constant of motion, then the projection of the system point is free to roam about the $\overline{q_1 p_1}$ plane corresponding to the surface $H = E$. However, if an additional integral of motion exists, then the system trajectory lies on the three-dimensional hypersurface which is the intersection of this constant and the energy surface. The intersection of this hypersurface and the $\overline{q_1 p_1}$ plane is a smooth curve. So if plots of trajectory points in the $\overline{q_1 p_1}$ plane show no systematic order, we conclude that the system has only one constant of motion, $H = E$. If, however, the trajectory points follow a smooth curve, we may conclude that an additional constant of the motion exists.

Three numerical readouts due to Hénon and Heiles are shown in Fig. 3.27. At lowest energy, $E = 0.0833$, the existence of a smooth curve in the $\overline{q_1 p_1}$ plane implies, to within computer accuracy, an additional constant of the motion. Each closed loop in these figures corresponds to one trajectory (that is, one set of initial data). As energy is increased, at $E = 0.125$, there is disintegration of the smooth curves. The random dots correspond to one trajectory. At slightly higher energy, $E = 0.167$, nearly all motion is chaotic, and the additional constant of motion is lost. Note in particular that these numerical data indicate a fairly abrupt transition to chaos.

3.8.8   Resonant Domains and the KAM Theorem

The loss of a constant of motion with the change in a parameter in the Hamiltonian, as in the preceding example, is well described as follows. With aid of the transformation

$$p_i = -(2m\omega_i J_i)^{1/2} \sin \theta_i$$

$$q_i = \left(\frac{2J_i}{m\omega_i} \right)^{1/2} \cos \theta_i \tag{8.51}$$

The Hamiltonian of a general two-dimensional anharmonic oscillator may be cast in the form

$$H(J_1, J_2, \theta_1, \theta_2) = H_0(J_1, J_2) + \lambda V(J_1, J_2, \theta_1, \theta_2) \tag{8.52}$$

where λ is a coupling constant. The potential V is periodic in θ_1 and in θ_2, whereas H_0 has a polynomial dependence on J_1 and J_2. At $\lambda = 0$, in accord with (8.39), we find[66] (for $i = 1, 2$)

$$\theta_I = \omega_i(J_1, J_2)t + a_i$$

$$\omega_1(J_1, J_2) = \frac{\partial H_0}{\partial J_i} \tag{8.53}$$

Before proceeding further, we introduce a more definite perturbing potential in the form of a multiply periodic series [see (8.38)]:

$$V = \sum_{n_1} \sum_{n_2} V_{n_1 n_2}(J_1, J_2) \cos(n_1 \theta_1 + n_2 \theta_2) \tag{8.54}$$

The sum runs over all positive and negative integers n_1, n_2. Note that this form of potential maintains periodicity in θ_1 and in θ_2.

Let us attempt the construction of a generating function $S(J_1', J_2', \theta_1, \theta_2)$, which renders J_1' and J_2' constants of the motion for sufficiently small values of the interaction parameter λ. We choose the form

$$S = J_1'\theta_1 + J_2'\theta_2 + \sum_{n_1} \sum_{n_2} B_{n_1 n_2} \sin(n_1 \theta_1 + n_2 \theta_2) \tag{8.55}$$

---

[66]Note that ω_i entering the present discussion is not relevant to the entire Hamiltonian (8.52) but only to the unperturbed term, H_0.

and determine the coefficients $B_{n_1 n_2}$ to give the desired cyclic property. With the preceding expression, we obtain

$$J_i = \frac{\partial S}{\partial \theta_i} = J'_i + \sum_{n_1} \sum_{n_2} n_i B_{n_1 n_2} \cos(n_1 \theta_1 + n_2 \theta_2) \qquad (8.56)$$

$$\theta'_i = \frac{\partial S}{\partial J'_i} \qquad (8.56a)$$

where J'_i, θ'_i are new canonical variables.

Substituting the potential (8.54) into the Hamiltonian (8.52) and first expanding the resulting form about $J_i = J'_i$, we obtain

$$H = H_0(J'_1, J'_2) + \frac{\partial H_0}{\partial J_1} \Delta J_1 + \frac{\partial H_0}{\partial J_2} \Delta J_2 + \cdots + \lambda V$$

With ΔJ_i inferred from (8.56) and keeping terms to $0(\lambda)$, we find

$$H = H_0(J'_1, J'_2) + \sum_{n_1} \sum_{n_2} [(n_1 \omega_1 + n_2 \omega_2) B_{n_1 n_2} + \lambda V_{n_1 n_2}] \cos(n_1 \theta_1 + n_2 \theta_2)$$

$$(8.57)$$

Thus, to lowest order in λ, J'_1 and J'_2 are constants, providing we set

$$B_{n_1 n_2} = -\frac{\lambda V_{n_1 n_2}}{n_1 \omega_1 + n_2 \omega_2} \qquad (8.58)$$

in which case

$$H = H_0(J'_1, J'_2) + 0(\lambda^2)$$

Expansions for the new action variables are obtained by substituting the coefficients (8.58) into (8.56). There results

$$J'_i = J_i + \sum_{n_1} \sum_{n_2} \frac{\lambda n_i V_{n_1 n_2}}{(n_1 \omega_1 + n_2 \omega_2)} \cos(n_1 \theta_1 + n_2 \theta_2) + 0(\lambda^2) \qquad (8.59)$$

It follows that the choice of S as given by (8.55) with coefficients (8.58) has the desired property of rendering J'_1 and J'_2 constants for sufficiently small λ. However, note that with (8.53) $\omega_i = \omega_i(J_1, J_2)$, and it is possible that $n_1 \omega_1 + n_2 \omega_2$ may vanish. In general, the series (8.59) diverges if

$$|n_1 \omega_1(J_1, J_2) + n_2 \omega_2(J_1, J_2)| \leq \lambda n_i V_{n_1 n_2} \qquad (8.60)$$

The region of phase space where (8.59) is obeyed is called the *resonance zone*, and $n_1 \omega_1 + n_2 \omega_2 = 0$ is the *resonance condition*.[67] In the resonance zone, J'_i as given by (8.59) ceases to be a well-defined constant and chaotic behavior ensues.

Suppose, in the expansion of J' given by (8.59), we exclude small regions in phase space about each resonance given by (8.60). In this event, we obtain

---

[67] This resonant condition is evidently equivalent to degeneracy of H_0 previously described in Section 8.3.

a well-behaved expansion for J_1'. For smooth potentials, the amplitudes $V_{n_1 n_2}$ decreases rapidly with increasing n_1 and n_2, and we may expect that, with larger n_1, n_2, smaller regions of phase space are excluded, thereby extending the domain of convergence of the series (8.59).

In three independent works, A. N. Kolmogorov,[68] V. I. Arnold,[69] and J. Moser[70] (KAM) established the following: In the limit $\lambda \to 0$, and providing H_0 has a nonzero Hessian (or, equivalently, that ω_i has a nonvanishing Jacobian)

$$\det \left| \frac{\partial \omega_i}{\partial J_j} \right| = \det \left| \frac{\partial^2 H_0}{\partial J_i \partial J_j} \right| \neq 0 \tag{8.61}$$

then the measure of excluded phase space approaches zero, or, equivalently, we may say that the invariant tori of H_0 are conserved.[71] In this event, we may expect orbits to remain well behaved in the limit $\lambda \to 0$ over most of phase space. Note that this theorem is consistent with (8.60). In the opposite case that λ grows away from zero, the theorem states that not all invariant tori of H_0 are destroyed. The ensuing motion of the system is a consequence of the remaining and lost invariant tori.

Two significant properties of resonance zones are as follows: Consider two points in phase space with arbitrarily small initial relative displacement. For chaotic motion, the subsequent displacement of trajectories stemming from these initial points grows exponentially with time.[72] This property is relevant to points within a resonant zone.[73] As this property maintains for arbitrarily small initial displacement, we may conclude that there is no finite-difference integration accuracy that suffices to retrace a trajectory back to its initial point in a resonant zone.

Second, it has been ascertained[74] that, as the number of resonances grow dense in phase space, integrals of the motion become nonanalytical or pathological, as do their intersection. As this intersection represents the system

---

[68] A. N. Kolmogorov in *Foundations of Mechanics*, R. Abraham (ed.), Benjamin, Menlo Park, Calif. (1967). See Appendix D.

[69] V. I. Arnold, *Russian Math. Surv. 18*, 9 (1963); *18*, 85 (1963).

[70] J. Moser, *Akad. Wiss. Gottingen II, Math. Physik Kl.* 1 (1962).

[71] For further discussion of this theorem, see G. M. Zaslavsky, *Chaos in Dynamical Systems*, Harwood Academic Publishers, New York (1985); and H. G. Schuster, *Deterministic Chaos*, Physik-Verlag, Weinheim (1984).

[72] The log of this exponential displacement is called the Liapunov exponent. For further discussion see: S. Chandra (ed.), *Chaos in Nonlinear Dynamical Systems*, SIAM, Philadelphia (1984).

[73] This behavior was demonstrated for the Hénon–Heiles system by C. H. Walker and J. Ford, *J. Math. Phys. 13*, 700 (1974).

[74] I. Prigogine, *From Being to Becoming*, W. H. Freeman, San Francisco (1980). This work includes a descriptive overview of irreversibility and a comprehensive reference list.

trajectory, it is evident that neither the trajectory nor its time reversal are well defined for this situation.

Problems

3.1. For the Hamiltonian given by (1.7) and potential of interaction given by (1.17), with $K = -|K|$, offer a geometrical argument that illustrates that orbits are bound for $E < 0$ and (conditionally) unbound for $E > 0$. *Hint*: Introduce the *effective potential*

$$V_{\text{eff}}(r) = V(r) + \frac{L^2}{2\mu r^2}$$

and make a sketch of $V_{\text{eff}}(r)$ versus r.

3.2. (a) Show that the scattering matrix

$$\overline{\overline{S}} = I - 2\overline{\alpha\alpha}$$

has the following properties:

$$\overline{\overline{S}} = \overline{\overline{S}}, \qquad \overline{\overline{S}} \cdot \overline{\overline{S}} = \overline{\overline{I}}$$

(b) What consequence do these properties have on the collision statement $\mathbf{g}' = \overline{\overline{S}}\mathbf{g}$?

3.3. Show that the external force $\mathbf{K}$ drops out in the derivation of (3.31).

3.4. This problem addresses the group property of the time-displacement operator $\hat{T}(t)$ introduced in Section 3.6 (where they were labeled $\hat{T}^n$) and in Section 8.1. Show that these operators satisfy the group property $\hat{T}(t_1)\hat{T}(t_2) = \hat{T}(t_1 + t_2)$.

Answer

Consider the form $\hat{T}(\Delta t_2)\hat{T}(\Delta t_1)\Omega_0$, where Δt_1 and Δt_2 are small time intervals and Ω_0 represents a volume in Γ-space. Let $(q_0, p_0) \in \Omega_0$. Integrating Hamilton's equations (1.1.20) over Δt_1 gives

$$q_1 \equiv q(\Delta t_1) = q_0 + H_p(q_0, p_0)\Delta t_1 \equiv q_0 + H_p(0)\Delta t_1$$

with a similar equation for p_1. We have written H_p for $\partial H/\partial p$. For the second displacement, we find

$$q_2 = q_1 + H_p(q_1, p_1)\Delta t_2$$

Substituting for (q_1, p_1) in H_p gives

$$q_2 = q_1 + H_p[q_0 + H_p(0)\Delta t_1, p_0 - H_q(0)\Delta t_1]\Delta t_2$$
$$q_2 = q_1 + H_p(0)\Delta t_2 + 0(\Delta t_1 \Delta t_2)$$

Dropping the last term and substituting for q_1 gives

$$q_2 = q_0 + H_p(0)(\Delta t_1 + \Delta t_2)$$

with a similar relation for p_2. These are the same forms that result for $\hat{T}(\Delta t_1 + \Delta t_2)\Omega_0$.

3.5. The value of viscosity for liquid D_2 at 20K is $\eta \simeq 370 \times 10^{-6}$ g/cm-s.[75] How does this value compare with that obtained from Maxwell's formula (4.33)? The diameter of a D_2 molecule may be taken to be $\simeq 3.5$ Å. Offer an explanation for this (good/bad) agreement.

3.6. Show that the following five integrals represent identical tensors. In these expressions, $\overline{\overline{G}}$ is a tensor independent of $\mathbf{c}$, and $F = F(c^2)$.

(1) $\displaystyle\int F\overline{\overline{\mathbf{cc}}}(\overset{\circ}{\overline{\overline{\mathbf{cc}}}} : \overline{\overline{G}})\,d\mathbf{c}$

(2) $\displaystyle\int F\overline{\overline{\mathbf{cc}}}(\overline{\overline{\mathbf{cc}}} : \overline{\overline{G}})\,d\mathbf{c}$

(3) $\displaystyle\frac{2}{15}\overset{\circ}{\overline{\overline{G}}}_s \int F c^4\,d\mathbf{c}$

(4) $\displaystyle\int F\overline{\overline{\mathbf{cc}}}(\overset{\circ}{\overline{\overline{\mathbf{cc}}}} : \overline{\overline{G}})\,d\mathbf{c}$

(5) $\displaystyle\frac{1}{5}\overset{\circ}{\overline{\overline{G}}}_s \int (\overline{\overline{\mathbf{cc}}} : \overline{\overline{\mathbf{cc}}}) F\,d\mathbf{c}$

We have written

$$\overset{\circ}{\overline{\overline{A}}} \equiv \overline{\overline{A}} - \frac{1}{3}(\text{Tr }\overline{\overline{A}})\overline{\overline{I}}$$

which is traceless and

$$\overset{\circ}{\overline{\overline{A}}}_s \equiv \frac{1}{2}\overline{\overline{(A + \tilde{A})}} - \frac{1}{3}(\text{Tr }\overline{\overline{A}})\overline{\overline{I}}$$

which is symmetric and traceless as well. The transpose of $\overline{\overline{A}}$ is written $\overline{\overline{\tilde{A}}}$.
Hint: See Appendix B, Section B.1. Note that if $\overline{\overline{A}}$ is symmetric then $\overset{\circ}{\overline{\overline{A}}} = \overset{\circ}{\overline{\overline{A}}}_s$.

3.7. (a) Show that the Sonine polynomials

$$S_m^{(n)}(x) = \sum_{p=0}^{n} \frac{(m+n)!(-x)^p}{(m+p)!(n-p)!p!}$$

occur as coefficients of expansion of the generating function

$$S(x) = (1-s)^{-m-1}e^{-xs/(1-s)}$$

where s is a positive number less than unity. That is, show that

$$S(x) = \sum_{n=0}^{\infty} S_m^{(n)}(x)s^n$$

3.8. Show that the coefficient α in (5.58) has the value $-15/4$.

Answer

First note that

$$\frac{F^0}{n}\,d\mathbf{c} = \frac{4}{\sqrt{\pi}}\xi^2\,d\xi e^{-\xi^2}$$

---

[75] H. M. Roder et al., NBS Technical Note 641 (1973).

With (5.56), we then obtain

$$\alpha_1 = \frac{4}{\sqrt{\pi}} \int_0^\infty e^{-\xi^2} \left(\xi^2 - \frac{5}{2} \right) \xi \cdot S_{3/2}^{(1)}(\xi^2) \xi \xi^2 \, d\xi$$

With the values of $S_m^{(n)}(x)$ given beneath (5.54a), we note

$$S_{3/2}^{(1)}(\xi^2) = \frac{5}{2} - \xi^2$$

Thus

$$\alpha_1 = -\frac{4}{\sqrt{\pi}} \int_0^\infty e^{-\xi^2} \xi^4 [S_{3/2}^{(1)}(\xi^2)]^2 \, d\xi$$

Let $x = \xi^2$ so that $2\xi^2 \, d\xi = \sqrt{x} \, dx$, whence

$$\alpha_1 = -\frac{2}{\sqrt{\pi}} \int_0^\infty e^{-x} x^{3/2} [S_{3/2}^{(1)}(x)]^2 \, dx$$

$$= -\frac{2}{\sqrt{\pi}} \frac{(5/2)!}{1!} = -\frac{2}{\sqrt{\pi}} \frac{5}{2} \cdot \frac{3}{2} \cdot \frac{1}{2} \sqrt{\pi}$$

$$= -\frac{15}{4}$$

3.9. Show that the coefficient β_1 in the equation preceding (5.67) has the value $\frac{5}{2}$.

Answer

From (5.64),

$$\beta_1 = \frac{4}{\sqrt{\pi}} \int \overset{\circ}{\xi\xi} : \overset{\circ}{\xi\xi} S_{5/2}^0(\xi^2) \xi^2 \, d\xi$$

$$\overset{\circ}{\xi\xi} : \overset{\circ}{\xi\xi} = \left(\xi_i \xi_j i - \frac{1}{3} \delta_{ij} \xi^2 \right) \left(\xi_i \xi_j - \frac{1}{3} \delta_{ij} \xi^2 \right)$$

$$= \xi^4 - \frac{1}{3} \xi^4 - \frac{1}{3} \xi^4 + \frac{1}{9} \cdot 3 \xi^4$$

$$= \frac{2}{3} \xi^4$$

Again, with $x \equiv \xi^2$, we obtain

$$\beta_1 = \frac{4}{3\sqrt{\pi}} \int_0^\infty e^{-x} [S_{5/2}^{(0)}(x)]^2 x^{5/2} \, dx$$

$$= \frac{4}{3\sqrt{\pi}} \left(\frac{5}{2} \right)! = \frac{4}{3\sqrt{\pi}} \cdot \frac{15}{8} \cdot \sqrt{\pi}$$

$$= \frac{5}{2}$$

3.10. Establish (5.61). That is, show that $A_{11} = -4n^2 \Omega^{(2,2)}$.

3.11. A vial of radioactive O^{19} breaks. Employing mean-free-path relations, estimate the velocity (cm/s) with which the O^{19} diffuses through air.

3.12. Show that the collision frequency $\nu(\xi)$ for interacting rigid spheres is given by (6.26).

3.13. Show that substitution of the Legendre expansion (7.21)

$$f(\mathbf{v}) = f_0(v) + \mu f_1(v)$$

into the left side of the equilibrium Boltzmann equation (7.22) yields the right side [keeping terms to $P_1(\mu)$]:

$$\frac{e\mathcal{E}}{m}\left[\mu\frac{df_0}{dv} + \frac{1}{3}\left(\frac{df_1}{dv} + \frac{2f_1}{v}\right)\right]$$

3.14. **(a)** Obtain an integral expression for the normalization constant K in the absolute Druyvesteyn distribution

$$f_0(x) = K(x+s)^s e^{-x}$$

$$x \equiv \frac{mv^2}{k_B T}$$

in terms of n, T, and m. *Hint*: Recall the normalization

$$\int f(\mathbf{v})\,d\mathbf{v} = n$$

and the connecting formula

$$f_1(v) = \frac{\tau e\mathcal{E}}{m}\frac{df_0}{dv}$$

(b) What does your expression for K reduce to in the limit of no electric field?

3.15. Establish the equality

$$\int g(v^2)\overline{\overline{\hat{\mathbf{v}}\hat{\mathbf{v}}}}\,d\mathbf{v} = \frac{\overline{\overline{I}}}{3}\int g(v^2)\,d\mathbf{v}$$

where $\hat{\mathbf{v}}$ is a unit vector, $\overline{\overline{I}}$ is the identity operator, and $g(v^2)$ is any scalar function. (A generalization of this result is given in Appendix B, Section B.1).

3.16. Show that the coarse-grained entropy $\eta(t)$ as given by (8.14) is minimum when $\Pi(\mathbf{z})$ is uniform. *Hint*: Let $\Pi(\mathbf{z})$ be uniform. Introduce a variation $\Pi \rightarrow \Pi' = \Pi \exp\kappa$, where κ is an arbitrary function of $\mathbf{z}$. With Π and Π' obeying the same normalization, obtain

$$\delta\eta = \int (\kappa e^\kappa + 1 - e^\kappa)\Pi(\mathbf{z})\,d\mathbf{z}$$

Then show that this form is greater than or equal to zero.

3.17. Show that for two-particle potential motion, periodic in r, the angle variable $\theta(t)$ satisfies the relation

$$\theta(t+T) = \theta(t) + A$$

where T is the period of r and A is a constant.

Answer

With $L_z = \mu r^2 \dot{\theta}$, we obtain

$$\theta(t) = \frac{L_z}{\mu} \int_0^t r^{-2}(t)\,dt$$

It follows that

$$\theta(t+T) - \theta(t) = \frac{L_2}{\mu} \int_t^{t+T} r^{-2}(t)\,dt \equiv A$$

Note in particular that

$$J_\theta = \int_{\theta(t)}^{\theta(t+T)} p_\theta\,d\theta = L_2 \int_{\theta(t)}^{\theta(t+T)} d\theta = L_z A$$

which is constant.

3.18. A two-dimensional harmonic oscillator has the Hamiltonian

$$H(1,2) = H_1 + H_2$$
$$H_1 = \frac{1}{2}(p_1^2 + \omega_1^2 q_1^2)$$
$$H_2 = \frac{1}{2}(p_2^2 + \omega_2^2 q_2^2)$$

Another constant of the motion, independent of H, is given by

$$A = H_1 - H_2$$

(a) What is the Poisson bracket of A and $H(1,2)$?

(b) Demonstrate a dynamical form that for degeneracy $\bar{n}_1\omega_1 = \bar{n}_2\omega_2$ gives a third constant independent of A and H, but that in the event of nondegeneracy, is not constant.

Answer (partial)

(b) Introduce the complex phase variable

$$z_j(t) = p_j(t) + i\omega_j q_j(t), \qquad j = 1, 2$$

The dynamical trajectory may be written

$$z_j(t) = z_j(0)e^{i\omega_j t}$$

Consider the form

$$B \equiv \frac{z_1^{\bar{n}_1}}{z_2^{\bar{n}_2}} = \frac{z_1(0)^{\bar{n}_1}\,e^{i\bar{n}_1\omega_1 t}}{z_2(0)^{\bar{n}_2}\,e^{i\bar{n}_2\omega_2 t}}$$

In the degenerate limit, we find that B is constant. Note that H and A may be written

$$H = \frac{1}{2}(z_1 z_1^* + z_2 z_2^*)$$
$$A = \frac{1}{2}(z_1 z_1^* - z_2 z_2^*)$$

With these relations, it is evident that the modulus of B is a function of H and A, but that the phase angle of B is independent of H and A.

3.19. Show that the differential measure of the cross-collision integral of the Boltzmann equation for a two-component fluid may be written

$$\sigma g \, d\Omega \, d\mathbf{v}_1 = \int_0^\infty \sigma \delta(\Delta E) \, d\mathbf{q} \, d\mathbf{v}_1$$

Here $\mathbf{q}$ is the momentum increment,

$$m_1(\mathbf{v}_1 - \mathbf{v}_1') = m_2(\mathbf{v}' - \mathbf{v}) \equiv \mathbf{q} = M\mathbf{u} = \mu(\mathbf{g}' - \mathbf{g})$$

$$M = m_1 + m_2, \qquad \mu = \frac{m_1 m_2}{M}$$

and ΔE is the energy increment,

$$\Delta E = \frac{\mu}{2}(g'^2 - g'^2)$$

The integration is over the scalar component of $\mathbf{q}$.
Recall

$$\int h(y)\delta[f(y) - a] \, dy = \left. \frac{h(y)}{|df/dy|} \right\}_{\substack{y=y_0 \\ f(y_0)=a}}$$

Answer

First note that

$$d\mathbf{g}' = g'^2 \, dg' \, d\Omega$$

(See Fig. 3.4) and recall

$$\delta\left(\frac{g'^2 - g^2}{2}\right) = \frac{1}{g'}\delta(g' - g)$$

It follows that

$$\sigma g \, d\Omega = \int_0^\infty \sigma \delta(g' - g) g' \, dg' \, d\Omega$$

$$= \int_0^\infty \sigma \delta\left(\frac{g'^2 - g^2}{2}\right) dg'$$

Now

$$d\mathbf{g}' = d(\mathbf{g}' - \mathbf{g}) = \frac{M}{\mu} \, d\mathbf{u}$$

so that

$$\sigma g \, d\Omega = \int_0^\infty \sigma \delta\left(\frac{g'^2 - g^2}{2}\right) \frac{M}{\mu} \, d\mathbf{u}$$

$$= \int_0^\infty \sigma \delta\left[\frac{\mu}{2}(g'^2 - g^2)\right] M \, d\mathbf{u}$$

$$= \int \sigma \delta\left[\frac{\mu}{2}(g'^2 - g^2)\right] d\mathbf{q}$$

Note that

$$\Delta E = \frac{\mu}{2}(g'^2 - g^2)$$

and we may write

$$\sigma g \, d\Omega \, d\mathbf{v}_1 = \int \sigma \delta(\Delta E) \, d\mathbf{v}_1 \, d\mathbf{q}$$

which was to be shown. Inserting this finding into the cross-collision integral gives the form [see (2.16)]

$$J(f \mid F) = \int (f'F_1' - fF_1)\sigma \delta(\Delta E) \, d\mathbf{v}_1 \, d\mathbf{q}$$

3.20. A point particle of mass m with speed v moves inside a spherical cavity of radius a with perfectly reflecting walls. Initial normal displacement of the particle trajectory from the origin (that is, impact parameter) is s.

(a) What are the constants of the motion for this system? Is angular momentum of the particle constant? Why?

(b) Consider the coordinate components of the energy surface in Γ-space for this system. Is the particle motion on this surface ergodic? Why? Describe the exact motion of the particle.

(c) How do your answers to the above change if the cavity is a rectangular parallelpiped?

Answers (partial)

(a) In addition to the kinetic energy of the particle, angular momentum, $\mathbf{L}$, is also a constant. The reason for this is that the force due to the wall is radial.

(b) Since there is a constant other than energy, the motion is nonergodic. The motion sweeps out an annulus that lies in a plane of constant orientation and has inner radius s.

3.21. (a) Employ (3.40) to calculate the change in entropy corresponding to the expansion process depicted in Fig. 3.14. Assume that the process occurs with both gases in Maxwellian states at fixed temperature T. Let the number of A and B molecules be N_A and N_B, respectively, with $N_A = N_B$, and call the total volume V.

(b) How does your answer change if labels in the molecules are removed (Gibb's paradox)?

3.22. Establish the following properties of the symmetric strain tensor (4.6a).

(a)

$$\overline{\overline{P}} : \overline{\overline{\nabla \mathbf{u}}} = \overline{\overline{P}} : \overline{\overline{\Lambda}}$$

(b) If $\overline{\overline{P}} = \overline{\overline{I}} p$, then $\overline{\overline{P}} : \overline{\overline{\Lambda}} = p \nabla \cdot \mathbf{u}$, where

$$p = \tfrac{1}{3} \operatorname{Tr} \overline{\overline{P}}$$

(c) Evaluate the Tr $\overline{\overline{\Lambda}}$.

3.23. Relevant to the Chapman–Enskog expansion [see (5.44)], show that

$$\int F^0 A(\xi^2)\xi^2 \, d\mathbf{c} = 0$$

3.24. Show that the functional operation (3.2)

$$\hat{I}(\phi) \equiv \int \hat{J}(f)\phi(\mathbf{v}) \, d\mathbf{v}$$

has the following properties:

(a) $\hat{I}(\phi) = \frac{1}{2}\int(\phi' + \phi_1' - \phi - \phi_1)ff_1g\sigma \, d\Omega \, d\mathbf{v} \, d\mathbf{v}_1$
(b) $\hat{I}(\phi) = \int(\phi' - \phi)ff_1g\sigma \, d\Omega \, d\mathbf{v} \, d\mathbf{v}_1$

3.25. Show that the stress tensor

$$\overline{\overline{S}} = 2\eta\left(\overline{\overline{\Lambda}} - \frac{1}{3}\overline{\overline{I}}\nabla\cdot\mathbf{u}\right) + \zeta\nabla\cdot\mathbf{u}$$

vanishes if $\mathbf{u} = \boldsymbol{\Omega}\times\mathbf{v}$, where $\boldsymbol{\Omega}$ is a constant angular velocity vector and $\overline{\overline{\Lambda}}$ is the symmetric strain tensor (4.6a).

3.26. (a) The equipartition theorem assigns $k_B T/2$ units of energy on the average to each degree of freedom for a system in equilibrium. Using this rule, obtain the value of c_V, the specific heat per molecule, for a gas of rigid diatomic molecules.

(b) For gaseous N_2 at room temperature, viscosity $\eta = 1.78\times 10^{-4}$ g/cm-s. Employing elementary mean-free-path estimates, obtain the value of thermal conductivity κ for a gas of N_2 at these conditions.

3.27. Consider a collection of N noninteracting particles confined to cubical box of edge length a, with perfectly reflecting walls. The speeds of the particles have values $\{v_{0i}\}$, $1 \le i \le N$. Express the motion $\{x_i(t)\}$ of this aggregate of particles in action-angle variables.

3.28. (a) Construct a simple kinetic model to show that the xx component of stress interior to a gas at rest is

$$p_{xx} = 2m\int d\mathbf{v}_\perp \int_0^\infty dv_x v_x^2 F(\mathbf{v})$$

where $\mathbf{v}_\perp = (v_y, v_z)$.

(b) The working definition of p_{xx} is that it is the force exerted on a unit area within the gas whose normal is parallel to the x axis. State a property of $F(\mathbf{v})$ at the surface of the test area that permits the preceding integral to be written

$$p_{xx} = m\int d\mathbf{v}_\perp \int_{-\infty}^\infty dv_x v_x^2 F(\mathbf{v}) = mn\langle v_x^2\rangle$$

3.29. Employing the continuity equation, together with the energy equation (3.31), derive the adiabatic law (5.23c).

Answer

Our starting equations are

$$\left(\frac{\partial}{\partial t} + \mathbf{u} \cdot \nabla\right) E_k + E_k \nabla \cdot \mathbf{u} + p \nabla \cdot \mathbf{u} = 0$$

$$\left(\frac{\partial}{\partial t} + \mathbf{u} \cdot \nabla\right) n + n \nabla \cdot \mathbf{u} = 0$$

Substituting

$$p = n k_B T = \frac{2}{3} E_k$$

into the first of these relations gives

$$\frac{D}{Dt}\left(\tfrac{3}{2}p\right) + \tfrac{5}{2}p \nabla \cdot \mathbf{u} = 0$$

whereas the continuity equation gives

$$\nabla \cdot \mathbf{u} = -\frac{D}{Dt}\ln n$$

Eliminating $\nabla \cdot \mathbf{u}$ from these latter two equations gives the desired result:

$$\frac{D}{Dt}\ln\left(\frac{P}{n^{5/3}}\right) = 0$$

3.30. Under what conditions is a dynamical system with N degrees of freedom integrable?

3.31. The orbits of a dynamical system with two degrees of freedom are given by

$$q_1 = q_1(0) \cos \omega_1 t$$
$$q_2 = q_2(0) \cos \omega_2 t$$

(a) If $\omega_2 = 2\omega_1$, what is the nature of the motion in the $q_1 - q_2$ plane?
(b) What is the minimum frequency associated with this motion?

3.32. A fluid is in a state of shear. A narrow slab of the fluid moves in the x direction with the velocity profile

$$v_x(y) = -\frac{a}{\tau}\xi e^{-w^2/2}$$

where a and τ are characteristic length and time intervals, respectively, and $w \equiv y/a$. The slab is centered at $w = 0$. The related xy component of stress in the fluid has the value

$$S_{xy} = 2.6\rho \left(\frac{a}{\tau}\right)^2 e^{-w^2/2}$$

where ρ is the mass density of the fluid. If $a = 2.4\,\overset{\circ}{A}$, and the thermal speed of molecules $C = 1.5a/\tau$, estimate the value of the mean free path of molecules, l, in the fluid. What are the assumptions of your estimate?

3.33. Employ the Hamilton–Jacobi equation (8.30) to solve for the orbit of a particle of mass m in the one-dimensional potential well

$$V(x) = \kappa \left(\frac{x^2}{2} + ax \right)$$

where a and κ are constants.

3.34. Repeating the development leading to the Fourier transform of electrical conductivity (4.72), derive a parallel expression for viscosity for a fluid in a state of shear. Identify all terms in your derivation. Show that your finding returns Maxwell's expression (4.33) in the mean-free-path estimate.

3.35. For attractive potentials (1.21a), the nondimensional inverse minimum displacement appears as

$$\bar{\beta}^2 = 1 + \left(\frac{\bar{\beta}}{b} \right)^N$$

Show that this equation has no positive solutions for $N \geq 2$, $b < 1$, and $E > 0$.

3.36. Consider the scattering of two particles with respective masses m_1 and m_2. The angle of scatter in the lab frame of particle 1 is the angle between $\mathbf{p}_1'$ and $\mathbf{p}_1$. Calling this angle θ_1 and the corresponding angle of scatter in the center-of-mass frame θ, show that

$$\tan \theta_1 = \frac{m_2 \sin \theta}{m_1 + m_2 \cos \theta}$$

3.37. Calling the cross section in the lab frame σ_L and that in the center-of-mass frame σ_C, obtain the equality

$$\sigma_L(\theta_1) \, d \cos \theta_1 = \sigma_C(\theta) \, d \cos \theta$$

3.38. What is $\sigma_C(\theta)$ for:

(a) Scattering of rigid spheres with respective diameters σ_{01} and σ_{02}?
(b) Scattering of Maxwell molecules of mass m?
(c) Coulomb scattering of particles with charge q and mass m?

3.39. Show that BY_1 [the first equation in the sequence (2.1.20)] implies the three conservation equations (3.14) (3.18), and (3.19). Note that the pressure tensor $\overline{\overline{p}}$ and heat flow vector $\mathbf{q}$ must be redefined to obtain these equations. See, for example, (2.1.34).

3.40. Show that the $\Omega^{(l,q)}$ cross section integrals (5.62) satisfy the equation

$$\frac{T \partial \Omega^{(l,q)}}{\partial T} + \left(q + \frac{3}{2} \right) \Omega^{(l,q)} = \Omega^{(l,q+1)}$$

3.41. (a) Equilibrium values of a fluid at rest are $\bar{n} = n_0$ and $\bar{T} = T_0$. Working with the Euler equations and assuming one-dimensional motion, obtain linear equations of motion for the perturbation variables n, u, and T, where, after a small disturbance, $\bar{n} = n_0 + n$, $\bar{u} = 0 + u$, and $\bar{T} = T_0 + T$.

(b) From your equations, show that the sound speed a is given by the isentropic derivative

$$a^2 = \frac{\partial p}{\partial \rho}$$

where ρ is mass density.

3.42. A bomb containing 1 mole of argon atoms at pressure $p = 3$ atm and temperature $T = 300K$ has a release valve of circular area $10^{-4}\pi \, cm^2$. The valve is open for $10^{-4}s$. The gas cools and then returns to the ambient temperature of 300 k.

Data: Atomic mass of argon $= 40$ amu. 1 amu $= 1.66 \times 10^{-24}$ g. 1 atm $= 10^6$ dyn/cm^2.

(a) Let the gas be *ideal* prior to the valve being open. What is the volume of the bomb in cubic centimeters?

(b) Write down the Maxwellian distribution function f for the gas prior to the valve opening. [The units of f are cm^{-3}(cm/s)$^{-3}$].

(c) Consider the flux J_+, of particles that pass through the valve opening. We may write

$$J_+ = n \langle v_x^+ \rangle.$$

The superscript denotes the fact that we are looking for the average velocity of particles in the $v_x > 0$ direction. Use this expression to obtain the mole fraction of particles that escape the bomb while the valve is open. Sketch the function $f(v_x)$ immediately after the valve is shut.

(d) Approximate the immediate drop in temperature, δT, of the bomb. *Hint*: Assume $\langle v^2 \rangle$ is constant and guess the value of $\langle v \rangle^2$ from your answer to part (c). Introduce the speed $C^2 \equiv 2k_B T/m$.

3.43. Let λ denote an eigenvalue of the $\hat{\mathcal{K}}$ operator defined by (6.29). Imagine circles of varying radii drawn about the origin in complex λ-space. What may be said about the number of eigenvalues of $\hat{\mathcal{K}}$ in each domain described by these circles?

3.44. In Section 2.2 an equation of state was given (2.2.53) for a liquid in equilibrium, which we now rewrite as

$$P = P_k + P_\Phi$$
$$P_K = nk_B T$$
$$P_\Phi = -\frac{n^2}{6} \int \Phi'(r)g(r)4\pi r^3 \, dr$$

Derive the latter expression for P_Φ from BY$_1$ (2.1.24). That is, take the appropriate moment of BY$_1$ to obtain the momentum equation and put the resulting collision integral in gradient form to identify P_Φ.

Answer

Operating on BY_1 with $\int d\mathbf{c} m \mathbf{c}$ and performing a parts integration ont he collision integral, we obtain

$$\rho\left(\frac{\partial}{\partial t}+\mathbf{u}\cdot\nabla\right)\mathbf{u}+\nabla\cdot\overline{\overline{P}}_K-N^2\int d\mathbf{c}_1\,d\mathbf{c}_2\,d\mathbf{x}_2\mathbf{G}_{12}f_2(1,2)=0$$

We may write this equation in the form

$$\rho\left(\frac{\partial}{\partial l}+\mathbf{u}\cdot\nabla\right)\mathbf{u}+\nabla\cdot(\overline{\overline{P}}_K+\overline{\overline{P}}_\Phi)=0$$

where we have set

$$\nabla\cdot\overline{\overline{P}}_\Phi=-N^2\int d\mathbf{c}_1\,d\mathbf{c}_2\,d\mathbf{x}_2\mathbf{G}_{12}f_2(1,2)$$

(Note that ∇ is written for ∇_1.) For a nonequilibrium fluid, we assume anisotropy in the fluid and write the following generalization of (2.2.44):

$$f_2(1,2)=f_0(c_1)f_0(c_2)g(\mathbf{x}_1,\mathbf{x}_2)$$

Substituting this relation into the preceding equation and integrating out the velocities gives

$$\nabla\cdot\overline{\overline{P}}_\Phi=\frac{N^2}{V^2}\int d\mathbf{x}_2\mathbf{G}_{12}(r)g(\mathbf{x}_1,\mathbf{x}_2)$$

Recalling the transformation introduced in Problem 2.14 from $\mathbf{x}_1,\mathbf{x}_2\to\mathbf{r},\mathbf{R}$, we write $g(\mathbf{x}_1,\mathbf{x}_2)\to g(\mathbf{r},\mathbf{R})$ and, with $\mathbf{r}=\mathbf{x}_2-\mathbf{x}_1$, set $d\mathbf{x}_2=d\mathbf{r}$. In equilibrium, the radial distribution function goes to the isotropic form $g(r)$, which causes the preceding integral to vanish. To obtain a finite result, we return to the form $g(\mathbf{r},\mathbf{R})$ and expand about $\mathbf{R}=\mathbf{x}_1$. There results

$$g(\mathbf{r},\mathbf{R})=g(\mathbf{r},\mathbf{x}_1)+\frac{\mathbf{r}}{2}\cdot\nabla_1 g(\mathbf{r},\mathbf{x}_1)+\cdots$$

(Note that $\mathbf{r}$ and $\mathbf{x}_1$ are new independent variables.) Further setting $\mathbf{G}_{12}(r)=\hat{\mathbf{r}}\,d\Phi(r)/dr$ and substituting these relations into the preceding equation gives

$$\nabla_1\cdot\overline{\overline{P}}_\Phi=-\frac{N^2}{V^2}\int d\mathbf{r}\Phi'(r)\hat{\mathbf{r}}g(\mathbf{r},\mathbf{x}_1)-\frac{N^2}{2V^2}\nabla_1\cdot\int d\mathbf{r}\Phi'(r)\frac{\overline{\overline{\mathbf{rr}}}}{r}g(\mathbf{r},\mathbf{x}_1)$$

As the fluid becomes isotropic, $g\to g(r)$ and the first integral vanishes. There remains

$$\overline{\overline{P}}_\Phi=-\frac{N^2}{2V^2}\int d\mathbf{r}\Phi'(r)\frac{\overline{\overline{\mathbf{rr}}}}{r}g(r)$$

Taking the trace and passing to the thermodynamic limit gives the desired result:

$$P_\Phi=\frac{1}{3}\,\mathrm{Tr}\,\overline{\overline{P}}_\Phi=-\frac{n^2}{6}\int_0^\infty dr4\pi r^3\Phi(r)g(r)$$

3.45. (a) Write down an expression for the force **F** on a small volume of fluid in terms of a closed surface integral of the pressure tensor $\overline{\overline{P}}$.

(b) What *form* does your expression for **F** reduce to if the fluid is in equilibrium and at rest.

Answer

(a) $\mathbf{F} = \oint \overline{\overline{P}} \cdot d\mathbf{S}$

(b) $\mathbf{F} = \oint p\overline{\overline{I}} \cdot d\mathbf{S} = 0$

3.46. (a) An ideal gas of N molecules is in equilibrium at temperature T for which $\langle \tilde{E} \rangle = \frac{3}{2}k_B T$ and $\langle (\tilde{E} - \langle \rangle \tilde{E})^2 \rangle = k_B T \langle \tilde{E} \rangle$, where $\tilde{E}$ is single-particle energy. Using the central-limit theorem, obtain the probability density $P(E, N)$ for this system, where E represents total energy.

(b) What are the assumptions that make this theorem relevant to this system?

(c) What is the variance σ^2 of the distribution you have found? In what manner does $\sigma/\langle E \rangle$ make $\langle E \rangle$ a *good thermodynamic variable* in the limit $N \to \infty$?

Answers

(a) From the central-limit theorem, we obtain

$$P(E, N) = \frac{1}{[2\pi N D(\tilde{E})]^{1/2}} \exp\{-\frac{[E - N\mathcal{E}(\tilde{E})]^2}{2N D(\tilde{E})}\}$$

Inserting the given information, we find

$$P(E, N) = \frac{1}{[3\pi N (k_B T)^2]^{1/2}} \exp\{-\frac{[E - \frac{3}{2}N k_B T]^2}{3N(k_B T)^2}\}$$

(b) Assumptions that permit use of the central-limit theorem are:

(i) $E = \sum_{i=1}^{N} \tilde{E}_i$, $N \gg 1$.

(ii) $\langle E_i \rangle$ are all equal.

(iii) $\langle D(\tilde{E}_i) \rangle$ are all equal.

(c) From our answer to part (a), we write

$$\langle E \rangle = \frac{3}{2} N k_B T$$

$$\sigma^2 = \frac{3}{2} N (k_B T)^2$$

so that

$$\frac{\sigma}{\langle e \rangle} = \frac{1}{\sqrt{(3/2)N}} \ll 1$$

Thus the Gaussian is sharply peaked about $E = \langle E \rangle$ so that E is effectively constant and therefore may be taken to be a good thermodynamic variable. *Note:* For the system at hand, E must be positive. However, from part (a) there is a finite probability that measurement find $E < 0$. The central-limit theorem addresses the case $N \gg 1$, in which case $\langle E \rangle$ grows large and $P(E < 0) \sim 0$.

3.47. (a) Obtain the expression for viscosity that the repulsive-core model (5.86) gives for $N = 4$ (Maxwell molecules). Set $K = V_0 a^4$, where V_0 and a are constant energy and length, respectively. [Hint: You should find $\eta(T) = \text{constant}$.]

(b) What value of viscosity does your formula give when applied to a gas of argon atoms at room temperature? Take $a = 1$ Å and $V_0 = 10$ eV. Give your answer in cgs units. [Hint: You should find $\eta(T) = AT^{-1/3}$.]

3.48. Employing fluid dynamical equations derive the following:

(a) $T \dfrac{D}{Dt} \sigma = \dfrac{1}{n} \left(\overline{\overline{\nabla \mathbf{u}}} : \overline{\overline{S}} - \nabla \cdot \mathbf{Q} \right)$

(b) $m \dfrac{D\mathbf{u}}{Dt} + \nabla h - T\nabla\sigma - \dfrac{1}{n} \nabla \cdot \overline{\overline{S}} = m\mathbf{K}$

where $m\mathbf{K}$ is an applied force field and σ, h, and n^{-1} are entropy, enthalpy, and volume per particle. The heat flow vector is $\mathbf{Q}$ (3.25).

(c) For $\sigma = \text{constant}$ and $\mathbf{K}$ conservative and $\overline{\overline{S}}$ symmetric, show that

$$\nabla \times \left(\dfrac{D\mathbf{u}}{Dt} \right) = 0$$

[where D/Dt is the conservative derivative given by (3.20a)]. From this result, show that

$$\oint_{C(t)} \mathbf{u} \cdot d\mathbf{l} = 0$$

where $C(t)$ denotes a closed loop moving with the fluid (Kelvin's theorem).[76]

Hint: Recall the energy equation (3.31), the defining relation for the stress tensor (4.4), and the first law of thermodynamics

$$d\tilde{E}_K = T\, d\sigma - p\, dn^{-1}$$

Enthalpy per particle, h, is given by

$$h = \bar{E}_K + \dfrac{p}{n}$$

and $\bar{E}_K$ denotes kinetic energy per particle.

3.49. Using the moment expansion (5.103), show that for the special case of Maxwell interaction [see (1.27)] the related collision integral (5.102) reduces to

$$J_{ij}^{(2)} = -\dfrac{3}{2} b_1 n a_{ij}^{(2)}$$

---

[76]For further discussion, see L. D. Landau and E. M. Lifshitz, *Fluid Mechanics*, Section 8, Addison-Wesley, Reading, Mass. (1959).

where

$$b_1 = \pi \int_0^\pi g\sigma \sin^2\theta \cos^2\theta\, d\theta$$

3.50. Determine if the following equations are reversible. That is, show that, if $y(t)$ is a solution, then so is $y(-t)$. (Primes denote time derivatives.)

(a) $y' = \int_0^a y(x)\, dx$ **(c)** $y' = \int_0^t y(x)\, dx$

(b) $y' = \int_0^a xy(x)\, dx$ **(d)** $y' = \int_0^a y(x - t)\, dx$

Answer

In part (a) we question whether $y(-t)$ is also a solution. Substituting $y(-t)$ for $y(t)$, we obtain

$$\frac{dy(-t)}{dt} = \int_0^a y(-x)\, dx$$

or, equivalently,

$$\frac{dy(\tau)}{d\tau} = \int_0^{-a} y(z)\, dz$$

which is not Eq. (a), so $y(-t)$ is not a solution and Eq. (a) is irreversible. Alternatively, we might have shown that the given equation is not invariant under the time-reversal transformation, $t \to t' = -t$. In a manner similar to the first, we find that (b) is irreversible, (c) is reversible, and (d) is irreversible. Note that differentiation of (a) gives a reversible equation.

3.51. Is the Boltzmann equation (2.14) reversible? That is, if $f = f(\mathbf{x}, \mathbf{v}, t)$ is a solution, is $f(\mathbf{x}, -\mathbf{v}, -t)$ also a solution?

3.52. Is Grad's second equation (2.6.12) reversible. That is, if $f_1(\mathbf{v}_1, t)$; $f_2(\mathbf{v}_1, \mathbf{v}_2, t)$ is a solution, is $f_1(-\mathbf{v}_1, -t)$; $f_2(-\mathbf{v}_1, \mathbf{v}_2, -t)$ also a solution? If your answer is yes, explain this result in light of the fact that Grad's second equation is approximate.

3.53. Derive an expression for the Hamiltonian of a freely moving rigid sphere of radius a and mass M in terms of the kinetic energy of its center of mass and the *spin* angular momentum $\mathbf{S}$. State explicitly what the variable S^2 is written for.

Answer

From Problem 1.7 we know that the Hamiltonian of the sphere may be written

$$H = \frac{p^2}{2M} + H_{rel}$$

where the subscript rel denotes motion relative to the center of mass. This relative motion is purely rotational, and for a rigid sphere we may write

$$H_{rel} = T = \frac{I}{2}(\dot\phi^2 \sin^2\theta + \dot\theta^2)$$

where I is the moment of inertia, which for a sphere is $\frac{2}{3}Ma^2$. Angles (θ, ϕ) specify the orientation of the rotational frequency vector of the sphere, ω, in terms of which $T = I\omega^2/2$. Taking derivatives of the preceding gives

$$\frac{\partial T}{\partial \dot{\theta}} = I\dot{\theta} = p_\theta$$

$$\frac{\partial T}{\partial \dot{\phi}} = I(\sin^2 \theta)\dot{\phi} = p_\phi$$

which gives the Hamiltonian

$$H_{rel} = \frac{p_\theta^2}{2I} + \frac{p_\phi^2}{2I \sin^2 \theta} = \frac{S^2}{2I}$$

3.54. This problem concerns "rough" and "smooth" hard spheres, where the quoted adjectives refer to the nature of surfaces of respective spheres. As noted in Section 2.3, smooth spheres do not exchange angular momentum in collision. Let all spheres have volume $\tau = 4\pi a^3/3$, mass M, and moment of inertia I. Consider a gas of N smooth spheres confined to a volume $V \gg N\tau$ in equilibrium at temperature T.

(a) Write down the Hamiltonian for this fluid.
(b) Write down an integral expression for the pair distribution function for this fluid, $f_2(\mathbf{x}_1, \mathbf{x}_2; \mathbf{S}_1, \mathbf{S}_2)$.
(c) If spheres in the fluid described above are rough, is your answer to part (a) maintained? If not, why not?

Answers

(a) With reference to Problem 3.53, we write

$$H(1, \ldots, N) = \sum_{i=1}^{n} \left(\frac{P_i^2}{2M} + \frac{S_i^2}{2I} \right) + \sum_{i<j}^{N} \theta(r_{ij} - 2a)$$

where

$$\mathbf{r}_{ij} = |\mathbf{x}_i - \mathbf{x}_j|$$

and

$$\theta(x) = \infty, \qquad \text{for } x \leq 0$$
$$\theta(x) = 0, \qquad \text{for } x > 0$$

(b) The equilibrium N-body distribution function (see Problem 2.5) for this fluid is given by

$$f_N(1, \ldots, N) = A_N \exp\left[-\frac{H(1, \ldots, N)}{k_B T} \right]$$

where A_N is a normalization constant.

Thus, with $r = r_{12}$, we obtain

$$f_2(\mathbf{P}_1, \mathbf{P}_2, \mathbf{S}_1, \mathbf{S}_2, r) = \int d3 \cdots dN \, A_N \exp\left[-\frac{H(1, \ldots, N)}{k_B T}\right]$$

Integration over $\mathbf{P}_1$ and $\mathbf{P}_2$ gives the desired result. See Appendix B, Section B.3.

(c) For rough spheres there is interaction between angular momentum of spheres in collision, and the above expression for H_N is not appropriate.

3.55. (a) Which dynamical systems corresponding to the following Hamiltonians satisfy the criterion for the KAM theorem?

(b) What is the consequence of the failure of a dynamical system to satisfy this theorem?

(c) Are any of the orbits corresponding to these systems homoclynic (see Problem 1.45)?

(i) *Harmonic oscillator*:

$$H = \frac{p^2}{2m} + \frac{kx^2}{2}$$

(ii) *Pendulum*:

$$H = \frac{p_\theta^2}{2ma^2} - mga \cos\theta$$

where $\theta = 0$ corresponds to the gravity direction.

(iii) *Anti Hénon–Heiles Hamiltonian*:

$$H = \frac{1}{2}(q_1^2 + p_1^2 + q_2^2 + p_2^2) + \frac{q_1^3}{3} + q_1 q_2^2$$

where we have written

$$q_i^2 \equiv k_i \bar{x}_i^2, \qquad p_i^2 \equiv \frac{\bar{p}_i^2}{m_i}$$

with $\bar{x}$ and $\bar{p}$ denoting physical displacement and momentum, respectively. *Hint*: In part (iii), introduce the change of variables

$$\begin{aligned} Q_1 = q_1 + q_2, \qquad & P_1 = p_1 + p_2 \\ Q_2 = q_1 - q_2, \qquad & P_2 = p_1 - p_2 \end{aligned}$$

to obtain

$$2H = H_1(Q_1, P_1) + H_2(Q_2, P_2)$$
$$H_1 = \frac{1}{2}[Q_1^2 + P_1^2] + \frac{1}{3}Q_1^3$$
$$H_2 = \frac{1}{2}[Q_2^2 + P_2^2] + \frac{1}{3}Q_2^3$$

3.56. Show that the Sonine polynomials $S_m^{(n)}(x)$ satisfy the equation (deleting subscripts and superscripts)

$$xS'' + (m + 1 - x)S' + nS = 0$$

3.57. With f_0 denoting the Maxwellian,

$$f_0(v, x) = \frac{1}{V} \frac{e^{-v^2/2C^2}}{(2\pi)^{3/2}C^3}$$

where $mC^2 \equiv k_B T$ and V is the volume of the fluid, show that:

(a) $\langle v \rangle = 3C^2$
(b) $\langle v \rangle^2 = \frac{8}{\pi}C^2$

3.58. Show that the generalized Gibbs entropy

$$H_\psi \equiv \int \psi(f_N) \ln \psi(f_N) \, d1 \ldots dN$$

is constant in time, providing the functional $\psi(f)$ has a bounded derivative.

Answer

First note that

$$\frac{D\psi \ln \psi}{Dt} = (1 - \ln \psi) \frac{d\psi}{df} \frac{Df}{Dt}$$

Operating on the Liouville equation with

$$\int d1 \cdots dN (1 + \ln \psi) \frac{d\psi}{df}$$

establishes the result.

3.59. (a) What is equilibrium distribution f_1 for a gas of N identical molecules of mass m and temperature T confined to a volume V? Assume that f_1 is normalized to unity.

(b) The gas described in part (a) adiabatically expands to a volume $2V$ and comes to equilibrium at the same temperature T. What is the new distribution, $\bar{f}_1$? Employing the Boltzmann $\mathcal{H}$ function, calculate the entropy change of the gas, ΔS, due to expansion.

Answers (partial)

(a) With $R \equiv k_B/m$,

$$f_1(x, v) = \frac{1}{V(2\pi RT)^{3/2}} e^{-v^2/2RT}$$

(b) Calculating

$$\Delta S = -k_B N \left[\int \bar{f}_1 \ln \bar{f}_1 \, dx \, dv - \int f_1 \ln f_1 \, dx \, dv \right]$$

gives the result

$$\Delta S = k_B N \ln 2$$

3.60. Three particles move on a plane and interact through conservative forces. In general, are there sufficient constants of motion to render the system integrable? Justify your answer.

3.61. The Liouville theorem[77] (concerning integrability of system) states that for a system with N degrees of freedom, if N independent constants, including the Hamiltonian, exist, each of which has zero Poisson brackets with each other, then the motion of the system may be reduced to quadratures. (See top p. 244.)

(a) Consider Problems 1.1 and 1.8. In each case verify the preceding theorem and reduce the second to quadrature.

(b) Do the 10 constants for the three-body problem (8.49) satisfy criteria for the Liouville theorem? How many of these constants do satisfy the criteria?

Hint: In Problem 1.1 note than $p_x(r, \theta)$ is constant. In Problem 1.8 consider motion relative to the center of mass.

Note: Previously we have noted that $2N$ constants are required to specify the trajectory in $\bar{\Gamma}$ space. The Liouville theorem states that N constants satisfying certain conditions, are sufficient to reduce the problem to quadratures. A quadrature is an integral which, for each degree of freedom, implies an added constant bringing the total to $2N$. In action-angle formalism, these $2N$ constants are $[J_i; \Theta_i(0)]$. See (8.24) et seq.

3.62. (a) Consider a force-free spatially homogenous fluid at rest. Employing the equipartition theorem, the conservation equation (3.19) and the definition of the coefficient of thermal conductivity, derive the *heat equation*,

$$\frac{\partial e_K}{\partial t} = \alpha \nabla^2 e_K$$

(b) What is your expression for α?

(c) What property of the structure of molecules in the fluid does your expression for α assume?

Answer (partial)

(b) $\alpha = \left(\dfrac{2\kappa}{3nk_B} \right)$

3.63. (a) The diffusion coefficient is written in terms of the autocorrelation function in (4.53). This relation is more formally written

$$D = \frac{1}{3} \lim_{\delta \to \infty} \int_0^\infty dt\, e^{-\delta t} \langle \mathbf{v}(0) \cdot \mathbf{v}(t) \rangle$$

Write down the corresponding autocorrelation function definition of the coefficient of viscosity, η [see (4.7) et seq.], related to shear force in the x direction due to y-momentum transported by x velocity components. In this construction introduce the momentum transfer component $\Gamma_{xy} \equiv p_x p_y / m$.

---

[77]For further discussion see, V. I. Arnold, *Mathematical Methods of Classical Mechanics*, Chapter 10, *ibid* (1978).

(b) Write down a component force equation in which this expression for η enters. What are the dimensions of the expression for η you have written down. [Check your answer with Maxwell's result (4.33)]

Answer (partial)

$$\eta = \lim_{\delta \to 0} \frac{1}{k_B T} \int_0^\infty dt\, e^{-\delta t} \langle \Gamma_{xy}(0)\Gamma_{xy}(t) \rangle$$

where the average is taken with respect to an equilibrium distribution.

Note: Numerical simulation calculation[78] of the velocity autocorrelation function has observed short-time exponential decay (over a few collision times) and a slower $t^{-3/2}$ long-time delay. This secular long-time decay of the autocorrelation function is labeled "long-time tails."[79]

[78] B. J. Adler, and T. E. Wainwright, *Phys. Rev. Lett. 18*, 988 (1967).
[79] M. H. Ernst, E. H. Hauge and J. M. J. van Leeuwen, *Phys. Rev. A4*, 2055 (1971).

CHAPTER 4

Assorted Kinetic Equations with Applications to Plasmas and Neutral Fluids

Introduction

The notion of a plasma was considered previously in Section 2.3 in describing fundamental plasma intervals and kinetic equations relevant to these intervals. The present chapter begins with applications of the Vlasov equation to an equilibrium charge-neutral plasma.

The plasma frequency and Debye length emerge in construction of the Fourier-transformed dielectric constant. The Debye parameter appears as a shielding length in the construction of the perturbed Coulomb potential due to the presence of an extraneous charge in the plasma. The dielectric constant further permits analysis of unstable and damped modes. Landau damping of an electric wave in a plasma is described, and the section concludes with the Nyquist analysis of unstable modes.

Various kinetic equations relevant to the description of a plasma are derived. Interrelations between these equations are obtained and illustrated in a flow-chart diagram (Fig. 4.9). The Fokker-Planck (FP) equation is obtained in expansion of the Boltzmann equation about grazing collisions. Equivalence between the FP and Landau equations is demonstrated. The Balescu-Lenard equation is obtained from an expansion of the hierarchy equations incorporating construction of the **E**-field auto correlation function. Divergence of plasma kinetic equations is discussed, and techniques for removing related singularities are reviewed.

The Krook-Bhatnager-Gross equation is applied to shock waves in a neutral fluid, and the chapter continues with a rederivation of the FP equation from the Chapman-Kolmogorov equation. This derivation illustrates the application of the FP equation to systems other than plasmas.

Due to its present-day widespread application, the final section of the chapter addresses the Monte-Carlo technique in kinetic theory. The discussion concludes with a flow chart for determination of the distribution function of a fluid immersed in an external force field.

4.1 Application of the Vlasov Equation to a Plasma

In this first section we employ the previously derived Vlasov equation in application to various wave and particle problems in plasma physics. Thus we consider a charge-neutral plasma comprised of electrons and ions in equilibrium at a given temperature. The plasma suffers an infinitesimal perturbation, and the resulting electric field and electron distribution are examined, which gives the very important plasma *response dielectric function*. When applied to perturbation in the form of an extraneous charge, this function yields the *Debye potential*, which reveals that the Coulomb field in a plasma is exponentially attenuated with a scale length equal to the Debye distance. The dielectric function is next applied to waves in a plasma, and the dispersion relation for long-wavelength longitudinal waves is found, which reveals that dispersion of waves grows with temperature. It is then shown that single-peaked electron distributions (in velocity space) are stable to infinitesimal perturbations.

The Vlasov equation is returned to and, working with a Laplace-Fourier transform, it is concluded that waves in a plasma experience (Landau) damping. An expression for the damping rate is obtained, and a physical description is included, which ascribes the damping of the electric-field wave in a plasma to increased kinetic energy of electrons.

The section concludes with a derivation of the Nyquist criterion, which, stemming from a theorem of complex analysis, offers a geometrical construction for the number of unstable modes in a plasma. Application of this result is made to multipeaked electron distributions.

4.1.1 Debye Potential and Dielectric Constant

We consider a charge-neutral plasma of electrons and ions in equilibrium at some temperature T. Ion $(+)$ and electron $(-)$ distribution functions are given by (assuming singly ionized atoms)

$$F_0^+ = \frac{n_0}{(2\pi R_+ T)^{3/2}} \exp\left[-\frac{v^2}{2R_+ T}\right] \tag{1.1a}$$

$$F_0^- = \frac{n_0}{(2\pi R_- T)^{3/2}} \exp\left[-\frac{v^2}{2R_- T}\right] \tag{1.1b}$$

$$R_\pm = k_B/m_\pm, \qquad \int F_0^\pm \, d\mathbf{v} = n_0$$

We assume that distribution functions satisfy the Vlasov equations [see (2.2.31)]:

$$\frac{DF_+}{Dt} + \frac{e}{m_+}\mathcal{E} \cdot \frac{\partial F_+}{\partial \mathbf{v}} = 0 \tag{1.2a}$$

$$\frac{DF_-}{Dt} - \frac{e}{m_-}\mathcal{E} \cdot \frac{\partial F_-}{\partial \mathbf{v}} = 0 \tag{1.2b}$$

where e is written for $|e|$ and we have set

$$\frac{D}{Dt} \equiv \frac{\partial}{\partial t} + \mathbf{v} \cdot \frac{\partial}{\partial \mathbf{x}}$$

Gauss's law links the distribution functions to $\mathcal{E}$.

$$\nabla \cdot \mathcal{E} = 4\pi e[n_+ - n_-] = 4\pi e\left[\int F_+\,d\mathbf{v} - \int F_-\,d\mathbf{v}\right] \tag{1.3}$$

The equilibrium value of the electric field is zero, and $n_\pm$ represent number densities.

We assume that the plasma suffers a small perturbation away from equilibrium. Variables assume the form[1]

$$F_- = F_0^- + \varepsilon g, \qquad F_+ = F_0^+ \tag{1.4}$$
$$\mathcal{E} = 0 + \varepsilon E$$

Due to their large mass, the ions are assumed to remain in their unperturbed state. Substituting the forms (1.4) into (1.2) gives the $O(\varepsilon)$ equations (deleting the superscript on F_0^-):

$$\frac{\partial g}{\partial t} + \mathbf{v} \cdot \frac{\partial}{\partial \mathbf{x}} g - \frac{e}{m} \mathbf{E} \cdot \frac{\partial F_0}{\partial \mathbf{v}} = 0 \tag{1.5a}$$

$$\nabla \cdot \mathbf{E} = -4\pi e \int g\,d\mathbf{v} \tag{1.5b}$$

Substituting the transforms[2]

$$g = \frac{1}{(2\pi)^4} \int d\mathbf{k} \int d\omega \bar{g}(\omega, \mathbf{k}, \mathbf{v}) e^{i(\mathbf{k}\cdot\mathbf{x} - \omega t)} \tag{1.6a}$$

$$\mathbf{E} = \frac{1}{(2\pi)^4} \int d\mathbf{k}\,d\omega \bar{\mathbf{E}}(\omega, \mathbf{k}) e^{i(\mathbf{k}\cdot\mathbf{x} - \omega t)} \tag{1.6b}$$

into (1.5), we find

$$(\omega - \mathbf{k} \cdot \mathbf{v})\bar{g} = i\frac{e}{m}\bar{\mathbf{E}} \cdot \frac{\partial F_0}{\partial \mathbf{v}} \tag{1.7}$$

---

[1] The present discourse is restricted to linearized theory. For further discussion, see R. C. Davidson, *Methods in Nonlinear Plasma Theory*, Academic Press, New York (1972).

[2] Here we are assuming a response that persists indefinitely.

We may decompose the $\bar{\mathbf{E}}$ vector as

$$\bar{\mathbf{E}} = \frac{\mathbf{k}(\mathbf{k} \cdot \bar{\mathbf{E}})}{k^2} + \frac{(\mathbf{k} \times \bar{\mathbf{E}}) \times \mathbf{k}}{k^2} \tag{1.8}$$

Assuming that only longitudinal waves emerge in the perturbation, (1.7) gives

$$\bar{g} = \frac{i(e/m[(\mathbf{k} \cdot \bar{\mathbf{E}})/k^2]\mathbf{k} \cdot (\partial F_0/\partial \mathbf{v})}{\omega - \mathbf{k} \cdot \mathbf{v}} \tag{1.9}$$

With

$$F_0 \equiv n_0 \tilde{f}_0(v)$$

we find

$$\frac{\partial F_0}{\partial \mathbf{v}} = n_0 \frac{\partial \tilde{f}_0}{\partial \mathbf{v}}$$

and (1.9) assumes the form

$$\bar{g} = \frac{i e n_0}{m} \frac{(\mathbf{k} \cdot \bar{\mathbf{E}})\mathbf{k}}{k^2(\omega - \mathbf{k} \cdot \mathbf{v})} \cdot \frac{\partial \tilde{f}_0}{\partial \mathbf{v}} \tag{1.10}$$

We assume that the perturbation is due to an extraneous electron. The corresponding charge density may be written

$$\rho = \rho_s + \rho_p \tag{1.11}$$

where

$$\rho_s = -e\delta(\mathbf{x} - \mathbf{v}_0 t) \tag{1.12}$$

represents the electron source and ρ_p denotes the charge density of the remaining plasma:

$$\rho_p = -e \int g \, d\mathbf{v}$$

The transform of the delta function (1.12) is

$$\bar{\rho}_s(k, \omega) = -e \int \delta(\mathbf{x} - \mathbf{v}_0 t)e^{-i(\mathbf{k} \cdot \mathbf{x} - \omega t)} \, d\mathbf{x} \, dt \tag{1.13}$$

Integrating first over $\mathbf{x}$, we obtain

$$\bar{\rho}_s(k, \omega) = -e \int e^{-i(\mathbf{k} \cdot \mathbf{v}_0 - \omega)t} \, dt$$

$$\bar{\rho}_s(k, \omega) = -2\pi e\delta(\omega - \mathbf{k} \cdot \mathbf{v}_0) \tag{1.14}$$

The generalization of Gauss' law to account for the extraneous test charge (1.12) appears as

$$\boldsymbol{\nabla} \cdot \mathbf{E} = 4\pi(\rho_s + \rho_p) = -4\pi e\delta(\mathbf{x} - \mathbf{v}_0 t) - 4\pi e \int g \, d\mathbf{v} \tag{1.15}$$

With (1.16) and (1.14), the transform of this equation becomes

$$i\mathbf{k} \cdot \bar{\mathbf{E}} = -e8\pi^2 \delta(\omega - \mathbf{k} \cdot \mathbf{v}_0) - 4\pi e \int \bar{g}\, d\mathbf{v} \qquad (1.16)$$

With (1.10), we obtain

$$i\mathbf{k} \cdot \bar{\mathbf{E}} = -e8\pi^2 \delta(\omega - \mathbf{k} \cdot \mathbf{v}_0) - i\frac{\omega_p^2 \mathbf{k} \cdot \bar{\mathbf{E}}}{k^2} \int \frac{\mathbf{k} \cdot \partial \tilde{f}_0 / \partial \mathbf{v}\, d\mathbf{v}}{\omega - \mathbf{k} \cdot \mathbf{v}} \qquad (1.17)$$

where ω_p represents the *plasma frequency*[3] (deleting the subscript on n_0):

$$\omega_p^2 = \frac{4\pi n e^2}{m_-} \qquad (1.18)$$

Solving (1.17) for $\mathbf{k} \cdot \bar{\mathbf{E}}$ gives

$$i\mathbf{k} \cdot \bar{\mathbf{E}} = \frac{-8\pi^2 e\delta(\omega - \mathbf{k} \cdot \mathbf{v}_0)}{1 + \frac{\omega_p^2}{k^2} \int \frac{\mathbf{k} \cdot \partial \tilde{f}_0 / \partial \mathbf{v}\, d\mathbf{v}}{\omega - \mathbf{k} \cdot \mathbf{v}}} \qquad (1.19)$$

The response of the plasma to the test charge may be given in terms of the dielectric response function defined by

$$\varepsilon(k, \omega) \equiv \bar{\rho}_s(k, \omega) / \bar{\rho}(k, \omega) \qquad (1.20)$$

The transform of (1.15) then gives

$$i\mathbf{k} \cdot \bar{\mathbf{E}} = \frac{4\pi \bar{\rho}_s(k, \omega)}{\varepsilon(k, \omega)} \qquad (1.21)$$

Rewriting (1.19) in the form [see (1.14)]

$$i\mathbf{k} \cdot \bar{\mathbf{E}} = \frac{4\pi \bar{\rho}_s(k, \omega)}{1 + \frac{\omega_p^2}{k^2} \int \frac{\mathbf{k} \cdot \partial \tilde{f}_0 / \partial \mathbf{v}\, d\mathbf{v}}{\omega - \mathbf{k} \cdot \mathbf{v}}} \qquad (1.22)$$

permits the identification

$$\boxed{\varepsilon(k, \omega) = 1 + \frac{\omega_p^2}{k^2} \int \frac{\mathbf{k} \cdot \partial \tilde{f}_0 / \partial \mathbf{v}\, d\mathbf{v}}{\omega - \mathbf{k} \cdot \mathbf{v}}} \qquad (1.23)$$

which is our explicit expression for the *plasma response dielectric function*.

The response electric field

With the decomposition (1.18), the transform integrals of the electric field assume the form

$$\mathbf{E}(\mathbf{x}, t) = \frac{1}{(2\pi)^4} \iint d\mathbf{k}\, d\omega\, e^{i(\mathbf{k} \cdot \mathbf{x} - \omega t)} \frac{\mathbf{k} \cdot \bar{\mathbf{E}}}{k^2} \mathbf{k} \qquad (1.24)$$

---

[3]Encountered previously in Section 2.3.9.

Substituting the expression (1.21) into (1.24) gives

$$\mathbf{E}(\mathbf{x}, t) = \frac{i}{(2\pi)^4} \iint \frac{d\mathbf{k}\, d\omega 8\pi^2 e\mathbf{k}\delta(\omega - \mathbf{k} \cdot \mathbf{v}_0)e^{i(\mathbf{k}\cdot\mathbf{x}-\omega t)}}{k^2 \varepsilon(k, \omega)} \tag{1.25}$$

Integrating first over ω gives

$$\mathbf{E}(\mathbf{x}, t) = \frac{2ie}{(2\pi)^2} \int \frac{\mathbf{k} e^{i\mathbf{k}\cdot(\mathbf{x}-\mathbf{v}_0 t)}\, d\mathbf{k}}{k^2 \varepsilon(k, \mathbf{k} \cdot \mathbf{v}_0)} \tag{1.26}$$

With the plasma dielectric constant (1.23), and $mC^2 = k_B T$, and

$$\tilde{f}_0 \equiv \frac{\exp(-v^2/2C^2)}{(2\pi)^{3/2}C^3}, \qquad \frac{\partial \tilde{f}_0}{\partial \mathbf{v}} = -\frac{\mathbf{v}}{C^2}\tilde{f}_0, \qquad \int \tilde{f}_0\, d\mathbf{v} = 1 \tag{1.27}$$

we find

$$\varepsilon(k, \mathbf{k} \cdot \mathbf{v}_0) = 1 + \frac{\omega_p^2}{C^2 k^2} \int \frac{\mathbf{k} \cdot \mathbf{v} \tilde{f}_0\, d\mathbf{v}}{\mathbf{k} \cdot (\mathbf{v} - \mathbf{v}_0)} \tag{1.28}$$

If the extraneous charge is stationary, $\mathbf{v}_0 = 0$, then

$$\varepsilon(k) = 1 + \frac{\omega_p^2}{C^2 k^2} \int \tilde{f}_0\, d\mathbf{v} = 1 + \frac{\omega_p^2}{C^2 k^2} \tag{1.29}$$

Thus (1.26) becomes

$$\mathbf{E}(\mathbf{x}) = \frac{2e}{(2\pi)^2} \int \frac{i\mathbf{k} e^{i\mathbf{k}\cdot\mathbf{x}}\, d\mathbf{k}}{k^2 + k_d^2} \tag{1.30}$$

Here we have introduced the Debye wave number

$$k_d^2 = \frac{\omega_p^2}{C^2} \tag{1.31}$$

Setting

$$\mathbf{E}(\mathbf{x}) = -\nabla\Phi = -\nabla \frac{1}{(2\pi)^3} \int \bar{\Phi}(k)e^{i\mathbf{k}\cdot\mathbf{x}}\, d\mathbf{k}$$

$$= -\frac{1}{(2\pi)^3} \int i\mathbf{k}\bar{\Phi}(k)e^{i\mathbf{k}\cdot\mathbf{x}}\, d\mathbf{k} \tag{1.32}$$

and comparing with (1.30) gives the transform

$$\bar{\Phi}(k) = -2e\frac{2\pi}{k^2 + k_d^2} \tag{1.33}$$

Solving for $\Phi(x)$ gives the response potential

$$\Phi(x) = \frac{1}{(2\pi)^3} \int \bar{\Phi}(k)e^{i\mathbf{k}\cdot\mathbf{x}}\, d\mathbf{k}$$

$$= \frac{-2e}{(2\pi)^2} \int \frac{e^{i\mathbf{k}\cdot\mathbf{x}}\, d\mathbf{k}}{k^2 + k_d^2} \tag{1.34}$$

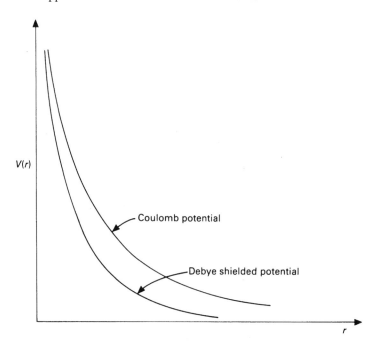

FIGURE 4.1. Bare Coulomb and Debye shielded potentials.

In spherical coordinates with $\mathbf{x}$ taken as the polar axis, we find

$$\int \frac{e^{i\mathbf{k}\cdot\mathbf{x}}d\mathbf{k}}{k^2 + k_d^2} = 2\pi \int_{-1}^{1} d\cos\theta \int_{0}^{\infty} \frac{dk k^2 e^{ikx\cos\theta}}{k^2 + k_d^2}$$

$$= \frac{2\pi}{x} \int_{0}^{\infty} \frac{2k \sin kx}{k^2 + k_d^2} dk = \frac{2\pi^2}{x} e^{-xk_d}$$

Substituting into (1.34) gives the desires result (writing r in place of x):

$$\Phi(r) = \frac{-e \exp(-k_d r)}{r},$$

$$V(r) = (-e)\Phi(r) = \frac{e^2}{r} \exp(-k_d r) \tag{1.35}$$

where $V(r)$ denotes potential energy (see Fig. 4.1). This is the Debye potential that surrounds a charge in a plasma. The distance

$$\lambda_d = \frac{1}{k_d}$$

is called the *Debye distance* and represents a shielding length of the Coulomb potential. [Recall (2.2.43).] With (1.18) and (1.31), we find

$$\lambda_d^2 = \frac{C^2}{\omega_p^2} = \frac{k_B T}{m\omega_p^2} = \frac{k_B T}{4\pi n e^2} \tag{1.36}$$

This is an important parameter in plasma physics. We note, in particular,

$$\lambda_d = 6.9 \left(\frac{T}{n}\right)^{1/2} \text{cm} \qquad (T \text{ in K})$$

$$\lambda_d = 740 \left(\frac{T}{n}\right)^{1/2} \text{cm} \qquad (T \text{ in eV}) \tag{1.37}$$

4.1.2 Waves, Instabilities, and Damping

Waves in a warm plasma

We next consider the response of a plasma to an arbitrary perturbation. Thus, omitting the point-charge delta function in (1.16) and (1.17), the result (1.21) may be rewritten

$$\varepsilon(k, \omega)\mathbf{k} \cdot \bar{\mathbf{E}} = 0 \tag{1.38}$$

The plasma dielectric constant $\varepsilon(k, \omega)$ is given by (1.23).

Equation (1.38) indicates that a finite longitudinal electric field will be present in the plasma, providing

$$\varepsilon(k, \omega) = 1 + \frac{\omega_p^2}{k^2} \int \frac{\mathbf{k} \cdot \partial \tilde{f}_0 / \partial \mathbf{v} \, d\mathbf{v}}{\omega - \mathbf{k} \cdot \mathbf{v}} = 0 \tag{1.39}$$

With (1.27), we obtain

$$1 + \frac{\omega_p^2}{C^2 k^2} \int \frac{\mathbf{k} \cdot \mathbf{v} \tilde{f}_0 \, d\mathbf{v}}{\mathbf{k} \cdot \mathbf{v} - \omega} = 0 \tag{1.40}$$

To reduce the integral, we choose $\mathbf{k}$ to lie along the Cartesian z axis. We may then integrate $\tilde{f}_0(v)$ over v_x and v_y. There results

$$\int_{-\infty}^{\infty} \int_{-\infty}^{\infty} dv_x \, dv_y \, \tilde{f}_0(v) = \frac{e^{-\mu^2/2}}{\sqrt{2\pi} C}$$

where[4]

$$\mu \equiv \frac{v_z}{C} \tag{1.41}$$

which permits the dispersion relation (1.40) to be written

$$1 + \frac{\omega_p^2}{C^2 k^2} \frac{1}{\sqrt{2\pi}} \int_{-\infty}^{\infty} \frac{d\mu \, \mu e^{-\mu^2/2}}{\mu - \beta} = 0 \tag{1.42}$$

Here we have set

$$\beta \equiv \frac{\omega}{Ck} \tag{1.42a}$$

---

[4]Recalling the nondimensional velocity ξ (3.5.28), we note $\xi_z^2 = \mu^2/2$.

We consider the limit $\beta \gg 1$ corresponding to $\omega/k/ \gg C$. Rewriting (1.42) as

$$1 = \frac{\omega_p^2}{C^2 k^2} \frac{1}{\sqrt{2\pi}} \int d\mu \frac{\mu}{\beta} e^{-\mu^2/2} \frac{1}{1 - (\mu/\beta)} \tag{1.43}$$

then permits the Taylor series expansion

$$1 = \frac{\omega_p^2}{C^2 k^2} \frac{1}{\sqrt{2\pi}} \int d\mu \frac{\mu}{\beta} e^{-\mu^2/2} \left[1 + \left(\frac{\mu}{\beta} \right) + \left(\frac{\mu}{\beta} \right)^2 + \left(\frac{\mu}{\beta} \right)^3 + \cdots \right] \tag{1.43a}$$

Only the even integrands contribute to the integration, and we obtain to terms of $0(\beta^{-4})$,

$$1 = \frac{\omega_p^2}{C^2 k^2} \left[\frac{1}{\beta^2} + \frac{3}{\beta^4} \right] \tag{1.44}$$

This equation may be rewritten

$$\omega^2 = \omega_p^2 + \frac{3\omega_p^2 (Ck)^2}{\omega^2} \tag{1.45}$$

Again expanding about small k gives

$$\boxed{\omega^2 = \omega_p^2 + 3C^2 k^2} \tag{1.46}$$

This is the dispersion relation that governs the propagation of longitudinal waves in a plasma at finite temperature and in the said $\omega - k$ domain.

A theorem for unstable plasma modes

The transform equation (1.6) indicates that the growth of waves in time (instability) corresponds to $\text{Im}(\omega) > 0$, or equivalently, $\text{Im}\beta > 0$. Thus, to examine unstable modes, we set

$$\beta = \alpha + i\tau, \qquad \tau > 0 \tag{1.47}$$

Our theorem states that an arbitrary single-peaked, one-dimensional equilibrium distribution $f_1(\mu)$ gives no unstable modes. Retracing steps leading to (1.42), we find that the generalized dispersion relation appears as [the one-dimensional velocity μ is defined by (1.41)]

$$1 = \frac{\omega_p^2}{C^2 k^2} \frac{1}{\sqrt{2\pi}} \int_{-\infty}^{\infty} \frac{d\mu \tilde{f}_1'}{\mu - \beta} \qquad f_1 = \frac{1}{\sqrt{2\pi C}} \tilde{f}_1 \tag{1.48}$$

where a prime denotes differentiation.[5] In a Maxwellian state, $\tilde{f}_1 = \exp(-\mu^2/2)$. Substituting the form of β given by (1.47) into (1.48) and

---

[5]Note that $\tilde{f}_1$ is the one-dimensional analog of $\tilde{f}_0$ as given in (1.27).

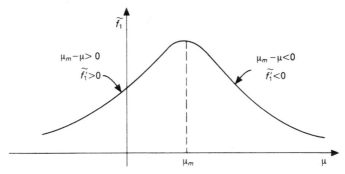

FIGURE 4.2. In either case, $(\mu_m - \mu)\tilde{f}_1' > 0$ for a single-peaked distribution.

rationalizing the integrand gives

$$1 = \frac{\beta_p^2}{\sqrt{2\pi}} \int_{-\infty}^{\infty} \frac{d\mu(\mu - \alpha + i\tau)\tilde{f}_1'}{(\mu - \alpha - i\tau)(\mu - \alpha + i\tau)}$$

$$1 = \frac{\beta_p^2}{\sqrt{2\pi}} \int_{-\infty}^{\infty} \frac{d\mu \tilde{f}_1'[(\mu - \alpha) + i\tau]}{(\mu - \alpha)^2 + \tau^2} \tag{1.49}$$

Here we have set

$$\beta_p^2 \equiv \left(\frac{\omega_p}{Ck}\right)^2 = \frac{1}{k^2\lambda_d^2} \tag{1.50}$$

Equation (1.49) gives the two equations

$$1 = \frac{\beta_p^2}{\sqrt{2\pi}} \int_{-\infty}^{\infty} \frac{d\mu \tilde{f}_1'(\mu - \alpha)}{(\mu - \alpha)^2 + \tau^2} \tag{1.51a}$$

$$0 = \frac{\beta_p^2 \tau}{\sqrt{2\pi}} \int \frac{d\mu \tilde{f}_1'}{(\mu - \alpha)^2 + \tau^2} \tag{1.51b}$$

Let $\tilde{f}_1(\mu_m)$ be the maximum value of $\tilde{f}_1$. Multiply (1.51b) by $\tau^{-1}(\alpha - \mu_m)$ and add the result to (1.51a). There results

$$1 = -\frac{\beta_p^2}{\sqrt{2\pi}} \int_{-\infty}^{\infty} \frac{d\mu(\mu_m - \mu)\tilde{f}_1'}{(\mu - \alpha)^2 + \tau^2} \tag{1.52}$$

If $\tilde{f}_1(\mu)$ has only a single maximum, then

$$(\mu_m - \mu)\tilde{f}_1'(\mu) > 0 \tag{1.53}$$

and the integrand of (1.52) is always positive, whence it has no solution. We may conclude that an infinitesimal perturbation of a single-peaked distribution results in no unstable modes (see Fig 4.2).

4.1.3   Landau Damping[6]

If we assume that the perturbation $g(t)$ initiates at $t = 0$, then $g(t < 0) = 0$ and $\mathbf{E}(t < 0) = 0$. In this case it is appropriate to work with the Laplace-fourier transform, and we write

$$
\begin{aligned}
\bar{g}(\omega, k) &= \int d\mathbf{x}\, e^{-i\mathbf{k}\cdot\mathbf{x}} \int_0^\infty dt\, e^{i\omega t} g(\mathbf{x}, \mathbf{v}, t) \\
\mathbf{E}(\omega, k) &= \int d\mathbf{x}\, e^{-i\mathbf{k}\cdot\mathbf{x}} \int_0^\infty dt\, e^{i\omega t} \mathbf{E}(\mathbf{x}, t)
\end{aligned}
\tag{1.54}
$$

The inverse of the time component of $\bar{g}(\omega)$ is given by[7]

$$
g(t) = \frac{1}{2\tau} \int_{i\omega_1-\infty}^{i\omega_1+\infty} \bar{g}(\omega) e^{-i\omega t}\, d\omega
\tag{1.55}
$$

with a similar expression for $\mathbf{E}(t)$. The line $\omega = i\omega_1$ in (1.55) lies above the singularities of $\bar{g}(\omega)$ in the ω plane. This condition ensures[8] that $g(t < 0) = 0$ (see Fig. 4.3).

Multiplying (1.5a) by the exponential operator in (1.54) gives

$$
\int d\mathbf{x} \int_0^\infty dt\, e^{i(\omega t - \mathbf{k}\cdot\mathbf{x})} \left(\frac{\partial g}{\partial t} + \mathbf{v} \cdot \frac{\partial}{\partial \mathbf{x}} g - \frac{e}{m} \mathbf{E} \cdot \frac{\partial F_0}{\partial \mathbf{v}} \right) = 0
\tag{1.56}
$$

Integrating the first two terms by parts and neglecting surface terms, we find

$$
\bar{g}(\omega, k) = \frac{\frac{ie}{m}\bar{\mathbf{E}} \cdot \frac{\partial F_0}{\partial \mathbf{v}} + i g(0)}{\omega - \mathbf{k}\cdot\mathbf{v}}
\tag{1.57}
$$

where the initial distribution $g(0)$ remains a function of $\mathbf{v}$. Inserting this expression into the divergence equation

$$
i\mathbf{k} \cdot \bar{\mathbf{E}} = -4\pi e \int \bar{g}\, d\mathbf{v}
$$

and recalling (1.8) gives

$$
-\mathbf{k} \cdot \bar{\mathbf{E}} = \frac{4\pi e \int \frac{g(0)\, d\mathbf{v}}{\omega - \mathbf{k}\cdot\mathbf{v}}}{\varepsilon(k, \omega)}
\tag{1.58}
$$

with $\varepsilon(k, \omega)$ given by (1.23).

If this form is substituted into the companion equation to (1.55) for the transform of the longitudinal component of $\mathbf{E}$, again we find that its time development is determined by the zeros of $\varepsilon(k, \omega)$ (that is, the poles of $\mathbf{k} \cdot \bar{\mathbf{E}}$).

---

[6]L. Landau, J. Phys. (USSR) *10*, 25 (1946).

[7]The Laplace transform was encountered previously in the discussion concerning the resolvent operator (Section 1.5.4).

[8]Consider the pole, $\mathrm{Im}\,\omega > \omega_1$. For $t < 0$, $\exp(-i\omega t)$ gives convergence in (1.55) for $\mathrm{Im}\,\omega > 0$. Thus, completion of the contour in (1.55) in the upper half-plane would give a finite contribution to $g(t < 0)$.

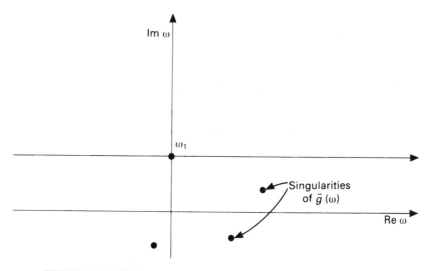

FIGURE 4.3. The line $\omega = i\omega_1$ lies above the singularities of $\bar{g}(\omega)$.

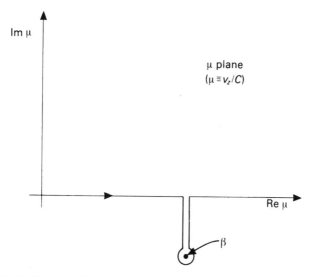

FIGURE 4.4. The analytic continuation of the integral in (1.42) for Im $\beta < 0$.

Here we envision that the contour of (1.55) shown in Fig. 4.3 is closed in a semicircle in the lower-half ω plane.

So once more we are led to the dispersion relation (1.42). With the Laplace transform (1.54), we see that convergence now demands that Im$\omega > 0$. Thus we realize that the integral in (1.42) is defined with Im$\beta > 0$.

However, with (1.55) we note that decaying modes correspond to Im$\beta < 0$. The integral in (1.42) for this case is evaluated by analytic continuation. This is effected through distortion of the contour of (1.42) to that shown in Fig. 4.4.

There results

$$1 = \frac{\omega_p^2}{C^2 k^2} \frac{1}{\sqrt{2\pi}} \left[\int_{-\infty}^{\infty} \frac{d\mu \, \mu e^{-\mu^2/2}}{\beta - \mu} + 2\pi i \, \tilde{f}_1'(\beta) \right] \qquad (1.59)$$

Where we have set $\tilde{f}_1 = \exp(-\mu^2/2)$. For weakly damped modes, β hovers just beneath the Re μ axis, and the pole in the integration is well represented by one-half the residue at that point. For Im $\beta \ll$ Re β, we obtain[9]

$$1 = \frac{\beta_p^2}{\sqrt{2\pi}} \left[\int_{-\infty}^{\infty} \frac{d\mu \, \mu e^{-\mu^2/2}}{\beta - \mu} + i\pi \, \tilde{f}_1'(\beta) \right] \qquad (1.60)$$

In this same limit, the expansion (1.43a) applies and (1.60) may be written

$$1 = \beta_p^2 \left[\frac{1}{\beta^2} + \frac{3}{\beta^4} + \cdots + i\sqrt{\frac{\pi}{2}} \, \tilde{f}_1'(\beta) \right] \qquad (1.61)$$

Assuming $\tilde{f}_1'(\beta) \ll 1$, in the first approximation we neglect the imaginary contribution, which again yields the solution (1.46), which we now write as

$$\bar{\beta}^2 = \beta_p^2 + 3 \qquad (1.62)$$

This is the first-order solution to (1.61). The second-order solution is constructed by reinserting this solution into the imaginary term in (1.61). We obtain

$$1 \simeq \frac{\beta_p^2}{\beta^2} + i\beta_p^2 \sqrt{\frac{\pi}{2}} \, \tilde{f}_1'(\bar{\beta}) \qquad (1.63)$$

or, equivalently,

$$\beta = \frac{\beta_p}{\left[1 - i\beta_p^2 \sqrt{\frac{\pi}{2}} \tilde{f}_1' \right]^{1/2}}$$

Expanding the radical about small $\tilde{f}_1'$ gives

$$\beta \simeq \beta_p \left[1 + i\frac{\beta_p^2}{2} \sqrt{\frac{\pi}{2}} \, \tilde{f}_1' \right] \qquad (1.64)$$

To within this approximation, we find

$$\text{Im} \, \beta = \beta_p^3 \sqrt{\frac{\pi}{8}} \, \tilde{f}_1'(\bar{\beta}) \qquad (1.65)$$

---

[9]This equation is exact for Im $\beta = 0$. We assume that it is also appropriate to the said limit with finite Im β. Note also that the integrals in (1.59) and (1.60) are principle values.

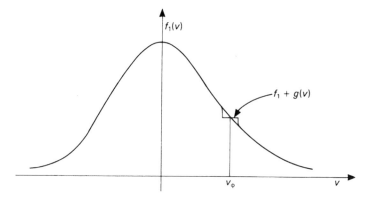

FIGURE 4.5. Variation in f_1 due to Landau damping in linear theory.

which, with the assumed Maxwellian form for $\tilde{f}_1$, gives the desired result:

$$\operatorname{Im}\omega = -\sqrt{\frac{\pi}{8}}e^{-3/2}\omega_p\beta_p^3 e^{-\beta_p^2/2} \equiv -\Gamma, \qquad \beta_p^2 = \frac{1}{k^2\lambda_d^2} \qquad (1.66)$$

The exponential form of the solution as given by (1.55) indicates that, at a given frequency, perturbation fields decay as $\exp -\Gamma t$, where Γ, the decay constant, is defined by (1.66).

The form of the solution as given by (1.65) suggests that damping will occur for $\tilde{f}_1' < 0$. Thus, with $\tilde{f}_1'$ negative for $\mu > 0$, we expect the perturbation to be damped for $v \equiv C\mu = v_\phi = \omega/k > 0$. Here v_ϕ denotes phase velocity. At these values, the $\mathbf{E}$ wave is damped and electric energy is lost to particle motion. This situation is depicted in Fig. 4.5, which illustrates the manner in which the perturbation alters the equilibrium distribution function. At the phase speed v_ϕ, the electric field loses energy, and electrons are driven to higher speeds, thereby depleting the region $v \lesssim v_\phi$ and increasing the region $v \gtrsim v_\phi$. For $v_\phi = \omega/k < 0$, $\beta_p < 0$. Since $\tilde{f}_1' > 0$ for $v < 0$, again (1.65) indicates wave damping.

Unstable modes

A brief recapitulation is in order: We have solved (1.60) for β through an approximation scheme in which its is assumed that $\operatorname{Im}\beta \ll \operatorname{Re}\beta$. The first-order iterate in this solution (1.62) gives the real part of β. This leads to $\operatorname{Im}\beta$ as given by (1.65).

From this result, we many conclude that damping results if $\beta_p^3\tilde{f}_1'(\bar{\beta}) < 0$. If the reverse of this inequality holds, then we may expect the mode at $\beta = \bar{\beta}$ to be unstable. (Unstable modes are discussed in the following section.)

For a single-peaked distribution, we find $\beta_p^3\tilde{f}_1'(\bar{\beta}) < 0$ for all $\bar{\beta}$, thereby recapturing our previously stated result that single-peaked equilibrium distribution functions are stable. A double-peaked distribution may or may not be unstable.

4.1.4 Nyquist Criterion

This criterion is a scheme for determining the number of unstable modes associated with a given unperturbed distribution function. The criterion stems from the observation that, for an arbitrary analytic function $\psi(z)$ with isolated poles

$$\frac{1}{2\pi i} \oint_C dz \frac{\psi'(z)}{\psi(z)} = N_z - N_p \tag{1.67}$$

where N_z denotes the number of zeros of $\psi(z)$ within the contour C and N_p denotes the number of singularities within C.

From (1.39), we infer that the zeros of the dielectric constant $\varepsilon(k, \omega)$ are solutions to the plasma dispersion relation. With (1.48), we may write

$$\varepsilon(\beta) = 1 + \frac{\omega_p^2}{C^2 k^2} \frac{1}{\sqrt{2\pi}} \int_{-\infty}^{\infty} \frac{d\mu \, \tilde{f}_1'}{\beta - \mu}$$

$$\varepsilon(\beta) = 1 + \beta_p^2 F(\beta) \tag{1.68}$$

where the function $F(\beta)$ is as implied.

It $\varepsilon(\beta)$ has no singularities in the upper-half β plane,[10] then, by (1.67),

$$\frac{1}{2\pi i} \int_C \frac{d\beta \varepsilon'(\beta)}{\varepsilon(\beta)} = N_z \tag{1.69}$$

represents the number of zeros $\varepsilon(\beta)$ has in the upper-half β plane. But by previous argument, each such value of β, with Im $\beta > 0$, corresponds to an unstable mode. The contour C includes the entire real β axis and an infinite semicircle in the upper-half β plane (see Fig. 4.6).

Convergence on the semicircular contour A demands[11] that $\beta \varepsilon'/\varepsilon \to 0$ with increasing β. With (1.68) we find that this is the case and we may proceed with the evaluation of (1.69).

Now

$$\frac{1}{2\pi i} \int_C \frac{d\beta \varepsilon'}{\varepsilon} = \frac{1}{2\pi i} \int_{\bar{C}} \frac{d\varepsilon}{\varepsilon} = N_z \tag{1.70}$$

In the ε plane, the function ε^{-1} has a simple pole at the origin. So the right equality in (1.70) indicates that N_z is equal to the number of times the contour $\bar{C}$ encircles the origin in the ε plane.

---

[10]It is conventional in the Nyquist analysis of a plasma to label $\varepsilon(\beta)$ as $H(\beta)$. The present description stays with $\varepsilon(\beta)$.

[11]See, for example, I. Sokolnikoff and R. Redheffer, *Mathematics of Physics and Modern Engineering*, 2nd ed., McGraw-Hill, New York (1968).

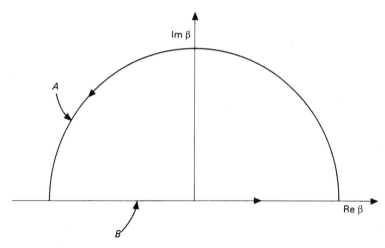

FIGURE 4.6. The contour C in (1.67) has a circular contribution A and a line contribution B.

We may use (1.68) to define a mapping from the β to the ε plane. Specifically, we wish to discern what curve $\bar{C}$ the contour C becomes in the ε plane. Let us perform this exercise for the single-peaked distribution described following (1.59).

As discussed above, the contour C for the β integration in (1.69) and (1.70) has two components. An infinite semicircle, which we have labeled A, and the real axis, labeled B (see Fig. 4.6).

With the single-peaked distribution inserted in $F(\beta)$, as defined in (1.68), we see that $F(\beta) \to 0$ on A.

On B, with β real, we recall (1.60), which may be rewritten as

$$F(\beta) = \mathcal{P} : F(\beta) + i\sqrt{\frac{\pi}{2}} \beta e^{-\beta^2/2} \qquad (1.71)$$

where $\mathcal{P}$: denotes "principal part."

Thus the dielectric constant (1.68) along B becomes

$$\varepsilon(\beta) = [1 + \beta_p^2 \mathcal{P} : F] + i\beta_p^2 \sqrt{\frac{\pi}{2}} \beta e^{-\beta^2/2} \qquad (1.72)$$

so that

$$\mathrm{Im}\varepsilon(\beta) = \beta_p^2 \sqrt{\frac{\pi}{2}} \beta e^{-\beta^2/2} \qquad (1.73)$$

which is an odd function of β. Furthermore, $\varepsilon(0) > 1$ and real. In this manner we obtain a sketch of the mapping of C into $\bar{C}$ as shown in Fig. 4.7.

So we find that the curve $\bar{C}$ in the ε plane does not encircle the origin for single-peaked distributions, and we again conclude that such equilibrium states are stable.

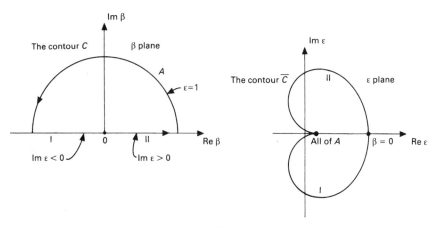

FIGURE 4.7. Inferred sketch of $\bar{C}$ for a single-peaked distribution

A Nyquist analysis for a double-peaked distribution was performed by J. D. Jackson[12] and was further described by T. H. Stix.[13] The results of this study are summarized in Fig 4.8. From these sketches, we see that the inner loop widens in the ε plane and drifts toward the origin as the peaks of $\tilde{f}_1$ separate. We may conclude that a sufficient separation of the two peaks is necessary for instability to occur. The related phenomena is called the *two-stream* instability.

4.2   Further Kinetic Equations of Plasmas and Neutral Fluids

In the discussion to follow, we find that certain limiting forms of the Boltzmann equation lead directly to kinetic equations relevant to neutral fluids and plasmas. The first equation derived in this scheme is the Krook-Bhatnager-Gross equation, appropriate to near-equilibrium states.

For gases dominated by long-range interactions or, equivalently, grazing collisions, we obtain the Fokker-Planck equation. With the aid of some tensor properties of the kernel of the Fokker-Planck equation, we obtain the Landau equation.

Returning to the BBKGY sequence, we finally obtain the Balescu-Lenard equation appropriate to the "long-time" limit of a plasma. This equation is found to include a shielding factor in its interaction integral representative of the Debye shielding discussed previously. In the limit of large number, the Balescu-Lenard equation is found to reduce to the Landau equation.

[12]J. D. Jackson, *J. Nuclear Energy C, 1*, 171 (1960).
[13]T. H. Stix, *The Theory of Plasma Waves*, McGraw-Hill, New York (1962).

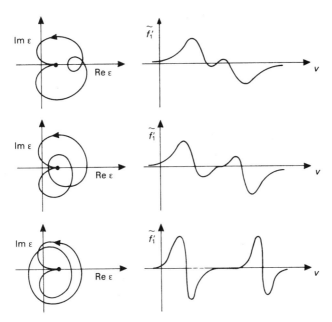

FIGURE 4.8. The loop $\bar{C}$ encircles the origin in the ε plane for sufficient separation of two maxima of $\tilde{f}_1$. Reprinted with permission of *J. Nuclear Energy C1*, J. D. Jackson, "Longitudinal Plasma Oscillations," copyright 1960, Pergamon Press, plc.

Logarithmic singularities of terms in plasma kinetic equations are discussed, and a review of various techniques of removing these singularities is included.

Connecting routes between the various kinetic equations derived in this section are graphically described in Fig 4.9.

4.2.1   Krook–Bhatnager–Gross Equation[14]

We recall the form of the Boltzmann collision integral (3.3.1):

$$\hat{J}(f_1) = \int f' f_1' \, d\mu_1 - \int f f_1 \, d\mu_1$$
$$d\mu_1 \equiv \sigma \, d\Omega g \, d\mathbf{v}_1$$

Near equilibrium, the fluid is close to a local Maxwellian state. The primed component of $\hat{J}(f)$, relevant to the after-collision interval, describes a situation closer to equilibrium than the unprimed component of $\hat{J}(f)$. Thus we set

$$\int d\mu_1 f_1' f' = \int d\mu_1 f_1^{0'} f^{0'} \tag{2.1}$$

[14]P. L. Bhatnagar, E. F. Gross, and M. Krook, *Phys. Rev. 94*, 511 (1954).

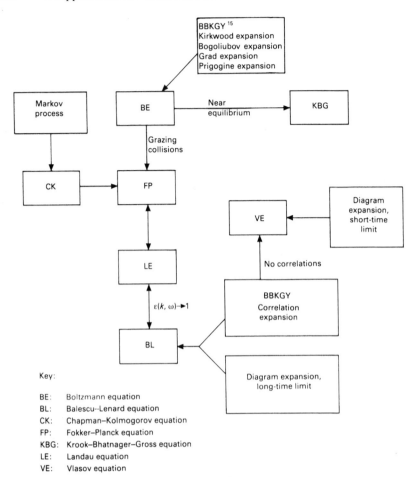

FIGURE 4.9. Relations among kinetic equations.

Since

$$\hat{J}(f^0) = \int d\mu_1 \left(f_1^{0\prime} f^{0\prime} - f_1^0 f^0 \right) = 0$$

we may write (2.1) as

$$\int d\mu_1 f_1' \simeq \int d\mu_1 f_1^0 f^0 \qquad (2.2)$$

[15]J. G. Kirkwood, *J. Chem. Phys.* **15**, 72 (1947). N. N. Bogoliubov, *Problems of a Dynamical Theory in Statistical Physics*, E. Gora (trans.), Providence College, Providence College, Providence, R.I. (1959). See also Section 2.4. H. Grad, *Hand. der Physik*, vol. XII, Springer Verlag, Berlin (1958). See also Section 2.5. I. Prigogine, *Non-Equilibrium Statistical Mechanics*, Wiley, New York (1962). See also Section 2.3.

Turning to the unprimed component of $\hat{J}(f)$, we note, in general,

$$\int d\mu_1 f_1 f = f \int d\mu_1 f_1 \qquad (2.3)$$

Next we recall that f^0 and f have the same first five moments. That is,

$$\int f^0 \begin{pmatrix} 1 \\ \mathbf{v} \\ v^2 \end{pmatrix} d\mathbf{v} = \int f \begin{pmatrix} 1 \\ \mathbf{v} \\ v^2 \end{pmatrix} d\mathbf{v} \qquad (2.4)$$

This common property motivates the final step of our derivation: in (2.3) we set

$$\int d\mu_1 f_1 = \int d\mu_1 f_1^0 \qquad (2.5)$$

With the relation (2.2) and substituting (2.51) into (2.3), we find

$$\hat{J}(f) = \left(\int d\mu_1 f_1^0 \right) \left[f^0 - f \right] \qquad (2.6)$$

which gives the *KBG equation*:

$$\boxed{\frac{Df}{Dt} = v(v) \left[f^0(v) - f(v) \right]} \qquad (2.7)$$

In this expression, we have written $v(v)$ for the *collision frequency*:

$$v(v) \equiv \iint \sigma \, d\Omega g \, d\mathbf{v}_1 f^0(v_1) \qquad (2.8)$$

Conservation equations

Owing to the moment equalities between f^0 and f as given by (2.4) and the defining relation (2.8), it may be shown that the KBG equation returns the conservation equations (3.3.14), (3.3.30), and (3.3.31). Specifically, consider the case that v is velocity independent.

$$N \int d\mathbf{v} \begin{pmatrix} 1 \\ m\mathbf{v} \\ \dfrac{mv^2}{2} \end{pmatrix} \frac{Df}{Dt} = \begin{pmatrix} \dfrac{\partial n}{\partial t} + \nabla \cdot n\mathbf{u} \\[4pt] \dfrac{\partial \rho \mathbf{u}}{\partial t} + \nabla \cdot \bar{\bar{p}} \\[4pt] \dfrac{\partial n e_\kappa}{\partial t} + \nabla \cdot \mathbf{q} \end{pmatrix}$$

$$= vN \int d\mathbf{v} \begin{pmatrix} 1 \\ m\mathbf{v} \\ \dfrac{mv^2}{2} \end{pmatrix} (f^0 - f) = \begin{pmatrix} 0 \\ 0 \\ 0 \end{pmatrix} \qquad (2.9)$$

It should be noted that, although the KBG equation appears simpler in form than the Boltzmann equation, it still maintains a strong nonlinear quality. That is, we note that the local Maxwellian, f^0, is a nonlinear function of the first five moments of f.

Time behavior of f

For a spatially homogeneous gas, the KBG equation becomes

$$\frac{\partial f}{\partial f} + \nu f = \nu f^0 \tag{2.10}$$

Multiplying through by the integrating factor $e^{\nu t}$ gives

$$\frac{\partial f e^{\nu t}}{\partial t} = \nu f^0 e^{\nu t}$$

Integrating, we find

$$f(t) = f(0)e^{-\nu t} + f^0 \left(1 - e^{-\nu t}\right) \tag{2.11}$$

Thus, the KBG equation parallels the conclusion of the $\mathcal{H}$ theorem and indicates that the fluid approaches a Maxwellian equilibrium state.

4.2.2   KBG Analysis of Shock Waves

We found previously (Problem 3.41) that relatively small disturbances in a fluid propagate at the sound speed. This process is isentropic. For disturbances of larger magnitude, the process becomes nonisentropic and the velocity of propagation exceeds the sound speed. Such phenomena are called *shock waves*.

A parameter important to the description of shock waves is the Mach number, M. This dimensionless parameter is the ration of the local fluid velocity u to the sonic velocity a:

$$M = \frac{u}{a}$$

If $M < 1$, the flow is called *subsonic*. If $M \simeq 1$, the flow is *transonic*. If $M > 1$, the flow is *supersonic*, and if $M \gg 1$, the flow is *hypersonic*.

Whereas fluid dynamic equations provide a good description of changes in macroscopic variables across a shock front,[16] a kinetic theory must be called on to examine corresponding changes in the distribution function. In the following analysis, we offer a brief description of application of the KBG equation to a one-dimensional shock wave. The fluid on either side of constant values of the macroscopic variables: ρ, u, and T.

We imagine a very long straight pipe filled with fluid that is at rest with a piston at one end. The piston undergoes a momentary acceleration carrying

---

[16]H. W. Liepman and A. Roshko, *Elements of Gas Dynamics*, Wiley, New York (1957).

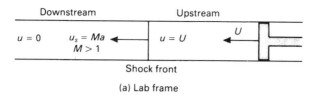

(a) Lab frame

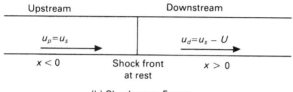

(b) Shock-wave Frame

FIGURE 4.10. Steady-state shock-wave configuration. Analysis is performed in frame (b) where the shock front is stationary.

it from zero to finite speed. The information that the piston has suffered this acceleration moves ahead of the piston in the form of a shock wave. In the subsequent steady-state configuration, the piston moves with constant velocity, pushing fluid ahead of it with the same speed, up to the shock front. Ahead of the shock front, information that the piston is moving has not yet reached the fluid and the fluid ahead of the shock remains at rest (see Fig. 4.10). It is conventional to view this steady-state configuration in the frame of the shock front. We call this speed u_p ("upstream").[17] The speed on the downstream side of the front we label u_d. Thus we have the boundary conditions

$$
\begin{aligned}
&\text{at } x = +\infty, \qquad u = u_d = \text{constant} \\
&\text{at } x = -\infty, \qquad u = u_p = \text{constant}
\end{aligned}
\tag{2.12}
$$

Since a steady-state condition maintains in this frame, the KBG equation reduces to

$$
v_x \frac{\partial}{\partial x} f = v \left(f^0 - f \right)
\tag{2.13}
$$

Assuming a Maxwellian distribution for velocities normal to the macroscopic flow, we set

$$
\begin{aligned}
f(\mathbf{x}, \mathbf{v}) &= g(x, v_x) \frac{1}{2\pi RT} \exp\left(-\frac{v_\perp^2}{2RT} \right) \equiv g f_\perp \\
f^0(\mathbf{x}, \mathbf{v}) &= g^0(x, v_x) \frac{1}{2\pi RT} \exp\left(-\frac{v_\perp^2}{2RT} \right) \equiv g^0 f_\perp
\end{aligned}
\tag{2.14}
$$

[17]These labels refer to flow in the frame of the shock wave. Labels are reversed in the lab frame (see Fig. 4.10).

Thus

$$\int d\mathbf{v}_\perp f(\mathbf{x}, \mathbf{v}) = g(x, v_x) \tag{2.15a}$$

and

$$nu_x = \int f v_x \, d\mathbf{v} = \int f_\perp \, d\mathbf{v}_\perp \int g u_x \, dv_x = \int g v_x \, dv_x$$

Substituting (2.14) into (2.13), with (2.15a) we find

$$\frac{\partial g}{\partial x} + \frac{v}{v} g = \frac{v}{v} g^0 \tag{2.16}$$

Here it is important to recall that g^0 is a local Maxwellian that contains actual values of macroscopic variables. Such variables change in value across the shock front.

Before examining (2.16), we more carefully consider the collision frequency ν. With Maxwell's expression for viscosity (3.4.33),

$$\eta = \frac{1}{3} \rho C l$$

and $l \simeq C/\nu, mC^2 = 3k_B T$, we find

$$\nu = \frac{nk_B T}{\eta}$$

Thus $\nu = \nu(n, T)$, and we may conclude that in the present application ν varies with x. This property must be kept in mind when integrating (2.16).

Multiplying (2.16) by the integrating factor

$$\exp \int_a^x \frac{v}{v} \, dx'$$

where a is an arbitrary length, gives

$$\frac{d}{dx}\left(g \exp \int_a^x \frac{v}{v} \, dx' \right) = \left(\exp \int_a^x \frac{v}{v} \, dx' \right) \frac{v}{v} g^0 \tag{2.16a}$$

· In the microscopic velocity domain: $v > 0$, we integrate (2.16a) over the interval $(-\infty, x)$ to obtain

$$\int_{-\infty}^x d\bar{x} \frac{d}{d\bar{x}}\left(g \exp \int_a^{\bar{x}} \frac{v}{v} dx' \right) = \int_{-\infty}^x d\bar{x} \left(\exp \int_a^{\bar{x}} \frac{v}{v} dx' \right) \frac{v}{v} g^0$$

$$g(x) \exp \int_a^x \frac{v}{n} dx' - g(-\infty) \exp \int_a^{-\infty} \frac{v}{v} dx' = \int_{-\infty}^x d\bar{x} \left(\exp \int_a^{\bar{x}} \frac{v}{v} dx' \right) \frac{v}{v} g^0$$

For $v > 0$, $\exp \int_a^{-\infty} (v/v)\, dx' \to 0$, and we obtain

$$g(x, v) = \left(\exp - \int_a^x \frac{v}{v} dx' \right) \int_{-\infty}^x d\bar{x} \left(\exp \int_a^{\bar{x}} \frac{v}{v} dx' \right) \frac{v}{v} g^0$$

which gives, for $v > 0$

$$g_+(x, v) \int_{-\infty}^x d\bar{x} \left(\exp - \int_{\bar{x}}^x \frac{v}{v} dx' \right) \frac{v}{v} g^0 \tag{2.17a}$$

For $v < 0$, we integrate (2.16a) over (∞, x). There results for $v < 0$,

$$g_-(x, v) \int_\infty^x d\bar{x} \left(\exp - \int_{\bar{x}}^x \frac{v}{v} dx' \right) \frac{v}{v} g^0 \tag{2.17b}$$

Number density and fluid velocity are contained in the equation

$$n(x)u(x) = \int_0^\infty dv\, v g_+(x, v) = \int_{-\infty}^0 dv\, v g_-(x, v) \tag{2.18}$$

Boundary conditions ar given by the relations

$$n_d u_d = \int_0^\infty g_+(v, +\infty)v\, dv + \int_{-a}^0 g_-(v, +\infty)v\, dv$$
$$n_p u_p = \int_0^\infty g_+(v, -\infty)v\, dv + \int_{-\infty}^0 g_-(v, -\infty)v\, dv \tag{2.19}$$

Since g^0 contains n, u, and T, which in turn are moments of g, (2.17)–(2.19) are integral equations for the distribution function. This formalism was applied by Leipmann et al.[18] to an argon gas. Equations were solved through numerical iteration, giving good agreement with results of the Navier-Stokes analysis at moderate Mach number ($M = 1.5$). At higher Mach number ($M = 5.0$), deviation from the Navier-Stokes equation was found to occur in the high-density side of the shock wave.[19]

4.2.3   The Fokker-Planck Equation[20]

Owing to the long-range nature of the Coulomb potential, we find that grazing collisions dominate in determining the kinetic properties of a plasma. The kinetic equation that best incorporates this collisional property is the Fokker-Planck equation. This equation may be obtained through expansion of the Boltzmann equation about small-angle, grazing collisions.

The cross section for the Coulomb interparticle force,

$$\mathbf{G}_{ij} = \frac{e^2}{r^3}(\mathbf{x}_i - \mathbf{x}_j) \tag{2.20}$$

[18]H. W. Liepmann, R. Narasimha, and M. T. Chahine, *Phys. Fluids 5*, 1313 (1962).
[19]For further discussion, see W. G. Vincenti and C. H. Kruger, Jr., *Introduction to Physical Gas Dynamics*, Wiley, New York, (1965).
[20]A. D. Fokker, *Ann. Physik 43*, 912 (1914); M. Planck, *Sitzungsber. Preuss. Akad. Wiss.*, p. 324 (1917). See also S. Chandrasekhar, *Revs. Mod. Phys. 15*, 1 (1943). The present derivation is due to H. Grad (unpublished).

is given by [see (3.1.31)]

$$go \, d \cos \theta = \frac{\kappa^2 \, d \cos \psi}{(g \cos \psi)^3} \tag{2.21}$$

$$\kappa \equiv \frac{2e^2}{m} \tag{2.21a}$$

The interparticle force given by (2.20) describes repulsion between two particles of charge e and reduce mass $m/2$. This description is relevant to a one-component plasma (OCP), which is a collection of like charges in a neutralizing background.[21] Substituting (2.21) into the Boltzmann collision integral gives

$$\hat{J}(f) = \kappa^2 \int (f_1' f' - f_1 f) \frac{d\phi \, d \cos \psi \, d\mathbf{v}_1}{(g \cos \psi)^3} \tag{2.22}$$

where, we recall, ψ and ϕ are the polar angles of the apsidal vector $\boldsymbol{\alpha}$.

With the kinematic relations

$$\begin{aligned} \Delta \mathbf{v}_1 &= \mathbf{v}_1' - \mathbf{v}_1 = -\boldsymbol{\alpha}(\boldsymbol{\alpha} \cdot \mathbf{g}) \\ \Delta \mathbf{v} &= \mathbf{v}' - \mathbf{v} = \boldsymbol{\alpha}(\boldsymbol{\alpha} \cdot \mathbf{g}) \end{aligned} \tag{2.23}$$

we conclude that for grazing collisions, $\boldsymbol{\alpha}(\boldsymbol{\alpha} \cdot \mathbf{g})$ may be taken as parameter of smallness. Consider first

$$f' \equiv f(\mathbf{v}') = f[\mathbf{v} + \boldsymbol{\alpha}(\boldsymbol{\alpha} \cdot \mathbf{g})]$$

Taylor series expanding this form about $\boldsymbol{\mathcal{E}} \equiv \boldsymbol{\alpha}(\boldsymbol{\alpha} \cdot \mathbf{g}) = 0$ gives

$$\begin{aligned} f' &= \exp(\boldsymbol{\mathcal{E}} \cdot \nabla)f = f + (\boldsymbol{\mathcal{E}} \cdot \nabla)f + \frac{1}{2}(\boldsymbol{\mathcal{E}} \cdot \nabla)^2 f + \cdots \\ &= f + (\boldsymbol{\alpha} \cdot \mathbf{g})(\boldsymbol{\alpha} \cdot \nabla)f + \frac{1}{2}(\boldsymbol{\alpha} \cdot \mathbf{g})^2(\boldsymbol{\alpha} \cdot \nabla)^2 f + \cdots \\ &= f + g \cos \psi (\boldsymbol{\alpha} \cdot \nabla)f + \frac{1}{2}g^2 \cos^2 \psi (\boldsymbol{\alpha} \cdot \nabla)^2 f + \cdots \end{aligned} \tag{2.24}$$

For f_1', we find

$$f_1' \equiv f(\mathbf{v}_1') = f[\mathbf{v}_1 + \boldsymbol{\alpha}(\boldsymbol{\alpha} \cdot \mathbf{g})]$$

$$f' = \exp(-\boldsymbol{\mathcal{E}} \cdot \nabla_1)f = f_1 - g \cos \psi (\boldsymbol{\alpha} \cdot \nabla_1)f_1 \tag{2.25}$$

$$+ \frac{1}{2}g^2 \cos^2 \psi (\boldsymbol{\alpha} \cdot \nabla_1)^2 f_1 + \cdots$$

In these expressions we have written

$$\nabla = \frac{\partial}{\partial \mathbf{v}} \qquad \nabla_1 = \frac{\partial}{\partial \mathbf{v}_1}$$

---

[21] An extensive review of the theory of one-component plasmas is given by S. Ichimaru, *Revs, Mod. Phys. 54*, 1017 (1982).

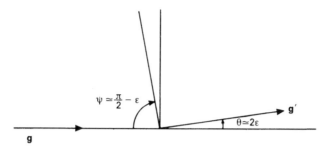

FIGURE 4.11. For grazing collisions, $\psi \simeq \frac{\pi}{2} - \varepsilon$, at which value $\cos \psi \simeq \varepsilon$.

The Cartesian components of the unit vector $\boldsymbol{\alpha}$ are

$$\boldsymbol{\alpha} = (\sin \psi, \cos \phi, \sin \psi \sin \phi, \cos \psi)$$

with $\mathbf{g}$ taken as the z axis [recall (3.2.26)].

Substituting (2.16) and (2.25) into the collision integral (2.22), keeping terms of $0(\varepsilon^2)$, and integrating over ϕ gives

$$\int_0^{2\pi} d\phi (f_1' f' - f_1 f)$$

$$= \pi \cos^2 \psi \left\{ 2\mathbf{g} \cdot (\boldsymbol{\nabla} - \boldsymbol{\nabla}_1) + \frac{1}{2}[\sin^2 \psi (\bar{\bar{I}} g^2 - \overline{\overline{\mathbf{gg}}}) + 2 \cos^2 \psi \overline{\overline{\mathbf{gg}}}] : \right.$$

$$\left. [\overline{\overline{\boldsymbol{\nabla}_1 \boldsymbol{\nabla}_1}} - 2\overline{\overline{\boldsymbol{\nabla} \boldsymbol{\nabla}_1}} + \overline{\overline{\boldsymbol{\nabla} \boldsymbol{\nabla}}}] \right\} f f_1 \qquad (2.26)$$

Here we have set

$$\int_0^{2\pi} d\phi g \boldsymbol{\alpha} \cdot \hat{\mathbf{D}} = 2\pi g \cos \psi \hat{D}_z = 2 \cos \psi \mathbf{g} \cdot \hat{\mathbf{D}}$$

where $\hat{\mathbf{D}}$ is written for $\boldsymbol{\nabla}$ or $\boldsymbol{\nabla}_1$. Integrating over $\cos \psi$, we encounter the forms

$$\int_0^1 \frac{d \cos \psi}{cos^3 \psi} \left(\cos^2 \psi, \cos^2 \psi \sin^2 \psi, \cos^4 \psi \right) \qquad (2.27)$$

At grazing collisions, $\psi = \pi/2 - \varepsilon$, where ε is an infinitesimal angle (see Fig. 4.11). This value of ψ corresponds to the lower limit in (2.27) and gives rise to the singularity:

$$\int_{\cos \psi,}^1 \frac{d \cos \psi}{\cos \psi} = -\ln \cos \left(\frac{\pi}{2} - \varepsilon \right)$$

$$= -\ln \sin \varepsilon$$

$$= \ln \varepsilon^{-1} \qquad (2.28)$$

For weakly coupled plasmas, we choose ε to be given by

$$\varepsilon = \Lambda = \frac{1}{4\pi n \lambda_d^3} \propto \frac{n^{1/2}}{T^3} \tag{2.28a}$$

The plasma parameter Λ was defined previously in (2.3.59). Having completed the angular integrations in (2.22), a final integration over $\mathbf{v}_1$ completes our derivation. Integration by parts reveals the form [keeping only the singular terms in (2.26)]

$$\hat{J}(f) = -\frac{\partial}{\partial \mathbf{v}} \cdot \boldsymbol{\alpha} f + \frac{1}{2} \frac{\overline{\overline{\partial^2}}}{\partial \mathbf{v} \partial \mathbf{v}} : \left(\overline{\overline{b}} f\right) \tag{2.29}$$

In this expression, $\boldsymbol{\alpha}$ represents the *friction coefficient* and $\overline{\overline{b}}$ the *diffusion coefficient*. They are given by

$$\boldsymbol{\alpha} = 2\pi \kappa^2 \ln \varepsilon^{-1} \int \frac{\mathbf{g}}{g^3} f_1 \, d\mathbf{v}_1 \tag{2.30a}$$

$$\overline{\overline{b}} = \pi \kappa^2 \ln \varepsilon^{-1} \int \left(\frac{g^2 \overline{\overline{I}} - \overline{\overline{\mathbf{g}\mathbf{g}}}}{g^3}\right) f_1 \, d\mathbf{v}_1 \tag{2.30b}$$

In the absence of an external force field, the Fokker-Planck equation then has the form

$$\boxed{\frac{\partial f}{\partial t} + \mathbf{v} \cdot \frac{\partial f}{\partial \mathbf{x}} + \frac{\partial}{\partial \mathbf{v}} \cdot \boldsymbol{\alpha} f - \frac{1}{2} \frac{\overline{\overline{\partial^2}}}{\partial \mathbf{v} \partial \mathbf{v}} : \overline{\overline{b}} f = 0} \tag{2.31}$$

To discover the physical significance of the $\boldsymbol{\alpha}$ and $\overline{\overline{b}}$ coefficients, we consider them separately and choose the simplest forms. Among the simpler forms we may choose for the friction coefficients $\boldsymbol{\alpha}$ is that corresponding to Rayleigh dissipation:

$$\boldsymbol{\alpha} = -\nu \mathbf{v} \tag{2.32}$$

With $\overline{\overline{b}} = 0$ and assuming spatial homogeneity, the Fokker-Planck equation (2.31) then reads

$$\frac{\partial f}{\partial t} - \nu \frac{\partial}{\partial \mathbf{v}} \cdot \mathbf{v} f = 0 \tag{2.33}$$

Choosing a local velocity frame, $\mathbf{v} = (v, 0, 0)$, and setting $h \equiv vf$, (2,23) becomes (with v and t as independent variables)

$$\frac{\partial h}{\partial t} - \mu v \frac{\partial}{\partial v} h = 0 \tag{2.33a}$$

The general solution for this equation is an arbitrary function of the solution to the characteristic equations

$$\frac{v \, dt}{1} = -\frac{dv}{v}$$

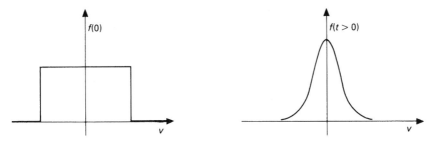

FIGURE 4.12. The friction coefficient causes particles to cluster near the origin in velocity space.

Thus h is constant on the curves $v_0 = v \exp vt$, and we write

$$h(v, t) = h(v_0, 0) = h(v \exp vt, 0)$$

We conclude that the friction coefficient (2.32) causes particles to slow down. Thus an initially square distribution in velocity space would tend to peak toward the origin (see Fig. 4.12).

The simplest form we may choose for the diffusion tensor $\overline{\overline{b}}$ is that of a diagonal matrix with equal elements. That is,

$$\overline{\overline{b}} = \overline{\overline{I}} b$$

With $a = 0$, the Fokker-Planck equation becomes

$$\frac{\partial f}{\partial t} = \frac{1}{2} b \nabla^2 f \tag{2.34}$$

where ∇^2 represents the Laplacian in velocity space. For the initial distribution given by

$$f(\mathbf{v}, 0) = V^{-1} \delta(\mathbf{v}) \tag{2.35}$$

the solution to the diffusion equation (2.34) is given by the Gaussian distribution[22]

$$f(\mathbf{v}, t) = \frac{V - 1}{(2\pi bt)^{3/2}} \exp\left(\frac{-v^2}{2bt}\right) \tag{2.36}$$

In these expressions, V represents the volume of the system. So the diffusion coefficient causes a distribution function initially peaked at the origin to flatten out (see Fig. 4.13).

Combining both the a and b coefficients gives (with $a = -\bar{a}v$ and again working in a local velocity frame)

$$\frac{\partial f}{\partial t} = \bar{a} \frac{\partial}{\partial v} vf + \frac{1}{2} b \frac{\partial^2 f}{\partial v^2} \tag{2.37}$$

---

[22]Encountered previously in (1.8.7) and (3.4.40).

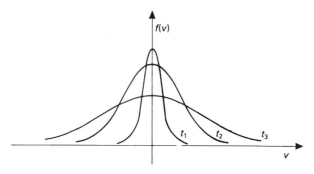

FIGURE 4.13. The diffusion coefficient causes a flattening of $f(v)$ in velocity space.

At equilibrium, this equation becomes

$$\frac{\partial}{\partial v}\left(\bar{a}vf + \frac{1}{2}b\frac{af}{av}\right) = 0$$

Thus

$$\bar{a}vf + \frac{1}{2}b\frac{af}{av} = A \qquad (2.38)$$

where A is independent of v. At $v = \infty$, we have

$$vf = 0 \qquad \frac{af}{av} = 0$$

It follows that the constant $A = 0$. Multiplying the resulting form of (2.38) through by the integrating factor

$$\exp \int \frac{2\bar{a}v}{b} dv$$

reduces the equation to

$$\frac{\partial}{\partial v}(e^{\bar{a}v^2/b} f) = 0$$

which gives the Maxwellian distribution

$$f = Be^{-\bar{a}v^2/b}$$

Thus incorporation of the friction coefficient that tends to peak $f(v)$ about the origin and the diffusion coefficient that tends to normalize $f(v)$ leads naturally to the Maxwellian distribution.

4.2.4   The Landau Equation[23]

The Fokker-Planck equation (2.31) may be cast in a more symmetric form known as the Landau equation. To demonstrate this transformation, we must first establish some tensor properties of the relative velocity

$$\mathbf{g} = \mathbf{v}_1 - \mathbf{v}$$

and the tensor

$$\overline{\overline{T}} \equiv g^2 \overline{\overline{I}} - \overline{\overline{\mathbf{gg}}} \tag{2.39}$$

which enters the Fokker-Planck diffusion coefficient $\overline{\overline{b}}$ given by (2.30b).
These are as follows:

$$\frac{\partial}{\partial v_\mu} g_\nu = -\delta_{\mu\nu} \tag{2.40}$$

$$\mathbf{\nabla} \cdot \mathbf{g} = \frac{\partial}{\partial v_\mu} g_\mu = -3 \tag{2.41}$$

$$\frac{\partial}{\partial v_\mu} g = \frac{\partial}{\partial v_\mu} (g^2)^{1/2} = \frac{1}{2}(g^2)^{-1/2} \frac{\partial}{\partial v_\mu} g_\nu g_\nu = \frac{2g_\nu}{2g} \frac{\partial g_\nu}{\partial v_\mu} = -\frac{g_\mu}{g} \tag{2.42}$$

$$\frac{\partial}{\partial v_\mu} g^2 = -2g_\nu \frac{\partial g_\nu}{\partial v_\mu} = -2g_\mu$$

If $f(g)$ is any scalar function, then

$$\frac{\partial f(g)}{\partial v_\mu} = -\frac{\partial f(g)}{\partial v_{1\mu}} \tag{2.43}$$

Next we turn to the calculation of $\mathbf{\nabla} \cdot \overline{\overline{T}}$.

$$\frac{\partial}{\partial v_\mu} T_{\mu\nu} = \frac{\partial}{\partial v_\mu} \delta_{\mu\nu} g^2 - \frac{\partial}{\partial v_\mu} g_\mu g_\nu$$

$$= \frac{\partial}{\partial v_\mu} g^2 - g_\mu \frac{\partial}{\partial v_\mu} g_\nu - g_\nu \frac{\partial}{\partial v_\mu} g_\mu$$

$$= -2g_\nu + g_\nu + 3g_\nu = 2g_\nu$$

Combining results, we have (see Problem 4.10

$$\mathbf{\nabla} \cdot \overline{\overline{T}} = 2\mathbf{g} \tag{2.44}$$

$$\mathbf{g} \cdot \overline{\overline{T}} = 0 \tag{2.45}$$

$$\mathbf{\nabla} \cdot f(g)\overline{\overline{T}} = f(g)\mathbf{\nabla} \cdot \overline{\overline{T}} \tag{2.46}$$

---

[23]L. Landau, Z. Eksp. i Teoret. Fiz. 7, 203 (1937)

An illustrative exercise involving these properties concerns the Fokker-Planck $\boldsymbol{\alpha}$ and $\bar{\bar{b}}$ coefficients (2.30). Let us establish the relation

$$\nabla \cdot \bar{\bar{b}} = \boldsymbol{\alpha} \tag{2.47}$$

or, equivalently,

$$\frac{\partial}{\partial v_\mu} K \int \frac{T_{\mu\nu}}{g^3} f(\mathbf{v}_1)\,d\mathbf{v}_1 = 2K \int \frac{g\nu}{g^3} f(\mathbf{v}_1)\,d\mathbf{v}_1 \tag{2.47a}$$

$$K \equiv \pi \kappa^2 \ln \varepsilon^{-1}$$

With (2.44) and (2.46), we find

$$\nabla \cdot \left(\frac{\bar{\bar{T}}}{g^3} \right) = \frac{1}{g^3} \nabla \cdot \bar{\bar{T}} = \frac{2\mathbf{g}}{g^3} \tag{2.48}$$

which establishes (2.47).

Finally, we turn to an explicit representation of $\bar{\bar{T}}$. Specifically, let us discover the matrix representation of $\bar{\bar{T}}$ in a form in which $\mathbf{g}$ is parallel to the z axis; that is $\mathbf{g} = (0, 0, g)$.

In general, we have

$$\bar{\bar{T}} = \begin{pmatrix} g^2 - g_x^2 & -g_x g_y & -g_x g_z \\ -g_y g_x & g^2 - g_y^2 & -g_y g_z \\ -g_z g_x & -g_z g_y & g^2 - g_z^2 \end{pmatrix}$$

In the said coordinate frame, $\bar{\bar{T}}$ is seen to reduce to the simple form

$$\bar{\bar{T}} = \begin{pmatrix} g^2 & 0 & 0 \\ 0 & g^2 & 0 \\ 0 & 0 & 0 \end{pmatrix} \tag{2.49}$$

This form will be returned to in our discussion of the Balescu-Lenard equation.

Symmetrization of the collision terms

We rewrite the Fokker-Planck equation in the form

$$\frac{1}{K}\frac{Df}{Dt} = -2\frac{\partial}{\partial v_\nu} f \int \frac{g\nu}{g^3} f_1\,d\mathbf{v} + \frac{1}{2}\frac{\partial}{\partial v_\nu}\frac{\partial}{\partial v_\mu} f \int \frac{T_{\mu\nu} f_1\,d\mathbf{v}_1}{g^3}$$

$$= \frac{\partial}{\partial v_\nu}\left[-2f \int \frac{g\nu}{g^3} f_1\,d\mathbf{v}_1 + \frac{1}{2}\frac{\partial}{\partial v_\mu} f \int \frac{T_{\mu\nu}}{g^3} f_1\,d\mathbf{v}_1 \right] \tag{2.50}$$

The second term within the brackets may be expanded to give

$$\frac{1}{2}\int d\mathbf{v}_1 f_1 \left(\frac{T_{\mu\nu}}{g^3}\frac{\partial}{\partial v_\mu} f + f\frac{\partial}{\partial v_\mu}\frac{T_{\mu\nu}}{g^3} \right) = \frac{1}{2}\int d\mathbf{v}_1 f_1 \left(\frac{T_{\mu\nu}}{g^3}\frac{\partial}{\partial v_\mu} f + \frac{2f g_\nu}{g^3} \right)$$

The second term in this expression combines with the first term within the brackets in (2.50), and we are left with

$$\frac{1}{K}\frac{DF}{Dt} = \frac{1}{2}\frac{\partial}{\partial v_\mu}\left(\int d\mathbf{v}_1 \frac{T_{\mu\nu}}{g^3} f_1 \frac{\partial}{\partial v_\mu} f - f \int f_1 \frac{\partial}{\partial v_\mu}\frac{T_{\mu\nu}}{g^3} d\mathbf{v}_1 \right)$$

The second term may be transformed as follows:

$$-\int f_1 \frac{\partial}{\partial v_\mu}\left(\frac{T_{\mu\nu}}{g^3}\right) d\mathbf{v}_1 = +\int f_1 \frac{\partial}{\partial v_{1\mu}}\left(\frac{T_{\mu\nu}}{g^3}\right) d\mathbf{v}_1$$

$$= \int \frac{\partial}{\partial v_{1\mu}}\left(\frac{_1 T_{\mu\nu}}{g^3}\right) d\mathbf{v}_1 - \int \frac{T_{\mu\nu}}{g^3}\frac{\partial f_1}{\partial v_{1\mu}} d\mathbf{v}_1$$

Dropping the first surface term and collecting remaining terms gives the *Landau* equation:

$$\boxed{\frac{2}{K}\frac{Df}{Dt} = \frac{\partial}{\partial u_\nu}\int d\mathbf{v}_1 \frac{T_{\mu\nu}}{g^3}\left[f_1 \frac{\partial f}{\partial v_\mu} - f \frac{\partial f_1}{\partial v_{1\mu}} \right]} \tag{2.51}$$

Thus we find an alternate symmetric form of the Fokker-Planck equation whose nonlinear collision integral more closely resembles that of the Boltzmann equation.

The question arises as to whether a Maxwellian distribution is an equilibrium solution of the Landau equation. Setting $f = A\exp(-v^2/2C^2)$, we find that, after differentiation with respect to v_μ and $v_{1\mu}$, the integrand contains the factor $\bar{\bar{T}}\cdot\mathbf{g}$, which by (2.45) vanishes. Thus we find that the Landau equation shares this additional property with the Boltzmann equation.

4.2.5   The Balescu–Lenard Equation[24]

In Section 2.3.9, we noted that the Balescu-Lenard equation is relevant to the time interval where shielding comes into play (see Fig. 2.14). This equation is now derived in a correlation expansion of the hierarchy relevant to a high-temperature rare-density plasma.

We return to the nondimensionalized BBKGY equation (2.2.12) and the BY_s write

$$\left(\frac{\partial}{\partial t} + \hat{\kappa}_s - a\hat{\Theta}_s\right) F_s = -\frac{\alpha}{\gamma}\hat{l}_s F_{s+1}$$

[24] A. Lenard, *Ann. Phys. 10*, 390 (1960); R. Balescu, *Phys. Fluids 3*, 52 (1960). The spectrum of the linearized Balescu-Lenard equation was examined by A. H. Merchant and R. L. Liboff, *J. Math. Phys. 14*, 119 (1973).

where κ_s is given by (2.2.22a), and we recall that the interaction term $\hat{I}_s F_{s+1}$ has the form

$$\hat{I}_s = \sum_{i=1}^{s} \frac{\partial}{\partial \mathbf{v}_i} \cdot \int d(s+1) \mathbf{G}_{i,s+1}$$

and $\mathbf{G}_{i,j}$ represents the two-particle interaction force for particle j on particle i.
    We examine a one-component rare plasma with

$$\frac{1}{\gamma} \equiv n_0 r_0^3 = 0(\varepsilon)$$

and weak coupling (or high temperature)

$$\alpha \equiv \frac{\Phi_0}{k_B T} = 0(\varepsilon)$$

where r_0 is the range of interaction [recall (2.2.9) and ε is a parameter of smallness. As particle coupling is small, the correlation expressions (2.2.21) for F_2 and F_3 are written

$$F_2(1, 2) = F_1(1)F_1(2) + \varepsilon C_2(1, 2)$$
$$F_3(1, 2, 3) = F_1(1)F_1(2)F_1(3) + \varepsilon \sum_{P(ijk)} F_1(i)C_2(j, k) + \varepsilon^2 C_3(1, 2, 3) \,(2.52)$$

The variables $P(ijk)$, we recall, denote permutations of (1,2,3).
    The equations BY_1 and BY_2 then appear as

$$\left(\frac{\partial}{\partial t} + \hat{K}_1 \right) F_1 = \varepsilon^2 \frac{\alpha}{\gamma} \hat{I}_1 \left[F_1(1)F_1(2) + \varepsilon C_2(1, 2) \right] \qquad (2.53a)$$

$$\left(\frac{\partial}{\partial t} + \hat{K}_1(1) + \hat{K}_1(2) - \varepsilon \alpha \Theta_2 \right) \left[F_1(1)F_1(2) + \varepsilon C_2(1, 2) \right]$$

$$= \varepsilon^2 \frac{\alpha}{\gamma} \hat{I}_2 \left[F_1(1)F_1(2)F_1(3) + \varepsilon \sum_{P} F_1(i)C_2(j, k) + \varepsilon^2 C_3(1, 2, 3) \right] \qquad (2.53b)$$

To terms of $O(\varepsilon)$, we find (returning to dimensional variables)

$$\left(\frac{\partial}{\partial t} + \hat{K}_1 \right) f_1(1) = 0 \qquad (2.54a)$$

$$\left[\frac{\partial}{\partial t} + \hat{K}_1(1) + \hat{K}_1(2) \right] C_2(1, 2) + \frac{\mathbf{G}_{1,2}}{m} \cdot \left(\frac{\partial}{\partial \mathbf{v}_1} - \frac{\partial}{\partial \mathbf{v}_2} \right) f_1(1) f_2(2) = 0 \qquad (2.54b)$$

The technique of solving these equations is as follows: (1) Solve (2.54b) with the aid of (2.54a) to obtain $C_2 = C_2(t)$. (2) Insert this solution into (2.53a) to obtain a kinetic equation that is valid to $O(\varepsilon^3)$.

The solution to (2.54a) is the streming solution, which contains only free-particle trajectories with $\dot{\mathbf{v}}_1 = \dot{\mathbf{v}}_2 = 0$. Thus, in this instance,

$$\frac{d}{dt}C_2(1,2) = \left(\frac{\partial}{\partial t} + \mathbf{v}_1 \cdot \frac{\partial}{\partial \mathbf{x}_1} + \mathbf{v}_2 \cdot \frac{\partial}{\partial \mathbf{x}_2}\right) C_2(1,2)$$

$$= \left[\frac{\partial}{\partial t} + \hat{K}_1(1) + \hat{K}_1(2)\right] C_2(1,2) \qquad (2.55)$$

It follows that (2.54b) may be rewritten

$$\frac{dC_2}{dt} = \frac{1}{m}\frac{\partial \Phi(1,2)}{\partial \mathbf{x}_1} \cdot \left(\frac{\partial}{\partial \mathbf{v}_1} - \frac{\partial}{\partial \mathbf{v}_2}\right) f_1(1) f_1(2)$$

$$\equiv A(1,2,t) \qquad (2.56)$$

Integrating this equation, we find

$$C_2(1,2,t) = \int_{-\infty}^{t} A(t')dt' + C_2(1,2,-\infty) \qquad (2.57)$$

If we assume that particles are uncorrelated at $t + -\infty$, then $C_2(-\infty)$, then $C_2(-\infty) = 0$. Changing variables to

$$\tau \equiv t - t', \qquad d\tau = -dt'$$

gives

$$C_2(1,2,t) = \int_0^{\infty} A(t-\tau)\,d\tau \qquad (2.58)$$

Proceeding with our solution prescription, we substitute (2.58) into (2.53a). There results

$$\left[\frac{\partial}{\partial t} + \hat{K}_1 - \frac{\alpha}{\gamma}\hat{I}_1 f_1(2)\right] f_1(1) = \frac{\alpha}{\gamma}\hat{I}_1 \int_0^{\infty} A(t-\tau)\,d\tau \qquad (2.59)$$

The left side of this equation is identical to that which appears on the left side of the Vlasov equation (2.2.30). Thus (2.59) may be rewritten (setting $f_1 \equiv f$)

$$\frac{\partial f}{\partial t} + \mathbf{v}_1 \cdot \frac{\partial}{\partial \mathbf{x}_1}f + \frac{e}{m}\mathcal{E} \cdot \frac{\partial f}{\partial \mathbf{v}_1} = \frac{1}{m}\frac{\partial}{\partial \mathbf{v}_1} \cdot \int d\mathbf{x}_2 \int d\mathbf{v}_2 \frac{\partial \Phi(x_{12})}{\partial \mathbf{x}_1}\left[\int_0^{\infty} A(t-\tau)\,d\tau\right] \qquad (2.60)$$

where $e\mathcal{E}$ is the Vlasov self-consistent force field. It the two-particle separation is $\mathbf{x}_{12}$ at t, then in the straight-line approximation, at $t - \tau$, its value is $\mathbf{x}_{12} - \mathbf{g}\tau$. Thus, with the identification (2.56), and writing $Df/Dt|_{VL}$ for the left side of (2.60), this equation becomes

$$\frac{Df}{Dt}\bigg|_{VL} = \frac{1}{m}\frac{\partial}{\partial \mathbf{v}_1} \cdot \int d\mathbf{x}_2 \int d\mathbf{v}_2 \left\{-e\mathbf{E}(\mathbf{x}_{12})\int_0^{\infty}\frac{1}{m}\left[-e\mathbf{E}(\mathbf{x}_{12}-\mathbf{g}\tau)\right.\right.$$

$$\left.\left. \cdot \left(\frac{\partial}{\partial \mathbf{v}_1} - \frac{\partial}{\partial \mathbf{v}_2}\right) f(1,t-\tau)f(2,t-\tau)d\tau\right]\right\} \qquad (2.61)$$

where $e\mathbf{E}$ here represents the two-particle force field.

Assuming that $f(1)f(2)$ is uniform over the $\mathbf{x}_2$, τ integration domain, we find

$$\left.\frac{Df}{Dt}\right|_{VL} = \left(\frac{e}{m}\right)^2 \frac{\partial}{\partial \mathbf{v}_1} \cdot \int d\mathbf{v}_2 \langle \overline{\overline{\mathbf{E}\mathbf{E}}}\rangle \cdot \left[f(2)\frac{\partial}{\partial \mathbf{v}_1}f(1) - f(1)\frac{\partial}{\partial \mathbf{v}_2}f(2)\right] \quad (2.62)$$

where

$$\langle \overline{\overline{\mathbf{E}\mathbf{E}}}\rangle = \int d\mathbf{x}_2 \int_0^\infty dt\,\overline{\mathbf{E}(\mathbf{x}_{12})\mathbf{E}(\mathbf{x}_{12} - \mathbf{g}\tau)} \quad (2.63)$$

represents the $\mathbf{E}$ field autocorrelation function in the straight-line approximation. We may rewrite (2.62) in a form that strongly resembles the Landau equation (2.50):

$$\left.\frac{Df}{Dt}\right|_{VL} = \frac{\partial}{\partial \mathbf{v}} \cdot \int \bar{\bar{Q}} \cdot \left(\frac{\partial}{\partial \mathbf{v}} - \frac{\partial}{\partial \mathbf{v}_1}\right) f(\mathbf{v})f(\mathbf{v}_1)\,d\mathbf{v}_1$$

$$\bar{\bar{Q}} \equiv \left(\frac{e}{m}\right)^2 \int d\mathbf{x}_1 \int_0^\infty d\tau\,\overline{\mathbf{E}(\mathbf{x} - \mathbf{x}_1)\mathbf{E}^*(\mathbf{x} - \mathbf{x}_1 - \mathbf{g}\tau)} \quad (2.64)$$

where we have changes notation: $\mathbf{x}_1, \mathbf{v}_1 \to \mathbf{x}, \mathbf{v}; \mathbf{x}_2, \mathbf{v}_2 \to \mathbf{x}_1, \mathbf{v}_1$. Note that the complex conjugate has no effect on the real $\mathbf{E}$ field.

Inserting the Fourier transform [see (1.32)]

$$e\mathbf{E}(\mathbf{x}) = -\nabla\Phi(\mathbf{x}) = \frac{1}{(2\pi)^3}\int d\mathbf{k}\,i\mathbf{k}e^{i\mathbf{k}\cdot\mathbf{x}}\bar{\Phi}(k) \quad (2.65)$$

into the autocorrelation function $\bar{\bar{Q}}$ gives

$$\bar{\bar{Q}} = \frac{1}{m^2}\int d\tau \int \frac{d\mathbf{k}\,\mathbf{k}}{(2\pi)^3}\int \frac{d\mathbf{k}'\,\mathbf{k}'}{(2\pi)^3}\bar{\Phi}(k)\bar{\Phi}^*(k')\int d\mathbf{x}_1 e^{i\mathbf{x}_1\cdot(\mathbf{k}'-\mathbf{k})}e^{i\mathbf{x}\cdot(\mathbf{k}'-\mathbf{k})}e^{i\mathbf{k}\cdot\mathbf{g}\tau} \quad (2.66)$$

With

$$\int d\mathbf{x}_1 e^{i\mathbf{x}_1\cdot(\mathbf{k}'-\mathbf{k})} = (2\pi)^3\delta(\mathbf{k}'-\mathbf{k})$$

integration of (2.66) over $\mathbf{k}'$ gives

$$\bar{\bar{Q}} = \frac{1}{m^2(2\pi)^3}\int d\mathbf{k}\,\overline{\overline{\mathbf{k}\mathbf{k}}}|\bar{\Phi}(k)|^2\int_{=}^{\infty} d\tau\,e^{i\mathbf{k}\cdot\mathbf{g}\tau}$$

which gives, finally,

$$\bar{\bar{Q}} = \frac{1}{8\pi^2 m^2}\int d\mathbf{k}\,\overline{\overline{\mathbf{k}\mathbf{k}}}\delta(\mathbf{k}\cdot\mathbf{g})|\bar{\Phi}(k)|^2 \quad (2.67)$$

Recalling (1.21), we reintroduce the dielectric constant $\varepsilon(k, \omega)$, obtained from linearized perturbation theory of the Vlasov equation, and set

$$\bar{\Phi}(k) \to \frac{\bar{\Phi}(k)}{\varepsilon(k, \omega)} \quad (2.68)$$

where $\varepsilon(k, \omega)$ is given by (1.23):

$$\varepsilon(k, \omega) = 1 + \frac{\omega_p^2}{k^2} \int \frac{\mathbf{k} \cdot \partial \tilde{f}_0/\partial \mathbf{v}\, d\mathbf{v}}{\omega - \mathbf{k} \cdot \mathbf{v}}$$

Substitution of the form of $\bar{\Phi}(k)$, so augmented, into (2.67) gives the Balescu-Lenard $\bar{\bar{Q}}$ tensor.

$$\bar{\bar{Q}}_{\mathrm{BL}} = \frac{1}{8\pi^2 m^2} \int \frac{d\mathbf{k}\, \overline{\overline{\mathbf{k}\mathbf{k}}}\, |\bar{\Phi}(k)|^2 \delta(\mathbf{k} \cdot \mathbf{g})}{\left|1 - \frac{\omega_p^2}{k^2} \int \frac{\mathbf{k} \cdot (\partial \tilde{f}_0/\partial \mathbf{v})\, d\mathbf{v}}{\mathbf{k} \cdot \mathbf{v} - \omega}\right|^2} \tag{2.69}$$

The Fourier transform $\bar{\Phi}(k)$ is that of the bar Coulomb potential energy [compare (1.33) with $k_d = 0$]:

$$\bar{\Phi}(k) = \frac{4\pi e^2}{k^2} \tag{2.70}$$

With $\bar{\bar{Q}}_{\mathrm{BL}}$ substituted into (2.64), we obtain the Balescu-Lenard equation:

$$\boxed{\left.\frac{Df}{Dt}\right|_{VL} = \frac{\partial}{\partial \mathbf{v}} \cdot \int \bar{\bar{Q}}_{\mathrm{BL}} \cdot \left(\frac{\partial}{\partial \mathbf{v}} - \frac{\partial}{\partial \mathbf{v}_1}\right) f(\mathbf{v}) f(\mathbf{v}_1)\, d\mathbf{v}_1} \tag{2.71}$$

Note that this equation includes the effects of shielding as contained in the $\bar{\bar{Q}}_{\mathrm{BL}}$ coefficient.

Reduction of $\bar{\bar{Q}}_{\mathrm{BL}}$ to $\bar{\bar{Q}}_{\mathrm{L}}$

An instructive example at this point concerns the reduction of $\bar{\bar{Q}}_{\mathrm{BL}}$ as given by (2.69) to the corresponding Landau form (2.51). First note that this equation may be written

$$\frac{Df}{Dt} = \frac{\partial}{\partial \mathbf{v}} \cdot \int \bar{\bar{Q}}_L \cdot \left(\frac{\partial}{\partial \mathbf{v}} - \frac{\partial}{\partial \mathbf{v}_1}\right) f_1 f\, d\mathbf{v}_1 \tag{2.72a}$$

$$\bar{\bar{Q}}_L = \frac{K}{2} \frac{\bar{\bar{T}}}{g^3} \tag{2.72b}$$

At $k \gg \lambda_d^{-1}$, $\varepsilon(k, \omega) \to 1$. In this limit, (2.68) and (2.69) give

$$\bar{\bar{Q}}_{\mathrm{BL}} \to \bar{\bar{Q}}'_{\mathrm{BL}} = \frac{2e^4}{m2} \int d\mathbf{k} \frac{\overline{\overline{\mathbf{k}\mathbf{k}}}}{k^4} \delta(\mathbf{k} \cdot \mathbf{g})$$

We obtain the Cartesian components of this matrix through evaluation of the integral in spherical coordinates. With $\mathbf{g}$ taken as the polar axis, there results

$$\bar{\bar{Q}}'_{\mathrm{BL}} = \frac{2e^4}{m^2} \int_0^\infty \frac{dk k^2}{k^4} \int_{-1}^1 d\cos\theta \int_0^{2\pi} d\phi\, \overline{\overline{\mathbf{k}\mathbf{k}}}\, \delta(kg\cos\theta)$$

$$= \frac{2e^4}{m^2} \int \frac{dk}{k} \int d\phi \int \frac{d\cos\theta}{g} \overline{\overline{\hat{k}\hat{k}}} \delta(\cos\theta)$$

Here $\hat{k}$ represents a unit k vector. The $\overline{\overline{\hat{k}\hat{k}}}$ matrix is given by

$$\overline{\overline{\hat{k}\hat{k}}} = \begin{pmatrix} \sin^2\theta\cos^2\phi & \sin^2\theta\sin\phi\cos\phi & \sin\theta\cos\theta\cos\phi \\ \sin^2\theta\cos\phi\sin\phi & \sin^2\theta\sin^2\phi & \sin\theta\cos\theta\sin\phi \\ \cos\theta\sin\theta\cos\phi & \cos\theta\sin\theta\sin\phi & \cos^2\theta \end{pmatrix}$$

Integration over $\cos\theta$ leaves

$$\int \overline{\overline{\hat{k}\hat{k}}} \delta(\cos\theta)\, d\cos\theta = \begin{pmatrix} \cos^2\phi & \sin\phi\cos\phi & 0 \\ \cos\phi\sin\phi & \sin^2\phi & 0 \\ 0 & 0 & 0 \end{pmatrix}$$

Completing the ϕ integration gives

$$\overline{\overline{Q}}'_{\mathrm{BL}} = \frac{2\pi e^4}{m^2} \int \frac{dk}{k} \frac{1}{g^3} \begin{pmatrix} g^2 & 0 & 0 \\ 0 & g^2 & 0 \\ 0 & 0 & 0 \end{pmatrix}$$

With the identification (2.49), the latter equation may be written

$$\overline{\overline{Q}}'_{\mathrm{BL}} = \frac{2\pi e^4}{m^2} \frac{\overline{\overline{T}}(g)}{g^3} \int_{k_d}^{k_0} \frac{dk}{k} \tag{2.73}$$

Here we have introduced the cutoff parameters

$$k_0^2 \equiv \left(\frac{k_B T}{e^2}\right)^2, \qquad k_d^2 \equiv 4\pi \frac{e^2 n}{k_B T} \tag{2.73a}$$

corresponding, respectively, to the wave number of closest approach and the Debye wave number, introduced previously in (1.31). It follows that

$$\frac{k_0}{k_d} = \frac{(k_B T)^{3/2}}{e^3 n^{1/2}} = 4\pi n\lambda_d^3 = \frac{1}{\Lambda} = \varepsilon^{-1} \tag{2.73b}$$

The plasma parameter Λ was introduced previously in (2.3.59). We recall that, for a weakly coupled plasma, such as is presently considered, Λ is a small parameter. We write

$$\int_{k_d}^{k_0} \frac{dk}{k} = \ln\varepsilon^{-1}$$

and $\overline{\overline{Q}}'_{\mathrm{BL}}$ reduces to the form

$$\overline{\overline{Q}}'_{\mathrm{BL}} = \left(\frac{2\pi e^4}{m^2} \ln\varepsilon^{-1}\right) \frac{\overline{\overline{T}}}{g^3} = \frac{K}{2}\frac{\overline{\overline{T}}}{g^3} = \overline{\overline{Q}}_L \tag{2.74}$$

We recognize this to be the Landau kernel given by (2.72b). This reduction was obtained in the limit $k \gg \lambda_d^{-1}$, $\varepsilon \to 1$, relevlant to a plasma with no shielding effects, such as described by the Landau or, equivalenty, the Fokker-Planck equations.

This finding brings our analysis around full circle. As illustrated in Fig 4.9, the Landau equation follows both from the Boltzmann equation (Section 2.3) and, as shown immediately above, from the Balescu-Lenard equation. This connecting loop indicates that, as previously described, the Boltzmann equation too, rests under the aegis of the BBKGY hierarchy.

4.2.6 Convergent Kinetic Equation

It is evident that the kinetic equations we have discussed relevant to plasma dynamics all suffer divergence. In deriving the Fokker-Planck equation, we found that the Boltzmann collision integral exhibits divergence at small-angle collisions or, equivalently, at large impact parameter collisions [see, for example, (2.28)]. When written in terms of impact parameter, this singularity has the form $\int ds/s$.

In the limit of large k, the Balescu-Lenard collision operator (2.69) exhibits a $\sim \int dk/k$ divergence. Since $\varepsilon(k, \omega) = 1$ in this limit, the Landau equation exhibits a like logarithmic divergence. At small k, $\bar{\bar{Q}}_{BL}$ is nonsingular, whereas the Landau equation remains logarithmically singular.

The regular behavior of $\bar{\bar{Q}}_{BL}$ at small k is due to the inclusion of dielectric shielding in its kernel, thereby lessening the effects of long-range collisions. Its remaining singularity at lark k reflects the fact that the equation does not incorporate effects of wide-angle collisions. This conclusion follows from our derivation of the Landau equation from the Boltzmann equation at small-angle collisions and the demonstration of the equivalence of the Landau and Balescu-Lenard equations at $\varepsilon(k, \omega) = 1$.

Resoultion of these divergent forms was considered by Hubbard,[25] Frieman and Book,[26] Gould and DeWitt,[27] and Aono,[28] among others.

We first write the convergent kinetic equation in the form

$$\frac{Df}{Dt} = I = I_0 + \tilde{I} - \tilde{I}_0 \qquad (2.75)$$

where the collision I terms denote the following:

$$I_0 = \text{Boltzmann collision integral}, \ s \leq s_0$$

$$\tilde{I} = \text{Balescu-Lenard integral}, \ k < 1/s_0$$

[25] J. Hubbard, *Proc. Roy. Soc. A261*, 371 (1961).
[26] E. A. Friedman and D. C. Book, *Phys. Fluids 6*, 1700 (1963).
[27] H. A. Gould and H. E. DeWitt, *Phys. Rev. 155*, 68 (1967).
[28] O. Aono, *Phys. Fluids 11*, 341 (1968).

$$\tilde{I}_0 = \text{renormalization term cancels singularities in } I_0 \text{ and } \tilde{I}$$

The inequalities on the right denote domains of convergence, with s_0 representing an intermediary value of the impact parameter.

We may write I_0 as an integral over impact parameter:

$$I_0 = \int_0^\infty B(s)\,ds \tag{2.75a}$$

For $s \gg s_0$, small-angle collisions are predominant and the Fokker-Planck expansion is relevant, which in turn yields the Landau equation (2.51). In developing this equation, the related singularity was absorbed in the coefficient K (2.47a). Rewriting this coefficient as an integral over the impact parameter,

$$K = \pi \kappa^2 \int \frac{ds}{s}$$

indicates that

$$B(s) \sim \frac{A}{s}$$

relevant to the large s, grazing-collision domain. The integral A is written for the Landau form:

$$A = \frac{\pi \kappa^2}{2} \int \frac{\partial}{\partial \mathbf{v}} \cdot \frac{\bar{\bar{T}}}{g^3} \cdot \left(\frac{\partial}{\partial \mathbf{v}} - \frac{\partial}{\partial \mathbf{v}_1} \right) f(\mathbf{v}) f(\mathbf{v}_1)\, d\mathbf{v}_1$$

Similarly, we write

$$\tilde{I} = \int G(k)k \tag{2.75b}$$

where $G(k)$ is the implied Balescu-Lenard form. For large k, $\varepsilon(\kappa, \omega) \to 1$ and, as shown above, the Balescu-Lenard equation reduces to the Landau equation. Thus, in this domain, $\tilde{I}$ suffers the same logarithmic singularity as the Landau collision form. That is , we may write

$$G(k) \sim \frac{A}{k}$$

For $\tilde{I}_0$ to cancel the s singularity in I_0 and the k singularity in $\tilde{I}$, it should have a dual form. This property may be satisfied with the aid of the Bessel function $J_1(x)$ by virture of its following property.

$$\int_0^\infty J_1(x)\,dx = 1 \; \begin{cases} ds \displaystyle\int_0^\infty J_1(ks)\,dk = \frac{ds}{s} \displaystyle\int_0^\infty J_1(ks)\,d(ks) = \frac{ds}{s} \\[2ex] dk \displaystyle\int_0^\infty J_1(ks)\,ds = \frac{dk}{k} \displaystyle\int_0^\infty J_1(ks)\,d(ks) = \frac{dk}{k} \end{cases}$$

Accordingly, we write

$$\bar{I}_0 = A \int_0^\infty \int_0^\infty J_1(ks)\,ds\,dk$$

$$= A \int_0^{s_0} ds \int_0^\infty dk J_1(ks) + \int_0^\infty dk \int_0^{s_0} ds J_1(ks)$$

so that

$$I = \int_0^{s_0} B(s)\, ds + \int_{s_0}^\infty \left[B(s) - A \int_0^\infty dk J_1(ks) \right] ds$$

$$+ \int_0^\infty \left[G(k) - A \int_0^{s_0} ds J_1(ks) \right] dk \qquad (2.76)$$

The integral over k in the first bracket leaves A/s, which cancels the $B(s)$ singularity for large s. In the singular domain of $G(k)$, that is, $ks_0 \gg 1$, the integral over s in the second bracket cancels the A/k singularity of $G(k)$. So we find that the combination integrals (2.76) gives a reasonable model for a convergent plasma kinetic equation.

4.2.7 Fokker–Planck Equation Revisited

Of the various kinetic equations described in Fig 4.9, the Boltzmann, KBG, and Fokker-Planck equations are relevant to neutral fluids as well as to plasmas. To exhibit this property for the Fokker-Planck equation, we revert to the Chapman-Komogorov equation (1.7.7) relevant to scattering processes homogeneous in time.

Let $\prod(\mathbf{v}, \Delta\mathbf{v})$ denote the probability that in the interval Δt, $\mathbf{v} \to \mathbf{v} + \Delta\mathbf{v}$. The Chapman-Kolmogorov equation (1.7.7) may then be written

$$f(\mathbf{v}, t) = \int f(\mathbf{v} - \Delta\mathbf{v}, t - \Delta\mathbf{v}) \prod(\mathbf{v} - \Delta\mathbf{v}, \Delta\mathbf{v})\, d\Delta\mathbf{v} \qquad (2.77)$$

If we assume that only small velocity changes $\Delta\mathbf{v}$ contribute to this integration then $\prod(\mathbf{v}-\Delta\mathbf{v}, \Delta\mathbf{v})$ will be sharply varying in its second variable but smoothly varying in its first variable. This permits expansion of the integrand in (2.77) about $\mathbf{v} - \Delta\mathbf{v} = \mathbf{v}$ and $t - \Delta t = t$. There results

$$f(\mathbf{v}, t) = \int d(\Delta\mathbf{v}) \left\{ f(\mathbf{v}, t) \prod(\mathbf{v}, \Delta\mathbf{v}) - \Delta t \frac{\partial f(\mathbf{v}, t)}{\partial t} \prod(\mathbf{v}, \Delta\mathbf{v}) \right. \qquad (2.78)$$

$$\left. - \Delta\mathbf{v} \cdot \frac{\partial f(\mathbf{v}, t) \prod(\mathbf{v}, \Delta\mathbf{v})}{\partial \mathbf{v}} + \frac{1}{2}\overline{\Delta\mathbf{v}\Delta\mathbf{v}} : \frac{\overline{\partial^2 [f(\mathbf{v}, t) \prod(\mathbf{v}, \Delta\mathbf{v})]}}{\partial\mathbf{v}\partial\mathbf{v}} \right\}$$

Dividing this equation through by Δt and then integrating over $\Delta\mathbf{v}$, with

$$\int \prod(\mathbf{v}, \Delta\mathbf{v})\, d\Delta\mathbf{v} = 1$$

we obtain

$$\frac{\partial f}{\partial t} = -\frac{\partial}{\partial \mathbf{v}} \cdot \left(\left\langle \frac{\Delta\mathbf{v}}{\Delta t} \right\rangle f \right) + \frac{1}{2} \frac{\overline{\partial^2}}{\partial\mathbf{v}\partial\mathbf{v}} : \left(\left\langle \frac{\overline{\Delta\mathbf{v}\Delta\mathbf{v}}}{\Delta t} \right\rangle f \right) \qquad (2.79)$$

Here we have set

$$\left\langle \frac{\Delta \mathbf{v}}{\Delta t} \right\rangle = \int \left(\frac{\Delta \mathbf{v}}{\Delta t} \right) \prod (\mathbf{v}, \Delta \mathbf{v}) \, d\Delta \mathbf{v}$$

$$\left\langle \overline{\frac{\Delta \mathbf{v} \Delta \mathbf{v}}{\Delta t}} \right\rangle = \int \left(\overline{\frac{\Delta \mathbf{v} \Delta \mathbf{v}}{\Delta t}} \right) \prod (\mathbf{v}, \Delta \mathbf{v}) \, d\Delta \mathbf{v} \tag{2.80}$$

Identifying $(\Delta \mathbf{v}/\Delta t)$ with the friction coefficient $\mathbf{a}$ and $\langle \Delta \mathbf{v} \Delta \mathbf{v}/\Delta t \rangle$ with the diffusion coefficient $\bar{\bar{b}}$, permits (2.79) to be recognized as the Fokker-Planck equation (2.31). However, the present derivation indicates that the Fokker-Planck equation has wider use than mere application to plasma physics.

A more tractable form of the average quantities (2.80) may be obtained as follows. With (2.23), we may write

$$\Delta \mathbf{v} = \alpha(\alpha \cdot \mathbf{g})$$

for velocity change due to collision. For the remaining factor in (2.80), we write

$$\prod (\mathbf{v}, \Delta \mathbf{v}) \, d(\Delta \mathbf{v}) = \Delta t \delta \nu \tag{2.81}$$

Here $\delta \nu$ represents the increment in collision frequency in which a $\mathbf{v}_1$ particle collides with a $\mathbf{v}$ particle, causing the change $\mathbf{v} \to \mathbf{v} + \Delta \mathbf{v}$. This factor is given by

$$\delta \nu = \sigma \, d\Omega g \, d\mathbf{v}_1 f(\mathbf{v}_1) \tag{2.82}$$

Completing the integral in (2.80) gives

$$\mathbf{a} = \left\langle \frac{\Delta \mathbf{v}}{\Delta t} \right\rangle = \iint \sigma \, d\Omega \, d\mathbf{v}_1 g\alpha(\alpha \cdot \mathbf{g}) f(\mathbf{v}_1) \tag{2.83a}$$

Repeating this contruction for the second integral in (2.80), we obtain

$$\bar{\bar{b}} = \left\langle \overline{\frac{\Delta \mathbf{v} \Delta \mathbf{v}}{\Delta t}} \right\rangle = \iint \sigma \, d\Omega \, d\mathbf{v}_1 g(\alpha \cdot \mathbf{g})^2 \overline{\alpha \alpha} f(\mathbf{v}_1) \tag{2.83b}$$

Let us employ these expressions to show that, for a plasma, $\mathbf{a}$ is given by (2.30a). With (2.21),

$$g\sigma \, d\Omega = \frac{\kappa^2 \, d \cos \psi \, d\phi}{(\mathbf{g} \cdot \alpha)^3} \tag{2.84}$$

Then (2.83a) becomes

$$\mathbf{a} = \iiint \frac{\kappa^2 \, d \cos \psi \, d\phi}{(g \cos \psi)^2} \alpha f(\mathbf{v}_1) \, d\mathbf{v}_1 \tag{2.85}$$

Again, choosing $\mathbf{g}$ as the polar axis for the ψ, ϕ integration gives the Cartesian components

$$\alpha = (\sin \psi \cos \phi, \sin \psi \sin \phi, \cos \psi)$$

Thus only the z component of $\mathbf{a}$ survives integration, and we may write

$$\mathbf{a} = 2\pi\kappa^2 \iint \frac{d\cos\psi}{\cos\psi} \frac{\mathbf{g}}{g^3} f(\mathbf{v}_1)\, d\mathbf{v}_1 \tag{2.86}$$

which with (2.28) returns (2.30a). The same identification follows for the diffusion coefficient $\bar{\bar{b}}$. Thus we find that the generalized Fokker-Planck equation (2.79) reduces to the plasma Fokker-Planck equation when the Coulomb cross section (2.84) is employed.

4.3 Monte Carlo Analysis in Kinetic Theory

In this concluding section of the present chapter we turn to a computational method in kinetic theory that has experienced wide use in the recent past. It is known as the Monte Carlo technique.

4.3.1 Master Equation

Primary use of this analysis addresses a spatially homogeneous fluid in an externally supported constant force field $\mathbf{K}$. The distribution function for this fluid, $f(\mathbf{v}, t)$, obeys the masterlike equation[29]

$$\frac{\partial f}{\partial t} + \mathbf{a} \cdot \frac{\partial f}{\partial \mathbf{v}} = \int d\mathbf{v}'[w(\mathbf{v}', \mathbf{v})f(\mathbf{v}', t) - w(\mathbf{v}, \mathbf{v}')f(\mathbf{v}, t)] \tag{3.1}$$

where $w(\mathbf{v}', \mathbf{v})$ represents the probability rate for scattering[30] from the velocity $\mathbf{v}'$ to $\mathbf{v}$ and $\mathbf{a} \equiv \mathbf{K}/m$.

An important parameter that enters this analysis is the total scattering rate from the velocity $\mathbf{v}$ to all other velocities:

$$\lambda(\mathbf{v}) = \int w(\mathbf{v}, \mathbf{v}')\, d\mathbf{v}' \tag{3.2}$$

To simplify (3.1), we introduce

$$\begin{aligned}
\tilde{w}(\mathbf{v}, \mathbf{v}') &\equiv w(\mathbf{v}, \mathbf{v}') + [\Gamma - \lambda(\mathbf{v})]\delta(\mathbf{v} - \mathbf{v}') \\
\tilde{w}(\mathbf{v}', \mathbf{v}) &\equiv w(\mathbf{v}', \mathbf{v}) + [\Gamma - \lambda(\mathbf{v}')]\delta(\mathbf{v}' - \mathbf{v})
\end{aligned} \tag{3.3}$$

---

[29]The master equation was encountered in Chapter 1 (1.7.20). It is used extensively in Chapter 5 for problems in quantum kinetic theory; see (5.2.28), (5.2.86), and (5.3.58).

[30]For inelastic scattering, we cannot assume $w(\mathbf{v}, \mathbf{v}') = w(\mathbf{v}', \mathbf{v})$. In an inelastic scattering, initial and final scattering constituents are not the same. Such is the case, for example, in ionizing collisions or collisions where an atom undergoes transition to an excited state.

where Γ is an arbitrary constant. The delta function component in these expressions permits no change in velocity in scattering and is referred to as a *self-scattering mechanism.*[31]

Note that the integration corresponding to (3.2) gives

$$\tilde{\lambda}(\mathbf{v}) = \int \tilde{w}(\mathbf{v}, \mathbf{v}')\, d\mathbf{v}' = \Gamma \tag{3.3a}$$

Substituting (3.3) into (3.1) gives

$$\frac{\partial f}{\partial t} + \mathbf{a} \cdot \frac{\partial f}{\partial \mathbf{v}} = \int d\mathbf{v}'\, \tilde{w}(\mathbf{v}', \mathbf{v}) f(\mathbf{v}', t) - \Gamma f(\mathbf{v}, t) \tag{3.4}$$

In order for the probability scattering rate $\tilde{w}$ to be positive, as the constant Γ is arbitrary, we choose

$$\Gamma = \max_{\mathbf{v}}[\lambda(\mathbf{v})] \tag{3.4a}$$

Equation (3.4) may be rewritten in the purely integral equation form

$$f(\mathbf{v}, t) = \iint d\mathbf{v}'\, d\mathbf{v}'' \int_0^t dt'\, f(\mathbf{v}', t)\tilde{w}(\mathbf{v}', \mathbf{v}'')\delta[\mathbf{v}' - \mathbf{v}'' - (t - t')\mathbf{a}]e^{-\Gamma(t-t')} \tag{3.5}$$

It is the interpretation of terms in (3.5) that forms the basis for the Monte Carlo analysis in kinetic theory

4.3.2 Equivalence of Master and Integral Equations

Prior to offering an interpretation of terms in the integral equation (3.5), let us demonstrate the equivalence of this equation to its precursor, the master equation (3.4). Towards these ends, we first set

$$\tau \equiv t - t'$$

and integrate (3.5) over $\mathbf{v}'$. There results

$$f(\mathbf{v}, t) = \int d\mathbf{v}' \int_0^t d\tau\, f(\mathbf{v}', t - \tau)\tilde{w}(\mathbf{v}', \mathbf{v} - \tau\mathbf{a})e^{-\Gamma\tau} \tag{3.6}$$

Now note that

$$\mathbf{a} \cdot \frac{\partial}{\partial \mathbf{v}}\tilde{w}(\mathbf{v}', \mathbf{v} - \tau\mathbf{a}) = -\frac{\partial}{\partial \tau}\tilde{w}(\mathbf{v}', \mathbf{v} - \tau\mathbf{a})$$

Operating on (3.6) with $\mathbf{a} \cdot (\partial/\partial\mathbf{v})$ then gives

$$\mathbf{a} \cdot \frac{\partial}{\partial \mathbf{v}} f(\mathbf{v}, t) = -\int d\mathbf{v}' \int_0^t d\tau\, f(\mathbf{v}', t - \tau)\left[\frac{\partial}{\partial \tau}\tilde{w}(\mathbf{v}', \mathbf{v} - \tau\mathbf{a})\right]e^{-\Gamma\tau}$$

---

[31]H. D. Rees, *J. Phys. Chem. Solids 30*, 643 (1969). For further discussion, see C. Jacoboni and L. Reggiani, *Revs. Mod. Phys. 55*, 645 (1983).

Integrating the right side of the latter equation by parts (over τ), we obtain

$$\mathbf{a} \cdot \frac{\partial}{\partial \mathbf{v}} f(\mathbf{v}, t) = - \int d\mathbf{v}' f(\mathbf{v}', 0) \tilde{w}(\mathbf{v}', \mathbf{v} - t\mathbf{a}) e^{-\Gamma \tau}$$

$$+ \int d\mathbf{v}' f(\mathbf{v}', t) \tilde{w}(\mathbf{v}', \mathbf{v}) - \int d\mathbf{v}' \int d\tau \, \tilde{w}(\mathbf{v}', \mathbf{v} - \tau \mathbf{a})$$

$$\times \left[\Gamma e^{-\Gamma \tau} f(\mathbf{v}', t - \tau) - e^{-\Gamma \tau} \frac{\partial}{\partial \tau} f(\mathbf{v}', t - \tau) \right]$$

$$\equiv \text{I} + \text{II} + \text{III} + \text{IV} \tag{3.7}$$

where the Roman numerals stand for corresponding respective integrals. Now recall that for any integrable function $g(t)$,

$$\frac{\partial}{\partial t} \int_0^t h(t - \tau) g(\tau) \, d\tau = h(0) g(t) + \int_0^t g(\tau) \frac{\partial}{\partial t} h(t - \tau) \, dt \tag{3.8}$$

We are now prepared to reconstruct (3.4). First we note that from (3.6)

$$\text{III} = -\Gamma f(\mathbf{v}, t)$$

Next we employ (3.8) with $h = f(\mathbf{v}', t)$ and $g = \tilde{w}(\mathbf{v}', \mathbf{v} - t)$ and $g = \tilde{w}(\mathbf{v}', \mathbf{v} - t\mathbf{a}) e^{-\Gamma \tau}$ and obtain

$$\text{I} + \text{IV} = - \frac{\partial}{\partial t} \iint_0^t f(\mathbf{v}', t - \tau) \tilde{w}(\mathbf{v}', \mathbf{v} - \tau \mathbf{a}) e^{\Gamma \tau} \, d\mathbf{v}' = - \frac{\partial}{\partial t} f(\mathbf{v}', t)$$

the last equality following from (3.6). Collecting terms (with II remaining intact) returns (3.4). Thus we find that (3.5) is equivalent to (3.4)

4.3.3 Interpretation of Terms

We wish to interpret the terms in (3.5). First we note that the probability of a particles of velocity $\mathbf{v}$ undergoing a collision in the time Δt is

$$P(\Delta t) = \tilde{\lambda}(\mathbf{v}) \Delta t$$

It follows that probability, $\bar{P}$, that the particles has no collision in the interval Δt is

$$\bar{P}(\Delta t) = 1 - \tilde{\lambda}(\mathbf{v}) \Delta t$$

Over a finite interval $t = N \, \Delta t$,

$$\bar{P}(t) = \left[1 - \tilde{\lambda}(\mathbf{v}_1) \Delta t \right] \left[1 - \tilde{\lambda}(\mathbf{v}_2) \Delta t \right] \cdots \left[1 - \tilde{\lambda}(\mathbf{v}_N) \Delta t \right] \tag{3.9}$$

where

$$\mathbf{v}_n = \mathbf{v}(n \Delta t) = \mathbf{v}(0) + \mathbf{a} \, n \Delta t \tag{3.9a}$$

and $n = 1, \ldots, N$. With (3.3a), the preceding becomes

$$\bar{P} = \left(1 - \frac{\Gamma t}{N} \right)^N$$

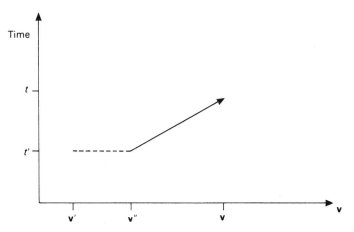

FIGURE 4.14. A particle undergoes a collision at t' and changes velocity from $\mathbf{v}'$ to $\mathbf{v}''(—)$. In the interval $(t - t')$, the particle undergoes free-flight acceleration from $\mathbf{v}''$ to $\mathbf{v}$ (—).

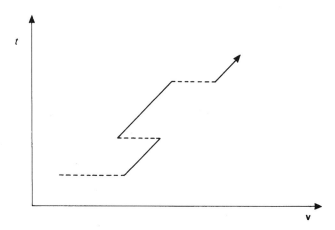

FIGURE 4.15. Three scattering free-flight events.

which in the limit $N \to \infty$ gives

$$\bar{P} = e^{-\Gamma t} \tag{3.10}$$

Thus, the exponential factor in (3.5) represents the probability that the particle drifts with no collision for the time $(t - t')$. The delta function tells us that during this collisionless flight the particle undergoes an acceleration due to the external field from velocity $\mathbf{v}''$ to $\mathbf{v}$. The factor $f(\mathbf{v}', t)\tilde{w}(\mathbf{v}', \mathbf{v}'')$ describes a particle with velocity $\mathbf{v}'$ being scattered to the velocity $\mathbf{v}''$. This series of events may be described in a $\mathbf{v} - \mathbf{t}$ diagram as shown in Fig. 4.14. The continuation of this collision-free flight process to three such events is shown in Fig. 4.15.

It is evident that the integration (3.5) carries this process over all $\mathbf{v}'$, $\mathbf{v}''$ and $0 \le t' \le t$. In the Monte Carlo analysis the distribution product $f(\mathbf{v})\Delta\mathbf{v}$ is constructed by measuring the time a typical trajectory spends in the interval $\Delta \mathbf{v}$ about the value $\mathbf{v}$, and we write

$$f(\mathbf{v})\Delta \mathbf{v} \propto \sum_{\mathbf{v}-\text{bin}} \Delta t \qquad (3.11)$$

The summation is over all time intervals Δt that the trajectory spends in the velocity interval, $\mathbf{v}, \mathbf{v} + \Delta \mathbf{v}$, which is termed a *velocity bin* ($\mathbf{v}$-bin).

4.3.4 Application of Random Numbers

As noted above, in this Monte Carlo process the computer follows the trajectory of a single particle. The intervals of free flight in this trajectory are related to a sequence of random numbers in the following manner. Consider that we have a distribution of random numbers, x, in the interval $(0, 1)$ with a uniform probability $P(x) = 1$. With (3.10), we note that the probability that the particle drifts for a time t and then suffers collision in the interval dt is equal to the product

$$(e^{-\Gamma t})\Gamma \, dt$$

With t also considered a random variable, we write

$$P(x) \, dx = \Gamma e^{-\Gamma t} \, dt = d(1 - e^{-\Gamma t})$$

Thus [recalling $P(x) = 1$] we find

$$t = -\frac{1}{\Gamma} \ln(1 - x) \qquad (3.12)$$

which is the desired mapping from the random variables x to the free flight intervals t.

Another step where random numbers come into play concerns the scattering rate $\tilde{w}(\mathbf{v}, \mathbf{v}') \, d\mathbf{v}'$ (dimensions, s^{-1}). Consider now that there is at our disposal a distribution of three sets of random numbers [all in the interval $(0,1)$]. Let us call a triplet of such values the vector $\mathbf{y}$. Viewing the final velocity $\mathbf{v}'$ as a random variable permits us to write

$$\tilde{w}(\mathbf{v}, \mathbf{v}') \, d\mathbf{v}' = P(\mathbf{y}) \, d\mathbf{y}$$

where again $P(\mathbf{y})$ is taken to be uniform. The preceding equation implies the relation

$$\mathbf{v}' = \mathbf{v}'(\mathbf{v}, \mathbf{y}) \qquad (3.13)$$

which effects a mapping from the random variables $(\mathbf{v}, \mathbf{y})$ to the random variables $\mathbf{v}'$. [As we are constructing $f(\mathbf{v})$, the velocity $\mathbf{v}$ is a known vector][32]

---

[32]It is this use of random numbers that motivates the gambling label of the analysis.

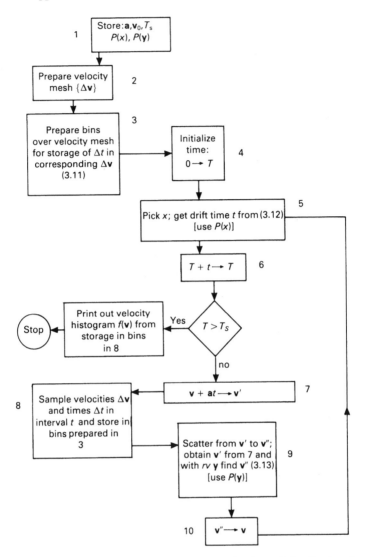

FIGURE 4.16. Flow chart for Monte Carlo determination of distribution function. Starting given values are constant acceleration α; starting velocity $\mathbf{v}_0$; random variable distributions $P(x)$, $P(\mathbf{y})$; final time T_s. Elapsed time of trajectory in any given loop is t.

4.3.5   Program for Evaluation of Distribution Function

Having interpreted the terms in (3.5) and described the manner in which random numbers enter the analysis, we are now prepared to construct a flow chart for the determination of the distribution function. Such a flow chart is shown in Fig. 4.16. The algorithm command $x \rightarrow y$ (also written $x = y$ in FORTRAN

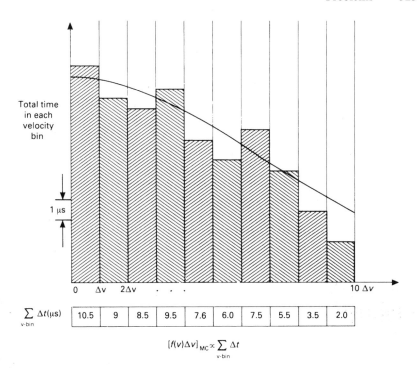

FIGURE 4.17. Typical bin storage histogram for one-dimensional Cartesian distribution component, $f(v)$, for force-free fluid. Velocity mesh is $10\Delta v$ units long. Smooth solid curve represents the distribution $\exp -(v/10)^2$.

and BASIC or $y := x$ in PASCAL) means "replace y by x." Recall also that the probability distributions $P(x)$ and $P(\mathbf{y})$ are uniform.

Note that the procedure displayed in Fig. 4.16 is directed at the construction of an equilibrium distribution. Furthermore, owing to the stochastic nature of these processes, it is assumed that $f(\mathbf{v})$ is independent of the starting velocity v_0 for sufficiently large T_s. If this is so, then the distributions $f(\mathbf{v}; v_0, T_s)$ and $f(\mathbf{v}; v_0 + \bar{\mathbf{v}}, T_s)$ are equal for arbitrary velocity $\bar{\mathbf{v}}$. To verify the equilibrium property of the distribution, we compare $f(\mathbf{v}; v_0, T_s)$ and $f(\mathbf{v}; v_0, T_s + \tau)$, which for equilibrium are equal for arbitrary time interval τ.

The manner in which $f(\mathbf{v})$ is constructed from time-interval storage in velocity bins is illustrated in Fig. 4.17.

Problems

4.1. (a) With $F(\beta)$ in (1.71) given by

$$F(\beta) = \frac{1}{\sqrt{2\pi}} \int_0^\infty \frac{d\mu\, \mu e^{-\mu^2/2}}{\mu - \beta}$$

show that

$$P : F(\beta) = 1 - \frac{2\beta^2 e^{-\beta^2/2}}{\sqrt{2\pi}} \int_{-\infty}^{\infty} dy e^{-y^2/2} \left[\frac{\sinh \beta y}{\beta y} \right]$$

Hint: Introduce the variable

$$y \equiv \mu - \beta$$

(b) With the preceding result, what is the explicit form of $\varepsilon(\beta)$ for a Maxwellian plasma?

(c) Establish the following rules concerning principle value:

$$P : \int_{-a}^{a} f_e(x) \, dx = 2 \lim_{\varepsilon \to 0} \int_{\varepsilon}^{a} f_e(x) \, dx$$

$$P : \int_{-a}^{a} f_o(x) \, dx = 0$$

where $f_e(x)$ and $f_o(x)$ are even and odd functions, respectively, and are regular in the open interval about the orgin.

4.2. What is the explicit form of the response dielectric function $\varepsilon(k, \mathbf{k} \cdot \mathbf{v}_0)$, as given by (1.28), for $\mathbf{v}_0 \neq 0$?

4.3. What does the static Debye potential $\Phi(r)$, given by (1.35), become for extraneous charge velocity $\mathbf{v}_0 \neq 0$?

4.4. Show that the derivation of the Landau decay constant Γ, preceding (1.66), remains valid providing $\Gamma \ll \omega_p$.

4.5. Can the contour in (1.69) be closed in the lower-half β plane? Explain your answer.

4.6. (a) Show that

$$\int f \ln f^0 \, d\mathbf{v} = \int f^0 \ln f^0 \, d\mathbf{v}$$

(b) With the preceding property, show that the KBG equation implies an $\mathcal{H}$ theorem.

Hint: Multiply both sides of the KBG equation by $f \, d\mathbf{v} \ln f$ and note that, with part (a),

$$\int (f^0 - f) \ln f \, d\mathbf{v} = \int (f^0 - f) \ln(f_0) \, d\mathbf{v}$$

4.7. A longitudinal electric wave with phase velocity $v_\phi = 0.8\omega_p/k$ and wavelength $\lambda = 1\mu m$ propagates through a plasma at $T = 20,000$ K and electron density $n = 4.2 \times 10^{18}$ cm^{-3}.

(a) What is the plasma frequency, ω_p, of this plasma?

(b) What is the Debye distance, λ_d, for this plasma?

(c) In how many seconds will the amplitude of the electric wave decay by the factor e^{-1} due to Landau damping?

4.8. Consider the following one-dimensional plasma distribution:

$$f_1(\mu) = 0, \qquad |\mu| > 1$$
$$f_1(\mu) = \frac{1}{2}, \qquad |\mu| \leq 1$$

where μ is the nondimensional velocity given by (1.41). Employ the Nyquist criterion to determine the number of unstable modes corresponding to this distribution.

4.9. Determine whether or not the Fokker-Planck equation (2.79) implies the conservation equations.

4.10. Establish the validity of (2.46),

$$\nabla \cdot f(g)\bar{\bar{T}} = f(g)\nabla \cdot \bar{\bar{T}}$$

where $f(g)$ has continuous first derivatives, the ∇ operator represents single-particle velocity differentiation, and $\bar{\bar{T}}$ is given by (2.39).

4.11. Show that the Fokker-Planck coefficients $\mathbf{a}$ and $\bar{\bar{b}}$ (2.30) satisfy the relation

$$\nabla \cdot \bar{\bar{b}} = \mathbf{a}$$

4.12. Show that the Maxwellian distribution reduces the Landau collision integral [see (2.51)] to zero.

4.13. Show that the matrix $\bar{\bar{T}}$ lies in a plane normal to $\mathbf{g}$. That is, show that

$$\mathbf{g} \cdot \bar{\bar{T}} = 0$$

4.14. Write down an explicit expression for the integrand $G(k)$ in (2.75b).

4.15. What is the physical significance of $\varepsilon(\omega, k) \to 1$ for a plasma, in the limit of large k. That is, what is the behavior of a wave in a plasma at such k?

4.16. Show that the electric-field auto correlation tensor in the straight-line approximation bay be written

$$\langle\overline{\overline{\mathbf{EE}}}\rangle = 2\pi e^2 \frac{\ln \varepsilon^{-1}}{g} \begin{pmatrix} 1 & 0 & 0 \\ 0 & 1 & 0 \\ 0 & 0 & 0 \end{pmatrix}$$

or, equivalently

$$\langle\overline{\overline{\mathbf{EE}}}\rangle = 2\pi e^2 \frac{\ln \varepsilon^{-1}}{g} \left(\frac{g^2 \bar{\bar{I}} - \overline{\mathbf{gg}}}{g^2} \right)$$

Hint: Recall that

$$\langle\overline{\overline{\mathbf{EE}}}\rangle = \int d\mathbf{x}_2 \int_0^\infty d\tau \overline{\mathbf{E}(\mathbf{x}_{12})\mathbf{E}}(\mathbf{x}_{12} - \mathbf{g}\tau)$$

Use cylindrical as shown in Fig 4.18. Introduce cutoff parameters in the s integration at the distance of closest approach, k_0^{-1}, and the Debye distance, λ_d.

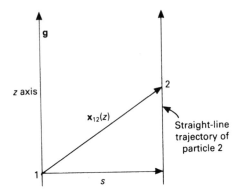

FIGURE 4.18. Configuration for Problem 4.16.

4.17. Relevant to the Monte-Carlo numerical procedure, write a flow chart corresponding to the output shown in Fig. 4.17. Assume three-dimensional scattering.

4.18. Determine if the KBG equation [(2.7),(2.8)] satisfies the conservation equations. Discuss your answer.

CHAPTER 5

Elements of Quantum Kinetic Theory

Introduction

This chapter begin with a review of some basic principles of quantum mechanics important to the formulation of quantum kinetic theory. In the following section the density operator is introduced and applied to problems involving beams of spinning particles. The Wigner distribution is described, and an equation of motion is derived for this function, which is found to reduce to the Liouville equation in the classical limit. The Wigner-Moyal equation and the notion of Weyl correspondence are described as well.

In the third section of the chapter, different manifestations of the KBG equation are applied respectively to the kinetic theory of photons interacting with a gas and electron transport in metals. In the former case, the canonical criterion for lasing is obtained from the relevant photon kinetic equation. In application to electron transport in metals, expressions are obtained for electrical and thermal conductivities and their ratio, the Lorentz number. The phenomenon of Thomas–Fermi screening is described, and forms for the related screening length and static potential are obtained. Dispersion relations for waves in quantum plasmas are discussed, and the close similarity between these equations and their classical counterparts is noted. A quantum modification of the Boltzmann equation due to Uehling and Uhlenbeck is introduced. This quasi-classical equation is found to give the correct quantum equilibrium distributions. The chapter continues with an overview of classical and quantum hierarchies, which includes a description of second quantization. With this formalism at hand, a quantum analysis of electrical conductivity is discussed, stemming from the Kubo formula derived earlier in Chapter 3. The chapter concludes with a brief introduction to the Green's function formalism relevant to quantum properties of many-body systems. Included in this description are diagrammatic representations of equations of motion for Green's functions. The discussion concludes with a derivation of the lifetime and energy of a particle interacting with its surroundings.

5.1   Basic Principles[1]

5.1.1   The Wave Function and Its Properties

In quantum mechanics, information concerning a given system is obtained from its wave function. Such information is derived from *expectation values*. Let

$$\psi(\mathbf{x}_1, \ldots, \mathbf{x}_N, t) \equiv \psi(\mathbf{x}^N, t)$$

be the wave function of a system comprised of N particles. Then, if $A(\mathbf{x}^N)$ is some dynamical property of the system, the expectation of A in the state $\psi(\mathbf{x}^N, t)$ is given by

$$\langle A \rangle = \int \psi^*(\mathbf{x}^N, t) \hat{A} \psi(\mathbf{x}^N, t) \, d\mathbf{x}^N \tag{1.1}$$

Here we have written $\hat{A}$ for the operator corresponding to the observable A. The quantity $\langle A \rangle$ is interpreted as an ensemble average.

In Dirac notation, (1.1) appears as

$$\langle A \rangle = \langle \psi | \hat{A} \psi \rangle \tag{1.2}$$

Another important manner in which information is derived from the wave function is through the *Born postulate*. This postulate indicates that the configurational probability of the system is given by

$$P(\mathbf{x}^N, t) = |\psi(\mathbf{x}^N, t)|^2 \tag{1.3}$$

This probability density is defined in the same manner as that obtained from the classical N particle distribution function $f_N(\mathbf{x}^N, \mathbf{p}^N, t)$:

$$P(\mathbf{x}^N, t) = \int f_N(\mathbf{x}^N, \mathbf{p}^N, t) \, d\mathbf{p}^N$$

The wave function evolves in time according to the *time-dependent Schrödinger equation*,

$$i\hbar \frac{\partial \psi}{\partial t} = \hat{H} \psi \tag{1.4}$$

where $\hat{H}$ is the Hamiltonian operator of the system.

---

[1] For further discussion, see P. A. M. Dirac, *The Principles of Quantum Mechanics*, 4th ed., Oxford, New York (1958); E. Merzbacher, *Quantum Mechanics*, 2nd ed., Wiley, New York (1970); R. L. Liboff, *Introductory Quantum Mechanics*, 4th ed., (2002), ibid.

Dynamic reversibility

Reversibility in quantum mechanics is given by the following description. With $\hat{H}(t) = \hat{H}(-t)$, we see that setting $t \to t' = -t$ and taking the complex conjugate of (1.4) gives

$$i\hbar \frac{\partial \psi^*(\mathbf{x}, -t)}{\partial t} = \hat{H} \psi^*(\mathbf{x}, -t)$$

This is the same equation as (1.4). We may conclude that if $\Psi(\mathbf{x}, t)$ is solution to the Schrödinger equation then $\Psi^*(\mathbf{x}, -t)$ is also a solution.

Time development of expectations

With (1.1) and (1.4), we obtain

$$i\hbar \frac{d\langle A \rangle}{dt} = \left\langle \left[\hat{A}, \hat{H}\right] + i\hbar \frac{\partial \hat{A}}{\partial t} \right\rangle \tag{1.5}$$

Here we have introduced the commutator

$$\left[\hat{A}, \hat{H}\right] \equiv \hat{A}\hat{H} - \hat{H}\hat{A} \tag{1.6}$$

The list of properties (1.1.26) obeyed by the Poisson brackets are also obeyed by the commutator.

Suppose $\hat{A}$ is not an explicit function of time. Then (1.5) indicates that if $\hat{A}$ commutes with $\hat{H}$ then the expectation of A is constant in time. This is the quantum mechanical analog of the classical theorem discussed in Chapter 1 which states that if a dynamical variable has a zero Poisson bracket with the Hamiltonian then it is a constant of the motion.

Measurement and operators

A fundamental postulate of quantum mechanics concerns measurement. Thus, if $\hat{A}$ is the operator corresponding to the observable A, then measured values of A, which we will call a, are eigenvalues of $\hat{A}$. The eigenvalue equation for $\hat{A}$ is given by

$$\hat{A}\varphi_a = a\varphi_a \tag{1.7}$$

where φ_a is the eigenfunction corresponding to the eigenvalue a.

The operator corresponding to energy is the Hamiltonian so that energy values of a given system are obtained from the time-independent Schrödinger equation:

$$\hat{H}\varphi_E = E\varphi_E \tag{1.8}$$

These eigenstates are called *stationary states* for the following reason.

The solution to the time-dependent Schrödinger equation (1.4) corresponding to the initial value $\psi(\mathbf{x}^N, 0)$ is given by (for time-independent $\hat{H}$)

$$\psi(\mathbf{x}^N, t) = e^{(-it\hat{H})/\hbar} \psi(\mathbf{x}^N, 0) \tag{1.9}$$

Suppose $\psi(\mathbf{x}^N, 0)$ is a solution to (1.8). Then

$$\psi_E(\mathbf{x}^N, t) = e^{(-it\hat{H})/\hbar} \varphi_E(\mathbf{x}^N)$$
$$= e^{(-itE)/\hbar} \varphi_E(\mathbf{x}^N) \tag{1.10}$$

Forming the expectation $\langle E \rangle$ as given by (1.2), we find

$$\langle E \rangle = \langle \psi_E | \hat{H} \psi_E \rangle = E \tag{1.11}$$

We may conclude that in a stationary state $\langle E \rangle$ is constant in time.

5.1.2   Commutators and Measurement

An important rule concerning measurement involves the commutator. Thus, if

$$[\hat{A}, \hat{B}] = 0$$

then $\hat{A}$ and $\hat{B}$ share common eigenstates.

As a simple example of this theorem, consider the configuration comprised of a single free particle whose Hamiltonian is

$$\hat{H} = \frac{\hat{p}^2}{2m}, \qquad \mathbf{p} = -i\hbar\nabla$$

It follows that

$$[\hat{H}, \hat{\mathbf{p}}] = 0$$

Eigenstates of $\hat{\mathbf{p}}$ and $\hat{H}$ are given by

$$\varphi_{\mathbf{k}}(\mathbf{x}) = \frac{1}{\sqrt{2\pi}} e^{i\mathbf{k}\cdot\mathbf{x}}$$
$$\varphi_E(\mathbf{x}) = A e^{i\mathbf{k}\cdot\mathbf{x}} + B e^{-i\mathbf{k}\cdot\mathbf{x}}$$

corresponding to

$$\hat{\mathbf{p}}\varphi_{\mathbf{k}} = \hbar\mathbf{k}\varphi_{\mathbf{k}}$$
$$\hat{H}\varphi_E = \frac{\hbar^2 k^2}{2m} \varphi_E$$

Thus $\hat{\mathbf{p}}$ and $\hat{H}$ share common eigenstates. That is, any eigenstate of the form $C \exp i\mathbf{k}\cdot\mathbf{x}$ is an eigenstate of both $\hat{\mathbf{p}}$ and $\hat{H}$.

Note, however, that $\varphi_E(\mathbf{x})$ as given above is an eigenstate of $\hat{H}$ but not of $\hat{\mathbf{p}}$. This property results from the degeneracy of free-particle energy eigenfunctions. A Venn diagram well illustrates this situation (see Fig. 5.1).

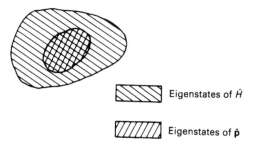

Eigenstates of $\hat{H}$

Eigenstates of $\hat{\mathbf{p}}$

FIGURE 5.1. Venn diagram depicting the eigenstates of $\hat{H}$ and $\hat{\mathbf{p}}$ for a free particle.

A simple example serves to illustrate the consequence on measurement that the commutator theorem has. Consider that the free particle is in the state

$$\psi_{\mathbf{k}}(\mathbf{x}) = A e^{i\mathbf{k}\cdot\mathbf{x}}$$

Measurement of the momentum of the particle in this state *is certain* to find the value $\mathbf{p} = \hbar\mathbf{k}$. Such measurement[2] leaves the system in the state $\Psi_{\mathbf{k}}(\mathbf{x})$. Subsequent measurement of E *is certain* to find the value $E = \hbar^2 k^2 / 2m$.

We conclude that there are states of a free particle in which E and $\mathbf{p}$ may be prescribed simultaneously. There is *no uncertainty* related to measurement of these observables in states that are common eigenstates of $\hat{H}$ and $\hat{\mathbf{p}}$.

Uncertainty in quantum mechanics

The uncertainty of measurement of an observable A in quantum mechanics is given by the variance of A [recall (1.8.2b)]:

$$(\Delta A)^2 = \langle (A - \langle A \rangle)^2 \rangle \tag{1.12}$$

For the example of the free particle considered above, in the common eigenstate $C \exp i\mathbf{k} \cdot \mathbf{x}$,

$$\Delta E = \Delta \mathbf{p} = 0$$

Clearly, this result is a consequence of the fact that for a free particle $[\hat{H}, \hat{\mathbf{p}}] = 0$. More generally, the *Robertson-Schrödinger theorem* tells us the following: Suppose two operators have a nonvanishing commutator. That is,

$$[\hat{A}, \hat{B}] = \hat{C} \tag{1.13}$$

The theorem then states that

$$\Delta A \Delta B \geq \frac{1}{2} |\langle C \rangle| \tag{1.14}$$

---

[2]Here we mean measurement in the ideal sense, that is, measurement that least perturbs the system.

The most famous application of this result follows from the fundamental commutator relation

$$[\hat{x}, \hat{p}] = i\hbar \tag{1.15}$$

Here we are writing x and p for parallel components of $\mathbf{x}, \mathbf{p}$. With the preceding relation, we conclude

$$\Delta x \Delta p \geq \frac{\hbar}{2} \tag{1.16}$$

That is, states *do not* exist in which x and p may be prescribed simultaneously.

Another important illustration of (1.14) concerns the Cartesian components of angular momentum, whose commutator relations are given by

$$\begin{aligned}
[\hat{J}_x, \hat{J}_y] &= i\hbar \hat{J}_z \\
[\hat{J}_y, \hat{J}_z] &= i\hbar \hat{J}_x \\
[\hat{J}_z, \hat{J}_x] &= i\hbar \hat{J}_y \\
[\hat{J}^2, \mathbf{J}] &= 0
\end{aligned} \tag{1.17}$$

From the first commutator we obtain

$$\Delta J_x \Delta J_y \geq \frac{\hbar}{2}|\langle J_z \rangle|$$

Thus, in general, J_x and J_y cannot be simultaneously designated in any quantum state. However, from the last of the four relations (1.17) we find, for example, that

$$\Delta J^2 \Delta J_z = 0$$

So states exist in which J^2 and J^z may be simultaneously designated.

5.1.3 Representations

Consider that the eigenstates of $\hat{A}$ comprise a discrete sequence. We may then write

$$\hat{A} |a_n\rangle = a_n |a_n\rangle \tag{1.18}$$

where n is an integer. In this notation, $|a_n\rangle$ represents the eigenstate of $\hat{A}$ corresponding to the eigenvalue a_n. The eigenstates $|a_n\rangle$ may be taken to be the basis of a Hilbert space $\mathcal{H}$. That is, $\{|a_n\rangle\}$ *spans* the Hilbert space $\mathcal{H}$. Let the operator $\hat{B}$ operate on elements of $\mathcal{H}$. We may represent $\hat{B}$ by a matrix with elements

$$B_{nn'} = \langle a_n| \hat{B} |a'_n\rangle$$

These are the matrix elements of $\hat{B}$ in a *representation that diagonalizes* $\hat{A}$. This description is due to the following. Consider the matrix elements of $\hat{A}$ in this same basis. With $\hat{A}$ taken to be Hermitian, its eigenstates are orthogonal, and we find

$$A_{nn'} = \langle a_n | \hat{A} | a'_n \rangle = a'_n \langle a_n | a'_n \rangle$$
$$A_{nn'} = a_n \delta_{nn'} \qquad (1.19)$$

which is a diagonal matrix.

Suppose the three operators $\hat{A}$, $\hat{B}$, and $\hat{C}$ compromise a *complete* set of commuting operators. This means that no other independent operator exists that commutes with $\hat{A}$, $\hat{B}$, and $\hat{C}$. Recalling the commutator theorem stated above, we may conclude that such commuting operators have common eigenstates, which may be written $|abc\rangle$. As such states cannot be further resolved, they are maximally informative and comprise quantum states of the system.

Consider, for example, the angular momentum operators (1.17). For a particle with spin $\hat{S}$ and orbital angular momentum $\hat{L}$, we write

$$\hat{J} = \hat{L} + \hat{S}, \qquad \hat{J}_z = \hat{L}_z + \hat{S}_z$$
$$\hat{J}^2 = \hat{L}^2 + \hat{S}^2 + 2\hat{L} \cdot \hat{S} \qquad (1.20)$$

Commuting operators for this system are

$$\{\hat{J}^2, \hat{J}_z, \hat{L}^2, \hat{S}^2\} \qquad (1.21)$$

Common eigenstates of these operators may be written $|jm_jls\rangle$ with eigenvalues

$$\begin{bmatrix} \hat{J}^2 \\ \hat{J}_z \\ \hat{L}^2 \\ \hat{S}^2 \end{bmatrix} |jm_jls\rangle = \hbar^2 \begin{pmatrix} j(j+1) \\ m_j/\hbar \\ l(l+1) \\ s(s+1) \end{pmatrix} |jm_jls\rangle \qquad (1.22)$$

It is quite clear that the four operators (1.21) are diagonal in a representation comprised of the basis $\{|jm_jls\rangle\}$.

5.1.4 Coordinate and Momentum Representations[3]

It is instructive at this point to include a brief review of the coordinate ($\hat{x}$) and momentum ($\hat{p}$) representations and their interplay. Thus let $|x'\rangle$ represent an eigenket of $\hat{x}$. Then with (1.18) we write

$$\hat{x}|x'\rangle = x'|x\rangle \qquad (1.23)$$

Eigenstates are orthogonal and obey the relation

$$\langle x'|x\rangle = \delta(x - x') \qquad (1.24)$$

---

[3]The relations developed here play an important role in Section 5.2.3 on the Wigner distribution and in Section 5.7 concerning the Green's function formalism.

appropriate to continuous eigenvalues spectra. Matrix elements of $\hat{x}$ are given by

$$\langle x|\hat{x}|x'\rangle = x'\langle x|x'\rangle = x'\delta(x-x') \tag{1.25}$$

The coordinate representation of the ket vector $|\psi\rangle$ is given by

$$\langle x'|\psi\rangle = \int \langle x'|x\rangle\langle x|\psi\rangle\,dx = \int \delta(x'-x)\psi(x)\,dx = \psi(x') \tag{1.26}$$

The matrix of the momentum operator $\hat{p}$ in the coordinate representation is

$$\langle x|\hat{p}|x'\rangle = -i\hbar\frac{\partial}{\partial x}\delta(x-x') \tag{1.27}$$

This relation allows us to calculate an explicit representation of the inner product $\langle x|p\rangle$:

$$p\langle x|p\rangle = \langle x|\hat{p}|p\rangle = \int dx'\langle x|\hat{p}|x'\rangle\langle x'|p\rangle$$

With (1.27), this integral reduces to

$$p\langle x|p\rangle = \int dx'\left[-i\hbar\frac{\partial}{\partial x}\delta(x-x')\right]\langle x'|p\rangle$$

$$= -i\hbar\frac{\partial}{\partial x}\int dx'\delta(x-x')\langle x'|p\rangle$$

$$= -i\hbar\frac{\partial}{\partial x}\langle x|p\rangle$$

Solving the differential equation gives

$$\langle x|p\rangle = \frac{1}{\sqrt{2\pi\hbar}}e^{ipx/\hbar} \tag{1.28}$$

Let us show that $\langle x|p\rangle$ is unitary. That is,

$$\int \langle p|x\rangle^*\langle p|x'\rangle\,dp = \int \langle x|p\rangle\langle p|x'\rangle\,dp$$

$$= \delta(x-x')$$

which indicates that $\langle x|p\rangle$ is unitary. We may employ the transfer matrix (1.28) to reestablish (1.27).

$$\langle x|\hat{p}|x'\rangle = \int \langle x|\hat{p}|p'\rangle\langle p|x'\rangle\,dp'$$

$$= \int p'\langle x|p'\rangle\langle p'|x'\rangle\,dp' = \int p'\frac{1}{\sqrt{2\pi\hbar}}e^{(ip'/\hbar)(x-x')}\,dp$$

$$= -i\hbar\frac{\partial}{\partial x}\left(\frac{1}{2\pi\hbar}\int_{-\infty}^{\infty}e^{(ip'/\hbar)(x-x')}\,dp'\right) = -\hbar\frac{\partial}{\partial x}\delta(x-x')$$

Here are some simple examples of the preceding formalism. First let us find the coordinate representation of $\hat{p}|\psi\rangle$.

$$\langle x|\hat{p}|\psi\rangle = \int_{-\infty}^{\infty} dx' \langle x|\hat{p}|x'\rangle\langle x'|\psi\rangle$$

$$= -i\hbar \int_{-\infty}^{\infty} dx' \frac{\partial}{\partial} \delta(x - x')\psi(x')$$

$$= -i\hbar \frac{\partial}{\partial x} \psi(x) \qquad (1.29)$$

Next consider the matrix elements of $[\hat{x}, \hat{p}]$ in the coordinate representation. First we calculate

$$\langle x|\hat{x}\hat{p}|x'\rangle = \int_{-\infty}^{\infty}\int_{-\infty}^{\infty}\int_{-\infty}^{\infty} dx''\, dp'\, dp \langle x|\hat{x}|x''\rangle\langle x''|p'\rangle\langle p'|\hat{p}|p\rangle\langle p|x'\rangle$$

$$= \frac{x}{2\pi\hbar} \int_{-\infty}^{\infty} dp\, p \exp[ip(x - x')/\hbar]$$

$$= -i\hbar x \frac{\partial}{\partial x}\delta(x - x')$$

Similarly, we find

$$-\langle x|\hat{p}\hat{x}|x'\rangle = -i\hbar x' \frac{\partial}{\partial x}\delta(x - x')$$

Combining results gives

$$\langle x|[\hat{x}, \hat{p}]|x'\rangle = -i\hbar(x - x')\frac{\partial}{\partial x}\delta(x - x')$$

$$= +i\hbar\delta(x - x') \qquad (1.30)$$

which is the continuous coordinate representation of the fundamental commutator relation (1.15).

5.1.5 Superposition Principle

Suppose a system is in an eigenstate $|a_n\rangle$ of the operator $\hat{A}$. If A is measured, we are certain to find the value a_n. Suppose the operator $\hat{B}$ corresponding to the observable B has a nonvanishing commutator with $\hat{A}$. That is,

$$[\hat{A}, \hat{B}] \neq 0$$

Then, with (1.14), outcome of measurement of B is uncertain. In this case, it is meaningful to ask of the probability that measurement of B finds the value b_n. The answer to this question is simply obtained by writing the given stage, $|a_n\rangle$, in a superposition of the eigenstates of $\hat{B}$:

$$|a_n\rangle = \sum_k |b_k\rangle\langle b_k|a_n\rangle \qquad (1.31)$$

Clearly, the coefficients $\langle b_k | a_n \rangle$ represent the *projections* of the state $|a_n\rangle$ onto the state $|b_n\rangle$. The probability $P(b_k)$ of finding b_k in measurement of B is given by the square of the projection:

$$Pb_k = |\langle b_x | a_n \rangle|^2 \tag{1.32}$$

The expansion (1.31) is the essential statement of the superposition principle. We may describe the state $|a_n\rangle$ as being partly in the various eigenstates of $\hat{B}$. The degree to which $|a_n\rangle$ contributes to one of the specific states of $\hat{B}$ is a measure of the probability that measurement of B will find the value corresponding to this specific state. The measure of this probability is given by (1.32).

An informative example of this principle concerns spin 1/2 particles, such as electrons. Eigenvalue equations for this case are given by

$$\hat{S}^2 \alpha_z = \frac{3}{4} \hbar^2 \alpha_z, \qquad \hat{S}_z \alpha_z = \frac{\hbar}{2} \alpha$$

$$\tag{1.33}$$

$$\hat{S}^2 \beta_z = \frac{3}{4} \hbar^2 \beta_z, \qquad \hat{S}_z \beta_z = -\frac{\hbar}{2} \beta_z$$

Working in a representation in which $\hat{S}^2$ and $\hat{S}_z$ are diagonal, we have the matrix forms

$$\hat{S}^2 = \frac{3}{4} \hbar^2 \begin{pmatrix} 1 & 0 \\ 0 & 1 \end{pmatrix}, \qquad \hat{S}_z = \frac{\hbar}{2} \begin{pmatrix} 1 & 0 \\ 0 & -1 \end{pmatrix}$$

$$\tag{1.34}$$

$$\alpha_z = \begin{pmatrix} 1 \\ 0 \end{pmatrix}, \qquad \beta_z = \begin{pmatrix} 0 \\ 1 \end{pmatrix}$$

Let a beam of electrons be polarized with spins in the $+x$ direction. Measurement of S_z is made. What values will be found and with what probabilities will these values occur?

In the given representation, eigenstates of $\hat{S}_x$ corresponding to $S_x = \hbar/2$ are given by

$$\alpha_x = \frac{1}{\sqrt{2}} \begin{pmatrix} 1 \\ 1 \end{pmatrix}$$

To answer the said problem, we must expand α_x in accord with the superposition statement (1.31). There results

$$\alpha_x = \frac{1}{\sqrt{2}} \alpha_z + \frac{1}{\sqrt{2}} \beta_z$$

Thus

$$P\left(S_z = +\frac{\hbar}{2}\right) = \frac{1}{2}, \qquad P\left(S_z = -\frac{\hbar}{2}\right) = \frac{1}{2}$$

It is equally likely that measurement finds either of the two values $S_z = \pm\hbar/2$.

5.1.6   Statistics and the Pauli Principle

In quantum mechanics, identical particles are indistinguishable. As a consequence, probability densities cannot change under exchange of particles. Consider, for example, the two-particle state $\psi(\mathbf{x}_1, \mathbf{x}_2)$. Then we must have

$$|\psi(\mathbf{x}_1, \mathbf{x}_2)|^2 = |\psi(\mathbf{x}_2, \mathbf{x}_1)|^2 \tag{1.35}$$

Consequently,

$$\psi(\mathbf{x}_1, \mathbf{x}_2) = \pm\psi(\mathbf{x}_2, \mathbf{x}_1) \tag{1.36}$$

The $+$ sign corresponds to a *symmetric wave function*, whereas the $-$ sign corresponds to an *antisymmetric wave function*.

The *Pauli principle* state the following. Particles with spin quantum number equation to one-half an odd integer are described by antisymmetric wave functions, whereas particles with integral spin quantum numbers are described by symmetric wave functions. Particles in the first class are called *fermions*, and particles in the second class are called *bosons*. Electrons and protons are examples of fermions. Photons, $\prod$, and K mesons are examples of bosons.

For a system comprised of N identical particles, properly symmetrized wave functions may be put in the form of *Slater determinants*. Let v_k denote the quantum numbers for the kth particle in the aggregate. For fermions, we write

$$\psi_A(1, \ldots, N) = \frac{1}{\sqrt{N!}} \sum_{P(v_1, v_2, \ldots, v_N)} (-1)^{|P|} \psi_{v_1}(1)\psi_{v_2}(2) \cdots \psi_{v_N}(N) \tag{1.37}$$

The summations run over the permutations of $(v_1 \cdots v_N)$. The symbol $|P|$ is zero or one depending on whether P is even or odd, respectively, and "1" represents $\mathbf{x}_1$, "2" represents $\mathbf{x}_2$, and so forth.

Since ψ_A is a determinant, it changes sign under exchange of two rows or two columns. So particle exchange in the wave function (1.37) carries a sign change.

Furthermore, ψ_A vanishes if two rows or columns are the same. This occurs it, for example, we set $v_1 = v_7$. Thus, in an aggregate of N fermions, no two particles can be in the same quantum state. This property of fermions is called the *Pauli exclusion principle*.

The wave function for an aggregate of N identical bosons may be written

$$\psi_S(1, \ldots, N) = \frac{1}{N!} \sum_{P(v_1, \ldots, v_2)} \psi_{v_1}(1) \cdots \psi_{v_N}(N) \tag{1.38}$$

This function remains unchanged under exchange of any two particles, so bosons do not obey the exclusion principle.

5.1.7   Heisenberg Picture

In addition to change in representation through transformation of basis in Hilbert space, as described in Section 1.3, we also speak of alternative "pictures" in quantum mechanics. In the Schrödinger picture wave functions obey the Schrödinger equation (1.4) and evolve in time through (1.9).

Consider the unitary operator

$$\hat{U} \equiv \exp\left(\frac{it\hat{H}}{\hbar}\right) \tag{1.39}$$

with properties

$$\hat{U}^\dagger = \hat{U}^{-1}$$
$$\hat{U}^\dagger \hat{U} = 1 \tag{1.40}$$

With $\hat{U}$ so defined, we may rewrite (1.9) as

$$\psi(t) = \hat{U}^\dagger(t)\psi(0) \tag{1.41}$$

In the *Heisenberg picture*, wave functions are related to their Schrödinger counterpart through the equation

$$\psi_H(t) = \hat{U}(t)\psi(t) \tag{1.42}$$

With (1.41), we obtain the important result

$$\psi_H(t) = \hat{U}(t)\psi(t) = \hat{U}(t)\hat{U}^{-1}(t)\psi(0) = \psi(0) = \psi_H(0) \tag{1.42a}$$

That is, in the Heisenberg picture wave functions remain fixed at their initial values.

Operators in the Heisenberg picture.

$$\hat{A}_H(t) = \hat{U}(t)\hat{A}U^{-1}(t) \tag{1.43}$$

are seen to vary in time. That is, an operator that is stationary in the Schrödinger picture varies in time in the Heisenberg picture.

An important result that emerges in the Heisenberg picture concerns the equation of motion for an operator. In the Schrödinger picture, this equation is given by (1.5). To obtain the corresponding equation in the Heisenberg picture, first we calculate the time derivative of (1.43).

$$\frac{d\hat{A}_H}{dt} = \frac{d\hat{U}}{dt}\hat{A}\hat{U}^{-1} + \hat{U}\hat{A}\frac{d\hat{U}^{-1}}{dt} + \hat{U}\frac{\partial\hat{A}}{\partial t}\hat{U}^{-1}$$

With (1.39), we find

$$-i\hbar\frac{d\hat{U}}{dt} = \hat{H}\hat{U}$$

$$-i\hbar\frac{d\hat{U}^\dagger}{dt} = \hat{H}\hat{U}^\dagger$$

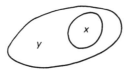

FIGURE 5.2. The whole system has the wave function $\psi(x, y)$. If $\psi(x, y) \neq \psi_1(x)\psi_2(y)$, then system X does not have a wave function. In this event, X is said to be in a *mixed state*.

Inserting these relations into the preceding equation gives the desired equation of motion:

$$i\hbar \frac{d\hat{A}_H}{dt} = [\hat{A}_H, \hat{H}] + i\hbar \frac{\partial \hat{A}_H}{\partial t} \qquad (1.44)$$

where we have written[4]

$$\frac{\partial \hat{A}_H}{\partial t} \equiv \hat{U} \frac{\partial \hat{A}}{\partial t} \hat{U}^{-1} \qquad (1.44a)$$

Equation (1.44) plays an important role in the interpretation of the corresponding equation of motion for the density operator described in Section 5.2.[5]

5.2   The Density Matrix

Mixed states

Consider a system that is coupled to an external environment, such as a gas of N particles maintained at a constant temperature through contact with a heat bath. If x denotes coordinates of the gas and y coordinates of its environment, then, whereas the closed composite of system plus environment has a self-contained Hamiltonian and wave function $\psi(x, y)$, this wave function does not, in general, fall into a product $\psi_1(x)\psi_2(y)$. Under such circumstances, we say that the system does not have a wave function (see Fig. 5.2). A system that does not have a wave function is said to be in a *mixed state*. A system that does have a wave function is said to be in a *pure state*.

It may also be the case that, owing to certain complexities of the system, less than complete knowledge of the state of the system is available. This situation arises for systems with a very large number of degrees of freedom, such as, for example, a mole of gas. The quantum state of such a system involves specification of $\sim 10^{23}$ momenta.

---

[4]For further discussion, see R. L. Liboff, *Found. Physics 17*, 981 (1987).

[5]Additional concepts in basic quantum mechanics appear below. Thus, for example, second quantization and closely allied Fock space are introduced in Section 5.5.

5.2.1   The Density Operator

For such cases as described above, in place of the wave function, we introduce the density operator $\hat{\rho}$. If A is some property of the system, the density operator determines the expectation of A through the relation

$$\langle A \rangle = \mathrm{Tr}(\hat{\rho}\hat{A}) \tag{2.1}$$

and

$$\mathrm{Tr}\,\hat{\rho} = 1 \tag{2.2}$$

Let us calculate the matrix elements of $\hat{\rho}$ for the case of a system whose wave function ψ is known. In this case we may write

$$\langle A \rangle = \langle \psi | \hat{A} \psi \rangle$$

Let the basis $\{|n\rangle\}$ span the Hilbert space containing ψ. We may expand ψ in this basis to obtain

$$|\psi\rangle = \sum_n |n\rangle\rangle n \mid \psi\rangle$$

Substituting this expansion into the preceding equation gives

$$\langle A \rangle = \sum_q \sum_n \langle \psi \mid q \rangle\langle q \mid \hat{A} \mid n \rangle\langle n \mid \psi \rangle$$

$$= \sum_q \sum_n \rho_{nq} A_{qn} = \mathrm{Tr}\,\hat{\rho}\hat{A} = \mathrm{Tr}\,\hat{A}\hat{\rho}$$

Have we have made the identification

$$\rho_{nq} = \langle q \mid \psi \rangle^* \langle n \mid \psi \rangle = a_q^* a_n \tag{2.3}$$

The coefficient a_n represents the projection of the state ψ onto the basis vector $|n\rangle$. The nth diagonal element of $\hat{\rho}$ is

$$\rho_{nn} = |\langle \psi \mid n \rangle|^2 = a_n^* a_n = P_n \tag{2.4}$$

which we recognize to be the probability P_n of finding the system in the state $|n\rangle$. Thus *the diagonal elements of $\hat{\rho}$ are probabilities* and must sum to 1.[6] This is the rationale for the property (2.2), $\mathrm{Tr}\,\hat{\rho} = 1$.

We note in passing that the representation of $\hat{\rho}$ given by (2.3) refers to the basis $\{|n\rangle\}$. Suppose these states are eigenstates of the Hamiltonian. As noted previously, we would say that ρ as given in (2.3) is written in a representation in which energy is diagonal or, more simply, the energy representation.

Now consider that the system is in a mixed state. Thus we may assume that the projections a_n are not determined quantities. In this case we define the

---

[6]The off-diagonal elements of $\hat{\rho}$ relevant to a given system have been shown to be related to the long-range order of the system (this phenomenon carries the acronym ODLRO). For further discussion, see C. N. Yang, *Revs. Mod. Phys. 34*, 694 (1962).

elements ρ_{nq} to be the *ensemble averages*:

$$\rho_{nq} = \overline{a_q^* a_n} \tag{2.5}$$

The diagonal element

$$\rho_{nn} = \overline{a_n^* a_n} \tag{2.6}$$

represents the probability that a system chosen at random from the ensemble is found in the nth state.

Equation of motion

Suppose again that a system is in a pure state and has the wave function ψ. Again, let $\hat{N}$ be a measurable property of the system with eigenstates $\{|n\rangle\}$. Expanding ψ in terms of the projections a_n gives

$$|\psi\rangle = \sum_n a_n(t)|n\rangle$$

From the Schrödinger equation for ψ, we obtain

$$i\hbar \sum_n \frac{\partial a_n}{\partial t}|n\rangle = \sum_n a_n \hat{H}|n\rangle$$

Operating on this equation from the left with $\langle l|$ gives

$$i\hbar \frac{\partial a_l}{\partial t} = \sum_n H_{ln} a_n, \qquad -i\hbar \frac{\partial a_l^*}{\partial t} = \sum_n H_{ln}^* a_n^* \tag{2.7}$$

We may use these relations to obtain an equation of motion for the matrix elements of $\hat{\rho}$

$$i\hbar \frac{\partial \rho_{ql}}{\partial t} = i\hbar \frac{\partial a_l^* a_q}{\partial t} = i\hbar \left(\frac{\partial a_l^*}{\partial t} a_q + a_l^* \frac{\partial a_q}{\partial t} \right) \tag{2.8}$$

Substituting the expressions (2.7) for the time derivatives of the projections a_n and setting $H_{lk}^* = H_{kl}$, together with forming the ensemble average, gives

$$i\hbar \frac{\partial \rho_{ql}}{\partial t} = \sum_k (H_{qk} \rho_{kl} - \rho_{qk} H_{kl}) \tag{2.9}$$

or, equivalently,

$$i\hbar \frac{\partial \hat{\rho}}{\partial t} = [\hat{H}, \hat{\rho}] \tag{2.10}$$

As $\hat{\rho}$ depends only on time [the projections a_n in the defining relation (2.5) are spartially independent], the partial derivative in the preceding equation is understood to be a total time derivative [as is the case for (2.12) et seq.].

In the Heisenberg picture, wavefunctions remain fixed [see (1.42a)] so that $\hat{\rho}_H$ is constant and we write

$$i\hbar \frac{d\hat{\rho}_H}{dt} = 0 \tag{2.11}$$

which, again, is in the form of the Liouville equation (1.4.7). Whereas these resemblances are very strong, we should not lose sight of the fact that in classical physics $f_N(\mathbf{x}^N, \mathbf{p}^N, t)$ gives information of both $\mathbf{x}^N$ and $\mathbf{p}^N$ at the time t. In the quantum domain, simultaneous specification of $\mathbf{x}^N$ and $\mathbf{p}^N$ cannot be made. However, soon after the emergence of quantum mechanics, E. P. Wigner developed a theory which gives a formal connection between the density matrix and the classical distribution function. We return to this topic in Section 2.2.

Some elementary examples

Given the form of the equation (2.9) some elementary solutions are evident. Thus, as with the classical Liouville equation, any operator of the form

$$\hat{\rho} = \hat{\rho}(\hat{H}) \tag{2.12}$$

gives

$$\frac{d\hat{\rho}}{dt} = [\hat{H}, \hat{\rho}] = 0$$

so that the form (2.10) may be termed an equilibrium distribution. Two important examples of this distribution are relevant to statistical mechanics:

1. *Canonical distribution*:

   $$\hat{\rho} = A e^{-\beta \hat{H}} \tag{2.13}$$

   This distribution is relevant to a system in equilibrium with a temperature bath at temperature $T \equiv 1/k_B \beta$.
2. *Microcanonical distribution*:

   $$\hat{\rho} = A\delta(\hat{H} - E) \tag{2.14}$$

This distribution is relevant to an isolated system of total energy E.

Let us consider these operators in the energy representation. Energy eigenstates obey the eigenvalue equation

$$\hat{H}|E_n\rangle = E_n|E_n\rangle \tag{2.15}$$

With these eigenstates at hand, diagonal elements of $\hat{\rho}$ in the canonical distribution are given by [recall (2.4)]

$$P(E_n) = \langle E_n|\hat{\rho}|E_n\rangle = A e^{-\beta E_n}$$

whereas, in the microcanonical distribution, we find

$$P(E_n) = A\delta(E_n - E)$$

All energy states are equally probable in this distribution.

Initially diagonal density matrix

Consider a quantum system whose density is initially diagonal:

$$\rho_{nl}(0) = P_n \delta_{nl} \tag{2.16}$$

The density matrix evolves according to (2.8). For the case at hand, we find that at $t = 0$,

$$i\hbar \frac{d\rho_{nl}}{dt}\bigg|_0 = \sum_k (H_{nk} P_k \delta_{kl} - P_n \delta_{nk} H_{kl})$$

Summing over k gives

$$i\hbar \frac{d\rho_{nl}}{dt}\bigg|_0 = H_{nl}(P_l - P_n) \tag{2.17}$$

For $n = l$, we find

$$\frac{d\rho_{ll}}{dt}\bigg|_0 = 0$$

So diagonal elements have zero slope in time initially. For $n \neq l$, (2.17) indicates that if $P_l \neq P_n$, off-diagonal elements of $\hat{\rho}$ develop and (2.16) cannot be termed an equilibrium distribution. If, however, all states are equally probable,[7] then (2.16) becomes

$$\rho_{nl} = P_0 \delta_{nl} \tag{2.18}$$

which gives $\partial\hat{\rho}/\partial t = 0$ so that (2.18) is a valid equilibrium distribution.

Random phases

An important case in which $\hat{\rho}$ is diagonal enters in the assumption of random phases. Consider the matrix elements of $\hat{\rho}$ (2.5):

$$\rho_{nm} = \overline{a_m^* a_n}$$

The indeterminacy of the state of the system may be manifest in a corresponding indeterminacy of the phases $\{\phi_n\}$ of the projections $\{a_n\}$. These phases are defined through

$$a_n = c_n e^{i\phi_n}$$

where c_n and ϕ_n are real. Consider the matrix element

$$\rho_{nm} = \overline{c_m^* c_n \exp[i(\phi_n - \phi_m)]} = \overline{c_m^* c_n [\cos(\phi_n - \phi_m) + i \sin(\phi_n - \phi_m)]}$$

If phases are random, then in averaging over the ensemble $\cos(\phi_n - \phi_m)$ occurs with positive value equally often as with negative value and similarly for $\sin(\phi_n - \phi_m)$, so that

$$\overline{\cos(\phi_n - \phi_m)} = \overline{\sin(\phi_n - \phi_m)} = 0$$

except when $n = m$. In this case

$$\rho_{nn} = \overline{c_n^* c_n \cos(\phi_n - \phi_n)} = \overline{c_n^* c_n}$$

It follows that for the case of random phases ρ is diagonal

---

[7]The condition of *equal a priori probabilities*.

5.2.2   The Pauli Equation

We wish to examine how an initially diagonal density matrix evolves under the influence of a perturbation. Matrix elements of $\hat{\rho}$ and the Hamiltonian of the system at $t < 0$ are given by

$$\rho_{nl} = \rho_{nl}(0)\delta_{nl}, \qquad (t < 0)$$

$$\hat{H} = \hat{H}_0, \qquad [\hat{H}_0, \hat{\rho}] = 0 \qquad (2.19)$$

The vanishing commutator ensures that the system is in equilibrium prior to $t = 0$.

A harmonic perturbation is turned on at $t = 0$ and the Hamiltonian of the system becomes

$$\hat{H} = \hat{H}_0 + \hat{H}', \qquad t \geq 0 \qquad (2.20)$$

The probability for transition from the nth to the kth state of the system, in short interval Δt after H' is turned on, is given by

$$w_{nk} = \frac{\Delta t |H'_{nk}|^2}{\hbar^2} \qquad (2.21)$$

Taylor-series expanding the density matrix about $t = 0$ gives

$$\rho(t) - \rho(0) = \left(\frac{\partial \rho}{\partial t}\right)_0 \Delta t + \left(\frac{\partial^2 \rho}{\partial t^2}\right)_0 \frac{(\Delta t)}{2} + \cdots$$

With (2.9), this equation may be rewritten

$$\rho(\Delta t) = \rho(0) = \frac{\Delta t}{i\hbar}[H, \rho(0)] - \frac{(\Delta t)^2}{2\hbar^2}[H, [H, \rho(0)]] + \cdots \qquad (2.22)$$

With the commutation property (2.19), (2.22) may be written

$$\frac{\rho(\Delta t) - \rho(0)}{\Delta t} = \frac{1}{i\hbar}[H', \rho(0)] - \frac{\Delta t}{2\hbar^2}[H_0, [H', \rho(0)]]$$

$$- \frac{\Delta t}{2\hbar^2}[H', [H', \rho(0)]] + \cdots$$

Taking diagonal elements and passing to the limit $\Delta t \to 0$ gives

$$\left.\frac{d\rho_{nn}}{dt}\right|_0 = \lim_{\Delta t \to 0} \left\{ \frac{1}{i\hbar}[H', \rho(0)]_{nn} - \frac{\Delta t}{2\hbar^2}[H_0, [H', \rho(0)]]_{nm} \right.$$

$$\left. - \frac{\Delta t}{2\hbar^2}[H', [H', \rho(0)]]_{nn} \right\} \qquad (2.23)$$

Consider first the leading term:

$$[H', \rho(0)]_{nn} = \sum_q [\langle n|H'|q\rangle\langle q|\rho(0)|n\rangle - \langle n|\rho(0)|q\rangle\langle q|H'|n\rangle]$$

$$= \sum_q [H'_{nq}\rho_{nn}(0)\delta_{qn} - \rho_{nn}(0)\delta_{nq}H'_{qn}]$$

$$= \rho_{nn}(0)[H'_{nn} - H'_{nn}] = 0$$

Similarly, we find

$$[H_0, [H', \rho(0)]]_{nn} = 0$$

With the first two elements on the right side of (2.23) vanishing, there remains

$$\frac{d\rho_{nn}(0)}{dt} = -\lim_{\Delta t \to 0} \frac{\Delta t}{2\hbar^2}[H'[H'\rho(0)]]_{nn}$$

$$= -\lim \frac{\Delta t}{2\hbar^2}\{H', [H', \rho(0) - \rho(0)H'] - [H'\rho(0) - \rho(0)H']H'\}_{nn}$$

$$= \lim \frac{\Delta t}{2\hbar^2}[-H'^2\rho(0) + H'\rho(0)H' + H'\rho(0)H' - \rho(0)H'^2]_{nn}$$

$$= \lim \frac{\Delta t}{2\hbar^2}[2H'\rho(0)H' - H'^2\rho(0) - \rho(0)H'^2]_{nn} \qquad (2.24)$$

Expanding products in terms of basis states gives

$$\frac{d\rho_{nn}}{dt} = \lim \frac{\Delta t}{2\hbar^2}\sum_k\sum_q [2\langle n|H'|k\rangle\langle k|\rho(0)|q\rangle\langle q|H'|n\rangle$$

$$- \langle n|H'|k\rangle\langle k|H'|q\rangle\langle q|\rho(0)|n\rangle$$

$$- \langle n|\rho(0)|q\rangle\langle q|H'|k\rangle\langle k|H'|n\rangle] \qquad (2.25)$$

With (2.19), we find

$$\frac{d\rho_{nn}}{dt} = \lim \frac{\Delta t}{2\hbar^2}\sum_k\sum_q [2H'_{nk}H'_{qn}\rho_{kk}(0)\delta_{kq}$$

$$- H'_{nk}H'_{kq}\rho_{qq}(0)\delta_{qn} - \rho_{nn}(0)\delta_{nq}H'_{qk}H'_{kn}] \qquad (2.26)$$

Summing over q gives

$$\frac{d\rho_{nn}}{dt} = \lim \frac{\Delta t}{2\hbar^2}\sum_k [2H'_{nk}H'_{kn}\rho_{kk}(0) - H'_{nk}H'_{kn}\rho_{nn}(0) - \rho_{nn}(0)H'_{nk}H'_{kn}]$$

$$= \lim \frac{\Delta t}{\hbar^2}\sum_k [|H'_{nk}|^2\rho_{kk}(0) - |H'_{nk}|^2\rho_{nn}(0)]$$

With (2.21), we obtain

$$\frac{d\rho_{nn}}{dt} - \sum_k w_{nk}[\rho_{kk}(0) - \rho_{nn}(0)] \qquad (2.27)$$

Since the left side of this equation represents the derivative of ρ_{nn} at $t = 0$, we may generalize this finding to the form

$$\boxed{\frac{d\rho_{nn}}{dt} = \sum_k (\rho_{kk} w_{kn} - \rho_{nn} w_{nk})} \qquad (2.28)$$

Here we have further made the identification $w_{kn} = w_{nk}$ corresponding to the hermiticity of $\hat{H}'$. The latter equation is called the *Pauli equation* or the *Master equation* [compare with (1.7.20)]. An $\mathcal{H}$ theorem for this equation is derived in Problem 5.48. Application of (2.28) follows.

Density matrix for spinning particles

Spin matrices for a spin $\frac{1}{2}$ particle are given by

$$\hat{S}_z = \frac{\hbar}{2}\begin{pmatrix} 1 & 0 \\ 0 & -1 \end{pmatrix}, \qquad \hat{S}_x = \frac{\hbar}{2}\begin{pmatrix} 0 & 1 \\ 1 & 0 \end{pmatrix}, \qquad \hat{S}_y = \frac{i\hbar}{2}\begin{pmatrix} 0 & -1 \\ 1 & 0 \end{pmatrix}$$
$$(2.29)$$

Once again we are working in a repesentation in which $\hat{S}^2$ and $\hat{S}_z$ are diagonal. These matrices may also be written

$$\hat{\mathbf{S}} = \frac{\hbar}{2}\hat{\boldsymbol{\sigma}} \qquad (2.30)$$

The components of $\boldsymbol{\sigma}$ are called the Pauli spin matrices. Thus

$$\hat{\sigma}_z = \begin{pmatrix} 1 & 0 \\ 0 & -1 \end{pmatrix}, \qquad \hat{\sigma}_x = \begin{pmatrix} 0 & 1 \\ 1 & 0 \end{pmatrix}, \qquad \hat{\sigma}_y = i\begin{pmatrix} 0 & -1 \\ 1 & 0 \end{pmatrix} \qquad (2.31)$$

Eigenvectors of $\hat{S}_z$ (or, equivalently, $\hat{\sigma}_z$) are given by (1.33). We repeat,

$$\alpha_z = \begin{pmatrix} 1 \\ 0 \end{pmatrix}, \qquad \beta_z = \begin{pmatrix} 0 \\ 1 \end{pmatrix} \qquad (2.32)$$

where

$$\hat{S}_z \alpha_z = \frac{\hbar}{2}\alpha_z, \qquad \hat{S}_z \beta_z = -\frac{\hbar}{2}\beta_z \qquad (2.33)$$

Consider a beam of electrons whose spins are isotropically polarized. The corresponding density matrix is

$$\hat{\rho} = \frac{1}{2}\begin{pmatrix} 1 & 0 \\ 0 & 1 \end{pmatrix} \qquad (2.34)$$

Since this is written in a representation where $\hat{S}_z$ is diagonal, diagonal elements of $\hat{\rho}$ give probabilties relevant to the components of S_z. For the matrix (2.34), these probabilities have the values

$$P\left(S_z = \pm\frac{\hbar}{2}\right) = \frac{1}{2}$$

Furthermore, we find

$$\langle S_x \rangle = \text{Tr}\, \hat{\rho}\hat{S}_x = 0$$
$$\langle S_y \rangle = \langle S_z \rangle = 0$$

These values are relevant to an isotropic beam.

Elementary application of the Pauli equation

Consider an electron beam with the density matrix

$$\hat{\rho} = \begin{pmatrix} a & 0 \\ 0 & 1-a \end{pmatrix}$$

The beam enters an infinitesimal region of space that contains a uniform **B** field that points in the z direction.

$$\mathbf{B} = (0, 0, B_0)$$

Let us find the z-component spin populations when the beam emerges from the field.

The perturbation Hamiltonian is given by

$$\hat{H}' = -\hat{\mu} \cdot \mathbf{B}$$

The electron's magnetic moment is given by

$$\hat{\mu} = -\mu_b \hat{\sigma}$$

where μ_b is the Bohr magneton. Thus we may write

$$\hat{H}' = \mu_b \hat{\sigma}_z B_0 = \mu_b B_0 \begin{pmatrix} 1 & 0 \\ 0 & -1 \end{pmatrix}$$

The Pauli equation for the case at hand gives

$$\frac{d\rho_{11}}{dt} = \sum_q w_{q1}(\rho_{qq} - \rho_{11}) = w_{21}(\rho_{22} - \rho_{11})$$

$$\frac{d\rho_{22}}{dt} = \sum_q w_{q2}(\rho_{qq} - \rho_{22}) = w_{12}(\rho_{11} - \rho_{22})$$

(2.35)

With (2.21), we have

$$w_{12} = w_{21} = \frac{\Delta t}{\hbar^2}|H'_{12}|^2$$

Since $\hat{\rho}$ is calculated in the basis (α, β), we find

$$H'_{12} = \mu_b B_0 \langle \alpha|\hat{\sigma}_z|\beta\rangle = -\mu_b B_0 \langle\alpha|\beta\rangle$$
$$H'_{12} = 0$$

Thus, with (2.36), we see that ρ_{11} and ρ_{22} remain constant under the given perturbation.

More generally, under an arbitrary perturbation, we note that the Pauli equation (2.28) indicates that the Tr $\hat{\rho}$ remains constant.

The projection representation

As noted previously, an ensemble description is relevant to a system in a mixed state. Let the states ψ of the ensemble be distributed with the probability P_ψ. In this event, we may write the density operator as the following projection sum over states of the ensemble:

$$\hat{\rho} = \sum_\psi |\psi\rangle P_\psi \langle\psi| \tag{2.36}$$

Consider, for example, that the probability of finding the energy E_n in measurement on a given system is P_n. For this case, (2.36) gives

$$\hat{\rho} = \sum_n |E_n\rangle P_n \langle E_n|$$

where $|E_n\rangle$ are energy eigenstates for the given system. As these states comprise and orthogonal sequence, it follow that $\hat{\rho}$ is diagonal with diagonal elements equal to P_n.

The projection representation of the density matrix given by (2.34) is

$$\hat{\rho} = |\alpha_z\rangle \frac{1}{2} \langle a_z| + |\beta_z\rangle \frac{1}{2} \langle \beta_z|$$

Again we find

$$S_z = \mathrm{Tr}\,\hat{\rho}\hat{S}_z = \langle\alpha_z|\hat{\rho}\hat{S}_z|\alpha_z\rangle + \langle\beta_z|\hat{\rho}\hat{S}_z|\beta_z\rangle$$
$$= \frac{1}{2}\langle\alpha_z|\hat{S}_z|\alpha_z\rangle + \frac{1}{2}\langle\beta_z|\hat{S}_z|\beta_z\rangle = 0$$

If a system is in a pure state, the wave function exists, and we write

$$\hat{\rho} = |\psi\rangle\langle\psi| \tag{2.37}$$

In a representation with basis $\{|n\rangle\}$, matrix elements of $\hat{\rho}$ are then given by

$$\rho_{nq} = \langle n|\psi\rangle\langle\psi|q\rangle = a_q^* a_n$$

in agreement with (2.3).

The projection representation comes into play again in Section 5.5 in consideration of generalized hierarchies.

5.2.3   The Wigner Distribution

Coordinate representation of $\hat{\rho}$

Another important form of the density matrix is found in the coordinate representation. The related density matrix appears as[8]

$$\rho(\mathbf{x}, \mathbf{x}') = \overline{\langle \mathbf{x}|\psi\rangle\langle\psi|\mathbf{x}'\rangle} = \overline{\psi(\mathbf{x})\psi^*(\mathbf{x}')} \tag{2.38}$$

($\mathbf{x}$ and $\mathbf{x}'$ are viewed as the indices of the matrix of $\hat{\rho}$.) The basic property (2.2) is written

$$\mathrm{Tr}\ \hat{\rho} = \int \delta(\mathbf{x} - \mathbf{x}')\rho(\mathbf{x}, \mathbf{x}')\,d\mathbf{x}\,d\mathbf{x}' = 1 \tag{2.39}$$

For the property (2.1), we write

$$\langle A\rangle = \mathrm{Tr}(\hat{\rho}\hat{A}) = \mathrm{Tr}(\hat{A}\hat{\rho}) = \int A(\mathbf{x}, \mathbf{x}')\rho(\mathbf{x}', \mathbf{x})\,d\mathbf{x}'\,d\mathbf{x} \tag{2.40}$$

where $A(\mathbf{x}, \mathbf{x}')$ is the coordinate representation of $\hat{A}$:

$$A(\mathbf{x}, \mathbf{x}') = \langle \mathbf{x}|\hat{A}|\mathbf{x}'\rangle \tag{2.41}$$

In a pure state, (2.40) becomes [recall (2.37)]

$$\mathrm{Tr}(\hat{A}\hat{\rho}) = \int \langle \mathbf{x}|\hat{A}|\mathbf{x}'\rangle\langle\mathbf{x}'|\psi\rangle\langle\psi|\mathbf{x}\rangle\,d\mathbf{x}'\,d\mathbf{x} \tag{2.42}$$

Here we have written $\langle \mathbf{x}'|\psi\rangle$ for the coordinate representation of ψ, more commonly written simply as $\psi(\mathbf{x}')$ (recall Section 1.3).

The last relation may be rewritten

$$\mathrm{Tr}(\hat{A}\hat{\rho}) = \int \langle\psi|\mathbf{x}\rangle\langle\mathbf{x}|\hat{A}|\mathbf{x}'\rangle\langle\mathbf{x}'|\psi\rangle\,d\mathbf{x}'\,d\mathbf{x} = \langle\psi|\hat{A}\psi\rangle = \langle A\rangle \tag{2.43}$$

One further preparatory remark is in order prior to our discussion of the Wigner distribution. This concerns the momentum representation of a wave function:

$$\varphi(\mathbf{p}) = \langle \mathbf{p}|\psi\rangle = \int \langle \mathbf{p}|\mathbf{x}\rangle\langle\mathbf{x}|\psi\rangle\,d\mathbf{x} \tag{2.44}$$

The transfer $\langle \mathbf{p}|\mathbf{x}\rangle$ is given by (1.28):[9]

$$\langle \mathbf{p}|\mathbf{x}\rangle = \frac{1}{(2\pi\hbar)^{3/2}}e^{-i\mathbf{p}\cdot\mathbf{x}/\hbar} \tag{2.44a}$$

---

[8]This is the form of $\hat{\rho}$ as originially put forth by J. von Neumann, *Mathematical Foundations of Quantum Mechanics*, R. T. Beyer, (trans.), Princeton University Press, Princeton, N.J. (1955).

[9]The relation (1.28) is relevant to one dimension, whereas (2.44a) is written in three dimensions.

Thus (2.44) becomes

$$\psi(\mathbf{p}) = \frac{1}{(2\pi\hbar)^{3/2}} \int e^{-i\mathbf{p}\cdot\mathbf{x}/\hbar} \psi(\mathbf{x})\,d\mathbf{x} \tag{2.45}$$

This function is the amplitude of the momentum probability density, $|\varphi(\mathbf{p})|^2$.

The Wigner distribution and its properties

To apply the following formalism to kinetic theory, we extend the analysis to N-particle systems. With $\hat{\rho}$ written in the coordinate repesentation, the Wigner distribution is given by[10]

$$F(\mathbf{x}^N, \mathbf{p}^N, t) = \left(\frac{1}{\pi\hbar}\right)^{3N} \int e^{2i\mathbf{p}^N\cdot\mathbf{y}^N/\hbar} \rho(\mathbf{x}^N_-, \mathbf{x}^N_+, t)\,d\mathbf{y}^N \tag{2.46}$$

where

$$\mathbf{x}^N_{\pm} \equiv \mathbf{x}^N \pm \mathbf{y}^N, \qquad \rho(\mathbf{x}^N_-, \mathbf{x}^N_+, t) = \langle \mathbf{x}^N_- | \hat{\rho} | \mathbf{x}^N_+ \rangle$$

Note that in a pure state (2.46) becomes

$$F(\mathbf{x}^N, \mathbf{p}^N, t) = \left(\frac{1}{\pi\hbar}\right)^{3N} \int e^{2i\mathbf{p}^N\cdot\mathbf{y}^N/\hbar} \psi^*(\mathbf{x}^N_+)\psi(\mathbf{x}^N_-)\,d\mathbf{y}^N \tag{2.47}$$

The inverse of (2.46) is given by

$$\rho(\mathbf{x}^N_+, \mathbf{x}^N_-, t) = \int F(\mathbf{x}^N, \mathbf{p}^N, t)\exp\left(\frac{-2i\mathbf{p}^N\cdot\mathbf{y}^N}{\hbar}\right)d\mathbf{p}^N \tag{2.48}$$

The Wigner distribution serves as a bridge between quantum and classical kinetic theory. Four key properties that demonstrate this connection are as follows.[11]

1. Integration over momenta gives the configurational probability density:

$$\int F(\mathbf{x}^N, \mathbf{p}^N, t)\,d\mathbf{p}^N = |\psi(\mathbf{x}^N, t)|^2 \tag{2.49}$$

2. Integration over spatial coordinates gives the momentum probability density:

$$\int F(\mathbf{x}^N, \mathbf{p}^N, t)\,d\mathbf{x}^N = |\varphi(\mathbf{p}^N, t)|^2 \tag{2.50}$$

---

[10]E. P. Wigner, *Phys. Rev. 40*, 749 (1932). In this paper we are informed that (2.46) is due to a joint efford of Wigner and L. Szilard. Note in particular that, as $F(\mathbf{x}, \mathbf{p}, t)$ is not positive definite, it cannot be view as a probability density.

[11]For further discussion, see *Studies in Statistical Mechanics*, vol I., J. de Boer and G. E. Uhlenbeck (eds.), North Holland, Amsterdam (1962); and B. Kursunoglu, *Modern Quantum Theory*, W. H. Freeman, San Francisco (1962).

3. The expectation of a dynamical function, $A(\mathbf{x}^N, \mathbf{p}^N)$, is given by

$$\langle A \rangle = \int A(\mathbf{x}^N, \mathbf{p}^N, t) F(\mathbf{x}^N, \mathbf{p}^N, t) \, d\mathbf{x}^N, \, d\mathbf{p}^N \qquad (2.51)$$

The preceding three properties indicate how the Wiegner distribution bears a striking resemblance to the classical N-particle joint probability distribution.

4. The derivation of a dynamical equation for $F(\mathbf{x}^N, \mathbf{p}^N, t)$ begins with Schrödinger's equation written for an N-body system with an interaction potential $V(\mathbf{x}^N)$:[12]

$$i\hbar \frac{\partial \psi(\mathbf{x}^N, t)}{\partial t} = \left[\frac{-\hbar^2}{2m} \sum_{k=1}^{N} \frac{\partial^2}{\partial x_k^2} + V(\mathbf{x}^N) \right] \psi(\mathbf{x}^N, t) \qquad (2.52)$$

Assuming the system to be in pure state, and with the identifications (2.52), (2.38), and (2.47) in mind, we obtain

$$\frac{\partial F}{\partial t} = \left(\frac{1}{\pi\hbar} \right)^{3N} \int d\mathbf{y}^N \exp\left(\frac{2i\mathbf{p}^N \cdot \mathbf{y}^N}{\hbar} \right)$$
$$\cdot \left\{ \frac{i\hbar}{2m} \sum_k \left[-\left(\frac{\partial^2 \psi_+^*}{\partial x_k^2} \right) \psi_- + \psi_+^* \left(\frac{\partial^2 \psi_-}{\partial x_k^2} \right) \right] \right.$$
$$\left. + \frac{i}{\hbar} [V(\mathbf{x}_+^N) - V(\mathbf{x}_-^N)] \rho(\mathbf{x}_-^N, \mathbf{x}_+^N) \right\} \qquad (2.53)$$

Next we replace $\mathbf{x}$ differentiation with $\mathbf{y}$ differentiation once and then perform a single parts integration over $\mathbf{y}$. Combining this operation with replacement of $\rho(\mathbf{x}_+^N, \mathbf{x}_-^N)$ with its inverse (2.48) and noting that

$$\frac{\partial \psi_\pm}{\partial x_k} = \pm \frac{\partial \psi_\pm}{\partial y_k}$$

gives

$$\boxed{\begin{aligned}
\frac{\partial F(\mathbf{x}^N, \mathbf{p}^N, t)}{\partial t} &= -\sum_k \frac{p_k}{m} \cdot \frac{\partial F}{\partial x_k} + \frac{i}{\hbar} \left(\frac{1}{\pi\hbar} \right)^{3N} \int \int d\mathbf{y}^N \, d\mathbf{p}'^N \\
&\times \exp\left[\frac{2i\mathbf{y}^N \cdot (\mathbf{p}^N - \mathbf{p}'^N)}{\hbar} \right] [V(\mathbf{x}_+^N) - V(\mathbf{x}_-^N)] F(\mathbf{x}^N, \mathbf{p}'^N, t)
\end{aligned}}$$

$$\qquad (2.54)$$

This is an integrodifferential equation for the Wigner distribution function. [The hierarchy equations for this distribution are given in (5.33).]

---

[12]Note that $\partial^2 / \partial x_k^2 = \nabla_k \cdot \nabla_k$.

5.2.4   Weyl Correspondence[13]

A note of caution is in order concerning the rule (2.51) for obtaining averages of dynamical variables: For consistency, this expression must agree with (2.1). Such agreement is quaranteed if the dynamical variable $A(\mathbf{x}^N, \mathbf{p}^N)$ in (2.51) is appriopriately related to the operator $\hat{A}$ in (2.1.)

Without loss in generality, we work in one dimension and introduce the Wigner counterpart of $\hat{A}$, written $(\hat{A})_W$, and given by the linear operation

$$(\hat{A})_W = 2 \int e^{2ip\bar{x}/\hbar} \langle x - \bar{x}|\hat{A}|x + \bar{x}\rangle \, d\bar{x} \tag{2.55}$$

where $(\hat{A})_W$ is a function of (x, p). Note that the matrix element of $\hat{A}$ in (2.55) is written in the coordinate representation. When written in the momentum representation, (2.55) appears as

$$(\hat{A})_W = 2 \int e^{2i\bar{p}x/\hbar} \langle p + \bar{p}|\hat{A}|p - \bar{p}\rangle \, d\bar{p} \tag{2.55a}$$

When the dynamical variable $A(x, p)$ in (2.51) satisfies the correspondence

$$(\hat{A})_W = A(x, p) \tag{2.56}$$

Weyl correspondence is obeyed and the expectation of $\hat{A}$ is given by (2.51).

Let us establish this rule. Again consider (2.51)

$$\langle \hat{A} \rangle = \frac{1}{\pi\hbar} \int \int \int dx \, dp \, dy \, A(x, p) e^{2ipy/\hbar} \rho(x_-, x_+) \tag{2.57}$$

Working in the coordinate representation, we insert (2.55) for A in (2.57). There results

$$\langle \hat{A} \rangle = \frac{2}{\pi\hbar} \int \int \int \int dx \, d\bar{x} \, dp \, dy \, e^{2ip(y+\bar{x})/\hbar} \langle x - \bar{x}|\hat{A}|x + \bar{x}\rangle \rho(x - y, x + y) \tag{2.58}$$

Integrating over p and recalling the delta-function normalization (2.3.11b), we obtain

$$\langle \hat{A} \rangle = \frac{2(2\pi)}{\pi\hbar} \left(\frac{\hbar}{2}\right) \int \int \int dx \, d\bar{x} \, dy \, \delta(y + \bar{x}) \langle x - \bar{x}|\hat{A}|x + \bar{x}\rangle \rho(x - y, x + y)$$

$$= 2 \int \int dx \, d\bar{x} \, \langle -x\bar{x}|\hat{A}|x + \bar{x}\rangle \langle x + \bar{x}|\hat{\rho}|x - \bar{x}\rangle$$

$$= \int \int dx_- \, dx_+ \langle x_-|\hat{A}|x_+\rangle \langle x_+|\hat{\rho}|x_+\rangle = \operatorname{Tr} \hat{A}\hat{\rho} = \operatorname{Tr} \hat{\rho}\hat{A} \tag{2.59}$$

---

[13]For further discussion, see N. L. Balazs and B. K. Jennings, *Phys. Repts, 104*, 347 (1984); and M. Hillery, R. F. O'Connell, M. O. Scully, and E. P. Wigner, *Phys. Repts. 106*, 121 (1984)

which is seen to agree with (2.1). Note that in the last step the change in variables $x, \bar{x} \rightarrow x_-, x_+$ carries a Jacobian of $\frac{1}{2}$. Thus we see that if the rule (2.56) is obeyed the average given by (2.51) agrees with that given by (2.1)

Correspondence forms

Here we wish to list some relations relevant to Weyl correspondence. First let us establish that operators of the form $f(\hat{x})$ or $g(\hat{p})$ obey Weyl correspondence. That is,

$$[f(\hat{x})]_W = f(x), \qquad [g(\hat{p})]_W = g(p) \tag{2.60}$$

Consider the first relation for $f(\hat{x})$.

$$\begin{aligned}
[f(\hat{x})]_W &= 2 \int e^{2ip\bar{x}/\hbar} \langle x - \bar{x}| f(\hat{x})|x + \bar{x}\rangle \, d\bar{x} \\
&= 2 \int e^{2ip\bar{x}/\hbar} f(x - \bar{x})\langle x - \bar{x}|x + \bar{x}\rangle \, d\bar{x} \\
&= 2 \int e^{2ip\bar{x}/\hbar} f(x - \bar{x})\delta(2\bar{x}) \, d\bar{x} \\
&= f(x)
\end{aligned}$$

A similar argument applies to $g(\hat{p})$. An important operator in this discussion is given by

$$\vec{0} = \frac{\overleftarrow{\partial}}{\partial x}\frac{\overrightarrow{\partial}}{\partial p} - \frac{\overleftarrow{\partial}}{\partial p}\frac{\overrightarrow{\partial}}{\partial x}$$

where $\overleftarrow{\partial}/\partial x$ operates to the left, and so forth.

With $[\hat{A}, \hat{B}]_\mp$ written for the commutator $(-)$ and anticommutator $(+)$, we may then write

$$([\hat{A}, \hat{B}]_-)_W = (A)_W \left(2i \sin\frac{\hbar}{2}\vec{0}\right)(B)_W \tag{2.61a}$$

$$([\hat{A}, \hat{B}]_+)_W = (A)_W \left(2\cos\frac{\hbar}{2}\vec{0}\right)(B)_W \tag{2.61b}$$

Thus, for example,

$$\left(\frac{1}{i\hbar}[\hat{x}, \hat{p}]_-\right)_W = x\left(\frac{2}{\hbar}\sin\frac{\hbar}{2}\vec{0}\right)p$$
$$= x\vec{0}p = 1$$

whereas

$$([\hat{x}, \hat{p}]_+)_W = 2x\left(\cos\frac{\hbar}{2}\vec{0}\right)p = 2xp$$

Thus

$$(\hat{x}\hat{p})_W = xp + \frac{i\hbar}{2}$$

We may conclude that the operator $\hat{x}\hat{p}$ does not obey Weyl correspondence, whereas the symmetric form $[\hat{x}, \hat{p}]_+$ does.

Stemming from (2.61), we write

$$(\hat{A}\hat{B})_W = (\hat{A})_W \left(\exp \frac{i\hbar\overleftrightarrow{0}}{2} \right) (\hat{B})_W \tag{2.62}$$

Here are some additional relations:

$$\langle \hat{A} \rangle = \text{Tr}(\hat{\rho}\hat{A}) = \int \frac{dx\, dp}{2\pi\hbar} (\hat{\rho})_W (\hat{A})_W \tag{2.63a}$$

$$\text{Tr}(\hat{A}^\dagger \hat{B}) = \int \frac{dx\, dp}{2\pi\hbar} (\hat{A})_W^* (\hat{B})_W \tag{2.63b}$$

$$(\hat{B}^\dagger)_W = (\hat{B})_W^* \tag{2.63c}$$

Let us establish (2.63c).

$$(\hat{B}^\dagger)_W = 2 \int e^{2ip\bar{x}/\hbar} \langle x - \bar{x}|\hat{B}^\dagger|x + \bar{x} \rangle\, d\bar{x}$$

$$= 2 \int e^{2ip\bar{x}/\hbar} \langle x + \bar{x}|\hat{B}^\dagger|x - \bar{x} \rangle^*\, d\bar{x}$$

$$= \left(2 \int e^{-2ip\bar{x}/\hbar} \langle x + \bar{x}|\hat{B}|x - \bar{x} \rangle\, d\bar{x} \right)^*$$

$$= (\hat{B})_W^*$$

In the last step we employed the change in variables: $\bar{x} \rightarrow -\bar{x}$. Additional properties of these correspondence froms are left to the problems.

The classical limit

Let us see how (2.54) reduces to the Liouville equation in the classical domain $\hbar = 0$. With the change of variables

$$\xi^N = \frac{2\mathbf{y}^N}{\hbar}$$

we may write

$$\mathbf{x}_\pm^N = \mathbf{x}^N \pm \frac{1}{2}\hbar\xi^N$$

and the integral term in (2.54) becomes

$$\frac{\partial F}{\partial t}\bigg|_{\text{Int}} = \frac{i}{\hbar} \left(\frac{1}{2\pi} \right)^{3N} \int d\xi^N\, d\mathbf{p}'^N$$

$$\times \exp[i\xi^N \cdot (\mathbf{p}^N - \mathbf{p}'^N)][V(\mathbf{x}_+^N) - V(\mathbf{x}_-^N)]F(\mathbf{x}^N, \mathbf{p}'^N, t) \tag{2.64}$$

In the limit $\hbar \to 0$, there results

$$\frac{V[\mathbf{x}^N + (\hbar \xi^N/2)] - V[\mathbf{x}^N - (\hbar \xi^N/2)]}{\hbar} \to \xi^N \cdot \frac{\partial}{\partial \mathbf{x}^N} V(\mathbf{x}^N)$$

and (2.64) reduces to

$$
\begin{aligned}
\frac{\partial F}{\partial t}\bigg|_{\text{Int}} &= \left(\frac{1}{2\pi}\right)^{3N} \frac{\partial}{\partial \mathbf{p}^N} \cdot \int \int \frac{\partial}{\partial \mathbf{x}^N} V(\mathbf{x}^N) \\
&\quad \times \exp[i\xi^N \cdot (\mathbf{p}^N - \mathbf{p}'^N)] F(\mathbf{x}^n, \mathbf{p}'^N, t) \, d\xi^N \, d\mathbf{p}'^N \\
&= \frac{\partial}{\partial \mathbf{p}^n} \cdot \int \frac{\partial}{\partial \mathbf{x}^N} V(\mathbf{x}^N) \delta(\mathbf{p}^N - \mathbf{p}'^N) F(\mathbf{x}^N, \mathbf{p}'^N, t) \, d\mathbf{p}'^N \\
&= \frac{\partial}{\partial \mathbf{x}^N} H(\mathbf{x}^N, \mathbf{p}^N) \cdot \frac{\partial}{\partial \mathbf{p}^N} F(\mathbf{x}^N, \mathbf{x}^N, t)
\end{aligned}
\tag{2.65}
$$

Combining (2.65) with (2.54) gives

$$\frac{\partial F}{\partial t} = [H, F]_C \tag{2.66}$$

where C denotes the classical Poisson bracket. We may conclude that the Wigner distribution function satisfies the Liouville equation in the classical domain $\hbar = 0$.

5.2.5 Wigner–Moyal Equation

The Wigner-Moyal equation in quantum kinetic theory is the analogue of the particle Liouville equation in kinetic theory. To derive this equation we examine the Wigner equation (2.54) for the case $N = 1$.

Three results

$$\frac{DF}{Dt} - \frac{i}{\hbar}\left(\frac{1}{\pi\hbar}\right)^3 \int d\mathbf{y}\, d\mathbf{p}' \exp\left(\frac{2i\mathbf{y}\cdot(\mathbf{p}-\mathbf{p}')}{\hbar}\right)[V(\mathbf{x}_+)-V(\mathbf{x}_-)]F(\mathbf{x}, \mathbf{p}', t) = 0 \tag{2.67}$$

where

$$\frac{DF}{Dt} \equiv \left(\frac{\partial}{\partial t} + \frac{\mathbf{p}}{m} \cdot \frac{\partial}{\partial \mathbf{x}}\right) F \tag{2.67a}$$

We wish to reduce (2.67) to a more concise form. Noting the relation

$$\exp\left(\mathbf{a} \cdot \frac{\partial}{\partial \mathbf{x}}\right) f(\mathbf{x}) = f(\mathbf{x} + \mathbf{a})$$

permits us to write

$$
\begin{aligned}
V(\mathbf{x}_+) - V(\mathbf{x}_-) &= V(\mathbf{x} + \mathbf{y}) - V(\mathbf{x} - \mathbf{y}) = [e^{\mathbf{y}\cdot(\partial/\partial\mathbf{x})} - e^{-\mathbf{y}\cdot(\partial/\partial\mathbf{x})}]V(\mathbf{x}) \\
&= -2i \sin\left(i\mathbf{y} \cdot \frac{\partial}{\partial \mathbf{x}}\right) V(\mathbf{x})
\end{aligned}
$$

Thus (2.67) becomes

$$\frac{DF}{Dt} - \frac{2}{\hbar}\left(\frac{1}{\pi\hbar}\right)^3 \int dy\, d\mathbf{p}'\, \exp\left(\frac{2i\mathbf{y}\cdot(\mathbf{p}-\mathbf{p}')}{\hbar}\right)$$
$$\times \sin\left(i\mathbf{y}\cdot\frac{\partial}{\partial\mathbf{x}}\right) V(\mathbf{x})F(\mathbf{x},\mathbf{p}',t) = 0 \qquad (2.68)$$

where $\partial/\partial\mathbf{x}$ operates on $V(\mathbf{x})$ only. To further reduce (2.68), we note the relation

$$\int e^{iyp'} i\mathbf{y}F(p')\,dp' = \int \left(\frac{\partial}{\partial p'}e^{iyp'}\right) F(p')\,dp'$$
$$= -\int e^{iyp'}\frac{\partial}{\partial p'}F(p')\,dp'$$

Thus, if $g(y)$ is a regular function, we obtain

$$\int e^{iyp'} g(i\mathbf{y})F(p')\,dp' = \int e^{iyp'} g\left(-\frac{\partial}{\partial p'}\right) F(p')\,dp'$$

Incorporating this property into (2.68) gives

$$\frac{DF}{Dt} - \frac{2}{\hbar}\left(\frac{1}{\pi\hbar}\right)^3 \int dy\, d\mathbf{p}'\, \exp\left(\frac{2i\mathbf{y}\cdot(\mathbf{p}-\mathbf{p}')}{\hbar}\right)$$
$$\times \sin\left(\frac{\hbar}{2}\frac{\partial}{\partial\mathbf{p}'}\cdot\frac{\partial}{\partial\mathbf{x}}\right) V(\mathbf{x})F(\mathbf{x},\mathbf{p},t) = 0 \qquad (2.69)$$

Recalling the delta function representation

$$\frac{1}{(2\pi)^3}\int d\mathbf{z}e^{i\mathbf{z}\cdot(\mathbf{p}-\mathbf{p}')} = \delta(\mathbf{p}-\mathbf{p}')$$

and setting $\mathbf{z} = 2\mathbf{y}/\hbar$ gives

$$\frac{DF}{Dt} - \frac{2}{\hbar}\left(\frac{1}{\pi\hbar}\right)^3 (2\pi)^3 \left(\frac{\hbar}{2}\right)^3 \int d\mathbf{p}'\delta(\mathbf{p}-\mathbf{p}')$$
$$\times \sin\left(\frac{\hbar}{2}\frac{\partial}{\partial\mathbf{p}'}\cdot\frac{\partial}{\partial\mathbf{x}}\right) V(\mathbf{x})F(\mathbf{x},\mathbf{p},t) = 0$$

Performing the final $\mathbf{p}'$ integration gives the desired result:

$$\left[\left(\frac{\partial}{\partial t}+\frac{\mathbf{p}}{m}\cdot\frac{\partial}{\partial\mathbf{x}}\right) - \frac{2}{\hbar}\sin\left(\frac{\hbar}{2}\frac{\partial}{\partial\mathbf{p}}\cdot\frac{\partial}{\partial\mathbf{x}}\right) V(\mathbf{x})\right] F(\mathbf{x},\mathbf{p},t) = 0 \qquad (2.70)$$

where, again, we recall that $\partial/\partial\mathbf{x}$ act only on $V(\mathbf{x})$. The relation (2.70) is sometimes called the Wigner-Moyal equation. It is instructive to note that for Hamiltonians of the form

$$H = \frac{p^2}{2m} + V(\mathbf{x})$$

(2.70) may be rewitten[14] (see Problem 5.8) as

$$\boxed{\frac{\partial F}{\partial t} + \frac{2}{\hbar} \sin \left[\frac{\hbar}{2} \left(\frac{\partial}{\partial \mathbf{x}} \cdot \frac{\partial}{\partial \mathbf{p}} - \frac{\partial}{\partial \mathbf{p}} \cdot \frac{\partial}{\partial \mathbf{x}} \right) \right] FH = 0} \qquad (2.71)$$

Here it is understood, as in its classical counterpart, that the right operators in the two products operate on H, whereas the left operators operate on F.

The classical limit of the Wigner-Moyal equation as given by (2.71) is particularly simple. Setting $\hbar \to 0$ and keeping the leading term in the expansion of sin[ ] returns the classical Liouville equation (2.66).

Note that all the above analysis can be generalized to an N-body problem with interaction potential $V(\mathbf{x}^N)$ by setting $\mathbf{x} \to \mathbf{x}^N$ and $\mathbf{p} \to \mathbf{p}^N$ [see (5.33)].

Let us return to the quantum one-particle Liouville equation (2.70). An ad hoc kinetic equation may be obtained from this relation by inserting an appropriate collision form on the right side which, in the Boltzmann formalism, is written $\hat{J}(F)$. In the KBG approximation, for example, we write

$$\frac{DF}{Dt} - \frac{2}{\hbar} \sin \left(\frac{\hbar}{2} \frac{\partial}{\partial \mathbf{p}} \cdot \frac{\partial}{\partial \mathbf{x}} \right) V(\mathbf{x}) F(\mathbf{x}, \mathbf{p}, t) = \nu [F_0(\mathbf{x}, \mathbf{p}, t) - F(\mathbf{x}, \mathbf{p}, t)]$$

$$(2.72)$$

where $F_0(\mathbf{x}, \mathbf{p}, t)$ is an appropriate equilibrium distribution and ν^{-1} is relaxation time.

An additional note is in order concerning the Wigner-Moyal equation (2.71). As $\mathbf{x}$ and $\mathbf{p}$ cannot be simultaneously specified in quantum mechanics, we expect the quantum distribution $F(\mathbf{x}, \mathbf{p}, t)$ to be defined for volumes in Γ-space, $\Omega \gtrsim \hbar^3$.

The Wigner-Moyal equation has found wide application in the analysis of charge carrier transport in a semiconductor.[15]

5.2.6 Homogeneous Limit: Pauli Equation Revisited

In this section we employ the equation of motion for the Wigner distribution function (2.70) to derive a kinetic equation for the matrix elements of the density matrix in the limit of spatial homogeneity. With this motivation and with reference to (2.46), we write the single-particle Wigner distribution function in momentum representation, which, together with the change in variable $2\mathbf{y} \to \mathbf{k}$, appears as

$$F(\mathbf{x}, \mathbf{p}, t) = \left(\frac{1}{\pi \hbar} \right)^3 \int d\mathbf{k} e^{i\mathbf{x} \cdot \mathbf{k}/\hbar} \left\langle \mathbf{p} + \frac{\mathbf{k}}{2} \left| \hat{\rho} \right| \mathbf{p} - \frac{\mathbf{k}}{2} \right\rangle \qquad (2.73)$$

---

[14]A hierarchy of equations for reduced Wigner distributions stemming from the Wigner-Moyal equation in N dimension is given in Section 5.5.4; see specifically (5.33).

[15]See, for example, *The Physics of Submicron Structures*, H. L. Grubin, K. Hess, G. J. Iafrate, and D. K. Ferry (eds.), Plenum, New York (1984).

Note that $\mathbf{k}$ has dimensions of momentum. Let us call

$$\left\langle \mathbf{p} + \frac{\mathbf{k}}{2} \middle| \hat{\rho} \middle| \mathbf{p} - \frac{\mathbf{k}}{2} \right\rangle \equiv \rho(\mathbf{k}, \mathbf{p}) \tag{2.74}$$

With these identifications, taking the Fourier transform of (2.70) (and recalling the convolution theorem) yields

$$\left(\frac{\partial}{\partial t} + \frac{i\mathbf{p}}{m} \cdot \frac{\mathbf{k}}{\hbar} \right) \rho(\mathbf{k}, \mathbf{p}) - \int d\mathbf{k}' \frac{2}{\hbar} \sin\left(\frac{i\mathbf{k}'}{2} \cdot \frac{\partial}{\partial \mathbf{p}} \right) \bar{V}(\mathbf{k}')\rho(\mathbf{k} - \mathbf{k}', \mathbf{p}) = 0 \tag{2.75}$$

where $\bar{V}(\mathbf{k})$ represents the Fourier tansform of the interaction potential $\mathbf{V}(\mathbf{x})$.

We wish to obtain an equation for the density matrix relevant to the homegeneous limit. This is performed through a process of iteration. In the homogeneous limit, the $\mathbf{k} = 0$ contribution to the integral (2.75) dominates, and we write

$$\frac{\partial}{\partial t} \rho(\mathbf{p}) = \int d\mathbf{k}' \hat{S}(\mathbf{k}') \bar{V}(\mathbf{k}') \rho(-\mathbf{k}', \mathbf{p}) \tag{2.76}$$

where

$$\rho(0, \mathbf{p}) \equiv \rho(\mathbf{p}) \tag{2.76a}$$

is written for the diagonal elements of the density matrix and $\hat{S}(\mathbf{k})$ represents the operator

$$\hat{S}(\mathbf{k}) \equiv \frac{2}{\hbar} \sin\left(\frac{i\mathbf{k}}{2} \cdot \frac{\partial}{\partial \mathbf{p}} \right) \tag{2.76b}$$

Changing variables $\mathbf{k}' \rightarrow -\mathbf{k}'$ in (2.76) gives [with $\bar{V}(-\mathbf{k}) = \bar{V}^*(\mathbf{k})$]

$$\frac{\partial}{\partial t} \rho(\mathbf{p}) = - \int d\mathbf{k}' \hat{S}(\mathbf{k}') \bar{V}^*(\mathbf{k}') \rho(\mathbf{k}', \mathbf{p}) \tag{2.77}$$

This equation contains two distributions, one relevant to $\mathbf{k} = 0$ and the other relevant to $\mathbf{k} \neq 0$. We assume that the $\mathbf{k} \neq 0$ distribution, $\rho(\mathbf{k}, \mathbf{p})$, varies in time as $\exp(i\omega t)$, where the interval ω^{-1} may be associated with characteristic times of the fluid. The time dependence of the $\mathbf{k} = 0$ distribution $\rho(\mathbf{p})$ is determined from its equation of motion (to be derived). For $\rho(\mathbf{k}, \mathbf{p})$, we return to (2.75) and write[16]

$$\rho(\mathbf{k}, \mathbf{p}) = \frac{1}{i[\omega + (\mathbf{p}/m) \cdot (\mathbf{k}/\hbar)]} \int d\mathbf{k}' \bar{V}(\mathbf{k}') \hat{S}(\mathbf{k}') \rho(\mathbf{k} - \mathbf{k}', \mathbf{p})$$

---

[16]Note that the time dependence of $\rho(\mathbf{p})$, as well as the ω dependence of $\rho(\mathbf{k}, \mathbf{p})$, is tacitly assumed.

Expanding the integrand about $\mathbf{k} = \mathbf{k}'$ and then keeping the first term with $\rho(\mathbf{k} - \mathbf{k}', \mathbf{p}) = \delta(\mathbf{k} - \mathbf{k}')\rho(\mathbf{p})$, we obtain

$$\rho(\mathbf{k}, \mathbf{p}) = \frac{1}{i[\omega + (\mathbf{p}/m) \cdot (\mathbf{k}/\hbar)]} \bar{V}(\mathbf{k})\hat{S}(\mathbf{k})\rho(\mathbf{p})$$

Inserting this value into (2.77) gives

$$\frac{\partial}{\partial t}\rho(\mathbf{p}) = -\int d\mathbf{k}' \hat{S}(\mathbf{k}')\left[\bar{V}^*(\mathbf{k}')\frac{1}{i[\omega + (\mathbf{p}/m) \cdot (\mathbf{k}'/\hbar)]} \bar{V}(\mathbf{k}')\hat{S}(\mathbf{k}')\rho(\mathbf{p})\right]$$

(2.78)

Now we note

$$\frac{\hbar}{2}\hat{S}(\mathbf{k}) = \sin\frac{i\mathbf{k}}{2}\cdot\frac{\partial}{\partial\mathbf{p}} = i\sinh\frac{\mathbf{k}}{2}\cdot\frac{\partial}{\partial\mathbf{p}}$$

$$= \frac{i}{2}\left[\exp\left(\frac{\mathbf{k}}{2}\cdot\frac{\partial}{\partial\mathbf{p}}\right) - \exp\left(-\frac{\mathbf{k}}{2}\cdot\frac{\partial}{\partial\mathbf{p}}\right)\right]$$

whence

$$\frac{\hbar}{2}\hat{S}(\mathbf{k})\rho(\mathbf{p}) = \frac{i}{2}\left[\rho\left(\mathbf{p} + \frac{\mathbf{k}}{2}\right) - \rho\left(\mathbf{p} - \frac{\mathbf{k}}{2}\right)\right]$$

(2.79)

Employing this relation in (2.78) gives

$$\frac{\partial}{\partial t}\rho(\mathbf{p}) = -\frac{2}{\hbar}\int d\mathbf{k}'|\bar{V}(\mathbf{k}')|^2 \hat{S}(\mathbf{k}')\frac{1}{i[\omega + (\mathbf{p}/m)\cdot(\mathbf{k}'/\hbar)]}$$

$$\times \frac{i}{2}\left[\rho\left(\mathbf{p} + \frac{\mathbf{k}'}{2}\right) - \rho\left(\mathbf{p} - \frac{\mathbf{k}'}{2}\right)\right]$$

(2.80a)

Performing the remaining $\hat{S}$ operation gives

$$\frac{\partial}{\partial t}\rho(\mathbf{p}) = \frac{1}{\hbar^2}\int d\mathbf{k}'|\bar{V}(\mathbf{k}')|^2 \left\{\frac{[\rho(\mathbf{p} + \mathbf{k}') - \rho(\mathbf{p})]}{i\left[\omega + \frac{\mathbf{k}'}{\hbar m}\cdot\left(\mathbf{p} + \frac{\mathbf{k}'}{2}\right)\right]} - \frac{[\rho(\mathbf{p}) - \rho(\mathbf{p} - \mathbf{k}')]}{i\left[\omega + \frac{\mathbf{k}'}{\hbar m}\cdot\left(\mathbf{p} - \frac{\mathbf{k}'}{2}\right)\right]}\right\}$$

(2.80b)

Changing variables in the second term, $\mathbf{k}' \to -\mathbf{k}'$, gives

$$\frac{\partial}{\partial t}\rho(\mathbf{p}) = \frac{1}{\hbar^2}\int d\mathbf{k}'|\bar{V}(\mathbf{k}')|^2[\rho(\mathbf{p} + \mathbf{k}') - \rho(\mathbf{p})]$$

$$\left\{\frac{1}{i\left[\omega + \frac{\mathbf{k}}{\hbar m}\cdot\left(\mathbf{p} + \frac{\mathbf{k}'}{2}\right)\right]} + \frac{1}{i\left[\omega - \frac{\mathbf{k}'}{\hbar m}\cdot\left(\mathbf{p} + \frac{\mathbf{k}'}{2}\right)\right]}\right\}$$

(2.81)

We may view $\mathbf{k}'$ as a momentum transfer in a scattering from the potential $V(\mathbf{x})$. The change in particle energy in this scattering is given by

$$\Delta E = \frac{(\mathbf{p} + \mathbf{k}')^2}{2m} - \frac{p^2}{2m} = \frac{\mathbf{k}'}{m}\cdot\left(\mathbf{p} + \frac{\mathbf{k}'}{2}\right)$$

(2.82)

Thus (2.81) may be written

$$\frac{\partial}{\partial t}\rho(\mathbf{p}) = \frac{1}{\hbar^2} \int d\mathbf{k}' |\bar{V}(\mathbf{k}')|^2 [\rho(\mathbf{p}+\mathbf{k}') - \rho(\mathbf{p})] \left[\frac{1}{i\left(\omega + \frac{\Delta E}{\hbar}\right)} + \frac{1}{i\left(\omega - \frac{\Delta E}{\hbar}\right)} \right]$$
(2.83)

As $\mathbf{k}'$ has been identified as momentum transfer in a collision, it follows that $\hbar/\Delta E$ is a measure of the corresponding collision time. Assuming this interval to be short compared to characteristic times of the fluid, we write

$$\frac{\hbar}{\Delta E} \ll \frac{1}{\omega}$$

To incorporate this property into the analysis, we set

$$i\omega \to \varepsilon > 0$$

where ε is a parameter of smallness. We may then write

$$\frac{1}{i\left(\omega \pm \frac{\Delta E}{\hbar}\right)} \to \frac{1}{i\left(-i\varepsilon \pm \frac{\Delta E}{\hbar}\right)} = \frac{\pm 1}{i\left(\frac{\Delta E}{\hbar} \mp i\varepsilon\right)}$$

Recall the representation (Plemelj's formula)

$$\lim_{\varepsilon \to 0} \left(\frac{1}{x \pm i\varepsilon} \right) = \mathcal{P}\left(\frac{1}{x} \right) \mp i\pi\delta(x)$$

where $\mathcal{P}$ denotes principal part. Inserting these relations into (2.83) gives the desired result:

$$\frac{\partial\rho(\mathbf{p})}{\partial t} = \frac{2\pi}{\hbar} \int d\mathbf{k}' |\bar{V}(\mathbf{k}')|^2 \, \delta(\Delta E)[\rho(\mathbf{p}+\mathbf{k}') - \rho(\mathbf{p})]$$
(2.84)

This completes our derivation of a closed kinetic equation for the diagonal matrix elements of $\hat{\rho}$ relevant to the homogeneous limit.[17]

The kinetic equation (2.84) may be cast in a more physically revealing form by virtue of the following observation. We note that the factor

$$w(\mathbf{p}, \mathbf{p}+\mathbf{k}') \equiv \frac{2\pi}{\hbar} |\bar{V}(\mathbf{k}')|^2 \, \delta(\Delta E)$$
(2.85)

that appears in (2.84) is an expression for the rate of scattering from $\mathbf{p}$ to $\mathbf{p}+\mathbf{k}'$ (or the reverse) (see Fig. 5.3).

With the relation (2.85) at hand, (2.84) may be identified with the *Pauli equation* (2.28). In the present case, the kinetic equation (2.85) is the momentum representation of the Pauli equation for which the scattering rate (2.85) is the continuum form of the discrete rate (2.21).

---

[17]For further discussion, see P. Carruthers and F. Zachariasen, *Rev. Mod. Phys.* **55**, 245 (1983); and W. Kohn and J. M. Luttinger, *Phys. Rev.* **108**, 590 (1957).

FIGURE 5.3. Diagrams for scattering by the potential V. For elastic interactions, the rate for collision and its reverse are equal.

Owing to the symmetric property of the scattering rate $w(\mathbf{p}, \mathbf{p} + \mathbf{k}')$, (2.84) may be written

$$\frac{\partial \rho(\mathbf{p})}{\partial t} = \int d\mathbf{k}'[w(\mathbf{p} + \mathbf{k}', \mathbf{p})\rho(\mathbf{p} + \mathbf{k}') - w(\mathbf{p}, \mathbf{p} + \mathbf{k}')\rho(\mathbf{p})] \qquad (2.86)$$

The first factor in the integrand contributes to scattering into momentum $\mathbf{p}$ states and represents again, whereas the second factor contributes to scattering out of momentum $\mathbf{p}$ states and represents a loss. The latter equation is returned to in Section 3.6 in the derivation of the relaxation time for charge carrier-phonon interactions.

5.3 Application of the KBG Equation to Quantum Systems

5.3.1 Equilibrium Distributions

One of the more significant problems in statistical mechanics concerns the distribution function for an appregate of particles in equilibrium at a given temperature T. There cases important in this study relate, respectively, to (1) classical particles, (2) bosons, and (3) fermions.

For aggregates of noninteracting particles, equilibrium distributions for these cases, giving mean occupation number per *energy level*, are given by[18]

$$f_0(E) = \frac{1}{e^{\beta(E-\mu)} + a} \qquad (3.1)$$

$$\beta \equiv (k_b T)^{-1}$$

where

$$\begin{aligned}
a &= 0 \quad \text{corresponds to classical particles} \\
a &= 1 \quad \text{corresponds to fermions} \qquad\qquad (3.2)\\
a &= -1 \text{ corresponds to bosons}
\end{aligned}$$

[18]For further discussion, see K. Huang, *Statistical Mechanics*. Wiley, New York (1967); and D. A. Mcquarrie, *Statistical Mechanics*, Harper & Row, New York (1976).

and the variable μ is the chemical potential.[19]

Note that quantum distributions reduce to the classical distribution in the limit

$$e^{\beta(E-\mu)} \gg 1 \tag{3.3}$$

or, more simply, $f_0(E) \ll 1$. That is, the classical distribution occurs when the mean occupation number per energy level is small compared to unity.[20]

The condition (3.3) may be shown to be equivalent to the criterion[21]

$$\lambda_d^3 \ll \frac{V}{N} \tag{3.4}$$

where

$$\lambda_d = \frac{h}{(2\pi m k_B T)^{1/2}} \tag{3.5}$$

is the *thermal de Broglie wavelength*, V is volume, and N is total number of particles. The condition (3.4) states that the classical picture results when the de Broglie wavelength is small compared to interparticle spacing.

We wish to apply the distributions (3.1) to two key problems: (1) an interacting gas of atoms and photons, and (2) electrons in a metal.

Planck distribution

Photons are bosons, so the related distribution is given by (3.1) with $a = -1$. Furthermore, for photons we may set[22] $\mu = 0$. With $E = \hbar\omega$, the appropriate distribution is given by

$$f_0(\omega) = \frac{1}{e^{\beta\hbar\omega} - 1} \tag{3.6}$$

We wish to obtain the number of photons per frequency interval, per volume, $F(\omega)$. The distribution (3.6) gives the mean number of photons per frequency mode. Thus we may set

$$F(\omega) = g(\omega) f(\omega) \tag{3.7}$$

In this equation, $g(\omega)$ represents the number of normal modes per frequency interval, per unit volume. An expression for the density of states, $g(\omega)$, may be

---

[19]The chemical potential enters the first law of thermodynamics for cases involving varying number of particles: $dE = T\,dS - P\,dV + \mu\,dN$. Here S represents entropy.

[20]For bosons with $\mu = 0$, the classical limit occurs for small β. See, for example, (3.68).

[21]R. K. Pathria, *Statistical Mechanics*, Pergamon Press. Elmsford, N. Y. (1978).

[22]This is a consequence of thermodynamics. Consider a gedanken process in which the number of photons in a monoenergetic field at frequency ω is doubled as $\omega \to \omega/2$. With $dE = dS = dV = 0$, and $dN > 0$, we obtain $\mu = 0$.

obtained in the following manner. First recall that the momentum of a photon of angular frequency ω is

$$p = \frac{\hbar \omega}{c} \tag{3.8}$$

It follows that the volume in momentum-coordinate phase space containing photons with frequencies in the interval ω, $\omega + d\omega$ is equal to the product of d^3x and the momentum volume corresponding to values in the interval p, $p + dp$ or, equivalently, $4\pi p^2 \, dp \, d^3x$. The number of states in the said interval is obtained by dividing this latter volume by the minimum of volume per state, h^3. Since photons have two possible polarizations, this minimum volume becomes $h^3/2$. Thus we may write

$$g(\omega) \, d\omega \, d^3x = \frac{4\pi p^2 \, dp \, d^3x}{h^3/2}$$
$$= \frac{\omega^2 \, d\omega \, d^3x}{\pi^2 c^3} \tag{3.9}$$

which gives the desired form for the density of states (Rayleigh's formula):

$$g(\omega) = \frac{\omega^2}{\pi^2 c^3} \tag{3.10}$$

Combining (3.7) and (3.10) gives the *Planck distribution*:

$$F_0(\omega) = \frac{\omega^2 / \pi^2 c^3}{e^{\beta \hbar \omega} - 1} \tag{3.11}$$

Here we have inserted a zero subscript to denote that (3.9) is a thermal equilibrium distribution.

An immediate result that follows from the Planck radiation law (3.11) is the total radiant energy density, $U(T)$, of a radiation field in equilibrium at the temperature T.

$$U(T) = \int F_0(\omega) \hbar \omega \, d\omega \tag{3.12}$$

Introducing the nondimensional parameter

$$x \equiv \beta \hbar \omega$$

gives

$$U(T) = T^4 \left(\frac{k_B^4}{\pi^2 c^3 \hbar^3} \int_0^\infty \frac{x^3 \, dx}{e^x - 1} \right)$$
$$U(T) = \left(\frac{\pi^2 k_B^4}{15 \hbar^3 c^3} \right) T^4 \tag{3.13}$$

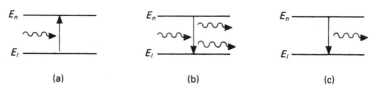

FIGURE 5.4. Three radiative collision processes are included in (3.18): (a) resonant absorption, (b) stimulated decay, (c) spontaneous emission.

[see (B2.23)]. The power radiated per unit area by the radiation field is given by

$$P = \frac{1}{4}cU(t) = \left(\frac{\pi^2 k_B^4}{60\hbar^3 c^2}\right) T^4$$

$$P = \sigma T^4 \tag{3.14}$$

Here c is the speed of light and σ if the Stefan-Boltzmann constant, which has the value

$$\sigma = 0.567 \times 10^{-4} \text{ erg/s} - \text{cm}^2 - K^4 \tag{3.15}$$

5.3.2 Photon Kinetic Equation

We consider a radiation field interacting with a gas of hydrogenic atoms whose excited states are characterized by the outer atomic electron. The density of the atoms in the ith excited state is written N_i.

In our kinetic model we discuss photons that are either entirely absorbed by an atom or give rise to stimulated decay of an atom. Furthermore, phontons may be generated through spontaneous decay of atoms. Thus the model does not included, for example, scattering of photons by atoms (see Fig 5.4).

The momentum of a photon of (angular) frequency ω is given by (3.8), which we now writes as

$$\mathbf{p} = \frac{\mathbf{s}\hbar\omega}{c} \tag{3.16}$$

where $\mathbf{s}$ is the unit vector

$$\mathbf{s} = \frac{\mathbf{p}}{p} \tag{3.17}$$

Our kinetic equation is then written[23]

$$\frac{\partial}{c\partial t}f(\mathbf{x}, \mathbf{s}, \omega) + \mathbf{s} \cdot \nabla f(\mathbf{x}, \mathbf{s}, \omega) = \sum_{l<n} \Delta_{ln} \tag{3.18}$$

---

[23]D. H. Sampson, *Radiative Contributions to Energy and Momentum Transport in a Gas*, Wiley-Interscience, New York (1965). See also D. M. Heffernan and R. L. Liboff, J. *Plasma Physics* 27, 473 (1982).

The collision term Δ_{ln} is defined below.

Let us first concentrate on the distribution function in (3.18). It is normalized with the differential measure (3.9). The quantity

$$\frac{2f\,d\mathbf{p}\,d\mathbf{x}}{h^3}$$

represents the number of photons in the phase volume $d\mathbf{p}\,d\mathbf{x}$ about the value $(\mathbf{p}, \mathbf{x})$. In the present description the distribution function f relates to photons progagating in the direction $\mathbf{p}$ with frequency $cp/\hbar$. As previously described, the factor 2 in the above expression is included to account for two possible polarizations at any given frequency.

Returning to (3.18), first we note that the double sum on l and n runs over all atomic states, and we have written

$$\Delta_{ln} \equiv N_n \sigma_{nl}(\omega)[1 + f(\omega)] - N_l \sigma_{ln}(\omega) f(\omega) \tag{3.19}$$

The first two terms in this expression represent spontaneous and induced decay, respectively. The third therm represents resonant absorption, and σ_{ln} is written for the absorption cross section at frequency ν corresponding to atomic transition from the lth to the nth state.

Writing g_l for the degeneracy of the lth atomic state and recalling the symmetry relation[24]

$$g_l \sigma_{ln} = g_n \sigma_{nl} \tag{3.20}$$

permits (3.18) to be written in the KBG form:

$$\frac{1}{c}\frac{Df(\omega)}{Dt} = \kappa(\bar{f}_0 - f) \tag{3.21}$$

The parameters κ and $\bar{f}_0$ are written for

$$\kappa \equiv \sum_{l<n} N_l \left(1 - \frac{g_l N_n}{g_n N_l}\right) \sigma_{ln} \tag{3.22}$$

$$\bar{f}_0 \equiv \frac{1}{\kappa} \sum_{l<n} N_n \frac{g_l}{g_n} \sigma_{ln} \tag{3.23}$$

The distribution $\bar{f}_0$ appears explicitly as

$$\bar{f} = \frac{\sum_{l<n} N_n (g_l/g_n)\sigma_{ln}}{\sum_{l<n} N_l[1 - (g_l N_n/g_n N_l)]} \tag{3.24}$$

Let us examine this distribution for the case of monochromatic radiation, which is resonant with two atomic states. That is

$$\hbar\omega = E_b - E_a \tag{3.25}$$

---

[24]H. Bethe and E. Salpeter, *Quantum Mechanics of One- and Two-Electron Atoms*, Springer, New York (1957).

Cross sections in (3.24) are peaked about his value of ω, and the expression (3.24) reduces to the simpler form

$$\bar{f}_0 = \frac{N_b(g_a/g_b)\sigma_{ab}}{N_a[1 - (g_a N_b/g_b N_a)]\sigma_{ab}}$$
$$\bar{f}_0 = \frac{1}{(N_a g_b/N_b g_a) - 1}$$

(3.26)

With atomic states taken to be in a Boltzmann distribution,

$$N_a = g_a e^{-\beta E_a}$$

(3.27)

the relation (3.26) is seen to return the Planck equilibrium distribution (3.11). Thus (3.23) may be interpreted as a generalized Planck distribution relevant to the dynamics of an electromagnetic field interacting with an atomic species.

Lasing criterion

Consider a laser beam of radiation interacting with a gas of atoms. Let the radiation be resonant with two atomic states as described by (3.25). In this case, the absorption coefficient κ given by (3.22) becomes

$$\kappa = N_a \left(1 - \frac{g_a N_b}{g_b N_a}\right) \sigma_{ab}$$

(3.28)

and $\bar{f}_0$ is given by (3.26).

In steady state, our kinetic equation (3.18) reduces to

$$\frac{\partial}{\partial s} f(s, \omega) + \kappa f(s, \omega) = \kappa \bar{f}_0(\omega)$$

(3.29)

Here we have written s for displacement in the direction of $\mathbf{s}$. The homogeneous solution to (3.29) gives the spatial dependence

$$f(s, \omega) = f(0, s)e^{-\kappa s}$$

(3.30)

It follows that the number of photons with the resonant frequency ω propagating in the direction $\mathbf{s}$ will grow, providing $\kappa < 0$. With (3.28), this gives the canoncial lasing criterion[25]

$$\frac{N_b}{g_b} > \frac{N_a}{g_a}$$

(3.31)

So, in fact, radiative gain demands something more than simply an excess of atoms in the higher lasing state than in the lower state.

------

[25] A. Yariv, *Quantum Electronics*, 2nd ed., Wiley, New York (1975).

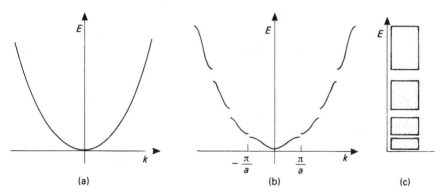

FIGURE 5.5. (a) Free-particle spectrum. (b) Energies corresponding to a periodic potential. Band energies occur at $k = n\pi/a$, where a is the lattice constant and n is an integer. (c) Related band-gap spectrum of energies.

5.3.3 Electron Transport in Metals

In Section 5.1, the following energy spectrum was obtained for free particles:

$$E(k) = \frac{\hbar^2 k^2}{2m}$$

When plotted as a function of k, this expression describes a parabola. If particles are not free, but propagate through a periodic potential, not all energies are allowed. The resulting $E(k)$ curve exhibits gaps that affect a band structure as illustrated in Fig 5.5.

The lattice of a metallic crystal presents a periodic potential to conduction electrons. Consequently, the energy spectrum of these electrons exhibits a bad structure containing forbidden and allowed bands. In a metal the band of highest energy is called the conduction band and is only partially filled. An applied electric field may therefore increase the energy of topmost electrons without violating the exclusion principle. The band structure of the metal sodium is shown in Fig. 5.6.

In semiconductors, the conduction band is empty at low temperature. However, due to the relatively small value of the energy gap separating valence and conduction bands, at higher temperatures, statse in the conduction band are populated, thereby increasing the conductivity of the sample. Vancancies left in the valence band due to promotion of electrons to the conduction band affect the presence of positive charge and are called holes. Such holes may act as charge carries and contribute to the net current, and we write for the conductivity

$$\sigma = |e|(n\mu_n + p\mu_p)$$

where n and p and hole concentrations (cm^{-3}), respectively, and μ_n and μ_p are related mobilities.

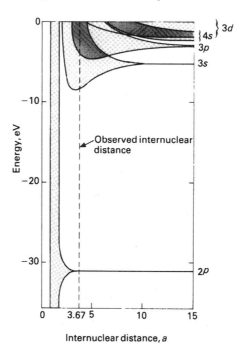

FIGURE 5.6. The band structure of sodium as a fuction of internuclear spacing. Note the overlapping of bands with (extreme) increase in density. Quantum states of atomic levels are also shown [J. C. Slater, *Rev. Mod. Phys.* **6**, 209 (1934)].

The quasi-classical description

For many cases of practical interest, a quasi-classical description of conduction in solid-state materials proves adequate. Since the presence of band gaps is a purely quantum mechanical effect, it is important to the self-consistency of a classical description that electron motion be confinded to a single band.

Let $K(k)$ denote the energy spectrum of a given band. Two parameters play an important role in the semiclassical model. These are the velocity,

$$\mathbf{v}(\mathbf{k}) = \frac{1}{\hbar} \nabla_{\mathbf{k}} E(k)$$

and the effective mass tensor

$$\left(\frac{1}{m^*}\right)_{\mu\nu} = \frac{\partial^2 E / \partial k_\mu \, \partial k_\nu}{\hbar^2}$$

For motion in one dimension, the effective mass tensor becomes a scalar, which is labeled m^*. For free-particle motion, m^* reduces to the classical mass, m.

As with our preceding analysis for photon kinetic theory, the following quantity is written for the number of electrons in the phase volume $d\mathbf{p}\,d\mathbf{x}$ (in the conduction band) about the phase point $(\mathbf{p}, \mathbf{x})$ (in the quasi-classical limit).

$$2f(\mathbf{p}, \mathbf{x})\frac{d\mathbf{p}\,d\mathbf{x}}{h^3} = \frac{2f\hbar^3\,d\mathbf{k}\,d\mathbf{x}}{h^3} = f(\mathbf{k}, \mathbf{x})\frac{d\mathbf{k}\,d\mathbf{x}}{4\pi^3} \qquad (3.32)$$

In (3.32), the factor 2 stems from the two possible spin projections relevant to electrons. Since $d\mathbf{k}\,d\mathbf{x}/4\pi^3$ represents a density of states, we may identify $f(\mathbf{k}, \mathbf{x})$ as the mean occupation number of electrons per state.

The Fermi-Dirac equilibrium distribution for an aggregate of electrons at temperature T is obtained from (3.1) and is given by

$$f_0[E(k)] = \frac{1}{e^{\beta[E(k)-\mu]} + 1} \qquad (3.33a)$$

With the relation

$$= \frac{2 \times 4\pi p^2\,dp}{h^3}\frac{4\pi(2m)^{3/2}E^{1/2}\,dE}{h^3} \qquad (3.33b)$$

the preceding two equations give

$$\frac{N}{V} = \int f_0(p)\frac{2\,d\mathbf{p}}{h^3} = \int_0^\infty \frac{4\pi(2m)^{3/2}E^{1/2}\,dE}{h^3(e^{\beta(E-\mu)} + 1)}$$

This relation serves to relate the chemical potential μ to the number density of N free electrons confined to the volume V. At $0\ K$, $\mu = E_F$, the Fermi energy, and the preceding relation gives

$$n = \frac{4\pi(2m)^{3/2}}{h^3}\int_0^{E_F} E^{1/2}\,dE$$

$$E_F = \frac{h^2}{2m}\left(\frac{3n}{8\pi}\right)^{2/3} = \frac{\hbar^2}{2m}(3\pi^2 n)^{2/3} \qquad (3.34)$$

where we have written n for particle number density. The preceding expression may be written in terms of the Fermi momentum p_F.

$$n = \int_0^{p_F} \frac{2\,d\mathbf{p}}{h^3} = \frac{2}{h^3} \times \frac{4}{3}\pi p_F^3$$

$$E_F = \frac{p_F^2}{2m} \qquad (3.34a)$$

Values of E_F for some typical metals are listed in Table 5.1.

Working with the distribution function $f(\mathbf{k}, \mathbf{x}, t)$, the KBG equation assumes the form[26]

$$\frac{\partial f}{\partial t} + \mathbf{v} \cdot \frac{\partial f}{\partial \mathbf{x}} + \dot{\mathbf{k}} \cdot \frac{\partial f}{\partial \mathbf{k}} = \frac{1}{\tau}(f_0 - f) \qquad (3.35)$$

---

[26]In the most solid-state works, this analysis is referred to as the relaxation-time approximation.

TABLE 5.1. Properties of Characteristic Conducting Metals: Resistivity, Relaxation Times and Fermi Values

Element	Z	$n(10^{22}/cm^3)$	$\sigma(\mu\Omega-cm)$ 273 K	$\tau(10^{-14}s)$	$E_F(eV)$	k_F $(\times10^8 cm^{-1})$	v_F $(\times10^8 cm/s)$
Li	1	4.70(78 K)	8.55	0.88	4.74	1.12	1.29
Na	1	2.65(5 K)	4.2	3.2	3.24	0.92	1.07
K	1	1.40(5 K)	6.1	4.1	2.12	0.75	0.86
Rb	1	1.15(5 K)	11.0	2.8	1.85	0.70	0.81
Cs	1	0.91(5 K)	18.8	2.1	1.59	0.65	0.75
Cu	1	8.47	1.56	2.7	7.00	1.36	1.57
Ag	1	5.86	1.51	4.0	5.49	1.20	1.39
Au	1	5.90	2.04	3.0	5.53	1.21	1.40
Be	2	24.7	2.8	0.51	14.3	1.94	2.25
Mg	2	8.61	3.9	1.1	7.08	1.36	1.58
Ca	2	4.61	3.43	2.2	4.69	1.11	1.28
Sr	2	3.55	23	0.44	3.92	1.02	1.18
Ba	2	3.15	60	0.19	3.64	0.98	1.13
Nb	1	5.56	15.2	0.42	5.32	1.18	1.37
Fe	2	17.0	8.9	0.24	11.1	1.71	1.98
Zn	2	13.2	5.5	0.49	9.47	1.58	1.83
Cd	2	9.27	6.8	0.56	7.47	1.40	1.62
Hg	2	8.65	Melted		7.13	1.37	1.58
Al	3	18.1	2.45	0.80	11.7	1.75	2.03
Ga	3	15.4	13.6	0.17	10.4	1.66	1.92
In	3	11.5	8.0	0.38	8.63	1.51	1.74
Ti	3	10.5	15	0.22	8.15	1.46	1.69
Sn	4	14.8	10.6	0.23	10.2	1.64	1.90
Pb	4	13.2	19.0	0.14	9.47	1.58	1.83
Bi	5	14.1	107	0.023	9.90	1.61	1.87
Sb	5	16.5	39	0.055	10.9	1.70	1.96

where

$$\hbar\dot{\mathbf{k}} = e\mathcal{E} \tag{3.35a}$$

and $\mathcal{E}$ represents electric field strength. The relaxation time τ is in general a function of $\mathbf{k}$ [see, for example, (3.81)]. Assuming steady state and spatial homogeneity, (3.34) reduces to

$$\frac{e\mathcal{E}}{\hbar} \cdot \frac{\partial f}{\partial \mathbf{k}} = \frac{1}{\tau}(f_0 - f) \tag{3.36}$$

For short relaxation times, we may set $\tau \to \varepsilon\tau$, where ε is a small bookkeeping parameter. Repeating the Chapman-Enskog expansion (3.5.8),

$$f = f^{(0)} + \varepsilon f^{(1)} + \varepsilon^2 f^{(2)} + \cdots \tag{3.37}$$

TABLE 5.2. Thermal Conductivities and Lorenz Numbers

| | 273 K | | 373 K | |
| | κ | $\kappa/\sigma T$ | κ | $\kappa/\sigma T$ |
Element	(W-cm/K)	(W-Ω/K^2)	(W-cm/K)	(W-Ω/K^2)
Li	0.71	2.22×10^{-8}	0.73	2.43×10^{-8}
Na	1.38	2.12		
K	1.0	2.23		
Rb	0.6	2.42		
Cu	3.85	2.20	3.82	2.29
Ag	4.18	2.31	4.17	2.38
Au	3.1	2.32	3.1	2.36
Be	2.3	2.36	1.7	2.42
Mg	1.5	2.14	1.5	2.25
Nb	0.52	2.90	0.54	2.78
Fe	0.80	2.61	0.73	2.88
Zn	1.13	2.28	1.1	2.30
Cd	1.0	2.49	1.0	
Al	2.38	2.14	2.30	2.19
In	0.88	2.58	0.80	2.60
Tl	0.5	2.75	0.45	2.75
Sn	0.64	2.48	0.60	2.54
Pb	0.38	2.64	0.35	2.53
Bi	0.09	3.53	0.08	3.35
Sb	0.18	2.57	0.17	2.69

Source: N. V. Ashcroft and N. D. Mermin, *Solid State Physics*, Holt, Rinehart, and Winston, New York (1976); and G. W. C. Kaye and T. H. Laby, *Table of Physical and Chemical Constants*, Longmans Green, London (1966).

and inserting into (3.36) and equating powers of ε gives

$$f^{(0)} = f_0$$

$$\frac{\tau e\mathcal{E}}{\hbar} \cdot \frac{\partial f^{(0)}}{\partial \mathbf{k}} = -f^{(1)}$$

$$\vdots$$

Thus, with (3.37), we find, to $0(\varepsilon)$,

$$f(\mathbf{k}) = f_0 - \frac{\tau e\mathcal{E}}{\hbar} \cdot \frac{\partial f_0}{\partial \mathbf{k}} \simeq f_0 \left(\mathbf{k} - \frac{\tau e\mathcal{E}}{\hbar} \right) \qquad (3.38)$$

Here we have recognized that the middle expression represents the first two terms in a Taylor series expansion of the function on the right about $\tau e\mathcal{E}/\hbar = 0$. This finding indicates that the Fermi sphere has been displaced by the amount $\tau e\mathcal{E}/\hbar$ in $\mathbf{k}$-space (see Fig. 5.7). It is evident from this figure that only electrons at the top of the Fermi sea contribute to conduction. Electrons sufficiently below the Fermi surface cannot be accelerated to nearby higher-lying levels. Such

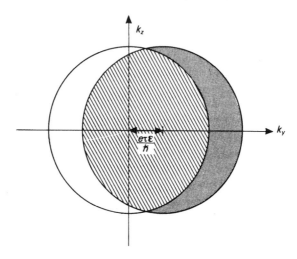

FIGURE 5.7. The Fermi sphere in **k** space suffers a displacement in the presence of an electric field. Electrons in the darkened area contribute to conduction.

promotion would violate the exclusion principle as these levels are already occupied.

We may use (3.38) to calculate conductivity. Current density is given by

$$\mathbf{J} = \frac{e}{4\pi^3} \int \mathbf{v}(\mathbf{k}) f(\mathbf{k}) \, d\mathbf{k}$$

Inserting (3.38) into this expression gives

$$\mathbf{J} = -\frac{e^2}{4\pi^3 \hbar} \int \tau(\mathbf{k}) \mathbf{v}(\mathbf{k}) \mathcal{E} \cdot \frac{\partial f_0}{\partial \mathbf{k}} \, d\mathbf{k} \qquad (3.39)$$

Here we have noted that f_0 has a vanishing first velocity moment. Let us picture the Fermi surface, $E = E_F$, in **k** space. If dS is an element of area on this surface, then a differential of volume $d\mathbf{k}$ generated by a normal, displacement $\delta\mathbf{k}$ is $|\delta\mathbf{k}| \, dS$.[27] The related change in energy is given by $dE = |\delta\mathbf{k}||\nabla_\mathbf{k} E|$. Thus we may write for the volume element

$$d\mathbf{k} = \frac{dS \, dE}{|\nabla_\mathbf{k} E|} \qquad (3.39a)$$

At 0 K, the Fermi-Dirac distribution (3.33) gives

$$\frac{\partial f_0}{\partial E} = -\delta(E - E_F) \qquad (3.39b)$$

Thus

$$\frac{\partial f_0}{\partial \mathbf{k}} = \frac{\partial f_0}{\partial E} \frac{\partial E}{\partial \mathbf{k}} = -\mathbf{v}\hbar \delta(E - E_F) \qquad (3.39c)$$

---

[27] A similar argument was given previously in Section 3.8.1.

and (3.39) becomes

$$\mathbf{J} = \frac{e^2}{4\pi^3} \int \tau(\mathbf{k})\mathbf{v}(\mathbf{k})\mathcal{E} \cdot \mathbf{v}(\mathbf{k})\delta(E - E_F)\frac{dE\,dS}{|\nabla_{\mathbf{k}}E|} \qquad (3.39\text{d})$$

Integrating over energy gives

$$\mathbf{J} = \frac{e^2}{4\pi^3} \int_{E=E_F} \frac{\tau(\mathbf{k})\mathbf{v}(\mathbf{k})\mathcal{E} \cdot \mathbf{v}(\mathbf{k})}{|\nabla_{\mathbf{k}}E|}\,dS \qquad (3.40)$$

Writing Ohm's law in tensoral form,

$$J\mu = \sigma_{\mu\nu}\mathcal{E}_\nu$$

and comparing with (3.40) implies the following for the *conductivity tensor*:[28]

$$\sigma_{\mu\nu} = \frac{e^2}{4\pi^3} \int_{E_F} = \frac{\tau(\mathbf{k})v_\mu(\mathbf{k})v_\nu(\mathbf{k})}{|\nabla_{\mathbf{k}}E|}\,dS \qquad (3.41)$$

Here, we recall, dS is an element of area in $\mathbf{k}$ space. Only electrons near the Fermi surface contribute to conductivity. (See Problem 5.52.)

Weidemann-Franz law

If a temperature gradient is present in the solid-state plasma, then the space gradient term in (3.35) is written

$$\mathbf{v} \cdot \frac{\partial f}{\partial \mathbf{x}} = \mathbf{v} \cdot \frac{\partial f}{\partial T}\frac{\partial T}{\partial \mathbf{x}}$$

The analog of (3.36) becomes

$$\frac{\partial f}{\partial T}\mathbf{v} \cdot \frac{\partial T}{\partial \mathbf{x}} = \frac{1}{\tau}(f_0 - f)$$

Writing the heat conductivity equation (3.4.3) in tensorial form,

$$Q\mu = -\kappa_{\mu\nu}\frac{\partial}{\partial x_\nu}T$$

and following the procedure that led to (3.41), we find

$$\kappa_{\mu\nu} = \frac{k_B^2 T}{12\pi} \int_{E=E_F} \frac{\tau(\mathbf{k})v_\mu(\mathbf{k})v_\nu(\mathbf{k})}{|\nabla_{\mathbf{k}}E|}\,dS \qquad (3.42)$$

Providing relaxation times in (3.41) and (3.42) are equal, we may conclude

$$\frac{\kappa}{\sigma T} = \frac{\pi^2}{3}\left(\frac{k_B}{e}\right)^2 \equiv L_0 = 2.445 \times 10^8 \text{W}-\Omega/\text{K}^2 \qquad (3.43)$$

---

[28]Electrical conductivity in Chapter 3 was labeled σ_c. A generalized tensor form for σ is given in Section 5.6. See also Section 7.1.2, 7.1.6, 7.2.3.

where L_0 is the Lorenz number (see Tables 5.1, 5.3). The relation (3.43) is the Weidemann-Franz law. An expression for the relaxation time $\tau(k)$ is obtained in Section 3.6.

5.3.4 Thomas–Fermi Screening

In this section we consider the repsonse to an extraneous electron placed in a charge-neutral degenerate Fermi gas. The analysis parallels the calculation (Section 4.1.1) relevant to a classical plasma. Converting from velocity to momentum, together with the transformation relevant to the quantum domain,

$$f_0(v)\,d\mathbf{v} \rightarrow 2f_0(p)\frac{d\mathbf{p}}{h^3}$$

permits (4.1.23) to be written (in the static limit) as

$$\varepsilon - 1 = -\frac{8\pi m e^2}{k^2} \int \frac{\mathbf{k}\cdot\partial f_0/\partial \mathbf{p}}{\mathbf{k}\cdot\mathbf{p}} \frac{d\mathbf{p}}{h^3} \tag{3.44}$$

Here we have written f_0 for the Fermi-Dirac distribution (3.33). With

$$\frac{\partial f_0}{\partial \mathbf{p}} = \frac{\mathbf{p}}{m}\frac{\partial f_0}{\partial E}$$

and

$$d\mathbf{p} = 2\pi (2m)^{3/2} E^{1/2}\,dE$$

(3.44) becomes

$$\varepsilon - 1 = -\frac{16\pi^2 e^2 (2m)^{3/2}}{k^2 h^3} \int \frac{\partial f_0}{\partial E} E^{1/2}\,dE \tag{3.45}$$

For sufficiently low temperature, again we write

$$\frac{\partial f_0}{\partial E} = -\delta(E - E_F)$$

and (3.45) may be written

$$\varepsilon - 1 = \frac{6\pi n e^2/E_F}{k^2} = \left(\frac{k_{TF}}{k}\right)^2 \tag{3.46}$$

where E_F is given by (3.34) and

$$k_{TF}^2 = \frac{6\pi n e^2}{E_F}, \qquad \lambda_{TF}^2 = \frac{1}{k_{TF}^2} \tag{3.46a}$$

are the Thomas–Fermi screening wave number and screening distance, respectively. It follows that the static dielectric constant for a degenerate plasma, in the quasi-classical limit, is given by

$$\varepsilon = 1 + \frac{k_{TF}^2}{k^2} \tag{3.47}$$

Repeating the analysis leading to (4.1.35) gives the static potential

$$\phi(r) = \frac{-e \exp(-k_{TF}r)}{r} \tag{3.48}$$

Whereas this form greatly resembles the classical result (4.1.35), a more detailed quantum mechanical description[29] indicates that, at relatively large distance, the interparticle potential oscillates as

$$\phi(r) \sim \frac{1}{r^3} \cos 2k_F r \tag{3.49}$$

Here k_F is written for the Fermi wave number, $E_F = \hbar^2 k_F^2/2m$.

Nonstatic limit

In the nonstatic, long-wavelength ($k^2 \ll \omega_p^2/v_F^2$) limit, we obtain[30] the following dispersion relation for a degenerate plasma:

$$\omega^2 = \omega_p^2 \left(1 + \frac{3}{5} \frac{v_F^2 k^2}{\omega_p^2} \right) \tag{3.50}$$

This relation bears a striking similarity to the parallel relation for long-wavelength waves in a classical plasma (4.1.46):

$$\omega^2 = \omega_p^2 \left(1 + \frac{3C^2 k^2}{\omega_p^2} \right) \tag{3.51}$$

However, as $T \to 0$, $C \to 0$ and (3.51) gives nonpropagating standing waves, whereas in the quantum domain v_F persists at $T = 0K$. Progagation of waves at 0 K in a degenerate plasma have been compared to the phenomenon of "zero sound" relevant to a fluid of uncharged fermions at 0 K (discussed in Section 4.2). Furthermore, longitudinal waves in a degenerate plasma (3.50) also suffer Landau damping.[31]

The similarity in relations of parallel quantities in classical and degenerate[32] plasmas is illustrated by the following:

---

[29] J. Linhard, *Kgl. Danske Videnskab. Selskab. Mat.-Fyd. Medd. 28*, No. 8 (1954). The behavior (3.49) is often referred to as Friedel oscillations.

[30] L. D. Landau, E. M. Lifshitz, and L. P. Pitaevskii, *Physical Kinetics*, Pergamon Press, Elmsford, N. Y. (1981).

[31] P. M. Platzman and P. A. Wolf, *Waves and Interactions in Solid State Plasmas*, Academic Press, New York (1973).

[32] Degenerate here refers to the quantum domain.

Classical

$$\frac{1}{k_d^2} = \lambda_d^2 = \frac{C^2}{\omega_p^2} = \frac{k_B T}{4\pi n e^2}, \qquad \left(\frac{\omega}{\omega_p}\right)^2 = 1 + \frac{3C^2 k^2}{\omega_p^2}, \qquad k^2 \ll \frac{\omega_p^2}{C^2}$$

$$(3.52)$$

Degenerate

$$\frac{1}{k_{TF}^2} = \lambda_{TF}^2 = \frac{1}{3}\frac{v_F^2}{\omega_p^2} = \frac{E_F}{6\pi n e^2}, \qquad \left(\frac{\omega}{\omega_p}\right)^2 = 1 + \frac{3}{5}\frac{v_F^2 k^2}{\omega_p^2}, \qquad k^2 \ll \frac{\omega_p^2}{v_F^2}$$

$$(3.53)$$

5.3.5   Mott Transition[33]

An instructive application of the Thomas–Fermi potential (3.48) comes into play in the derivation of the criteria for what is commonly called the Mott transition. Consider a conducting lattice with one free electron per atom, such as is the case with the alkali metals. The band structure typical to such conductors is shown in Fig 5.6. Let the crystal undergo an ideal expansion at 0 K. The question we wish to consider is, at which point in the expansion does the sample become an insulator? We will assume that the change of state, termed a Mott transition occurs when a conduction electron of the metal forms a bound state with an ion in the lattice.

At reasonably good expression for the potential of interaction between a conduction electron and an ion in a metal for the present model is given by the Thomas–Fermi potential (3.48) (now written as potential energy)

$$V(r) = -\frac{e^2 \exp(-k_{TF} r)}{r}$$

$$(3.54)$$

It has been shown[34] that an electron and ion interacting under this potential will permit a bound state providing

$$k_{TF} < \frac{1}{a_0}$$

$$(3.55)$$

where a_0 is the Bohr radius. With (3.46a), we may rewrite

$$k_{TF}^2 = \frac{6\pi n e^2}{E_F} = \frac{4}{a_0}\left(\frac{3n}{\pi}\right)^{1/3}$$

---

[33]Named for N. F. Mott.
[34]J. M. Blatt and V. F. Weisskopf, *Theoretical Nuclear Physics*, p. 55, Wiley, New York (1953).

The criterion (3.55) may then be rewritten

$$n < \left(\frac{\pi}{12}\right)^2 n_0 \qquad (3.56)$$

where

$$n_0 \equiv \frac{3}{4}\pi a_0^3$$

is the number density corresponding to one atom per "Bohr volume," $4\pi a_0^3/3$, which is relevant to a hydrogen-like lattice. Thus n represents both electron and ion densities. The relation (3.56) indicates that in the present crude model the Mott transition occurs for ion number density $\leq 0.1 n_0$.[35]

5.3.6   Relaxation Time for Charge-Carrier Phonon Interaction

For temperatures in excess of $\simeq 100K$, one of the more important scattering mechanisms of charge carriers in a semiconductor is that of phonon scattering.[36] Recall the phonons are quantized lattice vibrations. The dispersion relation for phonons contains two branches corresponding to *acoustic* and *optical* phonons, respectively.[37] For relatively small phonon wave number q, acoustic phonons satisfy the dispersion relation

$$\omega \simeq uq \qquad (3.57a)$$

where u is constant wave speed, whereas optical phonons in this same domain satisfy the relation

$$\omega \simeq \omega_0 \qquad (3.57b)$$

where ω_0 is a constant frequency.

For elemental semiconductors (for example, silicon and germanium), charge carriers interact with phonons through what is commonly termed the *deformation potential* interaction or, more simply, the *strain* interaction. Thus one speaks of strain-acoustic and strain-optical interactions.

---

[35] Whereas experiments are presently underway, the monatomic metallic phase of hydrogen has not been observed in the laboratory. It has been estimated that solid H_2 undergoes a transition to the metallic phase at approximately 3 Mbar. For further discussion, see B. I. Min, J. F. Jansen, and A. J. Freeman, *Phys. Rev. B. 33*, 6383 (1986); and D. Ceperely and B. Alder, *Phys. Rev. B. 36*, 2092 (1986).

[36] At lower temperature, scattering is dominated by interaction with impurity atoms in the lattice. For futher discussion, see K. Seeger, *Semiconductor Physics* 3rd ed., Springer-Verlag, New York (1985); and R. L. Liboff, *J. Phys. Chem. Solids 46*, 1327 (1985).

[37] For further discussion, see J. M. Ziman, *Principles of the Theory of Solids*, Cambridge, New York (1964).

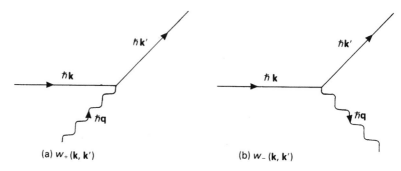

FIGURE 5.8. Diagrams corresponding to (a) absorption of a phonon, and (b) emission of a phonon.

In this section we derive an expression for the relaxation time $\tau(k)$, introduced in (3.35), appropriate to the acoustic strain interaction[38] for a semiconductor in the presence of an applied electric field $\mathcal{E}$.

Neglecting spin, the collision integral for this scattering proces may be written [compare with (2.86)]

$$\hat{J}(f) = \frac{V}{(2\pi)^3} \sum_{\pm} \int [w_{\pm}(\mathbf{k}', \mathbf{k}) f(\mathbf{k}') - w_{\pm}(\mathbf{k}, \mathbf{k}') f(\mathbf{k})] \, d\mathbf{k}' \qquad (3.58)$$

where V is the volume of the sample and with (3.32) we have written $\hbar\mathbf{k}$ for the momentum of charge carriers. The probability rate or charge carrier scattering from the $\mathbf{k}$ to $\mathbf{k}'$ state accompained by the absorption $(+)$ or emission $(-)$ of a phonon is written $w_{\pm}(\mathbf{k}, \mathbf{k}')$. We may associate diagrams with these interactions, as illustrated in Fig. 5.8.

The rate coefficient $w_{\pm}(\mathbf{k}, \mathbf{k}')$ may be written in terms of matrix elements of the perturbation Hamiltonian H', as follows [recall (2.85)]:

$$w_{\pm}(\mathbf{k}, \mathbf{k}') = |H'_{\mathbf{k}',\mathbf{k}}|^2 \frac{2\pi}{\hbar} \delta(\Delta E_{\pm}) \qquad (3.59)$$

Here $H'_{\mathbf{k}',\mathbf{k}}$ is the matrix element corresponding to the transition $\mathbf{k} \rightarrow \mathbf{k}'$. The delta function factor in (3.59) ensures conservation of energy where, with m written for effective mass of charge carriers,

$$\Delta E_{\pm} = \left[\frac{\hbar^2 k'^2}{2m} \mp \hbar\omega(q) \right] - \frac{\hbar^2 k^2}{2m} \qquad (3.60)$$

Our plan is to convert the right side of (3.58) to that of (3.35). To incorporate the anisotropy of the distribution function to the applied electric field, we recall

---

[38] It has been demonstrated that for cubic semiconductors the acoustic strain interaction dominates in the quasi-classical domain. See G. K. Schenter and R. L. Liboff, *J. Appl. Phys.* 62, 1977 (1987).

(3.7.21) and write

$$f(\mathbf{k}) = f_0(k) + \mu f_1(k) \tag{3.61}$$

where, again, μ is the cosine of the angle between $\mathcal{E}$ and $\mathbf{k}$. If we identify $f_0(k)$ as the equilibrium distribution, then by statistical balance (Section 3.3.9) we may set

$$w_\pm(\mathbf{k}, \mathbf{k}') f_0(k) = w_\pm(\mathbf{k}', \mathbf{k}) f_0(k') \tag{3.62}$$

Substituting (3.61) into (3.58) and incorporating (3.62), we obtain

$$\hat{J}(f) = \frac{V}{(2\pi)^3} \sum_p m \int [w_\pm(\mathbf{k}', \mathbf{k}) f_1(k')\mu' - w_p m(\mathbf{k}, \mathbf{k}') f_1(k)\mu]\, d\mathbf{k}'$$

$$= \frac{V}{(2\pi)^3} \sum_p m \int \left[\frac{f_0(k)}{f_0(k')} f_1(k')\mu' - f_1(k)\mu \right] w_p m(\mathbf{k}, \mathbf{k}')\, d\mathbf{k}' \tag{3.63}$$

In the high-energy limit, $E \gg \hbar\omega(q)$, we may further assume

$$f_0(k') \simeq f_0(k)$$
$$f_1(k') \simeq f_1(k) \tag{3.64}$$

Substituting these relations into (3.63) gives

$$\hat{J}(f) = -\frac{V}{(2\pi)^3} \sum_\pm \mu f_1(k) \int \left(1 - \frac{\mu'}{\mu} \right) w_\pm(\mathbf{k}, \mathbf{k}')\, d\mathbf{k}' \tag{3.65}$$

Recalling (3.61), we note

$$-\mu f_1(k) = f_0(k) - f(\mathbf{k})$$

Thus (3.65) gives the desired expression:

$$\frac{1}{\tau} = \frac{V}{(2\pi)^3} \sum_{(\pm)} \int \left(1 - \frac{\mu'}{\mu} \right) w_p m(\mathbf{k}, \mathbf{k}')\, d\mathbf{k}' \tag{3.66}$$

Explicit formula for $1/\tau$

To further reduce the integral in (3.66), we cite the specific form of the transition probability rate (3.59) for strain-acoustic interactions.[39]

$$\sum_{pm} w_\pm(\mathbf{k}, \mathbf{k}') = \frac{2\pi A q}{\hbar} [n_q \delta(\Delta E_+) + (n_q + 1)\delta(\Delta E_-)] \tag{3.67}$$

---

[39]E. M. Conwell in *High Field Transport in Semiconductors. Solid State Physics*, Vol. 9, Suppl., F. Seitz, D. Turnbull, and H. Ehrenreich (eds.), Academic Press, New York (1967).

where n_q is the Bose-Einstein distribution (3.6)[40] relevant to phonons, and $\Delta E_\pm$ are given by (3.60). The coefficient A is given by

$$A = \frac{E_1^2 \hbar}{2V\rho u} \tag{3.67a}$$

where ρ is crystal mass density and E_1/V is the shift in band edge[41] per unit dilation of the lattice. Note that E_1 has dimensions of energy and A that of energy squared times length.

For sufficiently large crystal temperature, we may write

$$k_B T \gg \hbar q u$$

This permits the Bose-Einstein distribution to be expaned[42] as

$$n_q = \frac{k_B T}{\hbar u q} - \frac{1}{2} + \cdots \gg 1 \tag{3.68}$$

which in turn allows (3.67) to be collapsed, and we obtain

$$\sum w_\pm(\mathbf{k}, \mathbf{k}') = \frac{2\pi A q}{\hbar} n_q [\delta(\Delta E_+) + \delta(\Delta E_-)] \tag{3.69}$$

We turn next to the delta functions $\delta(\Delta E_\pm)$. Substituting the dispersion relation (3.57a) into $\Delta E_\pm$ given by (3.60) permits us to write

$$
\begin{aligned}
\Delta E_\pm &= \frac{\hbar^2}{2m}[(\mathbf{k} \pm \mathbf{q})^2 - k^2] \mp \hbar q u \\
&= \frac{\hbar^2}{2m}[q^2 \pm 2KQ\bar{\mu}) \mp \hbar q u \\
&= \mp \frac{\hbar^2 k q}{m}\left[\pm \frac{q}{2k} - \bar{\mu} + \frac{mu}{\hbar k} \right]
\end{aligned}
$$

where we have written

$$\bar{\mu} \equiv \hat{\mathbf{k}} \cdot \hat{\mathbf{q}}$$

Taking the delta function of $\Delta E_p m$ gives

$$\delta(\Delta E_\pm) = \frac{2m}{\hbar^2}\frac{1}{2kq}\delta\left[\bar{\mu} - \left(\frac{mu}{\hbar k} \mp \frac{q}{2k}\right)\right] \tag{3.70}$$

Here we have recalled the delta-function relation

$$\delta(ax) = \frac{1}{|a|}\delta(x) \tag{3.71}$$

---

[40]That is, n_q (here) is written for $f_0[\omega(q)]$ in (3.6).

[41]For a one-dimensional lattice, band edgers occur at $k = n\pi/a$, where a is the lattice constant (see Fig. 5.5).

[42]Neglecting $\frac{1}{2}$ in (3.68) gives the phonon equipartition theorem: $k_B T = n_q \hbar\omega =$ energy/frequency modes.

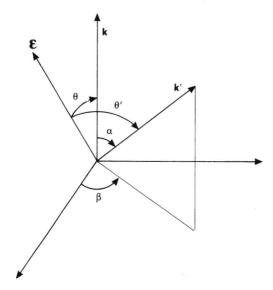

FIGURE 5.9. The angles θ, θ', α, and β for reduction of (3.66).

Finally, in our evaluation of the integral in (3.66), we seek a relation between μ and μ'. To these ends we refer to Fig. 5.9. Vectors are defined with respect to $\mathbf{k}$ taken as the polar axis. With this figure at hand, the cosine addition law gives

$$\mu' = \mu \cos \alpha + \sin \theta \sin \alpha \cos \beta \tag{3.72}$$

Furthermore,

$$q^2 = |\mathbf{k} - \mathbf{k}'|^2 = k^2 + k'^2 - 2kk' \cos \alpha \tag{3.73}$$

At high energy, $k \simeq k'$, and the preceding relation gives

$$\cos \alpha = 1 - \frac{q^2}{2k^2} \tag{3.74}$$

Substituting this relation into (3.72) and neglecting the $\cos \beta$ term owing to the $d\beta$ integration in (3.66), we obtain

$$\frac{\mu'}{\mu} = 1 - \frac{q^2}{2k^2} \tag{3.75}$$

With the relations (3.68) through (3.75) at hand, we return to (3.66) and write (with $d\mathbf{k}' = d\mathbf{q}$)

$$\frac{1}{\tau} = \frac{B}{k^3} \sum_{\pm} \int d\mathbf{q}\, q\delta \left[\bar{\mu} - \left(\frac{mu}{\hbar k} \mp \frac{q}{2k} \right) \right] \tag{3.76}$$

where

$$B \equiv \frac{mVAk_BT}{2(2\pi)^2\hbar^4 u} \tag{3.76a}$$

With **k** taken as the polar axis, we write

$$d\mathbf{p} = 2\pi q^2 \, dq \, d\bar{\mu}$$

and (3.76) gives

$$\frac{1}{\tau} = \frac{2\pi B}{k^3} \sum_{\pm} \int_0^{\infty} dq q^3 \int_{-1}^{1} d\bar{\mu} \, \delta\left[\bar{\mu} - \left(\frac{mu}{\hbar k} \mp \frac{q}{2k}\right)\right] \tag{3.77}$$

Integration over $\bar{\mu}$ leads to bounds on the dq integration, and (3.77) reduces to

$$\frac{1}{\tau} = \sum_{\pm} \frac{2\pi B}{k^3} \int_{q_1(\pm)}^{q_2(\pm)} dq q^3 \tag{3.78}$$

where $q(\pm)$ are the two solutions at the bounds of the inequalities

$$-1 \le \frac{mu}{\hbar k} \mp \frac{q}{2k} \le +1$$
$$q \ge 0 \tag{3.78a}$$

There results

$$q_1(\pm) = 0, \qquad q_2(\pm) = 2k \pm \frac{2mu}{\hbar} \tag{3.78b}$$

(see Problem 5.9). Substituting these values into (3.78) and integrating yields

$$\frac{1}{\tau} = \sum_{p} m \frac{8\pi B}{k^3 \hbar^4} (\hbar k \pm mu)^4$$

In the limit $u \ll \hbar k/m$, the preceding reduces to

$$\frac{1}{\tau} = 16\pi Bk$$

Substituting the value for A (3.67a) into B (3.76a) gives

$$\frac{1}{\tau} = \left(\frac{mk_BT}{\pi\hbar^3} \frac{E_1^2}{\rho u^2}\right) k \tag{3.79}$$

This result may be cast in a more revealing form if we identify

$$l = \frac{\pi\hbar^4 \rho u^2}{m^2 k_B T E_1^2} \tag{3.80}$$

as the acoustic phonon mean free path. Thus we may write

$$\frac{1}{\tau} = \frac{\hbar k/m}{l} \tag{3.81}$$

This finding is often employed in the literature.[43]

Here is a brief recapitulation of assumptions leading to (3.81): (1) $E \gg \hbar\omega(q)$, $k' \simeq k$, (2) $k_B T \gg \hbar\omega(q)$, (3) $\hbar k/m \gg u$. These inequalities are satisfied in most practical cases.

5.4   Quantum Modifications of the Boltzmann Equation

5.4.1   Quasi-Classical Boltzmann Equation

Normalizations

In the preceding analysis of photon (boson) and electron (fermion) kinetic theory a general rule was found describing the transformation of classical to quantum distribution functions. In the quasi-classical limit, it reads as follows:

$$f_{CL}(\mathbf{x}, \mathbf{v}) \, d\mathbf{x} \, d\mathbf{v} \rightarrow f_{QM}(\mathbf{x}, \mathbf{p}) \frac{d\mathbf{x} \, d\mathbf{p}}{h^2} \tag{4.1}$$

Here we recall that $d\mathbf{x} \, d\mathbf{p}/h^3$ represents the number of available states in the phase volume $d\mathbf{x} \, d\mathbf{p}$. Note that $f_{QM}(\mathbf{x}, \mathbf{p})$ is dimensionless. In the pure quantum domain, $\mathbf{x}$ and $\mathbf{p}$ may not be prescribed simultaneously, and the preceding rule becomes

$$f_{CL}(\mathbf{v}) \, d\mathbf{v} \rightarrow f_{QM}(\mathbf{p}) \frac{d\mathbf{p}}{h^3} \tag{4.2}$$

(This transformation was used previously in Section 3.4.) In our preceding examples, the density of states $d\mathbf{p}/h^3$ was increased by a degeneracy factor, which we now label G. For photons, $G = 2$, corresponding to two degrees of polarization, whereas for electrons, $G = 2$, corresponding to two spin projections. Thus, for photons, we found (3.9)

$$G \frac{d\mathbf{p}}{h^3} = \frac{2 \, d\mathbf{p}}{h^3} = \frac{\omega^2 \, d\omega}{\pi^2 c^3}$$

whereas, for electrons (3.33b),

$$G \frac{d\mathbf{p}}{h^3} = \frac{2 \, d\mathbf{p}}{h^3} = \frac{4\pi (2m)^{3/2} E^{1/2} \, dE}{h^3}$$

These notions play an important role in Uehling and Uhlenbeck's quantum generalization of the Boltzmann equation.[44]

---

[43] See, for example, L. Reggiani (ed.), *Hot-Electron Transport in Semiconductors*, Springer-Verlag, New York (1985).

[44] E. A. Uehling and G. E. Uhlenbeck, *Phys. Rev.* **43**, 552 (1933).

Collision integral

We concentrate on the collision integral in the Boltzmann equation (3.2.14) (reverting to dimensional form):

$$\hat{J}(f) = \int \int (f' f_1' - f f_1) g\sigma \, d\Omega \, dv_1$$

Here we recall that $f' f_1'$ relates to a collision in which particles are restored to velocity volumes dv about v and dv_1 about v_1, respectively. Suppose these particles are fermions. Then we must ensure that the exclusion principle is not violated when particles enter respective phase volumes. The number of available states in the volume $d\mathbf{x} \, d\mathbf{p}$ is

$$G\frac{d\mathbf{x} \, d\mathbf{p}}{h^3} = G\frac{m^3}{h^3} \, d\mathbf{x} \, dv \tag{4.3}$$

(Here we are writing $d\mathbf{x}$ for the collision-domain volume.) Thus, if $f = Gm^3/h^3$, the phase volume is filled and there can be no further reentry of fermions. This property can be incorported in the Boltzmann equation through the change in collision phase volume relevant to the term $f_1' f'$:

$$d\mathbf{x}_1 \, dv_1 \rightarrow \left[1 - \frac{h^3 f(v)}{Gm^3}\right]\left[1 - \frac{h^3 f(v_1)}{Gm^3}\right] d\mathbf{x}_1 \, dv_1 \tag{4.4}$$

A similar argument applies to the direct collision product $f f_1$. Inserting these changes into the derivation (3.2.1) et seq. yields the quasi-classical collision form

$$\hat{J}(f) = \int \int [f' f_1'(1+\xi f)(1+\xi f_1) - f f_1(1+\xi f')(1+\xi f_1')] g\sigma \, d\Omega \, dv_1 \tag{4.5}$$

Here we have labeled

$$\xi \equiv \delta \frac{h^3}{Gm^3} \tag{4.6a}$$

where

$$\delta = \begin{cases} +1 & \text{for bosons} \\ 0 & \text{for Boltzmann statistics} \\ -1 & \text{for fermions} \end{cases} \tag{4.6b}$$

The enhancement value $\delta = +1$ for bosons stems from a statistical affinity that identical bosons have to be in a common domain.[45]

---

[45]G. E. Uhlenbeck and L. Groper, *Phys. Rev.* **41**, 79 (1932). See also K. Huang, *Statistical Mechanics*, ibid.

Equilibrium distributions

The condition for statistical balance that follows from (4.5) is given by

$$\left(\frac{f}{1+\xi f}\right)\left(\frac{f_1}{1+\xi f_1}\right) = \left(\frac{f'}{1+\xi f'}\right)\left(\frac{f_1'}{1+\xi f_1'}\right) \tag{4.7a}$$

Taking the log of both sides reveals the collisional invariant:[46]

$$\ln\left(\frac{f}{1+\xi f}\right) = \beta(\mu - E) - \ln|\xi| \tag{4.7b}$$

Here $\beta \equiv (k_b T)^{-1}$, E is written for single-particle energy, and μ is a constant that may be identified with the chemical potential. Inverting the preceding relation gives

$$|\xi| f = \frac{1}{e^{\beta(E-\mu)} - \delta} = \frac{1}{e^{\beta(E-\mu)} \mp 1} \tag{4.7c}$$

With our definition of ξ and taking $|\delta| = 1$ gives the correct quantum distributions [Bose-Einstein $(-)$, Fermi-Dirac $(+)$]:

$$f_0 \, d\mathbf{v} = \frac{G \, d\mathbf{p}/h^3}{e^{\beta(E-\mu)} \mp 1} \tag{4.8}$$

In the following section, an equation closely allied to the Uehling-Uhlenbeck equation is discussed relevant to a Fermi liquid.

5.4.2   Kinetic Theory for Excitations in a Fermi Liquid

Elementary excitations

In quantum mechanics, a weakly excited state of a macroscopic body may be considered as an aggregate of individual *elementary excitations*. Such excitations are a consequence of a quantum mechanical description of the collective motion of the system and should not be identified with individual atoms.

Excitations behave like quasi-particles with definite momenta $\mathbf{p}$ and energy $\varepsilon(\mathbf{p})$. We may introduce a nondimensional distribution function for quasi-particles such that $f \, 2 \, d\mathbf{p}/h^3$ represents the number of quasi-particles per unit volume with momentum in the volume $d\mathbf{p}$. In the quasi-classical limit, f may be written as a function of both $\mathbf{x}$ and $\mathbf{p}$. In the present case, the quasi-classical domain is characterized as follows. Let L be the length over which f varies appreciably. Writing $k \sim 1/L$ for the equivalent wave number, the condition

---

[46]We may assume that the scattering potential is $\mathbf{x}$ dependent. It follows that $\mathbf{p}$ does not commute with the Hamiltonian and cannot be specified together with E in (4.7b).

becomes

$$\hbar k \ll p_F$$

where p_F is the Fermi momentum (3.34).

When the liquid suffers a small displacement from equilibrium, we write

$$f(\mathbf{x}, \mathbf{p}) = f_0(\mathbf{p}) + \delta f(\mathbf{x}, b\mathbf{p}) \qquad (4.9)$$

where $f_0(\mathbf{p})$ is the equilbrium distribution. In like manner, we write

$$\varepsilon(\mathbf{x}, \mathbf{p}) = \varepsilon_0(\mathbf{p}) + \delta\varepsilon(\mathbf{x}, \mathbf{p}) \qquad (4.10)$$

where $\varepsilon_0(p)$ corresponds to the equilibrium state. The change in excitation energy is given by

$$\delta\varepsilon(\mathbf{x}, \mathbf{p}) = \int K(\mathbf{p}, \mathbf{p}')\delta f(\mathbf{x}, \mathbf{p}')\frac{2\,d\mathbf{p}'}{h^3} \qquad (4.11)$$

where $K(\mathbf{p}, \mathbf{p}')$ is the quasi-particle interaction function.

Let the liquid occupy the volume V. For the total energy of the liquid we write UV so that U is energy density. The variation of U is then given by

$$\delta U = \int \varepsilon\delta f \frac{2\,d\mathbf{p}}{h^3} \qquad (4.12)$$

The quantity $\varepsilon(\mathbf{x}, \mathbf{p})$ plays the role of the Hamiltonian of a quasi-particle in the field of other particles, and its as well as (4.12) are employed below to construct an equation of motion for the distribution f.

Kinetic equation

A kinetic equation for quasi-particles in a Fermi liquid was presented by Landau in 1957.[47] It is given by[48]

$$\frac{\partial f}{\partial t} + \frac{\partial \varepsilon}{\partial \mathbf{p}} \cdot \frac{\partial f}{\partial \mathbf{x}} - \frac{\partial \varepsilon}{\partial \mathbf{x}} \cdot \frac{\partial f}{\partial \mathbf{p}} = \hat{J}(f)$$

---

[47] L. D. Landau, *J. Exp. Theor. Phys.* (USSR) 32, 59 (1957). Repringed in, D. Pines, *The Many-Body Problem*, W. A. Benjamin, Menlo Park, Calif. (1961).

[48] Note the resemblance between (4.13) and the Uehling-Uhlenbeck form (4.5). See also, F. Bloch, *Z. Physik 58*, 555 (1929); *59*, 208 (1930).

$$\hat{J}(f) = \int w(\mathbf{p}, \mathbf{p}_1; \mathbf{p}', \mathbf{p}_1')[f'f_1'(1-f)(1-f_1)] \tag{4.13}$$
$$- ff_1(1-f')(1-f_1')]\delta(\varepsilon + \varepsilon_1 - \varepsilon' - \varepsilon_1')$$
$$\delta(\mathbf{p} + \mathbf{p}_1 - \mathbf{p}' - \mathbf{p}_1')\,d\mathbf{p}_1\,d\mathbf{p}'\,d\mathbf{p}_1'$$

Delta functions indicate that momentum and energy are conserved over a collision. The function w gives the collision probability and obeys the symmetry relation

$$w(\mathbf{p}, \mathbf{p}_1; \mathbf{p}', \mathbf{p}_1') = w(\mathbf{p}', \mathbf{p}_1'; \mathbf{p}, \mathbf{p}_1)$$

That is, the probability of a collision and its reverse are equal.

As with the Uehling-Uhlenbeck collision term (4.5), the Landau form (4.13) likewise ensures that collisions satisfy the exclusion principle relevant to fermions. Note also that f in (4.13) is dimensionless.

The condition of statistical balance obtained from (4.13) again yields the Fermi distribution (3.33a).

Conservation equations

Equations of motion for macroscopic variables in a Fermi liquid are obtained by taking moments of (4.14). Integrating over momentum gives the continuity equation

$$\frac{\partial n}{\partial t} + \boldsymbol{\nabla} \cdot n\langle \mathbf{v} \rangle = 0 \tag{4.14}$$

where n represents quasi-particle number density,

$$n = \int f(\mathbf{x}, \mathbf{p}) \frac{2\,d\mathbf{p}}{h^3} \tag{4.15}$$

and

$$n\langle \mathbf{v} \rangle = \int f\mathbf{v} \frac{2\,d\mathbf{p}}{h^3} \tag{4.16}$$

is particle flux, where we have set

$$\mathbf{v} = \frac{\partial \varepsilon}{\partial \mathbf{p}} \tag{4.17}$$

for quasi-particle velocity.

Multiplying (4.13) by p_i, and integrating gives

$$\frac{\partial}{\partial t} n\langle p_i \rangle + \int p_i \left(\frac{\partial f}{\partial x_j} \frac{\partial \varepsilon}{\partial p_j} - \frac{\partial f}{\partial p_j} \frac{\partial \varepsilon}{\partial x_j} \right) \frac{2\,d\mathbf{p}}{h^3} = 0 \tag{4.18}$$

The integrand in this expression may be rewritten

$$\frac{\partial}{\partial x_j} \left(p_i \frac{\partial \varepsilon}{\partial p_j} f \right) + f \frac{\partial \varepsilon}{\partial x_i} - \frac{\partial}{\partial p_j} \left(p_i \frac{\partial \varepsilon}{\partial x_j} f \right) \tag{4.19}$$

Integration removes the last term. To transform the middle term, first, with (4.12), we write

$$\frac{\partial U}{\partial x_i} = \int \varepsilon \frac{\partial f}{\partial x_i} \frac{2\,d\mathbf{p}}{h^3} \tag{4.20}$$

It follows that

$$\int f \frac{\partial \varepsilon}{\partial x_i} \frac{2\,d\mathbf{p}}{h^3} = \frac{\partial}{\partial x_i}[n\langle \varepsilon \rangle - U] \tag{4.21}$$

where we have set

$$n\langle \varepsilon \rangle = \int \varepsilon f \frac{2\,d\mathbf{p}}{h^3} \tag{4.22}$$

The relation (4.21) permits the middle term in (4.11) to be recast and (4.18) becomes

$$\frac{\partial}{\partial t} n\langle p_i \rangle + \frac{\partial}{\partial x_j} \prod_{ij} = 0 \tag{4.23}$$

Here $\prod_{ij}$ represents the momentum flux (or stress) tensor

$$\prod_{ij} = n\langle p_i v_j \rangle + \delta_{ij}[n\langle \varepsilon \rangle - U] \tag{4.24}$$

Finally, we multiply (4.13) by ε and integrate. In like manner to (4.20), we write

$$\frac{\partial U}{\partial t} = \int \varepsilon \frac{\partial f}{\partial t} \frac{2\,d\mathbf{p}}{h^3} \tag{4.25}$$

There results

$$\frac{\partial U}{\partial t} + \nabla \cdot \mathbf{q} = 0 \tag{4.26}$$

where

$$\mathbf{q} = \int \varepsilon f \mathbf{v} \frac{2\,d\mathbf{p}}{h^3} \tag{4.27}$$

represents the energy flux vector.

In this manner, we obtain the classical-like[49] conservation equations relevant to a Fermi liquid:

$$\frac{\partial n}{\partial t} + \nabla \cdot n\langle \mathbf{v} \rangle = 0$$

$$\frac{\partial}{\partial t} n\langle \mathbf{p} \rangle + \nabla \cdot \overline{\prod} = 0 \tag{4.28}$$

---

[49] The similarity of macroscopic equations in classical and quantum physics is discussed in J. H. Irving and R. W. Zwanzig, *J. Chem. Phys. 19*, 1173 (1951).

$$\frac{\partial}{\partial t}U + \nabla \cdot \mathbf{q} = 0$$

In the quantum equilibrium state, all three flux vectors in (4.28) vanish.

Zero sound

We wish to apply the kinetic equation (4.13) to small perturbations away from equilibrium of a Fermi liquid at 0 K. If the frequency of the perturbation far exceeds the collision frequency of quasi-particles, then we may neglect the collision integral in (4.13). With this condition at hand, we obtain

$$\frac{\partial \delta f}{\partial t} + \frac{\partial \varepsilon_0}{\partial \mathbf{p}} \cdot \frac{\partial \delta f}{\partial \mathbf{x}} - \frac{\partial \delta \varepsilon \delta}{\partial \mathbf{x}} \cdot \frac{\partial f_0}{\partial \mathbf{p}} = 0 \qquad (4.29)$$

The quantitities δf and $\delta \varepsilon$ represent variations away from equilibrium, as in (4.9) and (4.10).

At $T = 0$ K, we have

$$\frac{\partial f_0}{\partial \mathbf{p}} = -\mathbf{n}\delta(p - p_F) = -\mathbf{v}\delta(\varepsilon - E_F) \qquad (4.30)$$

where $\mathbf{n}$ is a unit vector in the direction of $\mathbf{p}$. We seek a wave solution in the form

$$\delta f = \delta(\varepsilon - E_F)\alpha(\mathbf{n})e^{i(\mathbf{k}\cdot\mathbf{v}-\omega t)} \qquad (4.31)$$

where $\mathbf{n}$ is a unit vector in the direction of $\mathbf{p}$, and $\alpha(\mathbf{n})$, which is to be determined, represents displacement of the Fermi surface in the direction of $\mathbf{n}$. Combining the last three equations and recalling (4.11) for $\delta \varepsilon$, we find

$$(\omega - v_F \mathbf{n} \cdot \mathbf{k})\alpha(\mathbf{n}) = \mathbf{v} \cdot \mathbf{k} \int K(\mathbf{p}, \mathbf{p}')\alpha(\mathbf{n}')\delta(\varepsilon - E_F)\frac{2\,d\mathbf{p}'}{h^3}$$

where we have set

$$\mathbf{v}_F = \frac{\partial \varepsilon_0}{\partial \mathbf{p}}$$

Again transforming to surface area in $\mathbf{p}$-space [see equation following (3.39)],

$$d\mathbf{p} = \frac{dS_p\,d\varepsilon}{|\nabla_{\mathbf{p}}\varepsilon|} = \frac{dS_p\,d\varepsilon}{v} = \frac{p^2\,d\Omega\,d\varepsilon}{v}$$

permits the preceding equation to be written

$$(\omega - v_F \mathbf{n} \cdot \mathbf{k})\alpha(\mathbf{n}) = \mathbf{n} \cdot \mathbf{k}\frac{2p_F^2}{h^3}\int K(\bar\theta)\alpha(\mathbf{n}')\,d\Omega'$$

Here we have written $\bar\theta$ for the angle between $\mathbf{p}$ and $\mathbf{p}'$. With $\mathbf{k}$ taken as the polar axis and setting

$$s \equiv \frac{\omega}{kv_F}$$

gives

$$(s - \cos\theta)\alpha(\theta, \phi) = \cos\theta \int \bar{K}(\bar{\theta})\alpha(\phi', \phi')\frac{d\Omega'}{4\pi} \tag{4.32}$$

where we have set

$$\bar{K} \equiv \frac{2p_F^2}{h^3}4\pi K \tag{4.32a}$$

Let us examine the case that $\bar{K}$ is constant and equal to the value $\bar{K}_0$. We may then assume that the interaction integral in (4.32) is likewise independent of angle. This gives the solution

$$\alpha(\theta) = A\frac{\cos\theta}{s - \cos\theta} \tag{4.33}$$

where A is an arbitrary constant. The Fermi surface corresponding to this solution is elongated in the direction of wave propagation.

Substitution of the trial solution (4.33) into (4.32) gives the condition

$$\bar{K}_0 \int_0^\pi \frac{\cos\theta}{s - \cos\theta} \frac{2\pi\sin\theta\,d\theta}{4\pi} = 1$$

Integration gives

$$\frac{1}{2}s\ln\frac{s+1}{s-1} - 1 = \frac{1}{\bar{K}_0}$$

As s varies from 1 to infinity (corresponding to undamped waves), the left side of the last equation varies from infinity to zero. It follows that waves exist only for $\bar{K}_0 > 0$. In the limit of vanishing interactions, $\bar{K}_0 \to 0$ and s tends to unity corresponding to the zero sound speed $\omega/k = v_F$. For a moderately nonideal Fermi fluid, the zero sound speed is $v_F/\sqrt{3}$.[50]

5.4.3 H Theorem for Quasi-Classical Distribution

In this section we return to the quasi-classical Boltzmann equation (4.5) and present a derivation of the related H theorem. The appropriately generalized expression for entropy, $S = -k_B\mathcal{H}$, is given by[51,52]

$$\frac{S}{k_B} = \sum_i g_i[(f_i \pm 1)\ln(1 + f_i) - f_i\ln f_i] \tag{4.34}$$

In this expression, the sum is over all quantum states, g_i denotes of the ith state, and the $+$ sign is relevant to bosons and the $-$ sign is relevant to fermions.

[50] Yu. L. Klimontovich and V. P. Silin, *Zh. Eksper. Teor. Fiz.* **23**, 151 (1952).

[51] K. Huang, *Statistical Mechanics*, p. 196, Wiley, New York (1967).

[52] A. Akhiezer and S. Peletminskii, *Methods of Statistical Physics*, p. 134, Pergamon, Elmsford, N.Y. (1981).

In the quasi-classical limit, the sum over discrete quantum states reduces to an integral accordng to the rule

$$\sum_i g_i \rightarrow \frac{2V}{h^3} \int d\mathbf{p} \qquad (4.35)$$

Here we are assuming a spatially homogeneous system confined to the volume V. Employing the preceding relation, (4.34) becomes

$$\frac{S}{k_B} = \frac{2V}{h^3} \int d\mathbf{p}[(f \pm 1)\ln(1 \pm f) - f \ln f] \qquad (4.36)$$

With this expression at hand, we recall the generalized Boltzmann equation (4.5) and the relation (4.1):

$$\frac{DF}{Dt} = \hat{J}(f) \qquad (4.37)$$

$$\hat{J}(f) \equiv \int \frac{\sigma \, d\Omega g \, d\mathbf{p}_1}{h^3}[f'f_1'(1 \pm f)(1 \pm f_1) - ff_1(1 \pm f')(1 \pm f_1')] \quad (4.38)$$

The distribution functions in these equations are dimensionless.

Employing the relation

$$\frac{\partial}{\partial y}[\pm(a \pm f)\ln(a \pm f)] = [1 + \ln(a \pm f)]\frac{\partial f}{\partial y} \qquad (4.39)$$

where a is a constant, indicates that to obtain an equation of motion for S, as given by (4.36), we must multiply (4.37) from the left by

$$\frac{2V}{h^3} \int d\mathbf{p}\{-(1 + \ln f) + [1 + \ln(1 \pm f)]\}$$

There results

$$\frac{D}{Dt}\frac{S}{k_B} = \frac{2V}{h^3} \int d\mathbf{p} \ln \left(\frac{1 \pm f}{f}\right) \hat{J}(f) \qquad (4.40)$$

Recalling the $\hat{I}$ operator (3.3.2),

$$\hat{I}(\phi) \equiv \int \hat{J}(f)\phi(\mathbf{p}) \, d\mathbf{p} \qquad (4.41)$$

[with $\hat{J}(f)$ given by (4.38)] and repeating the symmetry operations (3.3.3 et seq.), again we find

$$\hat{I}(\phi) = \hat{I}(\phi_1) = -\hat{I}(\phi') = -\hat{I}(\phi_1') \qquad (4.42)$$

which gives

$$\hat{I}(\phi) = \frac{1}{4}[\hat{I}(\phi) + \hat{I}(\phi_1) - \hat{I}(\phi') - \hat{I}(\phi')]$$

$$\hat{I}(\phi) = \frac{1}{4}\hat{I}(\phi + \phi_1 - \phi' - \phi_1') \qquad (4.43)$$

Substituting this result into (4.40) gives

$$
\begin{aligned}
\frac{D}{Dt}\frac{S}{k_B} &= \frac{V}{2h^3}\int d\mathbf{p}\,\hat{J}(f)\left[\ln\left(\frac{1\pm f}{f}\right)+\ln\left(\frac{1\pm f_1}{f_1}\right)\right.\\
&\qquad\left.-\ln\left(\frac{1\pm f'}{f'}\right)-\ln\left(\frac{1\pm f_1'}{f_1'}\right)\right]\\
&= \frac{V}{2h^3}\int d\mathbf{p}\,\hat{J}(f)\left[\ln\left(\frac{1\pm f}{f}\right)\left(\frac{1\pm f_1}{f_1}\right)\left(\frac{f'}{1\pm f'}\right)\left(\frac{f_1'}{1\pm f_1'}\right)\right]
\end{aligned}
$$
(4.44)

Substituting (4.38) for $\hat{J}(f)$ and labeling

$$
\frac{d\mathbf{p}\,d\mathbf{p}_1\sigma\,d\Omega g}{h^3}\equiv d\mu
$$

we obtain

$$
\begin{aligned}
\frac{D}{Dt}\frac{S}{k_B} &= \frac{V}{2h^3}\int d\mu[f'f_1'(1\pm f)(1\pm f_1)-ff_1(1\pm f')(1\pm f_1')]\\
&\qquad-\ln\left[\left(\frac{1\pm f}{f}\right)\left(\frac{1\pm f_1}{f_1}\right)\left(\frac{f'}{1\pm f'}\right)\left(\frac{f_1'}{1\pm f_1'}\right)\right]
\end{aligned}
$$
(4.45)

Setting

$$
\begin{aligned}
f'f_1'(1\pm f)(1\pm f_1) &\equiv X\\
ff_1(1\pm f')(1\pm f_1') &\equiv Y
\end{aligned}
$$

permits (4.45) to be written as

$$
\frac{D}{Dt}\frac{S}{k_B}=\frac{V}{2h^3}\int d\mu(X-Y)\ln\frac{X}{Y}\geq 0
$$
(4.46)

which establishes the $\mathcal{H}$ theorem in the quasi-classical domain. Note also that the implied equilibrium solution, $X=Y$, returns (4.7), which gives the correct quantum equilibrium distributions (4.8).

5.5   Overview of Classical and Quantum Hierarchies[53]

5.5.1   Second Quantization and Fock Space

Consider the wave function relevant to N identical particles. In accord with the Pauli principle, the appropriately symmetrized wave function is given by

$$
|\mathbf{x}_1,\ldots,\mathbf{x}_N;N\rangle=\frac{1}{N!}\sum_{P(1,\ldots,N)}(\pm)^P|\mathbf{x}_1,\ldots,\mathbf{x}_N\rangle
$$
(5.1)

---

[53] An expanded presentation of these topics is given by G. K. Schenter, *Thesis*, Cornell University, Ithaca, N.Y. (1988). For further discussion, see L. E. Reichl, *A Modern Course in Statistical Physics*, University of Texas Press, Austin (1980).

The ket vector $|\mathbf{x}_1, \ldots, \mathbf{x}_N\rangle$ on the right side of (5.1) is of arbitrary symmetry. The sum is over $N!$ permutations of $(1, \ldots, N)$. The $+$ and $-$ signs are appropriate to bosons and fermions, respectively.

Let the wave function relevant to n_x particles at $\mathbf{x}$, n'_x particles at $\mathbf{x}'$, and so on, be written $|\ldots, n_\mathbf{x}, \ldots\rangle_F$. The subscript F denotes that this wave function exists in Fock space. This is a space whose independent variables are occupation numbers. The relation of this wave function to that given in (5.1) is

$$|\mathbf{x}_1, \ldots, \mathbf{x}_N; N\rangle = \sqrt{\frac{\prod_\mathbf{x} n_\mathbf{x}!}{N!}} |\ldots, n_\mathbf{x}, \ldots\rangle_F$$

$$N = \sum_\mathbf{x} n_\mathbf{x} \tag{5.2}$$

The only occupation numbers in the right ket vector of (5.2) which are nonzero are those with $\mathbf{x}$-values equal to those in the left ket vector. The wave function (5.2) may be generated by the field creation operator $\hat{\phi}^\dagger(\mathbf{x})$. Thus

$$|\mathbf{x}_1, \ldots, \mathbf{x}_N; N\rangle = \frac{1}{\sqrt{N!}} \hat{\phi}^\dagger(\mathbf{x}_N) \ldots, \hat{\phi}^\dagger(\mathbf{x}_1)|0\rangle_F \tag{5.3}$$

where $|0\rangle_F$ denotes the vacuum state. Note in particular that

$$\hat{\phi}^\dagger(\mathbf{x})|\ldots, n_\mathbf{x}, \ldots\rangle_F = \sqrt{1 \pm n_\mathbf{x}}|\ldots, (1 \pm n_\mathbf{x}), \ldots\rangle_F$$

$$\hat{\phi}(\mathbf{x})|\ldots, n_\mathbf{x}, \ldots\rangle_F = \sqrt{n_\mathbf{x}}|\ldots, (\mp 1 \pm n_\mathbf{x}), \ldots\rangle_F \tag{5.4}$$

The operator $\hat{\phi}^\dagger(\mathbf{x})$ creates a particle at $\mathbf{x}$, whereas $\phi(\mathbf{x})$ annihilates a particle at $\mathbf{x}$. These operators are appropriate to the representation called *second quantization*.[54] For fermions, $n_\mathbf{x} = 0$ or 1, whereas for bosons, $n_\mathbf{x} = 0, 1, 2, \ldots$

Commutation relations of field operators are given by

$$[\hat{\phi}(\mathbf{x})\hat{\phi}^\dagger(\mathbf{y}) \mp \hat{\phi}^\dagger(\mathbf{y})\hat{\phi}(\mathbf{x})] \equiv [\hat{\phi}(\mathbf{x}), \hat{\phi}^\dagger(\mathbf{y})]_\mp = \delta(\mathbf{x} - \mathbf{y}) \tag{5.5a}$$

where the upper sign corresponds to bosons and the lower sign to fermions. The wave functions of (5.4) carry the following normalizations:

$$\langle \mathbf{x}_1, \ldots, \mathbf{x}_N | \mathbf{x}'_1 \cdots \mathbf{x}'_N \rangle = \delta(\mathbf{x}_1 - \mathbf{x}'_1) \cdots \delta(\mathbf{x}_N - \mathbf{x}'_N) \tag{5.5b}$$

$$\langle \cdots n_\mathbf{x} \cdots | \cdots n'_\mathbf{x} \cdots \rangle_F = \cdots \delta_{n_\mathbf{x}, n'_\mathbf{x}} \cdots \tag{5.5c}$$

5.5.2 Classical and Quantum Distribution Functions

To this point in our discussion, we have encountered: (1) the classical distribution function, $f_N(\mathbf{x}^N, \mathbf{p}^N, t)$, (2) the quantum mechanical density

---

[54] An overview of quantum kinetic theory in second quantization is given in T.-J Lie and R. L. Liboff, *Annals of Physics* (1971).

matrix

$$\rho_N(\mathbf{x}^N, \mathbf{y}^N, t) = \langle \mathbf{x}^N | \hat{\rho} | \mathbf{y}^N \rangle \tag{5.6}$$

and (3) the quantum mechanical Wigner distribution, $F_N(\mathbf{x}^N, \mathbf{p}^N, t)$.

With the preceding description of second quantization operators $\hat{\phi}^\dagger(\mathbf{x})$ and $\hat{\phi}(\mathbf{x})$, we may introduce a fourth distribution-like form: (4) the N-particle F-space operator $\hat{G}_N(\mathbf{x}^N, \mathbf{y}^N, t)$, where F-space denotes Fock space. It is given by[55]

$$\hat{G}_N(\mathbf{x}^N, \mathbf{y}^N, t) = \hat{\phi}_H^\dagger(\mathbf{y}_1) \cdots \hat{\phi}_H^\dagger(\mathbf{y}_N) \hat{\phi}_H(\mathbf{x}_N) \cdots \hat{\phi}_H(\mathbf{x}_1) \tag{5.7}$$

where $\hat{\phi}_H(\mathbf{x})$ has been written for

$$\hat{\phi}_H(\mathbf{x}, t) = e^{(i\hat{H}t)/\hbar} \hat{\phi}(\mathbf{x}) e^{-(i\hat{H}t)/\hbar} \tag{5.7a}$$

which is the Heisenberg representation of $\hat{\phi}(\mathbf{x})$ and $\hat{H}$ is the corresponding N-body Hamiltonian in Fock space.

The equation of motion for $\hat{\phi}_H(\mathbf{x})$ is given by the Heisenberg equation of motion (1.44):[56]

$$i\hbar \frac{\partial \hat{\phi}_H}{\partial t} = [\hat{\phi}_H, \hat{H}] \tag{5.8}$$

It should be noted that, whereas the distributions f_N, ρ_n, F_N are defined with respect to a given number of N particles, $\hat{G}_N$ is defined for unbounded N.

To obtain a relation between the F-space operator $\hat{G}_s(\mathbf{x}^s, \mathbf{y}^s, t)$ and the density matrix $\rho_s(\mathbf{x}^s, \mathbf{y}^s, t)$, we first note that reduced density operators obey the relation

$$\rho_s = \mathrm{Tr}^{(N-s)} \hat{\rho}_N \tag{5.9}$$

where $\mathrm{Tr}^{(N-s)}$ denotes a trace over the $(N - s)$ dimensions $(s + 1, \ldots, N)$. Thus, for example in the coordinate representation,

$$\rho_s(\mathbf{x}^s, \mathbf{y}^s) = \int d\mathbf{x}_{s+1}\, d\mathbf{y}_{s+1} \cdots d\mathbf{x}_N\, d\mathbf{y}_N \delta(\mathbf{x}_{s+1} - \mathbf{y}_{s+1}) \cdots \delta(\mathbf{x}_N - \mathbf{y}_N) \rho_N(\mathbf{x}^N, \mathbf{y}^N)$$

The presence of the delta functions in this relation indicates that the integral is a trace operation.

At this point we introduce an s-particle observable equivalent operator. First we recall that the N-body potential is written

$$V_N = \sum_{i<j}^{N} \sum u_{ij}(|\mathbf{x}_i - \mathbf{x}_j|)$$

---

[55]This operator is closely allied to the N-body Green's function described in Section 5.7.

[56]Note that in (5.8), the partial time derivative is used to distinguish between $\mathbf{x}$ and t differentiation. It is not the same as that occurring on the right side of (1.44).

where u_{ij} are two-body potentials. Motivated by this form we introduce the s-particle operator

$$\hat{A}^{(s)} = \sum_{i_l < \cdots < i_s} \cdots \sum^{N} \hat{A}_{i_l \cdots i_s} \tag{5.10}$$

The time-dependent expectation of $\hat{A}^{(s)}$ is given by

$$\begin{aligned} \langle A^{(s)} \rangle_t &= \mathrm{Tr}^{(N)}[\hat{\rho}_N(t)\hat{A}^{(s)}] \\ &= \binom{N}{s} \mathrm{Tr}^{(s)} \hat{\rho}_s \hat{A}_{1\cdots s} \\ &= \binom{N}{s} \int d\mathbf{x}^s \, d\mathbf{y}^s \, \langle \mathbf{y}^s | \hat{A}_{1\cdots s} | \mathbf{x}^s \rangle \rho_s(\mathbf{x}^s, \mathbf{y}^s, t) \end{aligned} \tag{5.11}$$

Here we have taken advantage of the symmetry of the system so that $\hat{A}_{1\cdot s}$ operates only on the first s particles of the N-particle sample.

In Fock space, the analogy of (5.10) is given by

$$\hat{A}^{(s)} = \frac{1}{s!} \int d\mathbf{x}^s \, d\mathbf{y}^s \hat{\bar{G}}_s(\mathbf{x}^s, \mathbf{y}^s, 0) \langle \mathbf{y}^s | \hat{A}_{1\cdots s} | \mathbf{x}^s \rangle \tag{5.12}$$

Stemming from the first equality in (5.11), we write

$$\begin{aligned} \langle \hat{A}^{(s)} \rangle_t &= \mathrm{Tr}^{(N)}[\hat{\rho}_N(t)\hat{A}^{(s)}] \\ &= \mathrm{Tr}^{(N)}[\hat{\rho}_N(0)e^{i\hat{H}_N t/\hbar} \hat{A}^{(s)} e^{-i\hat{H}_N t/\hbar}] = \mathrm{Tr}^{(N)}[\hat{\rho}_N(0)\hat{A}_H^{(s)}] \end{aligned} \tag{5.13}$$

Here we have recalled that $\mathrm{Tr}[\hat{A}\hat{B}] = \mathrm{Tr}[\hat{B}\hat{A}]$. Inserting (5.12) into (5.13) and recalling (5.7) gives

$$\langle \hat{A}^{(s)} \rangle_t = \frac{1}{s!} \int d\mathbf{x}^s \, d\mathbf{y}^s \, \mathrm{Tr}^{(N)}[\hat{\rho}_N(0)\hat{\bar{G}}_s(\mathbf{x}^s, \mathbf{y}^s, t)]\langle \mathbf{y}^s | \hat{A}_{1\cdots s} | \mathbf{x}^s \rangle \tag{5.14}$$

Comparison of this latter result with (5.11) indicates that

$$\rho_s(\mathbf{x}^s, \mathbf{y}^s, t) = \frac{(N-s)!}{N!} \mathrm{Tr}^{(N)}[\hat{\rho}_N(0)\hat{\bar{G}}(\mathbf{x}^s, \mathbf{y}^s, t)] \tag{5.15}$$

which is the desired relation between the density matrix and $\hat{\bar{G}}_s$. Note, in particular, that $\hat{\bar{G}}_s$ determines the density matrix ρ_s, but the converse is not true.

As an elementary example, consider the problem of obtaining $\rho_1(\mathbf{x}, \mathbf{y}, 0)$ for a system of N particles in the pure state $|\ldots, n_\mathbf{x}, \ldots\rangle$. Writing $\hat{\rho}_N(0)$ in the projection representation (2.37), we obtain

$$\begin{aligned} \rho_1(\mathbf{x}, \mathbf{y}, 0) &= \frac{1}{N} \mathrm{Tr}^{(N)}[\hat{\rho}_N(0)\hat{\bar{G}}_1(\mathbf{x}, \mathbf{y}, 0)] \\ &= \frac{1}{N} \mathrm{Tr}^{(N)}[| \ldots, n_\mathbf{x}, \ldots\rangle\langle \ldots, n_\mathbf{x}, \ldots | \hat{\phi}_H^\dagger(\mathbf{y})\hat{\phi}_H(\mathbf{x})] \end{aligned}$$

$$= \frac{1}{N}[\langle \ldots, n_\mathbf{x}, \ldots | \hat{\phi}_H^\dagger(\mathbf{y}) \hat{\phi}_H(\mathbf{x}) | \ldots, n_\mathbf{x}, \ldots \rangle] \qquad (5.16)$$

A nonzero result occurs only if $\mathbf{x} = \mathbf{y}$. Thus we find

$$\rho_1(\mathbf{x}, \mathbf{y}) = \frac{n_\mathbf{x}}{N} \delta_{\mathbf{x}, \mathbf{y}}$$

This result has the proper normalization

$$\mathrm{Tr}\, \rho_1(\mathbf{x}, \mathbf{y}) = \mathrm{Tr}\left[\frac{n_\mathbf{x}}{N} \delta_{\mathbf{x}, \mathbf{y}} \right] = \sum_\mathbf{x} \frac{n_\mathbf{x}}{N} = 1$$

As a second example of this formalism, let us apply (5.14) to evaluate the average total two-body interaction potential

$$\hat{V} = \sum_{i>j}^{N} \sum \hat{u}(\hat{\mathbf{q}}_i - \hat{\mathbf{q}}_j)$$

Here $\hat{\mathbf{q}}_i$ represents the coordinate operator. Thus we must evaluate

$$\langle \mathbf{y}_1 \mathbf{y}_2 | \hat{u}(\hat{\mathbf{q}}_1 - \hat{\mathbf{q}}_2) | \mathbf{x}_1 \mathbf{x}_2 \rangle = \delta(\mathbf{y}_1 - \mathbf{x}_1)\delta(\mathbf{y}_2 - \mathbf{x}_2) u(\mathbf{x}_1 - \mathbf{x}_2)$$

Substituting this result into (5.14), and with reference to (5.7), we find

$$\langle V \rangle_t = \frac{1}{2!} \int \int \int \int dx_1\, dx_2\, dy_1\, dy_2\, \mathrm{Tr}^{(N)}$$
$$[\hat{\rho}_N(0)\hat{\phi}_H^\dagger(\mathbf{y}_1, t)\hat{\phi}_H^\dagger(\mathbf{y}_2, t)\hat{\phi}_H(\mathbf{x}_2, t)\hat{\phi}_H(\mathbf{x}_1, t)]$$
$$\times \delta(\mathbf{y}_1 - \mathbf{x}_1)\delta(\mathbf{y}_2 - \mathbf{x}_2) u(\mathbf{x}_1 - \mathbf{x}_2) \qquad (5.17)$$

Again, as in (5.16), let $\hat{\rho}_N(0)$ be relevant to a pure state. Substituting this form into (5.17) gives

$$\langle V \rangle = \frac{1}{2} dx_1\, dx_2 \langle \cdots, n_\mathbf{x}, \ldots |$$
$$\hat{\phi}_H^\dagger(\mathbf{x}_1, t)\hat{\phi}_H^\dagger(\mathbf{x}_2, t) u(\mathbf{x}_1 - \mathbf{x}_2)\hat{\phi}_H(\mathbf{x}_2, t)\hat{\phi}_H(\mathbf{x}_1, t) | \ldots, n_\mathbf{x}, \ldots \rangle (5.18)$$

This integral may be further reduced by setting $t = 0$. With (5.4) there results

$$\langle V \rangle = \frac{1}{2} \int \int dx_1\, dx_2\, n_{\mathbf{x}_1} n_{\mathbf{x}_2} u(\mathbf{x}_1 - \mathbf{x}_2) \qquad (5.19)$$

which has a self-evident interpretation. Had we not worked in a pure state, an additional probability factor would have appeared in the preceding integral [see (2.36)], together with a sum over $n_{\mathbf{x}_1}$ and $n_{\mathbf{x}_2}$.

Let us consider calculation of $\langle H \rangle_t$. With (5.13) we write

$$\langle H \rangle_t = \mathrm{Tr}[\hat{\rho}(t)\hat{H}] = \mathrm{Tr}[\hat{\rho}(0)e^{i\hat{H}t/\hbar}\hat{H}e^{-i\hat{H}t/\hbar}] = \mathrm{Tr}[\hat{\rho}(0)\hat{H}] = \langle H \rangle_0$$

Thus we find that $\langle H \rangle$ is constant in time.

5.5.3   Equations of Motion

We recall that the density operator and classical distribution function satisfy similar equations, the Liouville equation (1.4.7) and its quantum analog (2.11), [see (1.44)]

$$\frac{\partial f_N}{\partial t} = -[f_N, H] \tag{5.20}$$

$$i\hbar \frac{\partial \hat{\rho}_N}{\partial t} = -[\hat{\rho}_N, \hat{H}] \tag{5.21}$$

The classical Hamiltonian for our system is given by

$$H(1, \ldots, N) = \sum_{i=1}^{N} \frac{p_i^2}{2m} + \sum_{i<j}^{N} \sum u_{ij}(|\mathbf{x}_i - \mathbf{x}_j) \equiv K_N + V_N \tag{5.22}$$

where K_N and V_n are respective kinetic and potential energy terms. Substituting the quantum analog of (5.22) into (5.21) and forming matrix elements gives

$$\frac{\partial \rho_N(\mathbf{x}^N, \mathbf{y}^N)}{\partial t} = \frac{1}{i\hbar}\{[\hat{K}_N(\mathbf{x}^N) - \hat{K}_N^*(\mathbf{y}^N)] + [\hat{V}_N(\mathbf{x}^N) - \hat{V}_N^*(\mathbf{y}^N)]\}\rho_N(\mathbf{x}^N, \mathbf{y}^N) \tag{5.23}$$

The operators in (5.23) are give by

$$\hat{K}_N(\mathbf{x}^N) = \sum_{i=1}^{N} \hat{K}_i(\mathbf{x}_i)$$

$$\hat{K}_i(\mathbf{x}_i) = -\frac{\hbar^2}{2m}\nabla^2 \mathbf{x}_i \tag{5.24}$$

$$\hat{V}_N(\mathbf{x}^N) = \sum_{i<j} \sum u_{ij}(|\mathbf{x}_i - \mathbf{x}_j|)$$

The equation of motion for the Wigner distribution, F_N, is given by the Wigner-Moyal equation (2.70):

$$\frac{\partial F_N}{\partial t} = -\frac{\mathbf{p}^N}{m} \cdot \frac{\partial F_N}{\partial \mathbf{x}^N} + \frac{2}{\hbar}\sin\left(\frac{\hbar}{2}\frac{\partial}{\partial \mathbf{p}^N} \cdot \frac{\partial}{\partial \mathbf{x}^N}\right) V_N(\mathbf{x}^N)F_N \tag{5.25}$$

As noted previously, momentum and space derivatives in the sine function operate, respectively, on F_N and V_N.

To obtain the representation of the Hamiltonian (5.22) relevant to Fock space, we note the following second quantization rules: With $\hat{F}(\mathbf{x})$ and $L(\mathbf{x}, \mathbf{v})$ representing one- and two-particle functions, respectively, we write

$$\sum_{i=1}^{N} \hat{F}(\mathbf{x}_i) \rightarrow \int d\mathbf{x}\hat{\phi}_H^+(\mathbf{x})\hat{F}(\mathbf{x})\hat{\phi}_H(\mathbf{x})$$

$$\sum_{i<j}^{N} \sum L(\mathbf{x}_i, \mathbf{x}_j) \rightarrow \frac{1}{2}\int\int d\mathbf{x}\,d\mathbf{y}\hat{\phi}_H^+(\mathbf{x})\hat{\phi}_H^+(\mathbf{y})L(\mathbf{x}, \mathbf{y})\hat{\phi}_H(\mathbf{y})\hat{\phi}_H(\mathbf{x}) \tag{5.26}$$

where $\hat{\phi}_H(\mathbf{x})$ is the time-dependent operator (5.7a). There results

$$\hat{H} = \int \hat{\phi}_H^\dagger(\mathbf{x}) \hat{K}(\mathbf{x}) \hat{\phi}_H(\mathbf{x}) \, d\mathbf{x} + \frac{1}{2} \int \int \hat{\phi}_H^\dagger(\mathbf{x}) \hat{\phi}_H^\dagger(\mathbf{y}) u(\mathbf{x}, \mathbf{y}) \hat{\phi}_H(\mathbf{y}) \hat{\phi}_H(\mathbf{x}) \, d\mathbf{x} \, d\mathbf{y}$$
(5.27)

As an elementary example of the workings of these operators, consider that the double integral potential-energy operator in (5.27), which we label hQ, operates on the ket vector

$$|0 \ldots, 1_\mathbf{x}', 0 \ldots, 1\mathbf{y}', 0 \cdots\rangle \equiv |1_\mathbf{x}', 1_\mathbf{y}'\rangle$$

There results

$$\hat{Q}|1_\mathbf{x}', 1_\mathbf{y}'\rangle = \left[\frac{1}{2} u(\mathbf{x}', \mathbf{y}') + \frac{1}{2} u(\mathbf{y}', \mathbf{x}')\right]|1_\mathbf{x}', 1_\mathbf{y}'\rangle$$
$$= u(\mathbf{x}', \mathbf{y}')|1_\mathbf{x}', 1_\mathbf{y}'\rangle$$

We turn next to construction of equations of motion for $\hat{\phi}_H(\mathbf{x})$ and $\hat{\bar{G}}_N$. With the Hamiltonian (5.27) at hand, we employ (5.8) first to obtain an equation of motion for $\hat{\phi}_H(\mathbf{x})$. With the commutation relations (5.5), there results (see Problem 5.46)

$$i\hbar \, \hat{\phi}_H(\mathbf{x}) = \hat{K}(\mathbf{x}) \hat{\phi}_H(\mathbf{x}) + \int \hat{\phi}_H^\dagger(\mathbf{y}) u(\mathbf{x}, \mathbf{y}) \hat{\phi}_H(\mathbf{y}) \hat{\phi}_H(\mathbf{x}) \, d\mathbf{y} \qquad (5.28)$$

Employing this equation and the defining relation (5.7) gives the following equation of motion for $\hat{\bar{G}}_N(\mathbf{x}^N, \mathbf{y}^N, t)$ (see Problem 5.47):

$$\frac{\partial \hat{\bar{G}}_N}{\partial t} = \frac{1}{i\hbar}\{[\hat{K}_N(\mathbf{x}^N) - \hat{K}_N^*(\mathbf{y}^N)] + [\hat{V}_N(\mathbf{x}^N) - \hat{V}_N^*(\mathbf{y}^N)]\}\hat{\bar{G}}_N \qquad (5.29)$$

Whereas this equation is equivalent in form to (5.23), we should bear in mind that (5.29) is relevant to Fock space, where, as noted previously, N is unbounded.

5.5.4 Generalized Hierarchies

In the following, we write $\mathbf{z}_i$ for the single-particle state $(\mathbf{x}_i, \mathbf{p}_i)$. Thus

$$D_N = D_N(\mathbf{z}_1, \mathbf{z}_2, \ldots, \mathbf{z}_N) \equiv D_N(\mathbf{x}^N)$$

where D_N is written for any of the preceding distributions. Reduced distributions are given by

$$D_s(\mathbf{z}^s) = \int d\mu_{s+1} \cdots d\mu_N D_N(\mathbf{z}^N) \qquad (5.29a)$$

where the integration measure $d\mu$ is specific to the distribution at hand. In the following, we assume that $D_N(\mathbf{z}^N)$, as well as the Hamiltonian $H(\mathbf{z}^N)$, is symmetric with respect to $\mathbf{z}_i$, $\mathbf{z}_j$ interchange. Furthermore, we take

$$\int [D_N, \hat{K}_i] d\mu_i = 0 \tag{5.30a}$$

$$\int [D_N, \mu_{ij}] d\mu_i \, d\mu_j = 0 \tag{5.30b}$$

With the preceding relations at hand, hierarchies stemming from the Liouville equation (5.20), the density matrix equation (5.23), the Wigner–Moyal equation (5.25), and the equation for the Fock space operator (5.29) are obtained. We list first the hierarchy for the classical distributions:

$$\frac{\partial f_s}{\partial t} + \sum_{i=1}^{s} [f_s, K_i] + \sum_{i<j}^{x} [f_s, u_{ij}] + (N-s) \sum_{i=1}^{s} [f_{s+1}, u_{i,s+1}] \, dz_{s+1} = 0 \tag{5.31}$$

[compare with (2.1.20)]. This equation poses as a generic form for quantum heirarchy equations as well. Thus, with D_s representing any of the distribution functions considered, we write

$$\frac{\partial f_s}{\partial t} + \sum_{i=1}^{s} \beta_i D_s + \sum_{i<j}^{x} \gamma_{ij} D_s + \sum_{i=1}^{s} (N-s) \int [D_{s+1}, u_{i,s+1}] \, d\mu_{s+1} = 0 \tag{5.32}$$

The α_s, β_i, γ_{ij}, and $d\mu_{s+1}$ terms are listed in Table 5.3 relevant to the specific case at hand. Here are some comments concerning the various hierarchies in Table 5.3.

In the equation for $F_s(\mathbf{x}^s, \mathbf{p}^s)$, first note that

$$\left(\frac{\partial}{\partial \mathbf{p}_i} \cdot \frac{\partial}{\partial \mathbf{x}_i} + \frac{\partial}{\partial \mathbf{p}_j} \cdot \frac{\partial}{\partial \mathbf{x}_j} \right) u_{ij} = \left(\frac{\partial}{\partial \mathbf{p}_i} - \frac{\partial}{\partial \mathbf{x}_i} \right) \cdot \frac{\partial}{\partial \mathbf{x}_i} u_{ij}$$

This relation gives the correct correspondence between case (a) and (c) in the classical limit, $\hbar \rightarrow 0$. As in the example following (5.9), the delta functions in $d\mu_{s+1}$ stem from the trace operation in (5.9).

For a specific example from Table 5.3, we consider the hierarchy equations for the Wigner distribution. It appears as

$$\frac{\partial F_s}{\partial t} + \sum_{i=1}^{s} \frac{\mathbf{p}_i}{m} \cdot \frac{\partial F_s}{\partial \mathbf{x}_i} - \sum_{i<j}^{x} \frac{2}{\hbar} \sin \left[\frac{\hbar}{2} \left(\frac{\partial}{\partial \mathbf{p}_i} \cdot \frac{\partial}{\partial \mathbf{x}_i} + \frac{\partial}{\partial \mathbf{p}_j} \cdot \frac{\partial}{\partial \mathbf{x}_j} \right) \right]$$

$$\times u_{ij}(|\mathbf{x}_i - \mathbf{x}_j|) F_s(\mathbf{x}^s, \mathbf{p}^s) + \sum_{i=1}^{s} (N-s) \int [F_{s+1}, u_{i,s+1}] \, d\mathbf{x}_{s+1} \, d\mathbf{p}_{s+1} \tag{5.33}$$

As in (2.71), in the double sum in (5.33), the $\partial/\partial \mathbf{x}$ operators in the sine function operate only on u_{ij}.

In obtaining the hierarchy for $\hat{\tilde{G}}_s$, the equation of motion (5.29) was applied to $\hat{\tilde{G}}_s$ as given by (5.7).

TABLE 5.3. Elements of Classical and Quantum Hierarchies with Reference to the Generic Equation (5.32)

Case	D_s	β_i	γ_{ij}	$d\mu_{s+1}$
a	$f_s(\mathbf{x}^s, \mathbf{p}^s)$	$\dfrac{\mathbf{p}_i}{m} \cdot \dfrac{\partial}{\partial \mathbf{x}_i}$	$-\dfrac{\partial u_{ij}}{\partial \mathbf{x}_i} \cdot \left(\dfrac{\partial}{\partial \mathbf{p}_i} - \dfrac{\partial}{\partial \mathbf{p}_j} \right)$	$d\mathbf{x}_{s+1} d\mathbf{p}_{s+1}$
b	$\rho_s(\mathbf{x}^s, \mathbf{y}^s)$	$-\dfrac{1}{i\hbar}[\hat{K}_i(\mathbf{x}_i) - \hat{K}_i^*(\mathbf{y}_i)]$	$[\hat{u}_{ij}(\mathbf{x}_i, \mathbf{x}_j) - \hat{u}_{ij}^*(\mathbf{y}_i, \mathbf{y}_j)]$	$d\mathbf{x}_{x+1} d\mathbf{y}_{x+1} \delta(\mathbf{x}_{s+1} - \mathbf{y}_{s+1})$
c	$F_s(\mathbf{x}^s, \mathbf{p}^s)$	$\dfrac{\mathbf{p}_i}{m} \cdot \dfrac{\partial}{\partial \mathbf{x}_i}$	$-\dfrac{2}{\hbar} \sin\left[\dfrac{\hbar}{2} \left(\dfrac{\partial}{\partial \mathbf{p}_i} \cdot \dfrac{\partial}{\partial \mathbf{x}_i} + \dfrac{\partial}{\partial \mathbf{p}_j} \cdot \dfrac{\partial}{\partial \mathbf{x}_j} \right) \right] u_{ij}$	$d\mathbf{x}_{s+1} d\mathbf{p}_{s+1}$
d	$\hat{G}_s(\mathbf{x}^s, \mathbf{y}^s)$	$-\dfrac{1}{i\hbar}[\hat{K}_i(\mathbf{x}_i) - \hat{K}_i^*(\mathbf{y}_i)]$	$[\hat{u}_{ij}(\mathbf{x}_i, \mathbf{x}_j) - \hat{u}_{ij}^*(\mathbf{y}_i, \mathbf{y}_j)]$	$\dfrac{1}{N-s} d\mathbf{x}_{x+1} d\mathbf{y}_{x+1} \delta(\mathbf{x}_{s+1} - \mathbf{y}_{s+1})$

(a) Classical distribution; (b) density matrix; (c) Wigner distribution; (d) s-particle F-space operator

As noted above, the hierarchies for ρ_s and $\hat{\bar{G}}_s$ are nearly identical in form. However, as seen from the generic relation (5.32), at $s = N$ a closed equation for ρ_N results. This is not true for the Fock-space operator $\hat{\bar{G}}_s$. The particle number N is unbounded in this case, as are the number of equations in the hierarchy for this operator.

5.6   Kubo Formula Revisited

Having discussed second quantization in Section 5.1, we return to Kubo's formula (3.4.67) and develop a quantum expression for electrical conductivity.

5.6.1   Charge Density and Current

The Hamiltonian for a particle of charge e and mass m in an electromagnetic field is given by [recall (1.1.16)]

$$\hat{H} = \frac{1}{2m} \left[\hat{p} - \frac{e}{c}\mathbf{A}(\mathbf{x}, t) \right]^2 + e\Phi(\mathbf{x}, t) \tag{6.1}$$

Working in coordinate representation, together with the time-dependent Schrödinger equation, we obtain

$$\frac{\partial \rho}{\partial t} + \nabla \cdot \mathbf{J} = 0 \tag{6.2}$$

where

$$\rho = e\psi^*\psi \tag{6.3}$$

$$\mathbf{J} = \frac{e\hbar}{2mi}(\psi^*\nabla\psi - \psi\nabla\psi^*) - \frac{e^2}{mc}\mathbf{A}\psi^*\psi \tag{6.4}$$

We may view (6.1) as consisting of an unperturbed free-particle component $\hat{H}_0$ and perturbed component $\hat{V}$ so that

$$\hat{H} = \hat{H}_0 + \hat{V} \tag{6.5}$$

where

$$H_0 = \frac{p^2}{2m} \tag{6.5a}$$

$$\hat{V} = \frac{-e}{2mc}(\hat{p} \cdot \mathbf{A} + \mathbf{A} \cdot \hat{p}) + \frac{e^2}{2mc^2}A^2 + e\Phi \tag{6.5b}$$

In second quantized coordinate representation appropriate to Fock space, (6.5a,b) become

$$\hat{H}_0 = -\int \hat{\psi}^\dagger \frac{\hbar}{2m}\nabla^2\hat{\psi} \, d\mathbf{x} \tag{6.6a}$$

$$\hat{V} = \int \left\{ \frac{-e\hbar}{2mic} [\hat{\psi}^\dagger \nabla \cdot (\mathbf{A}\hat{\psi}) + \hat{\psi}^\dagger \mathbf{A} \cdot \nabla\hat{\psi}] + \frac{e^2}{2mc^2} A^2 \hat{\psi}^\dagger \hat{\psi} + e\Phi\hat{\psi}^\dagger\hat{\psi} \right\} d\mathbf{x}$$

$$(6.6b)$$

where

$$\hat{\psi}^\dagger = \hat{\psi}^\dagger(\mathbf{x}), \qquad \hat{\psi} = \hat{\psi}(\mathbf{x})$$

Recall that the wave functions for these operators obey the properties (5.4). Stemming from (6.4), we write

$$\hat{\mathbf{J}} = \hat{\mathbf{J}}^{(0)} - \frac{e^2}{mc} \mathbf{A}\hat{\psi}^\dagger\hat{\psi}$$

$$(6.7)$$

where

$$\hat{\mathbf{J}}^{(0)} = \frac{e\hbar}{2mi} [\hat{\psi}^\dagger(\nabla\hat{\psi}) - (\nabla\hat{\psi}^\dagger)\hat{\psi}]$$

$$(6.8)$$

This latter expression permits (6.6b) to be written

$$\hat{V} = \int \left(-\frac{1}{2}\hat{\mathbf{J}}^{(0)} \cdot \mathbf{A} + \frac{e^2}{2mc^2} A^2 \hat{\psi}^\dagger\hat{\psi} + e\Phi\hat{\psi} \right) d\mathbf{x}$$

$$(6.9)$$

5.6.2 Identifications for Kubo Formula

With these expressions at hand, we return to the Kubo formula and identify terms for the purpose of obtaining an expression for electrical conductivity. Employing (3.4.63) and (3.4.67) and assuming that $\langle B \rangle_0 \neq 0$, we write the quantum version of the Kubo formula:

$$\langle B \rangle - \langle B \rangle_0 = \int_0^t \frac{1}{i\hbar} \text{Tr}\{\hat{\rho}^{(0)}[\hat{H}, \hat{B}(t - t')]\} F(t')dt'$$

$$(6.10)$$

where $\hat{B}(t)$ is given by the quantum analog of (3.4.55),

$$\hat{B}(t) = \exp\left(\frac{-t}{i\hbar}[\hat{H}_0, \;\;] \right) \hat{B} \equiv \exp(-t\hat{S})\hat{B}$$

$$(6.11)$$

which serves to identify the superoperator $\hat{S}$.[57] The exponential commutator operator in (6.11) was defined previously by (1.4.20).

The following three relations allow (6.10) to be written in a more concise form. First we note that with $[\hat{\rho}^{(0)}, \hat{H}_0] = 0$ we may write

$$e^{-t\hat{S}}\hat{\rho}^{(0)} = \hat{\rho}^{(0)}$$

$$(6.12a)$$

The second relation is given by (see Problem 5.31)

$$\text{Tr}\{(e^{-t\hat{S}}\hat{A})(e^{-t\hat{S}}\hat{B}) = \text{Tr}\{\hat{A}\hat{B}\}$$

$$(6.12b)$$

---

[57]Superoperators operate on operators.

and the third by

$$e^{-t\hat{S}}[\hat{A}, \hat{B}] = [\hat{A}(t), \hat{B}(t)] \tag{6.12c}$$

Applying the latter three relations to (6.10), we obtain

$$\langle B \rangle - \langle B \rangle_0 = \int_0^t \frac{1}{i\hbar} \operatorname{Tr}\{\hat{\rho}^{(0)}[\hat{\tilde{H}}(t'), \hat{B}(t)]\} F(t')dt' \tag{6.13}$$

With the identifications

$$\hat{\tilde{H}}(t)F(t) \rightarrow 7 - haV(t)$$
$$\hat{B}(t) \rightarrow \hat{\mathbf{J}}(t) \tag{6.14}$$

(6.13) becomes

$$\langle \mathbf{J} \rangle - \langle \mathbf{J} \rangle_0 = \frac{1}{i\hbar} \int_0^t \operatorname{Tr}\{\rho^{(0)}[\hat{\mathbf{J}}(t), \hat{V}(t')]\}dt' \tag{6.15}$$

where $\langle \mathbf{L} \rangle_0$ represents current flow in equilibrium.

5.6.3 Electrical Conductivity

Substituting (6.7) through (6.9) for $\hat{\mathbf{J}}$ and $\hat{V}$ into the preceding expression, we find

$$\langle \mathbf{J} \rangle - \langle \mathbf{J} \rangle_0 = \frac{1}{i\hbar} \int_0^t dt' \operatorname{Tr}\left\{\hat{\rho}^{(0)}\left[\hat{\mathbf{J}}^{(0)}(\mathbf{x}, t) - \frac{e^2}{mc}\mathbf{A}(\mathbf{x}, t)\hat{\psi}^\dagger(\mathbf{x}, t)\hat{\psi}(\mathbf{x}, t)\right.\right.$$
$$\times \int d\mathbf{x}'\left(-\frac{1}{c}\hat{\mathbf{J}}^{(0)}(\mathbf{x}', t') \cdot \mathbf{A}(\mathbf{x}', t')\right.$$
$$+ \frac{e^2}{2mc}A^2(\mathbf{x}', t')\hat{\psi}^\dagger(\mathbf{x}', t')\hat{\psi}(\mathbf{x}', t')$$
$$\left.\left.\left. + e\Phi(\mathbf{x}', t')\hat{\psi}^\dagger(\mathbf{x}', t')\hat{\psi}(\mathbf{x}', t')\right)\right]\right\} \tag{6.16}$$

Working in the Coulomb gauge,

$$\nabla \cdot \mathbf{A} = 0, \qquad \Phi = 0 \tag{6.17a}$$

with

$$\mathcal{E} = -\frac{1}{c}\frac{\partial}{\partial t}\mathbf{A}, \qquad \mathbf{B} = \nabla \times \mathbf{A} \tag{6.17b}$$

and neglecting terms of $0(A^2)$, (6.16) reduces to

$$\langle \mathbf{J} \rangle - \langle \mathbf{J} \rangle_0 = \frac{i}{\hbar c} \int_0^t dt' \int d\mathbf{x}' \operatorname{Tr}\{\hat{\rho}^{(0)}[\hat{\mathbf{J}}^{(0)}(\mathbf{x}, t), \hat{J}_\beta^{(0)}(\mathbf{x}', t')]\} A_\beta(\mathbf{x}', t') \tag{6.18}$$

where subscript β represents Cartesian components and is summed from 1 to 3. We note that the time integral of the α component of (6.18) may be rewritten as

$$\langle J_\alpha \rangle - \langle J_\alpha \rangle_0 = \frac{i}{\hbar c} \int_{-\infty}^{\infty} dt' \int d\mathbf{x}' \theta(t - t') \operatorname{Tr}\{\cdots\}\theta(t) A_\beta(\mathbf{x}', t') \qquad (6.19)$$

where

$$\theta(t \geq 0) = 1 \quad \text{and} \quad \theta(t < 0) = 0$$

To bring (6.18) closer to the form of Ohm's law, we assume that

$$\frac{i}{\hbar c}\theta(t - t') \operatorname{Tr}\{\hat{\rho}^{(0)}[\hat{J}_\alpha^{(0)}(\mathbf{x}, t), \hat{J}_\beta^{(0)}(\mathbf{x}', t')]\} \equiv Q_{\alpha\beta}(\mathbf{x} - \mathbf{x}', t - t') \qquad (6.20a)$$

and that

$$\theta(t)\mathbf{A}(\mathbf{x}, t) = \mathbf{A}(\mathbf{x}, t) \qquad (6.20b)$$

Inserting these values into (6.19) gives

$$\langle J_\alpha \rangle(\mathbf{x}, t) - \langle J_\alpha \rangle(\mathbf{x}, t) = \int_{-\infty}^{\infty} dt' \int d\mathbf{x}' Q_{\alpha\beta}(\mathbf{x} - bx', \mathbf{t} - t') A_\beta(\mathbf{x}', t') \qquad (6.21)$$

We recognize this to be convolution integral, in which case the transform of (6.21) becomes

$$\langle J_\alpha \rangle(\omega, \mathbf{k}) = \langle J_\alpha \rangle_0(\omega, \mathbf{k}) + Q_{\alpha\beta}(\omega, \mathbf{k}) A_\beta(\omega, \mathbf{k}) \qquad (6.21a)$$

where

$$\mathbf{A}(\mathbf{x}, t) = \iint \frac{d\mathbf{k}\, d\omega}{(2\pi)^4} e^{-i\omega t + i\mathbf{k}\cdot\mathbf{x}} \mathbf{A}(\omega, \mathbf{k})$$

With (6.17b), we may write

$$\mathcal{E}(\omega, \mathbf{k}) = \frac{i\omega}{c}\mathbf{A}(\omega, \mathbf{k}) \qquad (6.22)$$

Substituting this transform into (6.21a) gives

$$\langle J_\alpha \rangle(\omega, \mathbf{k}) = \langle J_\alpha \rangle_0(\omega, \mathbf{k}) + \frac{c}{i\omega} Q_{\alpha\beta}(\omega, \mathbf{k}) \mathcal{E}_\beta(\omega, \mathbf{k}) \qquad (6.23)$$

The first term on the right may be further reduced as follows. Recalling (6.7), we write

$$\langle \mathbf{J} \rangle_0 = \operatorname{Tr}\{\hat{\rho}^{(0)}\hat{J}\} = \operatorname{Tr}\{\hat{\rho}_0 \hat{J}^{(0)}\} - \frac{e^2}{mc}\operatorname{Tr}\{\hat{\rho}_0 \mathbf{A}\hat{\psi}^\dagger \hat{\psi}\}$$

$$= \langle \mathbf{J}^{(0)} \rangle_0 - \frac{e^2}{mc}\operatorname{Tr}\{\hat{\rho}_0 \mathbf{A}\hat{\psi}^\dagger \hat{\psi}\} \qquad (6.24)$$

It is evident for (6.7) that $\langle J^0 \rangle_0$ is the equilibrium value of current with fields turned off. Without loss in generality, we may set this term equal to zero. Consider the remaining term in (6.24), which contains the form

$$\operatorname{Tr}\{\hat{\rho}_0 \mathbf{A}\hat{\psi}^\dagger \hat{\psi}\} \equiv n(\mathbf{x}, t)$$

where $n(\mathbf{x}, t)$ represents particle number density. The remaining term in (6.23) may then be written

$$\frac{e^2}{mc} n(\mathbf{x}, t)\mathbf{A}(\mathbf{x}; t)$$

The transform of this product is the convolution integral

$$\frac{e^2}{mc} \int\int \frac{d\mathbf{k}\, d\omega}{(2\pi)^4} n(\omega - \omega', \mathbf{k} - \mathbf{k}')\mathbf{A}(\omega', \mathbf{k}')$$

In the homogeneous limit

$$n(\omega - \omega', \mathbf{k} - \mathbf{k}') = (2\pi)^4 n_0 \delta(\omega - \omega')\delta(\mathbf{k} - \mathbf{k}')$$

Recalling (6.22), the preceding expression then reduces to

$$\frac{e^2}{mc} n_0 \mathbf{A}(\mathbf{k}, \omega) = \frac{e^2}{mic} n_0 \mathcal{E}(\omega, \mathbf{k}) \tag{6.25}$$

Substituting (6.24) in (6.23) with $\langle \mathbf{J}^{(0)}\rangle_0 = 0$ and the remaining term transferred as given by (6.25), we obtain

$$\langle J_\alpha\rangle(\omega, \mathbf{k}) = \left[\frac{c}{i\omega} Q_{\alpha\beta}(\omega, \mathbf{k}) - \delta_{\alpha\beta}\frac{e^2}{mic} n_0\right] \mathcal{E}_\beta(\omega, \mathbf{k}) \tag{6.26}$$

This relation is a tensor form of Ohm's law,

$$\langle \mathbf{J}\rangle = \bar{\bar{\sigma}}_{\alpha\beta}\mathcal{E} \tag{6.27}$$

[compare with (3.41)], and we may write

$$\sigma_{\alpha\beta} = \hat{\sigma}_{\alpha\beta} + \delta_{\alpha\beta}\sigma_0 \tag{6.28}$$

The terms in (6.26) may be identified by comparison with (6.24), with $Q_{\alpha\beta}$ given by (6.20a).

5.6.4 Reduction to Drude Conductivity

The relation (6.26) may be reduced to the Drude formula (3.4.14) as follows. First we neglect the off-diagonal $Q_{\alpha\beta}$ terms in (6.26). To incorporate collisions, we set

$$i\omega \to i\omega \to v \tag{6.29}$$

where, again, v represents collision frequency. In the dc limit, $\omega = 0$, and (6.28) reduces to

$$\sigma = \frac{e^2 n_0}{vm} \tag{6.30}$$

in agreement with (3.4.14) [as well as with (3.7.65)].[58]

---

[58] In these formulas, conductivity was labeled σ_c.

5.7   Elements of the Green's Function Formalism

The quantum mechanical Green's function comes into play in two classes of problems:

1. In constructing solutions to the Schrödinger equation for an isolated collection of few particles.[59]
2. In calculating average properties of many-body systems.[60]

As well be found below, for case 2, a significant property of the Green's function is that it gives few-particle properties of a many-body system. Out discussion begins with a brief review of topic 1.

5.7.1   Schrödinger Equation Green's Function

This Green's function enters in constructing solutions to equations of the form

$$\hat{R}\psi(\mathbf{x}, t) = I(\mathbf{x}, t) \tag{7.1}$$

where $\hat{R}$ typiclly is a linear differential operator and $I(\mathbf{x}, t)$ is a given inhomogeneous term. The Green's function corresponding to (7.1) satisfies the equation

$$\hat{R}(\mathbf{x}, t)G(\mathbf{x}, t; \mathbf{x}', t') = \delta(\mathbf{x} - \mathbf{x}')\delta(t - t') \tag{7.2}$$

If G is known, then the solution to (7.1) is given by

$$\psi(\mathbf{x}, t) = \int\int G(\mathbf{x}, t; \mathbf{x}', t')I(bx', t')\, d\mathbf{x}'\, dt' \tag{7.3}$$

which is readily seen to be a solution if this expression is substituted into (7.1).

Application of the Green's function to quantum theory is chiefly perturvative. Consider the Schrödinger equation for a single particle in a potential field $V(\mathbf{x})$:

$$\left(i\hbar\frac{\partial}{\partial t} + \frac{\hbar^2}{2m}\nabla^2\right)\psi(\mathbf{x}, t) = V(\mathbf{x})\psi(\mathbf{x}, t) \tag{7.4}$$

---

[59]See, for example, E. Marzbacher, *Quantum Mechanics*, 2nd ed., Wiley, New York (1970).

[60]For further discussion, see P. H. E. Meijer, *Quantum Statisical Mechanics*, Gordon and Breach, New York (1966); L. P. Kadanoff and G. Baym, *Quantum Statistical Mechanics*, W. A. Benjamin, Menlo Park, Calif., (1962); E. Em. Lifshitz an dL. P. PEtaevskii, *Physical Kinetics*, Pergamon Press, Elmsford, N.Y. (1981); and A. L. Fetter and J. D. Walecka, *Quantum Theory of Many Particle Systems*, McGraw-hill, New York (1978).

For weak interaction, the lowest-order solution to (7.4) is given by the free-particle equation

$$\left(i\hbar\frac{\partial}{\partial t} + \frac{\hbar^2}{2m}\nabla^2\right)\psi_0(\mathbf{x}, t) = 0 \qquad (7.5)$$

The second-order solution satisfies the equation

$$\left(i\hbar\frac{\partial}{\partial t} + \frac{\hbar^2}{2m}\nabla^2\right)\psi_1(\mathbf{x}, t) = V(\mathbf{x})\psi_0(\mathbf{x}, t) \qquad (7.6)$$

which has the structure of (7.1). Thus we may write

$$\left[i\hbar\frac{\partial}{\partial t} + \frac{\hbar^2}{2m}\nabla^2 \pm i\varepsilon\right]G_{\substack{R\\A}}^{(0)}(\mathbf{x}, t; \mathbf{x}', t') = \delta(\mathbf{x} - \mathbf{x}')\delta(t - t') \qquad (7.7)$$

where ε is an infinitesimal parameter of smallness that is evertually set equal to zero and, as we will see, ensures the boundedness of the transform of $G(\mathbf{x}, t)$. We have also written G_R, G_A for the retarded and advanced Green's functions and corresponding to $+i\varepsilon$ and $-i\varepsilon$ in (7.7), respectively. This terminology is motivated belwo. The zero superscript on $G_{\substack{R\\A}}^{(0)}$ indicates that it is the free-particle Green's function.

Transformed Green's function

The Fourier transform of $G_{\substack{R\\A}}^{(0)}$ is given by

$$G_{\substack{R\\A}}^{(0)}(x, t; x', t') = \iint\frac{d\omega}{2\pi}\frac{d\mathbf{k}}{(2\pi)^3}e^{-i\omega(t-t')_i\mathbf{k}\cdot(\mathbf{x}-\mathbf{x}')}\tilde{G}_{\substack{R\\A}}^{(0)}(\mathbf{k}, \omega) \qquad (7.8)$$

Taking the transform of (7.7) gives (where the limit $\varepsilon \to 0$ is understood)

$$\left(\hbar\omega - \frac{\hbar^2k^2}{2m} \pm i\varepsilon\right)\tilde{G}_{\substack{R\\A}}^{(0)}(\mathbf{k}, \omega) = 1 \qquad (7.9)$$

Separating solutions and subtracting, we obtain

$$[\tilde{G}_R^{(0)} - \tilde{G}_A^{(0)}] = \lim_{\varepsilon\to 0}\left[\frac{1}{\hbar\omega - (\hbar^2k^2/2m) + i\varepsilon} - \frac{1}{\hbar\omega - (\hbar^2k^2/2m) - i\varepsilon}\right] \qquad (7.10)$$

Now recall the relation [preceding (2.84)]

$$\lim_{\varepsilon\to 0}\frac{1}{x \pm i\varepsilon} = \mathcal{P}\left(\frac{1}{x}\right) \mp i\pi\delta(x) \qquad (7.11)$$

where $\mathcal{P}$ denotes principal part. Employing this relation in (7.10) gives

$$A^0(\mathbf{k}, \omega) \equiv i[\tilde{G}_R^{(0)} - \tilde{G}_A^{(0)}] = 2\pi\delta\left(\hbar\omega - \frac{\hbar^2k^2}{2m}\right) \qquad (7.12)$$

This form is generally known as the *spectral function* as it displays the dispersion relation of ω and k.

5.7.2   The s-Body Green's Function[61]

The s-body Green's function G_s, relevant to a many-body system is given by the following form written in second quantization:

$$i^s \hbar G_s(1, \ldots, s; 1', \ldots, s') = \langle \hat{T}[\hat{\phi}_H(1) \cdots \hat{\phi}_H(s) \hat{\phi}_H^\dagger(s') \cdots \hat{\phi}_H^\dagger(1')] \rangle \quad (7.13)$$

where

$$(j) \equiv (\mathbf{x}_j, t_j) \quad (7.13a)$$

and $\hat{T}$ is the time-ordering operator:

$$\hat{T}[\hat{\phi}_H(t_1)\hat{\phi}_H^\dagger(t_2)] = \begin{cases} \hat{\phi}_H(t_1)\hat{\phi}_H^\dagger(t_2) & t_1 > t_2 \\ \pm\hat{\phi}_H^\dagger(t_2)\hat{\phi}_H(t_1) & t_2 > t_1 \end{cases} \quad (7.14)$$

which ensures that operators in (7.13) are chronologically ordered with the earliest operator acting first. We recall that $\hat{\phi}_H$ satisfies the equation of motion (5.8). The signs $+$ and $-$ refer to bosons and fermions, repectively. The latter relation may alternatively be written

$$\hat{T}[\hat{\phi}_H(t_1)\hat{\phi}_H^\dagger(t_2)] = \theta(T_1 - t_2)\hat{\phi}(t_1)\hat{\phi}_H^\dagger(t_2) \pm \theta(t_2 - t_1)\hat{\phi}_H^\dagger(t_2)\hat{\phi}_H(t_1) \quad (7.15)$$

where $\theta(x)$ was introduced previously [beneath (6.19)]. Note that the dimensions of $\hbar G_s$ are (volume)$^{-s}$.

The average in (7.13) is written with respect to an N-body system in the Heisenberg representation and is given by (5.13):

$$\langle \hat{\Gamma} \rangle = \mathrm{Tr}^{(N)}[\hat{\rho}_N(0)\hat{\Gamma}] \quad (7.16a)$$

Working in Fock space, we obtain

$$\langle \hat{\Gamma} \rangle = \sum_{\mathbf{n},\mathbf{n}'} \langle \mathbf{n}| \hat{\rho}_N(0) |\mathbf{n}'\rangle \langle \mathbf{n}'| \hat{\Gamma} |\mathbf{n}\rangle \quad (7.16b)$$

where $|\mathbf{n}\rangle$ represents the many-body ket vector on the right side on (5.2).

5.7.3   Averages and the Green's Function

In this section we wish to obtain a relation between the average of a s-particle dynamical variable and the Green's function (7.13). Toward these ends, we note that the Fock-space operator, $\hat{\tilde{G}}_s$, defined by (5.7), is closely allied to the

---

[61] The many-body Green's function was originally formulated for relativistic quantum field theory where the use of multiple times is well motivated. See S. S. Schweber, *An Introduction to Relativisitic Quantum Field Theory*, Harper & Row, New York (1961). This analysis has found use in nonrelativisitic many-body theory as well. See, for example, H. Haken, *Quantum Field Theory of Solids*, North Holland, New York (1976); and C. Kettel, *Quantum Theory of Solids*, Wiley, New York (1963).

Green's function (7.13). The average of an s-particle dynamical variable is given by (5.14). This expression involves the form

$$\mathrm{Tr}^{(N)}[\hat{\rho}_N(0)\hat{\bar{G}}_s(\mathbf{x}^s, \mathbf{y}^s, t)] = \langle \hat{\bar{G}}_s(\mathbf{x}^s, \mathbf{y}^s, t) \rangle \qquad (7.17)$$

To relate this to the Green's function (7.13), first we make the notational change $\mathbf{x}_s = \mathbf{x}_s$, $\mathbf{y}_s = \mathbf{x}'_s$ and let variables include time so that (5.7) becomes

$$\hat{\bar{G}}_s(\mathbf{x}^s, \mathbf{x}'^s) = \hat{\phi}_H^\dagger(1')\cdots\hat{\phi}_H^\dagger(s')\hat{\phi}_H(s)\cdots\hat{\phi}_H(1) \qquad (7.18)$$

where again we have set $(j) = (\mathbf{x}_j, t_j)$ with

$$t'_1 = t'_2 = \cdots = t'_s = t_1 = t_2 = \cdots = t$$

The form (7.18) can be made congruent with (7.13) if in the latter we set

$$t'_1 > t'_2 > \cdots > t'_s > t_s > \cdots > t_1 = t$$

Let LIM represent the process

$$\lim_{\varepsilon \to 0} \begin{pmatrix} t'_i = t'_{i+1} + \varepsilon \\ t'_s = t_s + \varepsilon \\ t_i = t_{i-1} + \varepsilon \\ t_1 = t \end{pmatrix}$$

this limiting process brings the ordering of variables in (7.17) in agreement with those in (7.13), and we may write

$$\langle i^s\hbar \, \mathrm{LIM} \, \hat{\bar{G}}_s(\mathbf{x}^s, \mathbf{x}'^s, t) \rangle = i^s\hbar G_s(\mathbf{x}^s, \mathbf{x}'^s, t) \qquad (7.19)$$

Note in particular that averages of one-body observables such as particle momentum stem from $G_1(1, 1')$, whereas averages of two-body observables such as pair potential stem from $G_2(1, 2, 1', 2')$, and so on.

Combining the preceding results, we write [recall (5.14)]

$$\langle A^{(s)} \rangle = \frac{1}{s!} \int d\mathbf{x}^s \, d\mathbf{x}'^s i^s\hbar G_s(\mathbf{x}^s, \mathbf{x}'^s, t)\langle x'^s|A_1\cdots s|\mathbf{x}^s\rangle \qquad (7.20)$$

Note in particular that the s-dimensional summation in (5.10) is absorbed in the integral of (7.20).

Let us employ (7.20) to calculate (1) the total momentum and (2) the total potential energy of an N-particle system. For total momentum,

$$\hat{\mathbf{P}} = \sum_{i=1}^{N} \hat{\mathbf{p}}_i$$

and (7.20) gives

$$\langle \hat{\mathbf{P}} \rangle = \int d\mathbf{x}_1 \, d\mathbf{x}'_1 i\hbar G_1(\mathbf{x}_1, \mathbf{x}'_2, t)\langle \mathbf{x}'_1|\hat{\mathbf{p}}_1|\mathbf{x}_1\rangle \qquad (7.21)$$

The coordinate representation of $\hat{\mathbf{p}}_1$ is [recall (1.27)]

$$\langle \mathbf{x}'_1 | \hat{\mathbf{p}}_1 | \mathbf{x}_1 \rangle = -i\hbar \frac{\partial}{\partial \mathbf{x}_1} \delta(\mathbf{x}'_1 - \mathbf{x}_1)$$

Integrating (7.21) by parts we then obtain

$$\langle \hat{\mathbf{P}} \rangle = -(i\hbar)^2 \int d\mathbf{x}_1 \, d\mathbf{x}'_1 \left(\frac{\partial}{\partial \mathbf{x}_1} G_1(\mathbf{x}_1, \mathbf{x}'_1, t) \right) \delta(\mathbf{x}'_1 - \mathbf{x}_1)$$

$$= -(i\hbar)^2 \int d\mathbf{x}_1 \frac{\partial}{\partial \bar{\mathbf{x}}_1} G_1(\bar{\mathbf{x}}, \mathbf{x}_1, t) \bigg|_{\bar{\mathbf{x}}_1 = \mathbf{x}_1} \qquad (7.21a)$$

Our second example addresses total potential energy:

$$\hat{V} = \sum_{i<j}^{N} \sum u_{ij}(\hat{\mathbf{x}}_i, \hat{\mathbf{x}}_j)$$

Again employing (7.20) and dropping the subscripts on u_{ij}, we find (with $s = 2$)

$$\langle V \rangle = \frac{1}{2} \int d\mathbf{x}_1 \, d\mathbf{x}_2 \, d\mathbf{x}'_1 \, d\mathbf{x}'_2 i^2 \hbar G_2(\mathbf{x}_1, \mathbf{x}_2, \mathbf{x}'_1, \mathbf{x}'_2, t) \langle \mathbf{x}'_1 \mathbf{x}'_2 | u(\hat{\mathbf{x}}_1, \hat{\mathbf{x}}_2) | \mathbf{x}_1 \mathbf{x}_2 \rangle$$

$$= -\frac{\hbar}{2} \int d\mathbf{x}_1 \, d\mathbf{x}_2 \, d\mathbf{x}'_1 \, d\mathbf{x}'_2 G_2(\mathbf{x}_1, \mathbf{x}_2 \mathbf{x}'_1, \mathbf{x}'_2, t) u(\mathbf{x}'_1, \mathbf{x}'_2) \delta(\mathbf{x}'_1 - \mathbf{x}_1) \delta(\mathbf{x}'_2 - \mathbf{x}_2)$$

$$(7.22)$$

which reduces to

$$\langle V \rangle = -\frac{\hbar}{2} \int d\mathbf{x}_1 \, d\mathbf{x}_2 G(\mathbf{x}_1, \mathbf{x}_2, \mathbf{x}_1, \mathbf{x}_2, t) u(\mathbf{x}_1, \mathbf{x}_2)$$

Substituting the defining relation (7.19) into the preceding gives

$$\langle V \rangle = \frac{1}{2} \int d\mathbf{x}_1 \, d\mathbf{x}_2 \langle \hat{\phi}_H^\dagger(\mathbf{x}_1) \hat{\phi}_H^\dagger(\mathbf{x}_2) \hat{\phi}_H(\mathbf{x}_2) \hat{\phi}_H(\mathbf{x}_1) \rangle u(\mathbf{x}_1, \mathbf{x}_2) \qquad (7.23)$$

which agrees with our previous finding (5.18).

The preceding two example illustrate the manner in which the Green's function determines one- and two-particle properties of a many-body system.

Compilation of relations

One also defines the Green's functions (dropping the subscript on G_1)

$$G(1, 1') = G^>(1, 1'), \qquad \text{for } t_1 > t_{1'}$$
$$G(1, 1') = G^<(1, 1'), \qquad \text{for } t_1 < t_{1'} \qquad (7.24)$$

Note in particular that number density may be written

$$\langle n(\mathbf{x}, t) \rangle = \lim_{\varepsilon \to 0} \langle \hat{\phi}_H^\dagger(\mathbf{x}, t + \varepsilon) \hat{\phi}_H(\mathbf{x}, t) \rangle$$

$$= \pm i\hbar G^<(1, 1) \qquad (7.25)$$

Here is a compilation of relations among the assorted Green's functions introduced to this point.

$$G^{>}(1, 2) = \frac{1}{i\hbar}\langle \hat{\phi}_H(1)\hat{\phi}_H^{\dagger}(2)\rangle \tag{7.26a}$$

$$G^{<}(1, 2) = \pm\frac{1}{i\hbar}\langle \hat{\phi}_H^{\dagger}(2)\hat{\phi}_H(1)\rangle \tag{7.26b}$$

$$G(1, 2) = \theta(t_1 - t_2)G^{>}(1, 2) + \theta(t_2 - t_1)G^{<}(1, 2) \tag{7.26c}$$

$$G^{>}(1, 2) + G^{<}(1, 2) = \frac{1}{i\hbar}\langle [\hat{\phi}_H(1), \hat{\phi}_H^{\dagger}(2)]_{\pm}\rangle \tag{7.26d}$$

$$G^{>}(1, 2) - G^{<}(1, 2) = \frac{1}{i\hbar}\langle [\hat{\phi}_H(1), \hat{\phi}_H^{\dagger}(2)]_{\mp}\rangle \tag{7.26e}$$

$$G^{R}(1, 2) = \theta(t_1 - t_2)[G^{>}(1, 2) - G^{<}(1, 2)] \tag{7.27a}$$

$$G^{A}(1, 2) = \theta(t_2 - t_1)[G^{<}(1, 2) - G^{>}(1, 2)] \tag{7.27b}$$

$$G^{>}(1, 2) - G^{<}(1, 2) = G^{R}(1, 2) - G^{A}(1, 2) \tag{7.28}$$

5.7.4 The Quasi-Free Particle

In this and the following section we see how the Green's function is related to distribution functions of momentum. Consider the operators $\hat{\phi}(\mathbf{x})$ and $\hat{\phi}^{\dagger}(\mathbf{x})$ given in (5.4). Fourier analyzing these operators gives

$$\hat{\phi}(\mathbf{x}) = \int \frac{d\mathbf{k}}{(2\pi)^3}e^{i\mathbf{k}\cdot\mathbf{x}}\hat{a}(\mathbf{k}) \tag{7.29a}$$

$$\hat{\phi}^{\dagger}(\mathbf{x}) = \int \frac{d\mathbf{k}}{(2\pi)^3}e^{i\mathbf{k}\cdot\mathbf{x}}\hat{a}^{\dagger}(\mathbf{k}) \tag{7.29b}$$

where, with (5.5a), we write

$$[\hat{a}(\mathbf{x}), \hat{a}^{\dagger}(\mathbf{k}')]_{mp} = (2\pi)^3\delta(\mathbf{k} - \mathbf{k}') \tag{7.30}$$

The signs $(-, +)$ refer respectively to bosons and fermions.

Note that in this representation the Hamiltonian of a free particle may be written

$$\hat{H}_0 = \int \frac{d\mathbf{k}}{(2\pi)^3}\frac{\hbar^2 k^2}{2m}\hat{a}^{\dagger}(\mathbf{k})\hat{a}(\mathbf{k}) \tag{7.31}$$

For a *quasi-free particle*, we write

$$\hat{H} = \int \frac{d\mathbf{k}}{(2\pi)^3}E(\mathbf{k})\hat{a}^{\dagger}(\mathbf{k})\hat{a}(\mathbf{k}) \tag{7.32}$$

which is a good approximate Hamiltonian for the actual system.

Stemming from the Heisenberg equation of motion (5.8), the relation (7.32) gives (see Problem 5.32)

$$\hat{\phi}_H(\mathbf{x}, t) = \int \frac{d\mathbf{k}}{(2\pi)^3}e^{-(i/\hbar)E(\mathbf{k})t+i\mathbf{k}\cdot\mathbf{x}}\hat{a}(\mathbf{k}) \tag{7.33}$$

We recall that $\langle a^\dagger(\mathbf{k})a(\mathbf{k})\rangle$ represents the number of particles with momentum $\mathbf{k}$. For a spatially uniform system, we may write

$$\langle \hat{a}^\dagger(\mathbf{k})\hat{a}(\mathbf{k}')\rangle = (2\pi)^3\delta(\mathbf{k} - \mathbf{k}')f(\mathbf{k}) \tag{7.34a}$$

from which, with (7.30), we obtain

$$\langle \hat{a}(\mathbf{k})\hat{a}^\dagger(\mathbf{k}')\rangle = (2\pi)^3\delta(\mathbf{k} - \mathbf{k}')[1 \pm f(\mathbf{k})] \tag{7.34b}$$

where $f(\mathbf{k})$ represents the one-particle distribution function in $\mathbf{k}$ space. These findings are used below.

5.7.5   One-Body Green's Function

To obtain the related one-particle Green's function, we recall (7.13) with $s = 1$ and (7.15) to obtain

$$i\hbar G(\mathbf{x}, t; \mathbf{x}', t') = \theta(t - t')\langle\hat{\phi}_H(\mathbf{x}, t)\hat{\phi}_H^\dagger(\mathbf{x}', t')\rangle \pm (t' - t)\langle\hat{\phi}_H^\dagger(\mathbf{x}', t')\hat{\phi}_H(\mathbf{x}, t)\rangle \tag{7.35}$$

Substituting (7.33) into this equation, together with (7.34), we find

$$i\hbar G(\mathbf{x}, t; \mathbf{x}', t') = \int \frac{d\mathbf{k}}{(2\pi)^3} \exp\left\{ i\mathbf{k} \cdot (\mathbf{x} - \mathbf{x}') - \frac{i}{\hbar}E(\mathbf{k})(t - t') \right\}$$
$$\times \{\theta(t - t')[1 \pm f(\mathbf{k})] \pm \theta(t' - t)f(\mathbf{k})\} \tag{7.36}$$

Now note the repesentation of $\theta(t - t') \equiv \theta(\tau)$:

$$\theta(\pm\tau)e^{(-iE\tau)/\hbar} = \lim_{\varepsilon\to 0}(\pm i\hbar) \int \frac{d\omega}{2\pi} \frac{e^{i\omega\tau}}{\hbar\omega - E \pm i\varepsilon} \tag{7.37}$$

Substituting this representation into (7.36), we find

$$G(\mathbf{x}, t; \mathbf{x}', t') = \int\int \frac{d\mathbf{k}}{(2\pi)^3} \frac{d\omega}{2\pi} \exp\{i\mathbf{k}\cdot(\mathbf{x}-\mathbf{x}') = i\omega(t-t')\}G(\mathbf{k}, \omega) \tag{7.38}$$

$$G(\mathbf{k}, \omega) = \lim_{\varepsilon\to 0}\left[\frac{1 \pm f(\mathbf{k})}{\hbar\omega - E(\mathbf{k}) + i\varepsilon} \mp \frac{f(\mathbf{k})}{\hbar\omega - E(\mathbf{k}) - i\varepsilon} \right] \tag{7.39}$$

This equation gives the desired relation between the one-body Green's function $G(\mathbf{k}, \omega)$ and the one-body distribution function $f(\mathbf{k})$ with normalization.

$$V \int f(\mathbf{k})\frac{d\mathbf{k}}{(2\pi)^3} = N \tag{7.39a}$$

where N is total number of particles and V is volume.

5.7.6 Retarded and Advanced Green's Functions

Retarded and advanced one-body Green's functions are given by

$$G_R(1, 1') = \theta(t_1 - t_{1'})\frac{1}{i\hbar}\langle[\hat{\phi}_H(1), \hat{\phi}_H^\dagger(1')]_m p\rangle$$

$$G_A(1, 1') = \theta(t_1 - t_{1'})\left(-\frac{1}{i\hbar}\right)\langle[\hat{\phi}_H(1), \hat{\phi}_H^\dagger(1')]_m p\rangle \tag{7.40}$$

where, again, averages are with respect to an N-body system and are given by (7.16). For the quasi-free-particle problem considered above, we find

$$G_{\substack{R \\ A}}(\omega, \mathbf{k}) = \lim_{\varepsilon \to 0}\left(\frac{1}{\hbar\omega - E(\mathbf{k}) \pm i\varepsilon}\right) \tag{7.41}$$

which, for a free particle, agrees with (7.9) relevant to a single-particle system. Note that the Green's function (7.41) obeys the equation

$$[\hbar\omega - E(k)]G(k, \omega) = 1 \tag{7.42}$$

Again, for a free particle, with

$$E(k) = \frac{\hbar^2 k^2}{2m}$$

the Fourier inversion of (7.42) returns (7.7).

Let us return to the motivation for advanced and retarded terminology introduced above. With (7.37), we see that $+i\varepsilon$, $-i\varepsilon$ in the integral representation of the θ function corresponds to $t > 1'$ and $t < t'$, respectively, in the Green's function expression (7.35). If we revert back to (7.7), it is evident that t' may be associated with a source. Thus the Green's function in the interval $t < t'$ describes a response prior to the source being turned on and is called *advanced*. For the case $t > t'$, the Green's function obeys causality and the response is *retarded* in time with respect to the source.

5.7.7 Coupled Green's Function Equations

In this section we construct an equation of motion for the one-body Green's function, which, we will be seen is dependent on the two-body Green's function. The analysis begins with (5.28) [dropping the hat and H subscript on $\hat{\phi}_H(\mathbf{x}, t)$]; we write

$$\left(i\hbar\frac{\partial}{\partial t_1}\right)\phi(1) = \int d\mathbf{x}_2 \phi^\dagger(2)u(\mathbf{x}_1, \mathbf{x}_2)\phi(2)\phi(1)\bigg|_{t_2=t_1} \tag{7.43}$$

where, again, we are using the notation of (7.13a).

Now we note the relation (see Problem 5.34) relevant to differentiation of time-ordered products:

$$\frac{\partial}{\partial t_1}[\hat{T}\phi(1)\phi^\dagger(1')]$$

$$= \delta(1 - 1') \pm \theta(t_{1'} - t_1)\phi^\dagger(1')\frac{\partial\phi(1)}{\partial t} + \theta(t_1 - t_{1'})\frac{\partial\phi(1)}{\partial t_1}\phi^\dagger(1') \quad (7.44)$$

where again the $+$ sign refers to bosons and the $-$ sign to fermions, and we have set

$$\delta(1 - 1') = \delta(\mathbf{x} - \mathbf{x}')\delta(t - t')$$

Employing (7.43) for the time derivative terms in (7.44), after some rearrangement we obtain

$$\left(i\hbar\frac{\partial}{\partial t_1} + \hat{K}_1\right)[\hat{T}\phi(1)\phi^\dagger(1')] = i\hbar\delta(1 - 1')$$

$$\pm \int d\mathbf{x}_2\, u(\mathbf{x}_1, \mathbf{x}_2)[\hat{T}\phi(1)\phi(2)\phi^\dagger(2^+)\phi^\dagger(1')]\Big|_{t_2=t_1} \quad (7.45)$$

The time t_2^+ of $\phi^\dagger$ is infinitesimally greater than t_2 in order to maintain the order of operators in (7.43). From the defining relations (7.13), we write

$$i\hbar G_1(1, 1') = \langle \hat{T}\phi(1)\phi^\dagger(1')\rangle$$

$$-\hbar G_2(1, 2; 1', 2') = \langle \hat{T}\phi(1)\phi(2)\phi^\dagger(2')\phi^\dagger(1')\rangle \quad (7.46)$$

Constructing the average of (7.45) and substituting the preceding definitions gives the desired result:

$$\left(i\hbar\frac{\partial}{\partial t_1} + \hat{K}_1\right)G(1, 1') = \delta(1 - 1') \pm i\int d\mathbf{x}_2\, u(\mathbf{x}_1, \mathbf{x}_2)G(1, 2; 1', 2^+)\Big|_{t_1=t_2}$$
$$(7.47)$$

Note in particular that when particles become isolated, $u(\mathbf{x}_1, \mathbf{x}_2) \to 0$, and (7.47) returns (7.7) relevant to a free particle. Note further that carrying out the LIM process defined in (7.19) returns the first equation in the hierarchy for $\hat{\bar{G}}_s$, as given in Table 5.2 (see Problem 5.35).

5.7.8 Diagrams and Expansion Techniques

We continue this section with a brief discussion of diagrammatic representations of equations of motion for Green's functions. Toward these ends, (7.57) must be rewritten in a form that better lends itself to diagrammatic interpretation.

Consider first that the interaction in (7.47) is turned off. The resulting free-particle equation may then be written

$$\int G_0^{-1}(1, 1'')G_o(1'', 1')\,d1'' = \delta(1 - 1') \tag{7.48}$$

where $G_0(1, 1')$ is the solution to the free-particle equation and

$$G_0^{-1}(1, 1'') = \left(i\hbar \frac{\partial}{\partial t_1} + \hat{K}_1 \right) \delta(1 - 1'') \tag{7.49}$$

Let us further define

$$U(1, 2) \equiv u(\mathbf{x}_1, \mathbf{x}_2)\delta(t_1 - t_2) \tag{7.49a}$$

With the preceding identifications, (7.47) may be rewritten (see Problem 5.37)

$$\int d2\, G_0^{-1}(1, 2)G(2, 1') = \delta(1 - 1') \pm i \int d2\, U(1, 2)G(1, 2; 1'2^+) \tag{7.50}$$

Operating on this equation with $\int d1\, G_0(3, 1)$ gives

$$\iint d1\, d2\, G_0(3, 1)G_0^{-1}(1, 2)G(2, 1')$$

$$= \int d1\, G_0(3, 1)\delta(1 - 1') \pm i \int d1\, d2\, G_0(3, 1)U(1, 2)G(1, 2; 1'2^+)$$

which, with (7.48), gives the desired relation (exchanging 1 and 3):

$$G(1, 1') = G_0(1, 1') \pm i \iint d2\, d3\, G_0(1, 3)U(2, 3)G(3, 2; 1', 2^+) \tag{7.51}$$

This equation is the starting point for our description of diagrammatic representation of terms, which is described in Table 5.4. Note that in the box diagram a dashed line represents the interaction $U(1, 2)$ and implies integration. The direction of an arrow is associated with the order of arguments in the related Green's function. With these rules at hand, (7.51) has the following diagrammatic equivalent:

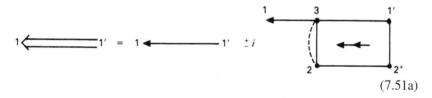

$$\tag{7.51a}$$

TABLE 5.4. Fundamental Green's Function Diagrams

Term	Diagram
$G(1; 1')$	
$G_0(1; 1')$	
$\int d1\, d2\, U(1, 2) G(1, 2; 1', 2^+)$	

The diagrammatic representation for the equation of motion for $G(1,2; 1',2')$ is given by

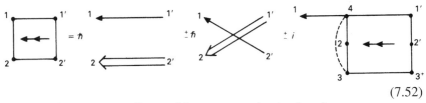

$$(7.52)$$

The equation corresponding to this representation is given by

$$G(1, 2; 1', 2') = \hbar G_0(1, 1') G(2, 2') \pm \hbar G_0(1, 2') G(2, 1')$$

$$\pm i \iint d3\, d4\, G_0(1, 4) U(3, 4) G(4, 2, 3; 1', 2', 3^+) \quad (7.52a)$$

The explicit equation of motion corresponding to the preceding equation [in analogy with the structure (7.47) is obtained by operating on (7.52a) with $\int d1\, G_0^{-1}(5, 1)$. There results

$$\int d1\, G_0^{-1}(5, 1) G(1, 2; 1', 2') = \hbar \delta(5 - 1') G(2, 2') \pm \hbar \delta(5 - 2') G(2, 1')$$

$$\pm i \int d3\, U(3, 5) G(5, 2; 1', 2', 3^+)$$

With (7.49), and changing $5 \rightarrow 1$, the preceding reduces to the desired equations of motion:

$$\left(i\hbar \frac{\partial}{\partial t_1} + \hat{K}_1 \right) G(1, 2; 1', 2') = \hbar \delta(1 - 1') G(2, 2') \pm \hbar \delta(1 - 2) G(2, 1')$$

$$\pm i \int d3 U(3, 1) G(1, 2, 3; 1', 2', 3^+)$$

Approximations

Let us return to (7.51a). For weak interaction it is consistent to expand the equations in powers of the interaction or, equivalently, expand the corresponding diagrms in terms of the dashed element. Thus, to lowest order, (7.51a) gives

$$1 \Longleftarrow 1' \ = \ 1 \longleftarrow 1' \qquad (7.53a)$$

With this simplification, the lowest-order form of (7.52) is given by

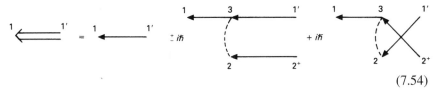

$$(7.53b)$$

Employing these diagrams in (7.51a) gives the following first-order diagrammatic equivalent for the equation of motion for $G(1, 1')$:

$$(7.54)$$

which has the topological equivalent

$$(7.54a)$$

In this latter representation, we have joined 2 and 2^+ as they represent the same space–time point. The equation corresponding to (7.54) is

$$G(1, 1') = G_0(1, 1') \pm i\hbar \int\int d2\,d3\,G_0(1, 3)G_0(3, 1')U(3, 2)G_0(2, 2^+)$$

$$+ i\hbar \int\int d2\,d3\,G_0(1, 3)G_0(3, 2)U(3, 2)G_0(2, 1') \qquad (7.55)$$

The full equation of motion, in diagrammatic form, for $G(1, 2, 3, 1', 2', 3')$ appears as

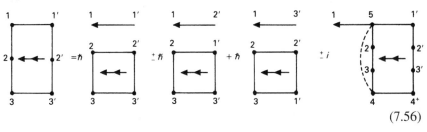

$$(7.56)$$

Furthermore, note that the integration in the final term in (7.56) over 4 and 5 leaves a function of $(1, 2, 3; 1', 2', 3')$ in accord with the left side of this equation. To lowest order, employing the decomposition (7.53b), (7.56) gives (keeping order of $1'$, $2'$, $3'$ on the right of each diagram).

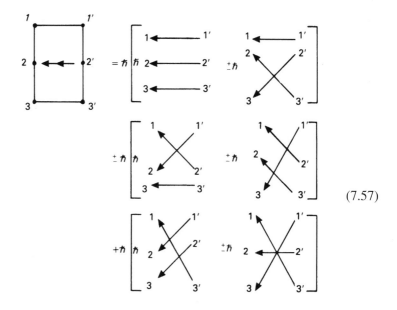

(7.57)

The corresponding functional relation may be written in the following determinant form:

$$G_0(1, 2, 3; 1', 2', 3') = \hbar^2 \begin{vmatrix} G_0(1, 1') & G_0(1, 2') & G_0(1, 3') \\ G_0(2, 1') & G_0(2, 2') & G_0(2, 3') \\ G_0(3, 1') & G_0(3, 2') & G_0(3, 3') \end{vmatrix}_{\pm}$$

(7.58)

It may be evident at this point that the equations of motion (7.56), (7.56), and (7.56) follow a general pattern given by the following rule: In the equation of motion for the N-body Green's function, the first N terms are products of the free one-body Green's function and the $(N-1)$-body Green's function with the $1'$ variable of $G_0(1, 1')$ exchanged in each product with one of the primed variables of the $(N-1)$-body Green's function. In the last term of the equation, the free one-body Green's function attaches to the $(N+1)$-body Green's function via interaction. This last term contains the factor $\pm i$. The sth term $(s = 1, \ldots, N)$ in the sum of products contains the factor $(\pm 1)^{s+1}\hbar$.

Following these rules of construction gives the following diagrammatic representation of the equation of motion of the N-body Green's function:[62]

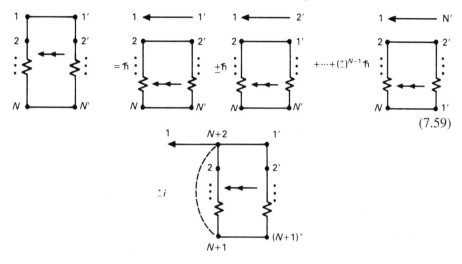

$$(7.59)$$

As noted previously, in each column of variables on the right side of the first N terms of the preceding equation, $1'$ is exchanged with an element of the right column of a box diagram with the order of the remaining variables maintained. Note also that the space–time coordinate 1 is preferred in the preceding equation due to the fact that this equation includes implicit differentiation with respect to $\mathbf{x}_1, t_1$.

5.8 Spectral Function for Electron–Phonon Interactions

5.8.1 Hamiltonian

We conclude this section with a derivation of an approximate form of the spectral function [see (7.12)] relevant to electrons interacting with phonons in a lattice (described previously in Section 3.6). The first step in this derivation is the construction of the related Green's function which in turn is obtained from its equation of motion. With (5.8) and (7.13) we see that formulation of this equation stems from the Hamiltonian of the system. Thus, with $\hat{b}^{\dagger}(\mathbf{q})$ representing the creation operator of a phonon with momentum $\hbar\mathbf{q}$ and energy

---

[62] An alternative development of (7.56) follows from Wick's theorem, which addresses the appropriate expansion of time-ordered product of operators. G. C. Wick, *Phys. Rev.* **80**, 268 (1950). See also A. L. Fetter and J. D. Walecka, *Quantum Theory of Many Particle Systems*, McGraw-Hill, New York (1971).

$\hbar\omega(\mathbf{q})$, (7.32) is extended to read

$$\hat{H} = \int \frac{d\mathbf{k}}{(2\pi)^3} E(\mathbf{k})\hat{a}^\dagger(\mathbf{k})\hat{a}(\mathbf{k}) + \int \frac{d\mathbf{q}}{(2\pi)^3}\hbar\omega(\mathbf{q})\hat{b}^\dagger(\mathbf{q})\hat{b}(\mathbf{q})$$
$$+ \int \frac{d\mathbf{k}}{(2\pi)^3} \frac{d\mathbf{q}}{(2\pi)^3}[C(\mathbf{q})\hat{a}^\dagger(\mathbf{k} + \mathbf{q})\hat{a}(\mathbf{k})\hat{b}(\mathbf{q})$$
$$+ C^*(\mathbf{q})\hat{a}^\dagger(\mathbf{k})\hat{a}(\mathbf{k} + \mathbf{q})\hat{b}^\dagger(\mathbf{q})] \tag{8.1}$$

The coefficient $C(\mathbf{q})$ represents the strength of the electron–phonon interaction.

The interpretation of the interaction terms in (8.1) is as follows. Consider, for example, the third integral on the right side of the equation. In this interaction, a phonon of momentum $\hbar\mathbf{q}$ is annihilated in scattering with an electron, which in turn suffers momentum change from $\hbar(\mathbf{k})$ to $\hbar(\mathbf{k}+\mathbf{q})$. A diagram corresponding to this process was sketched previously in Fig. 5.7a.

In the Heisenberg picture [recall (5.7a)], we write

$$\hat{a}_H(\mathbf{k}, t) = e^{(i\hat{H}t)/\hbar}\hat{a}(\mathbf{k})e^{-(i\hat{H}t)/\hbar} \tag{8.2}$$

This operator has the equation of motion [recall (5.8)]

$$i\hbar\frac{\partial}{\partial t}\hat{a}_H(\mathbf{k}, t) = [\hat{a}_H(\mathbf{k}, t), \hat{H}] \tag{8.2a}$$

with parallel relations held by $\hat{a}_H^\dagger$, $\hat{b}_H$, and $\hat{b}_H^\dagger$. As $\hat{a}(\mathbf{k})$ relates to fermions and $\hat{b}(\mathbf{q})$ to bosons, recalling (7.30), we write

$$[\hat{a}(\mathbf{k}), \hat{a}^\dagger(\mathbf{k}')]_+ = (2\pi)^3\delta(\mathbf{k} - \mathbf{k}') \tag{8.3a}$$
$$[\hat{b}(\mathbf{k}), \hat{b}^\dagger(\mathbf{q}')]_- = (2\pi)^3\delta(\mathbf{q} - \mathbf{q}') \tag{8.3b}$$

All other commutators are equal to zero.

Combining the preceding relations and recalling the expansion

$$[\hat{A}, \hat{B}\hat{C}]_- = [\hat{A}, \hat{B}]_\pm\hat{C} \mp \hat{B}[\hat{A}, \hat{C}]_\pm$$

gives the following four equations of motion:

$$i\hbar\frac{\partial}{\partial t}\hat{a}_H(\mathbf{k}, t) = E(\mathbf{k})\hat{a}_H(\mathbf{k}, t) + \int \frac{d\mathbf{q}}{(2\pi)^3}[C(\mathbf{q})\hat{a}_H(\mathbf{k} - \mathbf{q}, t)\hat{b}_H(\mathbf{q}, t)$$
$$+ C^*(\mathbf{q})\hat{a}_H(\mathbf{k} + \mathbf{q}, t)\hat{b}_H^\dagger(\mathbf{q}, t)]$$
$$i\hbar\frac{\partial}{\partial t'}\hat{a}_H^\dagger(\mathbf{k}', t') = -E(\mathbf{k}')\hat{a}_H^\dagger(\mathbf{k}', t') - \int \frac{d\mathbf{q}}{(2\pi)^3}[C(\mathbf{q})\hat{a}_H^\dagger(\mathbf{k}' + \mathbf{q}, t')\hat{b}_H(\mathbf{q}, t')$$
$$+ C^*(\mathbf{q})\hat{a}_H(\mathbf{k}' - \mathbf{q}, t')\hat{b}_H^\dagger(\mathbf{q}, t')] \tag{8.4}$$
$$i\hbar\frac{\partial}{\partial t}\hat{b}_H(\mathbf{q}, t) = \hbar\omega(\mathbf{q})\hat{b}_H(\mathbf{q}, t) + C^*(\mathbf{q})\int \frac{d\mathbf{k}}{(2\pi)^3}\hat{a}_H^\dagger(\mathbf{k}, t)\hat{a}_H(\mathbf{k} + \mathbf{q}, t)$$
$$i\hbar\frac{\partial}{\partial t'}\hat{b}_H^\dagger(\mathbf{q}, t') = -\hbar\omega(\mathbf{q})\hat{b}_H^\dagger(\mathbf{q}, t') - C(\mathbf{q})\int \frac{d\mathbf{k}}{(2\pi)^3}\hat{a}_H^\dagger(\mathbf{k} + \mathbf{q}, t')\hat{a}_H(\mathbf{k}, t')$$

5.8.2 Green's Function Equations of Motion

As in 7.13 we introduce [dropping the tilda notation of (7.8)]

$$i\hbar G(\mathbf{k}, t; \mathbf{k}', t') = \langle \hat{T}[\hat{a}_H(\mathbf{k}, t)\hat{a}_H^\dagger(\mathbf{k}', t')] \rangle \tag{8.5}$$

With the property (7.44), we obtain the equation of motion

$$i\hbar \frac{\partial}{\partial t} G(\mathbf{k}, t; \mathbf{k}', t') = (2\pi)^3 \delta(\mathbf{k} - \mathbf{k}') \delta(t - t') + E(\mathbf{k}) G(\mathbf{k}, t; \mathbf{k}', t')$$

$$+ \int \frac{d\mathbf{q}}{(2\pi)^3} [C^*(\mathbf{q}) \Xi'(\mathbf{k} + \mathbf{q}, t; \mathbf{k}', t'; \mathbf{q}, t)$$

$$+ C(\mathbf{q}) \Xi(\mathbf{k} - \mathbf{q}, t; \mathbf{k}', t'; \mathbf{q}, t)] \tag{8.6}$$

where

$$i\hbar \Xi(\mathbf{k}, t; \mathbf{k}', t'; \mathbf{q}, \tau) \equiv \langle \hat{T}[\hat{a}_H(\mathbf{k}, t)\hat{a}_H^\dagger(\mathbf{k}', t')\hat{b}_H(\mathbf{q}, \tau)] \rangle$$
$$i\hbar \Xi'(\mathbf{k}, t; \mathbf{k}', t'; \mathbf{q}, \tau) \equiv \langle \hat{T}[\hat{a}_H(\mathbf{k}, t)\hat{a}_H^\dagger(\mathbf{k}', t')\hat{b}_H^\dagger(\mathbf{q}, \tau)] \rangle \tag{8.7}$$

With (8.6) we note that in order to determine $G(\mathbf{k}, t; \mathbf{k}', t')$ it is necessary to know Ξ and Ξ'. Equations of motion for these variables follow from the defining relations (8.7) and the equations of motion (8.2a). There results

$$i\hbar \frac{\partial}{\partial \tau} \Xi'(\mathbf{k}, t; \mathbf{k}', t'; \mathbf{q}, \tau) = -\hbar \omega(\mathbf{q}) \Xi'(\mathbf{k}, t; \mathbf{k}', t'; \mathbf{q}, \tau) \tag{8.7a}$$

$$+ i \int \frac{d\mathbf{K}}{(2\pi)^3} C(\mathbf{q}) G_2(\mathbf{k}, t, \mathbf{K}, \tau; \mathbf{k}', t', \mathbf{K} + \mathbf{q}, \tau^+)$$

$$i\hbar \frac{\partial}{\partial \tau} \Xi(\mathbf{k}, t; \mathbf{k}', t'; \mathbf{q}, \tau) = \hbar \omega(\mathbf{q}) \Xi(\mathbf{k}, t; \mathbf{k}', t'; \mathbf{q}, \tau) \tag{8.7b}$$

$$- i \int \frac{d\mathbf{K}}{(2\pi)^3} C^*(\mathbf{q}) G_2(\mathbf{k}, t, \mathbf{K}+\mathbf{q}, \tau; \mathbf{k}', t', \mathbf{K}, \tau^+)$$

For differentiation with respect to t, we find

$$i\hbar \frac{\partial}{\partial t} \Xi(\mathbf{k}, t; \mathbf{k}', t'; \mathbf{q}, \tau)$$

$$= (2\pi)^3 \delta(\mathbf{k} - \mathbf{k}') \delta(t - t') \langle \hat{b}_H(\mathbf{q}, \tau) \rangle + E(\mathbf{k}) \Xi(\mathbf{k}, t; \mathbf{k}', t'; \mathbf{q}, \tau)$$

$$+ \int \frac{d\mathbf{Q}}{(2\pi)^3} \left\{ \frac{C^*(\mathbf{Q})}{i\hbar} \langle \hat{T}[\hat{a}_H(\mathbf{k} + \mathbf{Q}, t)\hat{a}_H^\dagger(\mathbf{k}', t')\hat{b}_H^\dagger(\mathbf{Q}, t)\hat{b}_H(\mathbf{q}, \tau)] \rangle \right.$$

$$\left. + \frac{C(\mathbf{Q})}{i\hbar} \langle \hat{T}[\hat{a}_H(\mathbf{k} - \mathbf{Q}, t)\hat{a}_H^\dagger(\mathbf{k}', t')\hat{b}_H(\mathbf{Q}, t')\hat{b}_H(\mathbf{q}, \tau)] \rangle \right\} \tag{8.8a}$$

The equation of motion for $\Xi'(\mathbf{k}, t; \mathbf{k}', t'; \mathbf{q}, \tau)$ is the same as the preceding with $\hat{b}_H(\mathbf{q}, \tau)$ replaced by $\hat{b}_H^\dagger(\mathbf{q}, \tau)$. [We will refer to this implicit equation as (8.8b).] Thus the equations of motion for Ξ and Ξ', (8.7) and (8.8), involve additional unknowns and, with (8.6), do not comprise a closed system. To remedy this situation, we set $t = \tau$ and assume a homogeneous phonon distribution and write

$$\langle \hat{b}_H^\dagger(\mathbf{q}')\hat{b}_H(\mathbf{q})\rangle = (2\pi)^3\delta(\mathbf{q}-\mathbf{q}')n(\mathbf{q})$$
$$\langle \hat{b}_H(\mathbf{q})\hat{b}_H^\dagger(\mathbf{q}')\rangle = (2\pi)^3\delta(\mathbf{q}-\mathbf{q}')[n(\mathbf{q})+1]$$
$$\langle \hat{b}_H(\mathbf{q})\hat{b}_H(\mathbf{q}')\rangle = 0 \qquad (8.9)$$
$$\langle \hat{b}_H^\dagger(\mathbf{q})\hat{b}_H^\dagger(\mathbf{q}')\rangle = 0$$
$$\langle \hat{b}_H(\mathbf{q})\rangle = \langle \hat{b}_H^\dagger(\mathbf{q}')\rangle$$

where $n(\mathbf{q})$ is phonon number density. Assuming further that the time-ordering terms in (8.7) factor into products such as

$$\langle \hat{T}[\hat{a}_H(\mathbf{k}+\mathbf{Q},t)\hat{a}_H^\dagger(\mathbf{k}',t')\hat{b}_H^\dagger(\mathbf{Q},t)\hat{b}_H(\mathbf{q},t)]\rangle$$
$$= \langle \hat{T}[\hat{a}_H(\mathbf{k}+\mathbf{Q},t)\hat{a}_H^\dagger(\mathbf{k}',t')]\rangle\langle \hat{b}_H^\dagger(\mathbf{Q},t)\hat{b}_H(\mathbf{q},t)\rangle \qquad (8.10)$$

and neglecting the two-body Green's functions in (8.7), we obtain

$$i\hbar\frac{D}{Dt}\Xi(\mathbf{k}-\mathbf{q},t;\mathbf{k}',t';\mathbf{q},t)$$
$$\simeq \Xi(\mathbf{k}-\mathbf{q},t;\mathbf{k}',t';\mathbf{q},t)[\hbar\omega(\mathbf{q})+E(\mathbf{k}-\mathbf{q})]+C^*(\mathbf{q})n(\mathbf{q})G(\mathbf{k},t;\mathbf{k}',t')$$
$$(8.11\text{a})$$

$$i\hbar\frac{D}{Dt}\Xi'(\mathbf{k}+\mathbf{q},t;\mathbf{k}',t';\mathbf{q},t)$$
$$\simeq \Xi'(\mathbf{k}+\mathbf{q},t;\mathbf{k}',t';\mathbf{q},t)[-\hbar\omega(\mathbf{q})+E(\mathbf{k}+\mathbf{q})]+C(\mathbf{q})[n(\mathbf{q})+1]G(\mathbf{k},t;\mathbf{k}',t')$$
$$(8.11\text{b})$$

Here we have set

$$\frac{D}{Dt} \equiv \frac{\partial}{\partial t} + \frac{\partial}{\partial \tau}$$

and, after differentiation, set $t=\tau$.

Introducing the time transformation

$$G(\mathbf{k},t;\mathbf{k}',t') = \int \frac{d\omega}{2\pi}e^{-i\omega(t-t')}G(\mathbf{k},\mathbf{k}',\omega)$$
$$G(\mathbf{k},\mathbf{k}',\omega) = \int d(t-t')e^{i\omega(t-t')}G(\mathbf{k},t,\mathbf{k}',t')$$

into (8.6) and (8.11), we obtain

$$\hbar\omega G(\mathbf{k},\mathbf{k}',\omega) = (2\pi)^3\delta(\mathbf{k}-\mathbf{k}') + E(\mathbf{k})G(\mathbf{k},\mathbf{k}',\omega)$$
$$+ \int \frac{d\mathbf{q}}{(2\pi)^3}[C^*(\mathbf{q})\Xi'(\mathbf{k}+\mathbf{q},\mathbf{k}',\mathbf{q},\omega)$$
$$+ C(\mathbf{q})\Xi(\mathbf{k}-\mathbf{q},\mathbf{k}',\mathbf{q},\omega)] \qquad (8.12\text{a})$$
$$\hbar\omega\Xi(\mathbf{k}-\mathbf{q},\mathbf{k}',\mathbf{q},\omega) \simeq [\hbar\omega(\mathbf{q})+E(\mathbf{k}-\mathbf{q})]\Xi(\mathbf{k}-\mathbf{q},\mathbf{k}',\mathbf{q},\omega)$$
$$+ C^*(\mathbf{q})n(\mathbf{q})G(\mathbf{k},\mathbf{k}',\omega) \qquad (8.12\text{b})$$

$$\hbar\omega \, \Xi'(\mathbf{k} + \mathbf{q}, \mathbf{k}', \mathbf{q}, \omega) \simeq [-\hbar\omega(\mathbf{q}) + E(\mathbf{k} + \mathbf{q})] \Xi'(\mathbf{k} + \mathbf{q}, \mathbf{k}', \mathbf{q}, \omega)$$
$$+ C(\mathbf{q})[n(\mathbf{q}) + 1] G(\mathbf{k}, \mathbf{k}', \omega) \qquad (8.12c)$$

5.8.3   The Spectral Function

These equations comprise three equations for G, Ξ, Ξ'. Solving for G, we obtain

$$\left[\hbar\omega - E(\mathbf{k}) - \sum(\mathbf{k}, \omega) \right] G(\mathbf{k}, \mathbf{k}', \omega) = (2\pi)^3 \delta(\mathbf{k} - \mathbf{k}') \qquad (8.13)$$

where

$$\sum(\mathbf{k}, \omega) \equiv \int \frac{d\mathbf{q}}{(2\pi)^3} |C(\mathbf{q})|^2 \qquad (8.13a)$$
$$\times \left[\frac{n(\mathbf{q}) + 1}{\hbar\omega - E(\mathbf{k} + \mathbf{q}) + \hbar\omega(\mathbf{q})} + \frac{n(\mathbf{q})}{\hbar\omega - E(\mathbf{k} - \mathbf{q}) - \hbar\omega(\mathbf{q})} \right]$$

In the homogeneous limit, we integrate (8.13) over $\mathbf{k}'$ to obtain

$$\left[\hbar\omega - E(\mathbf{k}) - \sum(\mathbf{k}, \omega) \right] G(\mathbf{k}, \omega) = 1 \qquad (8.14)$$

In analogy with (7.39), we write

$$G(\mathbf{k}, \omega) = \frac{1 \pm f(\mathbf{k})}{\hbar\omega - E(\mathbf{k}) - \sum(\mathbf{k}, \omega + i\eta)} \mp \frac{f(\mathbf{k})}{\hbar\omega - E(\mathbf{k}) - \sum(\mathbf{k}, \omega - i\eta)} \qquad (8.15)$$

and note that in the limit $\eta \to 0$, $G(\mathbf{k}, \omega)$, as given by (8.15), is a solution to (8.14). Furthermore, in the limit that interaction is turned off, $C = 0$, whence $\sum = 0$ and

$$G(\mathbf{k}, \omega) \to G_0(\mathbf{k}, \omega)$$

the free-particle Green's function.

We now make the identification

$$\lim_{\eta \to 0} \sum(\mathbf{k}, \omega \pm i\eta) = \Delta(\mathbf{k}, \omega) \mp i\Gamma(\mathbf{k}, \omega) \qquad (8.16)$$

where Δ and Γ are real parameters and $\Gamma \geq 0$.

With (8.16), (8.15) may be written

$$G(\mathbf{k}, \omega) = \frac{1 \pm f(\mathbf{k}, \mathbf{q})}{\hbar\omega - E(\mathbf{k}) - \Delta(\mathbf{k}, \omega + i\Gamma(\mathbf{k}, \omega)}$$
$$\mp \frac{1 \pm f(\mathbf{k})}{\hbar\omega - E(\mathbf{k}) - \Delta(\mathbf{k}, \omega) - i\Gamma(\mathbf{k}, \omega)} \qquad (8.17)$$

In analogy with (7.41), we write

$$G_{\substack{R \\ A}}(k, \omega) = \frac{1}{\hbar\omega - E(\mathbf{k}, \omega) - \Delta(\mathbf{k}, \omega) \pm i\Gamma(\mathbf{k}, \omega)} \qquad (8.18)$$

which, in the limit $C \to 0$, returns the free particle functions $G_R^{(0)} {}_A$ as given by (7.41). With (8.18), (8.17) may be written

$$G(\mathbf{k}, \omega) = [1 \pm f(\mathbf{k})]G_R(k, \omega) \mp f(\mathbf{k})G_A(\mathbf{k}, \omega) \qquad (8.19)$$

Explicit time dependence

To revert to the time-dependent Green's function, we write

$$G(\mathbf{k}, t) = \int \frac{d\omega}{2\pi} e^{-i\omega t} G(\mathbf{k}, \omega)$$

Noting that $G_R(\mathbf{k}, \omega)$ has no poles in the upper-half ω plane and that $G_A(\mathbf{k}, \omega)$ has no poles in the lower-half ω plane, we write

$$G(\mathbf{k}, t) = \theta(t) \oint_{LH} \frac{d\omega}{2\pi} e^{-i\omega t} [1 \pm f(\mathbf{k})]G_R(\mathbf{k}, \omega)$$
$$+ \theta(-t) \oint_{UH} \frac{d\omega}{2\pi} e^{-i\omega t} [\mp f(\mathbf{k})]G_A(\mathbf{k}, \omega) \qquad (8.20)$$

Here we have written LH to denote a path that includes the real ω axis and a great semicircle in the lower-half ω plane, whereas UH includes a great semicircle in the upper-half ω plane.

To recapture the spectral function, with the property

$$\oint_{LH} \frac{d\omega}{2\pi} G_A(\mathbf{k}, \omega) = \oint_{UH} \frac{d\omega}{2\pi} G_R(\mathbf{k}, \omega) = 0$$

we write

$$G(\mathbf{k}, t) = \theta(t) \oint_{LH} \frac{d\omega}{2\pi} e^{-i\omega t} [1 \pm f(\mathbf{k})][G_R(\mathbf{k}, \omega) - G_A(\mathbf{k}, \omega)]$$
$$= \theta(-t) \oint_{UH} \frac{d\omega}{2\pi} e^{-i\omega t} [\pm f(\mathbf{k})][G_R(\mathbf{k}, \omega) - G_A(\mathbf{k}, \omega)] \qquad (8.21)$$

With (7.27), this expression permits the identification

$$G^>(\mathbf{k}, \omega) = [1 \pm f(\mathbf{k})][G_R(\mathbf{k}, \omega) - G_A(\mathbf{k}, \omega)]$$
$$G^<(\mathbf{k}, \omega) = \pm f(\mathbf{k})[G_R(\mathbf{k}, \omega) - G_A(\mathbf{k}, \omega)] \qquad (8.22)$$

Furthermore, we note that (8.21) implies that the time dependence of $G(\mathbf{k}, \omega)$ is contained in the spectral function [recall (7.12)]

$$A(\mathbf{k}, \omega) = i[G_R(\mathbf{k}, \omega) - G_A(\mathbf{k}, \omega)] \qquad (8.23)$$

5.8.4 Lorentzian Form

We conclude this discussion with a derivation of an explicit expression for the spectral function (8.23). From (8.18), we write

$$i[G_R(\mathbf{k}, \omega) - G_A(\mathbf{k}, \omega)] = i \left[\frac{1}{\hbar\omega - E(\mathbf{k}) - \Delta(\mathbf{k}, \omega) + i\Gamma(\mathbf{k}, \omega)} \right.$$

$$\left. -\frac{1}{\hbar\omega - E(\mathbf{k}) - \Delta(\mathbf{k}, \omega) - i\Gamma(\mathbf{k}, \omega)} \right]$$

which gives the main result of the preceding analysis:

$$A(\mathbf{k}, \omega) = \frac{2\Gamma(\mathbf{k}, \omega)}{[\hbar\omega - E(\mathbf{k}) - \Delta(\mathbf{k}, \omega)]^2 + \Gamma^2(\mathbf{k}, \omega)} \qquad (8.24)$$

This Lorentzian form is the desired generalization of the free-particle spectral function (7.12). Note that with (8.18a), $\sum(\mathbf{k}, \omega)$ [or, equivalently, $\Delta(\mathbf{k}, \omega)$ and $\Gamma(\mathbf{k}, \omega)$] contains the specifics of the phonon–electron interaction for the problem at hand. Thus, to recapture the spectral function appropriate to no interaction, we consider the limit

$$\lim_{\Gamma, \Delta \to 0} \left[\frac{2\Gamma(\mathbf{k}, \omega)}{[\hbar\omega - E(\mathbf{k}) - \Delta]^2 + \Gamma^2(\mathbf{k}, \omega)} \right] = 2\pi \delta[\hbar\omega - E(\mathbf{k})] \qquad (8.25)$$

which agrees with our preceding free-particle expression (7.12), with the free-particle energy $\hbar^2 k^2 / 2m$ replaced by the quasi-free-particle energy $E(\mathbf{k})$.

To gain deeper physical insight into these results, first we substitute the general form (8.24) into the Green's function (8.21) to obtain

$$G(\mathbf{k}, t) = \theta(t) \int_{\text{LH}} \frac{d\omega}{2\pi i} e^{-i\omega t} [1 \pm f(\mathbf{k})] A(\mathbf{k}, \omega)$$

$$+ \theta(-t) \int_{\text{UH}} \frac{d\omega}{2\pi i} e^{-i\omega t} [\pm f(\mathbf{k}) A(\mathbf{k}, \omega)] \qquad (8.26)$$

Again consider the limit of no interaction. With (8.25) substituted in (8.26) and integrating over ω, we find

$$G_0(\mathbf{k}, t) = \frac{\theta(t)}{i\hbar} \left(\exp -i \frac{E(\mathbf{k})t}{\hbar} \right) [1 \pm f(\mathbf{k})]$$

$$+ \frac{\theta(-t)}{i\hbar} \left(\exp -i \frac{E(\mathbf{k})t}{\hbar} \right) [\pm f(\mathbf{k})] \qquad (8.27)$$

We note that this form is identical to that obtained from the Fourier inversion of (7.36). Thus (8.27) is a representation of the quasi-free-particle Green's function.

For further interpretation of the variables Δ and Γ in the interaction $\sum$ [see (8.16), let us assume that both these variables are frequency independent. In this event

$$A(\mathbf{k}, \omega) = \frac{2\Gamma(\mathbf{k})}{[\hbar\omega - E(\mathbf{k}) - \Delta(\mathbf{k})]^2 + \Gamma^2(\mathbf{k})} \qquad (8.28)$$

With

$$z^2 \equiv \frac{[\hbar\omega - E(\mathbf{k}) - \Delta(\mathbf{k})]^2}{\hbar^2}$$

the inverse frequency transform of $A(\mathbf{k}, \omega)$ is written

$$A(\mathbf{k}, t) = \int_{-\infty}^{\infty} \frac{d\omega}{2\pi} e^{-i\omega t} A(\mathbf{k}, \omega) = \frac{1}{\pi\hbar} \left(\exp -i \frac{(E + \Delta)t}{\hbar} \right) \int_{-\infty}^{\infty} \frac{\gamma e^{-izt} dz}{z^2 + \gamma^2}$$

(8.29)

where $\gamma \equiv \Gamma/\hbar$. Thus

$$A(\mathbf{k}, \omega) = \frac{1}{\pi\hbar} \left(\exp -i \frac{(E + \Delta)t}{\hbar} \right) \int_{-\infty}^{\infty} \frac{\gamma e^{-izt} dz}{z^2 + \gamma^2}$$

(8.30)

Consider the integral

$$I \equiv \frac{1}{\pi} \int_{-\infty}^{\infty} \frac{\gamma e^{-izt} dz}{z^2 + \gamma^2} = \frac{1}{\pi} \int \frac{\gamma e^{-izt} dz}{(z - i\gamma)(z + i\gamma)}$$

$$= \int \frac{1}{2\pi i} \left(\frac{1}{z - i\gamma} - \frac{1}{z + i\gamma} \right) e^{-izt} dz$$

$$= \oint_{\substack{\text{UH} \\ (t<0)}} \frac{1}{2\pi i} \frac{e^{-izt}}{z - i\gamma} dz - \oint_{\substack{\text{LH} \\ (t>0)}} \frac{1}{2\pi i} \frac{e^{-izt}}{z - i\gamma} dz$$

That is, the first integral in the last equality is evaluated on UH [defined below (8.20) with $t < 0$ (for convergence), whereas the second integral is defined on LH with $t > 0$. Thus we obtain

$$A(\mathbf{k}, t) = \frac{1}{\hbar} \left(\exp -i \frac{(E + \Delta)t}{\hbar} \right) [\theta(-t)e^{\Gamma t/\hbar} + \theta(t)e^{-\Gamma t/\hbar}]$$

(8.31)

Substituting this result into (8.26) and recalling the property $\theta(t)\theta(-t) = 0$, we obtain

$$G(\mathbf{k}, t) = \frac{\theta(t)}{i\hbar} [1 \pm f(\mathbf{k})] \left(\exp -i \frac{(E + \Delta)t}{\hbar} \right) e^{-\Gamma t/\hbar}$$

$$\pm \theta(-t) f(\mathbf{k}) \left(\exp -i \frac{(E + \Delta)t}{\hbar} \right) e^{\Gamma t/\hbar}$$

(8.32)

5.8.5 Lifetime and Energy of a Quasi-Free Particle

To discover the meaning of the expression (8.32), we assume that the system at hand is in the pure state $|\Phi\rangle$. Consider that at time $t \equiv 0$ an electron with momentum $\hbar\mathbf{k}$ is added to the system. The initial state of the new system is then given by

$$|\phi_{\mathbf{k}}(0)\rangle = \hat{a}_H^\dagger(\mathbf{k}, t = 0) |\Phi\rangle$$

(8.33a)

With the electron now contained in the system, we view it as a quasi-free particle with energy $E(\mathbf{k}) + \Delta(\mathbf{k})$. At $t > 0$, (8.33a) becomes

$$|\phi_{\mathbf{k}}(t)\rangle = \hat{a}_H^\dagger(\mathbf{k}, t) |\Phi\rangle$$

(8.33b)

Now we ask the question, what is the probability density for finding the particle with momentum $\hbar\mathbf{k}'$ at $t > 0$ if it had momentum $\hbar\mathbf{k}$ at $t = 0$? This probability is given by

$$P(t) = |\langle \psi_{\mathbf{k}'}(t) \mid \psi_{\mathbf{k}}(0) \rangle|^2$$
$$= |\langle \Phi | \hat{a}_H(\mathbf{k}', t) \hat{a}_H^\dagger(\mathbf{k}, 0)|\Phi\rangle|^2 \tag{8.34}$$

With (8.5), we may then write

$$P(t) = |i\hbar G(\mathbf{k}', t; \mathbf{k}, 0)|^2 \tag{8.35}$$

For a system homogeneous in space and time, we write

$$G(\mathbf{k}', t'; \mathbf{k}, t) = (2\pi)^2 \delta(\mathbf{k}' - \mathbf{k}) G(\mathbf{k}, t' - t) \tag{8.36}$$

Substituting $G(\mathbf{k}, t)$ as given by (8.32) into this equation, for $t > 0$, the latter two equations give

$$P(t) = \left|(2\pi)^3 \delta(\mathbf{k} - \mathbf{k}')[1 \pm f(\mathbf{k})] \exp\left\{-\frac{i}{\hbar}[E(\mathbf{k}) + \Delta(\mathbf{k})]t - \frac{\Gamma(\mathbf{k})t}{\hbar}\right\}\right|^2$$
$$= e^{-2\Gamma(\mathbf{k})t/\hbar} P(0) \tag{8.37}$$

With $\mathbf{k} = \mathbf{k}'$, $P(t)$ gives the desired expression for the probability density that the particle remains with momentum $\hbar\mathbf{k}$ at time t. In this sense, the lifetime of our quasi-particle is given by $\hbar/2\Gamma$.

With reference to (8.17), we may conclude that the poles of the Green's function in the complex ω-plane give the lifetime $\hbar/2\Gamma$ and energy $E + \Delta$ of a particle interacting with its surroundings.

Problems

5.1. (a) Integrate Heisenberg's equation of motion (1.44) for a free-particle Hamiltonian to obtain $\hat{q}(t)$ and $\hat{p}(t)$ as a function of $\hat{q}(0)$ and $\hat{p}(0)$.
   (b) Show that in this case

$$[\hat{q}(t), \hat{q}(0)] = -\frac{i\hbar}{m}t$$

Answer (partial)

Inserting

$$\hat{H} = \frac{\hat{p}^2}{2m}$$

into (1.44), we find

$$i\hbar\frac{d\hat{p}}{dt} = 0, \qquad \hat{p} = \hat{p}(0)$$

$$i\hbar\frac{d\hat{q}}{dt} = [\hat{q}, \hat{H}] = i\hbar\frac{\hat{p}}{m}$$

$$\hat{q}(t) = \hat{q}(0) + \frac{\hat{p}(0)t}{m}$$

5.2. A homogeneous beam of electrons has the density matrix

$$\hat{p} = \begin{pmatrix} a & 0 \\ 0 & 1-a \end{pmatrix}$$

in a representation where $\hat{S}_z$ and $\hat{S}^2$ are diagonal. If $\langle S_z \rangle = 0.75\hbar$, what is the value of a?

5.3. (a) If a density matrix is initially diagonal, what are the values of the derivatives $\partial p_{nn}/\partial t$ at $t = 0$?

(b) Is your answer to part (a) consistent with the Pauli equation (2.28)? If not, explain this discrepancy.

5.4. (a) Show that, in a pure state, $\hat{p}^2 = \hat{p}$.

(b) Show that for spinors this property is obeyed by the matrix

$$\hat{p} = \begin{pmatrix} \cos^2\theta & e^{i\phi_1}\sin\theta\cos\theta \\ e^{-i\phi_2}\sin\theta\cos\theta & \sin^2\theta \end{pmatrix}$$

providing the phases ϕ_1 and ϕ_2 are equal.

5.5. In a representation where $\hat{S}^2$ and $\hat{S}_z$ are diagonal, the density matrix for a neutron in a homogeneous beam is

$$\hat{p} = \begin{pmatrix} \frac{3}{4} & 0 \\ 0 & \frac{1}{4} \end{pmatrix}$$

The beam enters a region of space containing a magnetic field $\mathbf{B} = (0, B_0, 0)$.

(a) What are the values of $\langle S_x \rangle$, $\langle S_y \rangle$, and $\langle S_z \rangle$ before the beam enters the $\mathbf{B}$ field?

(b) The Hamiltonian for a neutron in a $\mathbf{B}$ field may be taken to be of the form

$$\hat{H} = -\hat{\mu}_n \cdot \mathbf{B}$$

where $\hat{\mu}_n$ is the magnetic moment of the neutron. Employing the Pauli equation (2.28), determine the *probabilities* of finding $S_z = +(\hbar/2)$ and $S_z = -(\hbar/2)$ after the beam leaves the $\mathbf{B}$-field domain. Data:

$$\hat{\mu}_n = g\left(\frac{\mu_N}{\hbar}\right)\hat{\mathbf{S}}$$

where

$$\mu_N = 0.505 \times 10^{-23}\,\text{erg/gauss}$$

and

$$g = 2(-1.91)$$

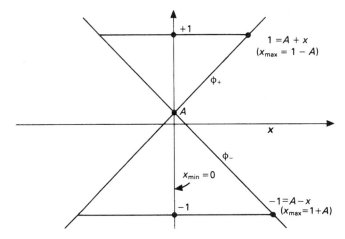

FIGURE 5.10. Graphical solution to Problem 5.9. Recall that the relevant domain is $x \geq 0$.

5.6. Show that integration over momentum of the Wigner distribution gives the spatial probability density. That is, establish the equality (2.49).

5.7. Show that the integral of the Wigner distribution function over coordinate space gives the momentum probability density. That is, establish (2.50).

5.8. Show that the Wigner–Moyal equation (2.70) goes to the form (2.71) for the canonical Hamiltonian given between these two equations.

5.9. Show that the inequalities (3.78a) relevant to the relaxation-time integral (3.78) imply the solutions (3.78b). *Hint*: Employ a graphical technique.

Answer

First label

$$x \equiv \frac{q}{2k} \geq 0$$
$$A \equiv \frac{mu}{\hbar k}$$

We must find the minimum and maximum of x (that is, q) corresponding to

$$-1 \leq \phi_\pm \leq +1$$

where

$$\phi_\pm \equiv A \pm x$$

This function is plotted against x in Fig. 5.10, from which we see that

$$x_{min} = 0 \Rightarrow q_{min} = 0$$
$$x_{max} = 1 \pm A \Rightarrow$$
$$q_{max} = 2k(1 \pm A) \simeq 2k$$

5.10. Employing the expression for the relaxation time $\tau(\mathbf{k})$ obtained in Section 3.6 and assuming free-particle motion near the Fermi surface, with (3.40) obtain an expression for electrical conductivity σ, in terms of mean-free path, l, and $k_F = p_F/\hbar$, the Fermi wave vector.

Answer

Substituting

$$dS = k_F^2 d\Omega$$

$$E \simeq \frac{\hbar^2 k^2}{2m}$$

together with (3.81)

$$\tau(k) = \frac{lm}{\hbar k}$$

into (3.40) gives

$$J = \frac{e^2 l}{4\pi^3 \hbar} \int \frac{\mathbf{v}\mathcal{E}\cdot\mathbf{v}}{v^2} k_F^2 d\Omega$$

$$= \frac{e^2 l k_F^2}{4\pi^3 \hbar} \mathcal{E} \cdot \int \overline{\overline{\hat{\mathbf{v}}\hat{\mathbf{v}}}} d\Omega$$

Recalling Problem 3.15 (see also Appendix B, Section B.1) we write

$$\mathcal{E} \cdot \int \overline{\overline{\hat{\mathbf{v}}\hat{\mathbf{v}}}} d\Omega = \mathcal{E} \cdot \frac{\overline{\overline{I}}}{3} 4\pi = \frac{4\pi}{3} \mathcal{E}$$

which gives the desired result:

$$\sigma = \frac{e^2 l k_F^2}{3\pi^2 \hbar}$$

5.11. Employing the rule (2.51) show that the first three moments of the Wigner–Moyal equation (2.71) return the classical fluid dynamical equations (3.3.14), (3.3.18), and (3.3.19).

5.12. Show that the s-particle density matrix (5.15) may be written as

$$\rho_s(\mathbf{x}^s, \mathbf{y}^s, t) = \frac{(N-s)!}{N!} \mathrm{Tr}^{(N)}[\hat{\rho}_N(t)\hat{\tilde{G}}_s(\mathbf{x}^s, \mathbf{y}^s, 0)]$$

5.13. Employing results of Problem 5.12, show that in a pure state, with $\hat{\rho}_N = |\mathbf{n}\rangle\langle\mathbf{n}|$:

(a) $N\rho_1(\mathbf{x}, \mathbf{y}) = \delta_{xy} n_x$

(b) $N(N-1)\rho_2(\mathbf{x}_1, \mathbf{x}_2, \mathbf{y}_1, \mathbf{y}_2) = (\delta_{x_1 y_1}\delta_{x_2 y_2} \pm \delta_{x_1 y_2}\delta_{x_2 y_1}) \begin{bmatrix} n_{x_1}(n_{x_2} - \delta_{x_1 x_2}) \\ n_{x_1} n_{x_2} \end{bmatrix}$

In part (b), the top term and $+$ sign are relevant to bosons and the bottom term and $-$ sign to fermions.

(c) Show that (for $\mathbf{x}_1 \neq \mathbf{x}_2$)

$$\rho_2(\mathbf{x}_1, \mathbf{x}_2, \mathbf{y}_1, \mathbf{y}_2) = \begin{vmatrix} \rho_1(\mathbf{x}_1, \mathbf{y}_1) & \rho_1(\mathbf{x}_1, \mathbf{y}_2) \\ \rho_1(\mathbf{x}_2, \mathbf{y}_2) & \rho_1(\mathbf{x}_2, \mathbf{y}_2) \end{vmatrix}_{\pm}$$

where $(+, -)$ are relevant to bosons and fermions, respectively. *Hint*: For, part (b) note that for fermions we may write

$$n^2 = n$$
$$n(1 - n) = 0$$
$$n(2 - n) = n$$

5.14. Show that the Fock-space Hamiltonian (5.27) remains invariant under the change

$$\hat{\phi}_H(\mathbf{x}) \to \hat{\phi}(\mathbf{x})$$

Hint: Note that, in $\hat{U}$, $\hat{H}$ is the Fock-space Hamiltonian in the Schrödinger representation.

5.15. Obtain a quasi-classical kinetic equation valid to $0(\hbar^3)$ from the Wigner–Moyal equation (2.71).

5.16. Write down the specific form of the first equation (BY_1) in the hierarchy (3.33) for the Wigner distributions. In what limit does this relation become a closed equation?

5.17. (a) The thermodynamic properties of a system comprised of N molecules in equilibrium at temperature T and confined to a fixed volume V is contained in the *partition function*[63]

$$Z_N = \sum_r g_r e^{-\beta H_r(1,...,N)}$$

where the sum on r is over all states of the system, and g_r is the degeneracy of the rth state. Establish the following relations:

$$\langle E \rangle = -\frac{\partial \ln Z}{\partial \beta}, \qquad \beta P = \frac{\partial \ln Z}{\partial V}$$

$$S = -k_B \beta^2 \frac{\partial}{\partial \beta} \left(\frac{1}{\beta} \ln Z \right)$$

(b) The *grand partition function*, Q, is relevant to an equilibrium fluid under the same constraints that hold for Z_N but whose total number of molecules may vary. It is given by

$$Q = \sum_N z^N Z_N$$

---

[63]More generally, $Z_N = \text{Tr} \exp(-\beta \hat{H})$. The expression given for Z_N in the problem assumes a representation in which $\hat{H}$ is diagonal.

where

$$z = \exp \beta \mu$$

is fugacity and μ is chemical potential. Show that

$$\langle N \rangle = z \frac{\partial}{\partial z} \ln Q$$

$$(\Delta N)^2 \equiv \langle (N - \langle N \rangle)^2 \rangle = z \frac{\partial}{\partial z} \langle N \rangle$$

5.18. Consider a *two-dimensional* gas of noninteracting spin 1/2 fermions confined to an area A at temperature T and chemical potential μ.

   (a) What is $\ln Q$ for this system written as a sum over single-particle momenta p? ($\ln Q$ is defined in Problem 5.17.)
   (b) Convert the sum in your answer to part (a) to an integral employing the following rule:

$$\sum_p \rightarrow \frac{gA}{h^2} \int d^2 p$$

   where g is spin degeneracy.
   (c) Obtain an expression for the average number of particles $\langle N \rangle$ as a function of fugacity z and inverse temperature $\beta \equiv (k_B T)^{-1}$.
   (d) Obtain an expression for the Fermi energy E_F from your answer to part (c) in terms of $n_0 \equiv \langle N \rangle_0 / A$, where the zero subscript denotes zero temperature. Recall that $E_F = \mu\,(0\ \text{K})$.

Answers

   (a)
$$\ln Q = \sum_p \ln(1 + z e^{-\beta p^2/2m})$$

   (b)
$$\ln Q = \frac{2A}{h^2} \int_0^\infty 2\pi p\, dp\, \ln(1 + z e^{-\beta p^2/2m})$$

   (c)
$$\langle N \rangle = \frac{mA}{\pi \beta \hbar^2} \int_0^\infty \frac{z e^{-x} dx}{(1 + z e^{-x})}$$

$$\langle N \rangle = \frac{mA}{\pi \hbar^2 \beta} \ln(1 + z)$$

   (d)
$$\langle N \rangle_{T \to 0} \equiv \langle N \rangle_0 = \lim_{\beta \to \infty} \frac{mA}{\pi \hbar^2} \frac{\ln(1 + e^{-\beta E_F})}{\beta}$$

$$= \frac{mA}{\pi \hbar^2} E_F$$

Thus we find

$$E_F = \frac{\pi \hbar^2 n_0}{m}$$

5.19. Show that in a *mixed state*

$$\Delta A \Delta B \geq \frac{1}{2} |\langle [\hat{A}, \hat{B}] \rangle|$$

where ΔA is the uncertainly

$$(\Delta A)^2 = \langle (\hat{A} - \langle \hat{A} \rangle)^2 \rangle$$

and $\hat{A}$ and $\hat{B}$ are heritian.

Answer

Define

$$S \equiv \hat{A} + iz\hat{B}$$
$$\hat{S}^+ = \hat{A} - iz\hat{B}$$

where z is a real variable. It follows that

$$\hat{S}^+\hat{S} = \hat{A}^2 + iz[\hat{A}, \hat{B}] + z^2\hat{B}^2$$

For a mixed state, with (2.36), we write

$$\hat{\rho} = \sum_\mu P_\mu |\psi_\mu\rangle\langle\psi_\mu|$$

so that

$$\langle \hat{S}^+\hat{S}\rangle = \sum_\mu P_\mu \langle \hat{S}\psi_\mu | \hat{S}\psi_\mu\rangle \geq 0$$

Choosing the value

$$z = -\langle i[\hat{A}, \hat{B}]\rangle/2\langle\hat{B}^2\rangle$$

we find

$$\langle \hat{S}^+\hat{S}\rangle = -\frac{\langle i[\hat{A}, \hat{B}]\rangle^2}{4\langle\hat{B}^2\rangle} + \langle\hat{A}^2\rangle \geq 0$$

Now let $\hat{A} \to \hat{A} - \langle\hat{A}\rangle$ and $\hat{B} \to \hat{B} - \langle\hat{B}\rangle$. Then

$$\Delta A \Delta B \geq \frac{1}{2}|\langle[\hat{A}, \hat{B}]\rangle|$$

5.20. Show that the uncertainty relation $\Delta x \Delta p \geq \hbar/2$ is consistent with the Wigner distribution function.

Answer

From Problem 5.19, we obtain

$$\Delta x \Delta p \geq \frac{1}{2}|\langle[x, p]\rangle| = \frac{\hbar}{2}$$

where

$$(\Delta x)^2 = \langle x^2\rangle - \langle x\rangle^2$$

However, as established in (2.61a), both x^2 and x obey Weyl correspondence so that averages may be written with respect to the Wigner distribution function as given by (2.51).

5.21. A system is in an eigenstate of its Hamiltonian. Show that the Wigner distribution is constant in time for this state.

Answer

For an eigenstate, with (1.10), we write

$$\psi(x, t) = \varphi(x)E^{-itE/\hbar}$$

Substitution of this form into (2.47) indicates that the Wigner distribution is constant in time for an eigenstate of the Hamiltonian. Note that this conclusion also follows from (2.51):

$$\langle A \rangle = \int A(x, p)F(x, p, t)\,dx\,dp$$

For a stationary state, $\langle A \rangle$ is constant, so $F(x, p, t)$ must be stationary.

5.22. Show that the Wigner distribution is real.

Answer

We refer to the defining relation (2.46). Deleting vector notation, we write

$$F(x, p, t) \propto \int e^{2ipy/\hbar}\rho(x_-, x_+, t)\,dy$$

Taking the complex conjugate, we obtain

$$F^*(x, p, t) \propto \int e^{-2ipy/\hbar}\rho^*(x_-, x_+, t)\,dy$$
$$\rho^*(x_-, x_+, t) = \langle x_-| \hat{\rho} |x_+\rangle^*$$
$$= \langle x_+| \hat{\rho} |x_-\rangle$$
$$= \langle x + y| \hat{\rho} |x - y\rangle$$

Changing variables in the above integral expression for $F^*(x, p, t)$ from $y \to -y$ gives $F^* = F$.

5.23. A student argues that the Wigner distribution function is not consistent with quantum mechanics for the following reasons. In the absence of interactins, the equation of motion (2.54) reduces to

$$\left(\frac{\partial}{\partial t} + \frac{\mathbf{p}}{m} \cdot \frac{\partial}{\partial \mathbf{x}}\right) F(\mathbf{x}, \mathbf{p}, t) = 0$$

As was discussed in Chapter 1, the general solution to this equation is given by

$$F = F\left(\mathbf{x} - \frac{\mathbf{p}}{m}t - \mathbf{x}_0, \mathbf{p}_0\right)$$

In particular, we may take F to be a delta function for which $F \neq 0$ only on the system trajectory, in which case $\mathbf{x}$ and $\mathbf{p}$ may be specified simultaneously in violation of quantum mechanics. Is the student's argument sound? Why?

Answer

It was established in Problem 5.20 that the Wigner formalism is consistent with the uncertainty relation between **x** and **p**. This finding disavows solutions (2.54) that imply a classical trajectory.

5.24. Show that

$$([\hat{x} + f(\hat{p})]^2)_W = [x + f(p)]^2$$

Answer

First note that

$$\frac{1}{2}[\hat{A}, \hat{A}]_+ = \hat{A}^2$$

Then employ (2.61b)

$$\left(\frac{1}{2}[\hat{A}, \hat{A}]_+\right)_W = (\hat{A})_W \left(\cos \frac{\hbar}{2}\overleftrightarrow{0}\right)(\hat{A})_W = (\hat{A})_W^2$$

5.25. Show that the Wigner distributions given by (2.46) is normalized to unity.

Answer

Working in one dimension, we find

$$\iint F(x, p)\, dx\, dp = \frac{1}{\pi\hbar}\iiint dy\, dx\, dp\, e^{2ipy/\hbar}\, \langle x_- | \rho | x_+\rangle$$

Integrating over p gives

$$\iint F(x, p) dx\, dp = \iint dx\, dy\, \delta(y)\, \langle x_- | \hat{\rho} | x_+\rangle$$

$$= \int dx\, \langle x | \hat{\rho} | x\rangle = \mathrm{Tr}\, \hat{\rho} = 1$$

5.26. In quantum mechanics the continuity equation is given by

$$\frac{\partial \rho}{\partial t} + \nabla \cdot \mathbf{J} = 0$$

where ρ is particle density

$$\rho = \psi^* \psi$$

and **J** is particle current

$$\mathbf{J} = \frac{\hbar}{2mi}(\psi^*\nabla\psi - \psi\nabla\psi^*)$$

Working in a pure state, show that **J** and ρ may be written

$$\mathbf{J} = \int d\mathbf{p}\, \frac{\mathbf{p}}{m} F(\mathbf{x}, \mathbf{p})$$

$$\rho = \int d\mathbf{p}\, F(\mathbf{x}, \mathbf{p})$$

where $F(\mathbf{x}, \mathbf{p})$ is the Wigner distribution function.

5.27. Show that the z component of angular momentum

$$\hat{L}_x = \hat{x}\hat{p}_y - \hat{y}\hat{p}_x$$

obeys Weyl corrrespondence.

5.28. Employing the projection representation (2.36), derive the following expressions for the density matrix $\rho(x, x')$ for a canonical distribution [see (2.13)] of (a) free particles of mass m, and (b) particles of mass m confined to a one-dimensional box of length L.

(a) $\rho(x, x') = \dfrac{1}{Z} \displaystyle\int e^{-\beta \hbar^2 k^2 / 2m} e^{ik(x-x')}\, dk$

(b) $\rho(x, x') = \dfrac{1}{Z} \displaystyle\sum_{n=1}^{\infty} e^{-\beta E_n} \varphi_n(x)\varphi_n(x')$

where

$$E_n = n^2 E_1, \qquad E_1 = \frac{1}{2m}\left(\frac{\pi\hbar}{L}\right)^2$$

$$\varphi(x) = \sqrt{\frac{2}{L}}\, \sin\frac{n\pi x}{L}$$

and Z is a normalization factor that ensures that $\operatorname{Tr}\hat{\rho} = 1$ (that is, the partition function).

5.29. Write down the eigenvalue equation that the Fock-space ket vector given on the right side of (5.2) obeys.

5.30. The quantum mechanical s-body Green's function is given by (7.13) and (7.14). With $s = 1$ we obtain

$$G_1(\mathbf{x}, t, \mathbf{y}, t') = \frac{1}{i\hbar}\theta(t - t')\langle \hat{\phi}_H(\mathbf{x}, t)\hat{\phi}_H^{\dagger}(\mathbf{y}, t')\rangle$$

$$\pm \frac{1}{i\hbar}\theta(t' - t)\langle \hat{\phi}_H^{\dagger}(\mathbf{y}, t')\hat{\phi}_H(\mathbf{x}, t)\rangle$$

With $t' = t + \varepsilon$, where ε is an infinitesimal, show that

$$G(\mathbf{x}, \mathbf{y}, t) = \pm\frac{1}{i\hbar}\langle \hat{\hat{G}}_1(\mathbf{x}, \mathbf{y}, t)\rangle$$

where the F-space operator $\hat{\hat{G}}_1$ is given by (5.7).

5.31. (a) Show that

$$e^{-i\hat{\hat{S}}}\hat{A} = e^{-(t/i\hbar)\hat{H}_0}\hat{A}e^{(t/i\hbar)\hat{H}_0}$$

where $\hat{\hat{S}}$ is defined by (6.11).

(b) With the preceding result, establish (6.12b), that is,

$$\operatorname{Tr}\{(e^{-\hat{\hat{S}}}\hat{A})(e^{-i\hat{\hat{S}}}\hat{B})\} = \operatorname{Tr}\{\hat{A}\hat{B}\}$$

5.32. Employing the fundamanetal time-development equation (5.8), show that (7.33) follows from (7.32).

5.33. If the operator $\hat{R}$ in (7.1) is nonlinear, is the Green's function integral (7.3) still a solution of (7.1)? Explain your answer.

5.34. Employing the relation (7.15) for a time-ordered product, establish the differentiation rule (7.44).

5.35. Show that (7.47) coupling the first two Green's functions may be transformed to the first of the hierarchy for the $\hat{G}_s$ functions as given in Table 5.3. The specifics of this tranformation are given in the text.

5.36. This problem addresses *composite bosons*; that is, bosons comprised of two coupled fermions. Consider such a two-component composite system with respective creation operators $\hat{a}^\dagger$ and $\hat{b}^\dagger$ with the model Hamiltonian[64]

$$\hat{H} = E_0 \sum_i (\hat{a}_i^\dagger \hat{a}_i + \hat{b}_i^\dagger \hat{b}_i) + g \sum_i \sum_j \hat{b}_i^\dagger (\hat{a}_i^\dagger \hat{a}_j \hat{b}_j) \qquad (P36.1)$$

where E_0 and g are constants and $\hat{a}_i$ and $\hat{b}_i$ both satisfy fermion anticommutation relations. Let

$$\hat{A} \equiv \frac{1}{\sqrt{N_0}} \sum_i \hat{a}_i \hat{b}_i, \qquad A^\dagger \equiv \frac{1}{\sqrt{N_0}} \sum_i \hat{b}_i^\dagger \hat{a}_i^\dagger \qquad (P36.2)$$

and

$$\hat{H}_0 \equiv E_0 \sum_i (\hat{a}_i^\dagger \hat{a}_i + \hat{b}_i^\dagger \hat{b}_i) \qquad (P36.3)$$

where N_0 is the number of free-particle states.

(a) Show that (P36.1) may be written

$$\hat{H} = \hat{H}_0 + g N_0 \hat{A}^\dagger \hat{A} \qquad (P36.4)$$

(b) Show that

$$[\hat{A}^\dagger, \hat{A}] = \hat{I} - \frac{\hat{N}}{N_0} \qquad (P36.5)$$

where $\hat{N}$ is the operator $\hat{H}_0/E_0$.

(c) Given the above findings, what may be said of the bose quality of composite bosons with strong coupling ($g \gg E_0$)?

Answer (partial)

(c) The noncanonical form of (P36.5) indicates that $\hat{A}^\dagger$ is not a pure creation operator and in this sense invalidates the interpretation of $\hat{A}^\dagger A$ in (P36.4) as representing a quasi-particle number operator. Thus we may conclude

---

[64]For further discussion on this topic, see A. K. Kerman, *Annals of Physics 12*, 300 (1961).

that to within the consistency of the stated model the composite boson does not satisfy boson properties.

5.37. Show that (7.50) is equivalent to (7.47).

5.38. Obtain the explicit time-dependent equation of motion for $G(1, 2, 3; 1', 2', 3')$ [in a form parallel to (7.47)] corresponding to (7.56).

5.39. Write down the diagrammatic equation of motion for the four-particle Green's function $G(1, 2, 3, 4; 1', 2', 3', 4')$.

5.40. Write down the lowest-order diagrammatic expansion for the equation of motion for $G(1, 2; 1', 2')$, as given by (7.52), in the topological form given by (7.54a).

Answer

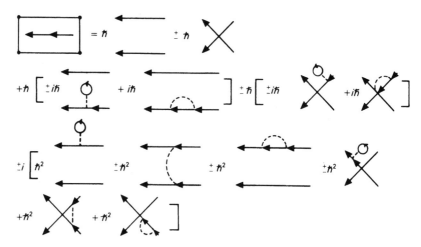

The last six diagrams in this expansion stem from the three-body interaction term in (7.52). Consider, for example, reorientation of the term

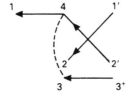

Maintaining the format of (7.52) gives the topologically equivalent form

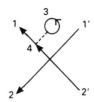

which is the third from the last diagram in the given expansion.

5.41. The Green's function for a quasi-free electron propagating through a medium is given by (with I denoting a constant length)

$$G(\mathbf{k}, \omega) = \left[\frac{i\theta(k - k_F)}{\hbar\omega - \left(\frac{\hbar^2 k^2}{2m} + \Delta \right) + \frac{i\hbar^2 k}{2ml}} \right]$$

(a) What is the single-particle distribution for the medium? What is its significance?

(b) What is the energy of the quasi-free particle?

(c) What is the lifetime of the quasi-free particle? Why is this so?

(d) What is the minimum excitation energy of a quasi-particle described by the Green's function above?

5.42. (a) Show that

$$\hat{T}[\hat{\phi}_H(t_1)\hat{\phi}_H^\dagger(t_2)] = \pm\hat{T}[\hat{\phi}_H^\dagger(t_2)\hat{\phi}_H(t_1)]$$

(b) Does validity of the preceding equation depend on the commutation properties on $\hat{\phi}_H$ and $\hat{\phi}_H^\dagger$?

5.43. (a) Write the correlation function for velocity operator $\overline{\langle \mathbf{v}(t)\mathbf{v}(0)\rangle}$, corresponding to particles of mass m, in terms of an integral over Heisenberg operators $\hat{a}_H^\dagger(\mathbf{k}, t)$, $a_H(\mathbf{k}, t)$.

(b) Show that this correlation function may be written in terms of a two-particle Green's function.

Answer

(a) $$\overline{\langle \hat{v}(t)\hat{v}(0)\rangle} = \int\int \frac{d\mathbf{k}_1 d\mathbf{k}_2}{(2\pi)^3(2\pi)^3} \frac{\hbar}{m^2}$$
$$\times \overline{\mathbf{k}_1 \mathbf{k}_2} \langle \hat{a}_H^\dagger(\mathbf{k}_1, t)\hat{a}_H(\mathbf{k}_1, t)\hat{a}_H^\dagger(\mathbf{k}_2, 0)\hat{a}_H(\mathbf{k}_2, 0)\rangle$$

(b) First recall

$$i^2\hbar G_2(1, 2; 1', 2') = \langle \hat{T}[\hat{a}_H(1)\hat{a}_H(2)\hat{a}_H^\dagger(2')\hat{a}_H^\dagger(1')]\rangle$$

with the results of Problem 5.42, the right side of the preceding equation may be rewritten as

$$\text{RHS} = \langle \hat{T}[\hat{a}_H(1)\hat{a}_H^\dagger(1')\hat{a}_H(2)\hat{a}_H^\dagger(2')]\rangle$$
$$= \langle \hat{T}[\hat{a}_H^\dagger(1')\hat{a}_H(1)\hat{a}_H^\dagger(2')\hat{a}_H(2)]\rangle$$

Now choose $1' = (\mathbf{k}_1, t^+)$, $1 = (\mathbf{k}_1, t)$, $2' = (\mathbf{k}_2, 0^+)$, $2 = (\mathbf{k}_2, 0)$ to to obtain

$$\overline{\langle \hat{v}(t)\hat{v}(0)\rangle} = \int\int \frac{d\mathbf{k}_1 d\mathbf{k}_2}{(2\pi)^3(2\pi)^3} \frac{\hbar}{m^2}$$
$$\times \overline{\mathbf{k}_1 \mathbf{k}_2} i^2\hbar G_2[(\mathbf{k}_1, t), (\mathbf{k}_2, 0); (\mathbf{k}_1, t^+), (\mathbf{k}_2, 0^+)]$$

which is the desired relation.

5.44. We have seen that in both the Chapman–Enskog expansion [see (3.5.37)] and in the study of electron–phonon interaction [see (3.58)] integral equations of the following form typically emerge (for elastic interactions):

$$R(\mathbf{v}) = \int [f(\mathbf{v}) - f(\mathbf{v}')]w(\mathbf{v}, \mathbf{v}')d\mathbf{v}'$$

$$R(\mathbf{v}) = \hat{K} f(\mathbf{v}) \qquad\qquad (P44.1)$$

These relations define the integral operator $\hat{K}$. With $R(\mathbf{v})$ taken to be a known function, the preceding is a linear inhomogeneous integral equation. Let us work in Dirac notation with the inner product given by

$$\langle \phi \mid \psi \rangle \equiv \int \phi(\mathbf{v})\psi(\mathbf{v})\,d\mathbf{v}$$

(We assume that all functions are real.) Given that w is a symmetric kernel:

(a) Show that $\hat{K}$ is self-adjoint. That is,

$$\langle \phi | \hat{K}\psi \rangle = \langle \hat{K}\phi \mid \psi \rangle = \langle \psi \mid \hat{K}\phi \rangle$$

(b) Show that $\hat{K}$ is a nonnegative operator. That is,

$$\langle \phi \mid \hat{K}\phi \rangle \geq 0$$

(c) Consider the equation

$$\langle \phi \mid \hat{K}\phi \rangle = \langle \phi \mid R \rangle \qquad\qquad (P44.2)$$

Not all solutions to this equation are solutions to (P44.1). (Two distinct vectors can have the same projection onto a third vector.) Show that of all solutions, $\phi(\mathbf{v})$, to (P44.2), the solution that satisfies (P44.1) maximizes $\langle \phi \mid \hat{K}\phi \rangle$.

(d) Show that solutions to (P44.1) render the functional

$$I[\phi] \equiv \frac{\langle \phi \mid \hat{K}\phi \rangle}{[\langle \phi \mid R \rangle]^2} \qquad\qquad (P44.3)$$

minimum.

Answer

(a) We may easily establish

$$\langle \phi \mid \hat{K}\psi \rangle = \frac{1}{2} \int\int [\phi(\mathbf{v}) - \phi(\mathbf{v}')]w(\mathbf{v}, \mathbf{v}')[\psi(\mathbf{v}) - \psi(\mathbf{v}')]\,d\mathbf{v}\,d\mathbf{v}'$$

from which the self-adjoint property of $\hat{K}$ follows. Setting $\psi = \phi$ into the preceding representation [with $w(\mathbf{v}, \mathbf{v}')$ positive and multiplicative] establishes part (b).

(c) Let $\phi(\mathbf{v})$ be a solution to (P44.1) and let $\psi(\mathbf{v})$ be a solution to (P44.2) but not to (P44.1). Then

$$0 \leq \langle (\phi - \psi) \mid \hat{K}(\phi - \psi) \rangle$$

$$= \langle \phi \mid \hat{K}\phi \rangle + \langle \psi | \hat{K}\psi \rangle - \langle \phi \mid \hat{K}\psi \rangle - \langle \psi \mid \hat{K}\phi \rangle$$
$$= \langle \phi \mid \hat{K}\phi \rangle + \langle \psi | \hat{K}\psi \rangle - 2\langle \psi \mid \hat{K}\phi \rangle$$
$$= \langle \phi \mid \hat{K}\phi \rangle + \langle \psi | \hat{K}\psi \rangle - 2\langle \psi \mid R \rangle$$
$$= \langle \phi \mid \hat{K}\phi \rangle + \langle \psi | \hat{K}\psi \rangle$$

Thus

$$\langle \phi \mid \hat{K}\phi \rangle \geq \langle \psi \mid \hat{K}\psi \rangle$$

which was to be shown.

(d) To show this, we consider the functional variation

$$\delta I[\phi] = I[\phi + \delta\phi] - I[\phi]$$
$$= \frac{\langle \delta\phi \mid \hat{K}\phi \rangle + \langle \phi \mid \hat{K}\delta\phi \rangle}{(\langle \phi \mid R \rangle)^2} - \frac{2\langle \phi \mid \hat{K}\phi \rangle\langle \delta\phi \mid R \rangle}{(\langle \phi \mid R \rangle)^3}$$
$$= \frac{2}{(\langle \phi \mid R \rangle)^3}[\langle \delta\phi \mid \hat{K}\phi \rangle\langle \phi \mid R \rangle - \langle \phi \mid \hat{K}\phi \rangle\langle \delta\phi \mid R \rangle]$$

Thus, for $\hat{I}[\phi]$ to be stationary, we must have

$$\langle \delta\phi \mid \hat{K}\phi \rangle\langle \phi \mid R \rangle = \langle \delta\phi \mid R \rangle\langle \phi \mid \hat{K}\phi \rangle$$

Viewing this equation in function space, with $|\delta\phi\rangle$ an arbitrary infinitesial element of the space, the preceding equality implies that

$$|\hat{K}\phi\rangle\langle \phi \mid R \rangle = |R\rangle \, \langle \phi \mid \hat{K}\phi \rangle \tag{P44.4}$$

We wish to show that this equation implies that

$$|\hat{K}\phi\rangle = c\,|R\rangle \tag{P44.5}$$

where c is a constant. First note that substituting (P44.5) into (P44.4) gives an equality. Now assume that (P44.4) does *not* imply (P44.5). Then we may write

$$\hat{K}\,|\phi\rangle = a\,|R\rangle + |\delta\rangle \tag{P44.6}$$

where $\langle \delta \mid R \rangle = 0$. Substituting this form into (P44.4) gives

$$|\delta\rangle \, \langle \phi \mid R \rangle = |R\rangle \, \langle \phi \mid \delta \rangle$$

which implies that $\langle \delta \mid R \rangle \neq 0$, whence the assumption (P44.6) is inconsistent, and we may conclude that (P44.4) implies (P44.5). Inserting the solution (P44.5) into (P44.3) indicates that $I[\phi]$ is insensitive to the constant c, and we may conclude that $I[\phi]$ is stationary when $\hat{K}\,|\phi\rangle = |R\rangle$, in which case

$$I[\phi] = \frac{1}{\langle \phi \mid \hat{K}\phi \rangle} \tag{P44.7}$$

As we have shown, $\langle \phi \mid \hat{K}\phi \rangle$ is maximum for $\hat{K}\,|\phi\rangle = |R\rangle$, so we conclude that (P44.7) is minimum, which was to be shown. *Note*: The fact that (P44.3) is minimum affords a means of obtaining an approximate solution to (P44.1): Choose a *trial function* comprised of known functions and a number of arbitrary parameters. Insert this trial solution into (P44.3) and vary the parameters until a minimum of this form is obtained. If the trial solution was a good guess, then the solution so obtained is a good approximate solution to (P44.1).[65]

5.45. Show that

$$\hat{\rho}(t) = \hat{U}^\dagger \hat{\rho}(0)\hat{U}$$

where $\hat{U}$ is given by (1.39).

Answer

Working in a pure state, we write

$$\hat{\rho}(t) = |t\rangle\langle t|$$

Recalling (1.41), it follows that

$$\hat{\rho}(t) = \hat{U}^\dagger |0\rangle\langle 0| \hat{U} = \hat{U}^\dagger \hat{\rho}(0)\hat{U}$$

5.46. Employing the Fock-space Hamiltonian (5.27), derive the equation of motion (5.28).

Answer

With (5.8), we write

$$i\hbar\frac{\partial}{\partial t}\hat{\phi}_H(\mathbf{x}) = [\hat{\phi}_H, \hat{H}]$$

Consider the kinetic energy term in $\hat{H}$ given in (5.27). Forming the commutator gives

$$\hat{\phi}_H(\mathbf{x})\int d\mathbf{x}'\hat{\phi}_H^\dagger(\mathbf{x}')\hat{K}(\mathbf{x}')\hat{\phi}_H(\mathbf{x}') - \int d\mathbf{x}'\hat{\phi}_H^\dagger(\mathbf{x}')\hat{K}(\mathbf{x}')\hat{\phi}_H(\mathbf{x}')\hat{\phi}_H(\mathbf{x})$$

$$= \int d\mathbf{x}'[\hat{\phi}_H(\mathbf{x})\hat{\phi}_H^\dagger(\mathbf{x}') - \hat{\phi}_H^\dagger(\mathbf{x}')\hat{\phi}_H(\mathbf{x})]\hat{K}(\mathbf{x}')\hat{\phi}_H(\mathbf{x}')$$

$$= \int d\mathbf{x}'\delta(\mathbf{x} - \mathbf{x}')\hat{K}(\mathbf{x}')\hat{\phi}_H(\mathbf{x}') = \hat{K}(\mathbf{x})\hat{\phi}_H(\mathbf{x})$$

which is the first term in the right side of (5.28). A similar construction gives the potential integral of (5.28).

5.47. Obtain the equation of motion for $\hat{G}_N$ given by (5.29) and expressions for $\hat{K}_N(\mathbf{x}^N)$ and $\hat{V}_N(\mathbf{x}^N)$.

---

[65]This technique has been employed in the calculation of electrical and thermal conductivity in metals. For further discussion, see A. Haug, *Theoretical Solid State Physics*, vol. 2, Pergamon, Elmsford, N.Y. (1972).

Answer

Consider the case $N = 2$. Dropping the subscript on $\hat{\phi}_H$, we write

$$i\hbar\frac{\partial}{\partial t}\hat{\tilde{G}}_2 = i\hbar[\dot{\hat{\phi}}^\dagger(\mathbf{y}_1)\hat{\phi}^\dagger(\mathbf{y}_2)\hat{\phi}(\mathbf{x}_2)\hat{\phi}(\mathbf{x}_1)$$

$$+ \hat{\phi}^\dagger(\mathbf{y}_1)\dot{\hat{\phi}}^\dagger(\mathbf{y}_2)\hat{\phi}(\mathbf{x}_2)\hat{\phi}(\mathbf{x}_1)$$

$$+ \hat{\phi}^\dagger(\mathbf{y}_1)\hat{\phi}^\dagger(\mathbf{y}_2) + \dot{\hat{\phi}}(\mathbf{x}_2)\hat{\phi}(\mathbf{x}_1)$$

$$+ \hat{\phi}^\dagger(\mathbf{y}_1)\hat{\phi}^\dagger(\mathbf{y}_2)\hat{\phi}(\mathbf{x}_2)\dot{\hat{\phi}}(\mathbf{x}_1)]$$

With (5.28), we obtain

$$i\hbar\frac{\partial}{\partial t}\hat{\tilde{G}}_2 = [-\hat{K}^*(\mathbf{y}_1) - \hat{Q}^*(\mathbf{y}_1) - \hat{K}^*(\mathbf{y}_2) - \hat{Q}^*(\mathbf{y}_2)$$

$$+ \hat{K}^*(\mathbf{x}_1) + \hat{Q}^*(\mathbf{x}_1) + \hat{K}^*(\mathbf{x}_2) + \hat{Q}^*(\mathbf{x}_2)]\hat{\tilde{G}}_2$$

where $\hat{Q}(\mathbf{x})$ is written for the integral operator in (5.28). Generalizing this finding indicates that

$$\hat{K}_N(\mathbf{x}^N) = \sum_{i=1}^{N} \hat{K}(\mathbf{x}_i)$$

$$\hat{V}_N(\mathbf{x}^N) = \sum_{i=1}^{N} \int \hat{\phi}^\dagger(\mathbf{x}')u(\mathbf{x}_i, \mathbf{x}')\hat{\phi}(\mathbf{x}')d\mathbf{x}'$$

5.48. Show that the master equation (2.28) implies the $\mathcal{H}$ theorem.

Answer

We label $\rho_{nn} \equiv P_n$ and define

$$\mathcal{H} = \sum_n P_n \ln P_n$$

Operating on (2.28) with

$$\sum_n (1 + \ln P_n)$$

gives

$$\frac{d\mathcal{H}}{dt} = \sum_k \sum_n (1 + \ln P_n)(P_k w_{kn} - P_n w_{nk})$$

With

$$\sum_n \sum_k P_k w_{kn} = \sum_n \sum_k P_n w_{nk}$$

the preceding reduces to

$$\frac{d\mathcal{H}}{dt} = \sum_k \sum_n \ln P_n(P_k w_{kn} - P_n w_{nk})$$

To obtain the $\mathcal{H}$-theorem, first exchange indices in the second term on the right. This gives

$$\frac{d\mathcal{H}}{dt} = \sum_k \sum_n P_k w_{kn} \ln\left(\frac{P_n}{P_k}\right)$$

Now we exchange indices in the first term on the right to obtain

$$\frac{d\mathcal{H}}{dt} = \sum_k \sum_n P_n w_{nk} \ln\left(\frac{P_k}{P_n}\right)$$

Adding the latter two equations and setting $w_{nk} = w_{kn}$ gives

$$\frac{d\mathcal{H}}{dt} = \frac{1}{2} \sum_k \sum_n w_{nk}(P_k - P_n) \ln\left(\frac{P_n}{P_k}\right)$$

Noting that [recall (3.3.49)]

$$(P_k - P_n) \ln\left(\frac{P_n}{P_k}\right) \leq 0$$

gives the desired result

$$\frac{d\mathcal{H}}{dt} \leq 0$$

5.49. (a) Show that the Master equation (2.28) implies that $\mathrm{Tr}\,\hat{\rho}$ is constant in time.

(b) What is the physical significance of this result?

(c) What property of the perturbing Hamiltonian does you proof of part (a) depend on?

5.50. (a) A system comprises N identical, interacting particles. Write down relation (5.1) and (5.2) in the energy representation, in which n_E represents the number of particles with energy E, and E_i denotes the energy of the ith particle.

(b) What is the coordinate representation of the (unsymmetrized) wavefunctions for this system if the particles are non-interacting? Give your answer in terms of the single-particle energy states, $\varphi_{E_i}(\mathbf{x}_i)$.

(c) What is your answer to part (b) if it is known that all particles of the system are in the ground state?

(d) Under the conditions of part (b), what is the coordinate representation of the Fock-space vector, $|\ldots, n_E, \ldots\rangle_F$?

Answer (partial)

(a)
$$|E_1, \ldots, E_n; N\rangle = \frac{1}{N!} \sum_P (\pm)^P \, |E_1, \ldots, E_N\rangle$$

$$|E_1, \ldots, E_n; N\rangle = \sqrt{\frac{\prod_E n_E}{N!}} \, |\ldots, n_E, \ldots\rangle_F$$

$$\sum_E n_E = N$$

(b) $\qquad \langle \mathbf{x}_1, \ldots, \mathbf{x}_N \mid \mathbf{E}_1, \ldots, \mathbf{E}_N \rangle = \varphi_{E_1}(\mathbf{x}_1)\varphi_{E_2}(\mathbf{x}_2)\cdots\varphi_{E_N}(\mathbf{x}_N)$

5.51. Again consider a system of N interacting, identical particles. Wavefunctions for this system exist in a Hilbert space spanned by products of 1-body functions, $\{\varphi_{n_i}(\mathbf{x}_i)\}$. In this representation, field operators, $\phi(\mathbf{x})$, $\phi^\dagger(\mathbf{x})$ may be expanded in the form

$$\phi(\mathbf{x}) = \sum_n a_n \varphi_n(\mathbf{x})$$

(In the preceding and following, hats over operators are omitted.)

(a) What is the representation of the number operator N in this representation?

(b) What are the companion commutator relations to (7.30)?

(c) Employing this representation, write down the form of the N-body Hamiltonian (5.22) in second quantization.

(d) For which basis functions $\varphi_n(\mathbf{x})$, does the kinetic energy term in your answer to the preceding question, reduce to a diagonal sum?

Answer

(a) $\qquad\qquad\qquad N = \sum_n a_{nn}^\dagger a_n$

(b) $\qquad\qquad\qquad [a_n, a_{n'}^{\ \dagger}]_\mp = \delta_{nn'}$

(c) $H = \sum_n \sum_{n'} \int d\mathbf{x}\, a_n^{\ \dagger}\varphi_n^{\ \dagger}\dfrac{p^2}{2m}a_{n'}\varphi_{n'}$

$\qquad + \sum_n \sum_{n'} \sum_\lambda \sum_{\lambda'} \int d\mathbf{x}\, d\mathbf{x}'\, a_n^{\ \dagger}\varphi_n^{\ \dagger}(\mathbf{x})a_{n'}^{\ \dagger}\varphi_{n'}^{\ \dagger}(\mathbf{x}')u(\mathbf{x}, \mathbf{x}')a_\lambda \varphi_\lambda(\mathbf{x})a_{\lambda'}\varphi_{\lambda'}(\mathbf{x}')$

(d) Plane waves.

5.52. Show that if the collision time, $\tau(\mathbf{k})$ is constant, then the expression for conductivity (3.41) reduces to

$$\sigma = \frac{e^2 l}{12\pi^3\hbar} S_F$$

where $l = \tau v_F$ is mean fee path on the Fermi surface, and $S_F = 4\pi k_F^2$.

Answer

We return to (3.40) and assume that the electric field is in the x-direction. It follows that

$$\varepsilon^{-1}[\mathbf{v}(\mathcal{E} \cdot \mathbf{v})]_x = v_x^2 = \frac{1}{3}v^2$$

Note further that

$$|\nabla_k E|_{E_F} = \frac{\hbar k_F}{m} = v_F$$

There results

$$\sigma = \frac{e^2 \tau v_F}{12\pi^3\hbar} S_F = \frac{e^2 l}{12\pi^3\hbar} S_F$$

which was to be shown.

5.53. Consider a dense fluid with zero isothermal compressibility (recall Problem 2.23).

(a) What is the normalization condition of the radial distribution function under these constraints?

(b) Establish a relation between the radial distribution function, $g(\mathbf{r})$, and the quantum mechanical two-body wavefunction, $\psi_2(\mathbf{x}_1, \mathbf{x}_2)$.

(c) What is the physical meaning of $g(\mathbf{r})$? What does your statement reduce to for an isotropic fluid?

(d) Write down an expression for the number of pairs of particles in the volume element $\Delta v \ll V$, in terms of $g(\mathbf{r})$, where V is the volume of the whole system.

Answer (partial)

(b) We note that the function $|\psi_2(\mathbf{x}_1, \mathbf{x}_2)|^2$ is a joint probability density. It follows that the normalization of this wavefunction is given by

$$\int |\psi_2(\mathbf{x}_1, \mathbf{x}_2)|^2 d\mathbf{x}_1 \, d\mathbf{x}_2 = 1$$

With the unit Jacobian change in variables (see Problem 2.14)

$$\mathbf{r} = \mathbf{x}_1 - \mathbf{x}_2$$
$$2\mathbf{R} = \mathbf{x}_1 + \mathbf{x}_2$$

the preceding normalization becomes

$$\int |\psi_2(\mathbf{r}, \mathbf{R})|^2 d\mathbf{r} \, d\mathbf{R} = 1$$

The radial distribution function is then given by

$$g(\mathbf{r}) = V \int |\psi_2(\mathbf{r}, \mathbf{R})|^2 \, d\mathbf{R}$$

which gives the correct normalization (under the said conditions)

$$\int g(\mathbf{r}) \, d\mathbf{r} = V$$

(c) The factor $g(\mathbf{r}) \, d\mathbf{r} / V$ gives the probability of finding a pair of particles with one particle at the origin and the other at $\mathbf{r} + d\mathbf{r}$, where $d\mathbf{r}$ represents a volume element. For an isotropic fluid, the preceding probability factor is given by $g(r) 4\pi r^2 dr / V$ which represents the probability of finding a pair of particles with one of the particles at the origin and the other in the sperical differential shell of volume $4\pi r^2 dr$.

(d) The number of pairs of particles, ΔN_2, in ΔV is given by

$$\Delta N_2 = N^2 \frac{\Delta V}{V} g(r')$$

where r' is a radial measure of ΔV.

CHAPTER 6

Relativistic Kinetic Theory

Introduction

This concluding chapter addresses elements of relativistic kinetic theory.[1] This component of kinetic theory plays a significant role in areas of astrophysics, free-electron lasing, and certain approaches to controlled thermonuclear fusion. The chapter begins with elementary notions of relativity including postulates, Lorentz transformation, covariance, and invariance. Hamilton's equations are written in covariant form. With these preliminaries at hand, the discussion turns to a covariant description of relativistic kinetic theory. This formalism is employed to obtain a covariant one-particle Liouville equation, which when integrated yields the relativistic Vlasov equation for a plasma in an electromagnetic field. A covariant Drude formulation of Ohm's law is given in relation to a relativistic monoenergetic beam. The following section addresses Lorentz invariants in kinetic theory and a compilation of these forms is presented in a table. The chapter continues with a derivation of the relativistic Maxwellian. Relativity in non-Cartesian coordinates is discussed in the concluding section. Here the reader is introduced to the metric tensor and the notion of contra and covariant tensors. Lagrange's equations are derived, from which a one-particle Liouville equation is obtained relevant to non-Cartesian space–time.

---

[1]For further discussion, See S. R. de Groot, W. A. van Leewen, and Ch. G. van Weert, *Relativistic Kinetic Theory: Principles and Applications*, North-Holland, New York (1980); and P. G. Bergman, *Introduction to the Theory of Relativity*, Prentice-Hall, Englewood Cliffs, N.J. (1955).

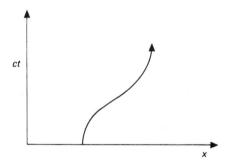

FIGURE 6.1. The world line of a particle moving in one dimension.

6.1 Preliminaries

6.1.1 *Postulates*

There are two postulates of special relativity. These are as follows:

1. The laws of physics are invariant under inertial transformations.
2. The speed of light is independent of the motion of the source.

The first postulate states that the result of an experiment in a given inertial frame of reference is independent of the constant translation motion of the system as a whole. An inertial frame is one in which a mass at rest experiences no force. Thus there is no absolute frame in the universe with respect to which motion of an arbitrary inertial frame is uniquely defined. Only relative motion between frames is relevant.[2]

Concerning the second postulate, consider a light source fixed in a frame S. The frame moves relative to the observer in a frame S' with speed v. The observer measures the speed of light c, independent of the speed v. This situation is evidently equivalent to one in which S' moves relative to S with speed v. Thus the speed of light is independent of the motion of the receiver, as well as that of the source. This conclusion is alien to our intuitive picture of either wave or particle motion.

6.1.2 *Events, World Lines, and the Light Cone*

Einstein defined an *event* as a point in space–time coordinates. The locus of events of a particle is called the *world line* of the particle. For one-dimensional motion, the world line of a particle is a curve in (x, ct) space, where c is the speed of light (see Fig. 6.1).

Of particular interest in the study of relativity is the concept of the *light cone*. This is the world line of the leading edge of a light wave stemming from

---

[2]Unless otherwise stated, frames referred to in this description are inertial.

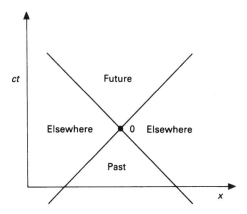

FIGURE 6.2. Light pulse is initiated at 0. Points in the domain marked "elsewhere" cannot be reached from 0 at speed less than c.

a source switched on at a given instant at a given location. The notion of past and future may be defined with respect to the light cone (see Fig. 6.2).

6.1.3   Four-Vectors

An event at a given location $\mathbf{x}$ and time t may be described by the four-dimensional vector[3]

$$\bar{\mathbf{x}} = (\mathbf{x}, ict) \tag{1.1}$$

which is called a *four-vector*. The momentum four-vector is given by

$$\bar{\mathbf{p}} = \left(\mathbf{p}, \frac{iE}{c} \right) \tag{1.2}$$

Here we have written

$$E = \gamma mc^2 = mc^2 + T \tag{1.3}$$

for the total energy of a particle (in the absence of potential) of *rest mass m*, with kinetic energy T. The parameter γ is written for [recall (1.1.19a)]

$$\gamma^2 \equiv \frac{1}{1 - \beta^2}, \qquad \beta \equiv \frac{v}{c} \tag{1.4}$$

Note that γ increases monotonically from 1 to $\beta = 0$ to ∞ at $\beta = 1$. The relativistic momentum three-vector is given by

$$\mathbf{p} = \gamma m \mathbf{v}, \qquad \mathbf{v} = \frac{d\mathbf{x}}{dt} \tag{1.5}$$

---

[3] An alternative formalism in terms of the metric tensor avoids use of imaginary time. See Section 6.4.

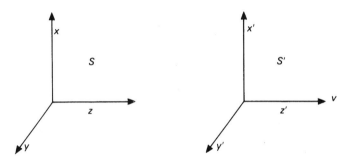

FIGURE 6.3. The frame S' moves at velocity v with respect to the frame S.

where t is time measured in the lab frame.

Let us write p_μ for the components of the four-vector $\bar{\mathbf{p}}$. Thus

$$p_4 = \frac{iE}{c}, \qquad -p_4^2 c^2 = E^2 = p^2 c^2 + m^2 c^4 \tag{1.6}$$

In this notation we may write

$$p_\mu = m u_\mu \tag{1.7}$$

where u_μ is the velocity four-vector,

$$u_\mu = \frac{p_\mu}{m} = \frac{dx_\mu}{d\tau} \equiv \dot{x}_\mu \tag{1.8}$$

and τ denotes *proper time*. This is the time measured on a clock attached to the moving particle (described in Section 1.5).

Three other important four-vectors are[4]

$$\bar{\mathbf{J}} = (\mathbf{J}, ic\rho) \tag{1.9a}$$

$$\bar{\mathbf{A}} = (\mathbf{A}, i\Phi) \tag{1.9b}$$

$$\bar{\mathbf{k}} = \left(\mathbf{k}, \frac{i\omega}{c}\right) \tag{1.9c}$$

In these expressions, $\mathbf{J}$ is current density, ρ is charge density, $\mathbf{A}$ is vector potential, Φ is scalar potential, $\mathbf{k}$ is a wave vector, and ω is frequency.

6.1.4   Lorentz Transformation

Consider that a frame S' moves at constant speed v in the z direction with respect to a frame S, as shown in Fig. 6.3. It is readily shown from the two postulates above that if A_μ is a four-vector in S an observer in S' observes the components

$$A'_\mu = L_{\mu\nu} A_\nu \tag{1.10}$$

---

[4]The manner in which electric and magnetic fields are related to vector and scalar potentials is described in Problem 1.22. See also Section 2.3.

where

$$L = \begin{pmatrix} 1 & 0 & 0 & 0 \\ 0 & 1 & 0 & 0 \\ 0 & 0 & \gamma & i\gamma\beta \\ 0 & 0 & -i\gamma\beta & \gamma \end{pmatrix} \tag{1.11}$$

Note that L is orthogonal; that is

$$\tilde{L} = L^{-1} \tag{1.12}$$

where $\tilde{L}$ is the transpose of L. Note further that

$$\det L = 1 \tag{1.13}$$

(see Problem 6.1). Thus L effects a rotation in complex four-dimensional space. Consequently, the "length" of a four-vector is preserved under a *Lorentz* transformation. That is

$$A'_\mu A'_\mu = A_\mu A_\mu \tag{1.14}$$

Let us show this formally:

$$A'_\mu A'_\mu = L_{\mu\nu} A_\nu L_{\mu\lambda} A_\lambda = L_{\mu\nu} L_{\mu\lambda} A_\nu A_\lambda = L^{-1}_{\nu\mu} L_{\mu\lambda} A_\nu A_\lambda = \delta_{\nu\lambda} A_\nu A_\lambda = A_\lambda A_\lambda$$

For example, consider the length of the momentum four-vector,

$$p_\mu p_\mu = \gamma^2 m^2 v^2 - \gamma^2 m^2 c^2 = -m^2 \gamma^2 c^2 (1 - \beta^2) = -m^2 c^2 \tag{1.15a}$$

This returns the useful relation

$$E^2 = c^2 p^2 + m^2 c^4 \tag{1.15b}$$

For $\bar{x}$, we find

$$x_\mu x_\mu = x^2 - c^2 t^2 \tag{1.16}$$

Such entities, which remain invariant under a Lorentz transformation, are called *Lorentz invariants*. Note in particular that the Lorentz invariant $x_\mu x_\mu$ as given by (1.16) is a reiteration of the second postulate stated above that the speed of light is c in all frames.

The Lorentz matrices (1.11) have the following group property.

$$\hat{L}(\beta_1)\hat{L}(\beta_2) = \hat{L}(\beta_{12}) \tag{1.17}$$

where

$$\beta_{12} = \frac{\beta_1 + \beta_2}{1 + \beta_1\beta_2} \tag{1.18}$$

(see Problem 6.2). Note that (1.18) precludes the speed of any object from exceeding c. Consider that one of the frames moves with relative speed c, so that, for example, $\beta_1 = 1$. Then (1.18) returns the value $\beta_{12} = 1$. Furthermore, in that $L(0) = I$, the identity operator, and $L^{-1} = \tilde{L}$, we see that the Lorentz transformations comprise a group, called the *Lorentz group*.

6.1.5   Length Contraction, Time Dilation, and Proper Time

With relative interframe motion again confined to the z direction, Lorentz transformation of the even four-vector gives

$$\begin{pmatrix} z' \\ ict' \end{pmatrix} = \gamma \begin{pmatrix} 1 & i\beta \\ -i\beta & 1 \end{pmatrix} \begin{pmatrix} z' \\ ict \end{pmatrix} \tag{1.19}$$

We find

$$z' = \gamma(z - \beta ct) \tag{1.20a}$$
$$ict' = i\gamma(-\beta z + ct) \tag{1.20b}$$

Let a rod of given length $\Delta z'$ lie fixed in a frame that we label S'. The frame S' moves with speed βc relative to the frame S. At a given instant, the length of the rod is measured in the S frame. This means that the locations of the ends of the rods are measured simultaneously in S. Calling the length so measured $\Delta z = z_b - z_a$, with $t_b = t_a$ (1.20a), gives

$$\Delta z = \frac{1}{\gamma} \Delta z' \leq \Delta z' \tag{1.21}$$

The rod moving past the frame S is measured to be shortened by an observer in S. This is the phenomenon of *length contraction*.

Consider next a clock that is at a fixed location in the moving S' frame ($z_b' = z_a'$). To find the manner in which intervals τ' on this clock are observed in S, we write the inverse of (1.20b):

$$ict = i\gamma(\beta z' + ct')$$

There results

$$\tau = \gamma \tau' \geq \tau' \tag{1.22}$$

Thus an observer in S concludes that intervals on his clock, τ, are longer than those on the S' clock or that the S' clock is "running slow."

An important parameter in relativity is that of *proper time*. The proper time of a particle is the time measured on a clock that moves with the particle. Thus, if t denotes time in the lab frame and τ proper time, we write

$$dt = \gamma \, d\tau \tag{1.23}$$

Comparison with (1.22) reveals that we have identified proper time with the single clock fixed in S'.

6.1.6   Covariance, Hamiltonian, and Hamilton's Equations

It is important to the first postulate that laws of physics be written in a manner that guarantees that invariance under Lorentz transformation. Relations so written are said to be *covariant*.[5] For example, the covariant forms of Hamilton's equations are given by[6]

$$\frac{dx_\lambda}{d\tau} = \frac{\partial \tilde{H}}{\partial p_\lambda}, \qquad \frac{dp_\lambda}{d\tau} = \frac{\partial \tilde{H}}{\partial x_\lambda} \tag{1.24}$$

For a free particle, the covariant Hamiltonian is given by

$$\tilde{H} = \frac{p_\mu p_\mu}{2m} \tag{1.25}$$

For a charged particle in an electromagnetic field,

$$\tilde{H} = \frac{[p_\mu - (e/c)A_\mu][p_\mu - (e/c)A_\mu]}{2m} \tag{1.26}$$

where A_μ is the four-vector potential (1.9b).

In three-vector form, the relativistic Hamiltonian is given by (1.1.19). When working with this form, Hamilton's equations must be written with respect to the lab time t, as in (1.1.12)[7] (see Problem 6.3).

6.1.7   Criterion for Relativistic Analysis

Prior to our discussion of relativistic kinetic theory, let us ask the question, when is it necessary to use this formalism? The answer to this question is given in terms of a simple rule of thumb. Compare the rest-mass energy of particles whose kinetic theory is being studied to their kinetic energy. Thus, with (1.3), the criterion for nonrelativistic theory is written

$$T = (\gamma - 1)mc^2 \ll mc^2 \tag{1.27a}$$

or, equivalently,

$$1 \le \gamma \ll 2 \tag{1.27b}$$

or, more simply, $v \ll c$. The left inequality in (1.27b) stems from the definition of γ.

---

[5]Not to be confused with covariant and contravariant vectors discussed in Section 6.4.

[6]For further discussion, see H. Goldstein, *Classical Mechanics*, 2nd ed., Addison-Wesley, Reading, Mass. (1981).

[7]See, for example, A. O. Barut, *Electrodynamics and Classical Theory of Fields and Particles*, Dover Publications, New York (1980).

6.2   Covariant Kinetic Formulation

6.2.1   Distribution Function[8]

We consider a distribution $\mathcal{F}(x_\mu, p_\mu)$ with the following properties:

$$\mathcal{F} \geq 0 \tag{2.1a}$$

$$\mathcal{F} \to 0 \text{ as } p_\mu \to \infty \tag{2.1b}$$

It has the normalization

$$\int_{\Sigma_x} \int \mathcal{F} u_\mu d\sigma_\mu d^4 p = \text{constant} \tag{2.2}$$

where Σ_x is a hypersurface in four-dimensional $\bar{x}$-space, with differential element $d\sigma_\mu$.

The definition of $\mathcal{F}(x_\mu p_\mu)$ is as follows. The product

$$\mathcal{F}(x_\mu, p_\mu) u_\mu d\sigma_\mu d^4 p \tag{2.3a}$$

represents the probability that the world line of a particle intersects the hypersurface element $d\sigma_\mu$ at x_μ about the point p_μ. Furthermore, the product

$$\mathcal{F}(x_\mu, p_\mu) \dot{p}_\mu d\tilde{\sigma}_\mu d^4 x \tag{2.3b}$$

represents the probability that the world line of a particle intersects the hypersurface element $d\tilde{\sigma}_\mu$ at p_μ about the point x_μ, where $d\tilde{\sigma}_\mu$ represents an element of hypersurface in four-dimensional $\tilde{p}$-space. As in (1.8), the dot in (2.3b) represents differentiation with respect to proper time.

In (2.3a), the inner product

$$u_\mu d\sigma_\mu = u_1(dx_2\ dx_3\ dx_4) + \cdots + u_4(dx_1\ dx_2\ dx_3) \tag{2.4}$$

where, for example, $dx_2\ dx_3\ dx_4$ is the element of the hypersurface $d\sigma_\mu$ parallel to u_1.

The relation of $\mathcal{F}(x_\mu, p_\mu)$ to the classical distribution $f(\mathbf{x}, \mathbf{p}, t)$ is given by

$$f(x_\mu, \mathbf{p}) = \int_{-\infty}^{\infty} \mathcal{F}\, dp_4 \tag{2.5}$$

With the normalization condition (2.2), we write[9]

$$\int_{\Sigma_x} \int_1 \int_2 \int_3 \int_4 \mathcal{F} u_\mu\, d\sigma_\mu\, d^3 p\, dp_4 = \text{constant}$$

---

[8]B. Kursunoglu, *Nuclear Fusion*, 1 (1961). This formulation takes the four components of p_μ to be independent. An alternative description that directly incorporates the constraint (1.15a) was given by Y. Klimontovich, *JETP*, 37, 524 (1960).

[9]Choosing Σ_x to be a plane normal to the time axis (2.6) gives $\int \gamma f\, d^3x\, d^3p = $ constant, which in the nonrelativistic limit returns the classical normalization $\int f\, d^3x\, d^3p = $ constant.

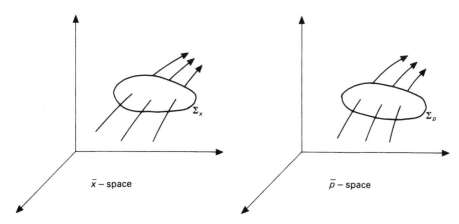

FIGURE 6.4. A three-dimensional depiction of four-dimensional $\tilde{x}$ and $\tilde{p}$ spaces and closed Σ_x and Σ_p hypersurfaces. The net number of world lines entering the closed hypersurfaces Σ_x and Σ_p is zero.

$$= \int_{\Sigma_x} \int_1 \int_2 \int_3 f u_\mu \, d\sigma_\mu \, d^3 p = \text{constant} \qquad (2.6)$$

6.2.2 One-Particle Liouville Equation

Consider that the hypersurface Σ_x in the normalization (2.2) is a closed hypersurface. In this case the net number of world lines entering Σ_x is zero. With the definition (2.3a), we find

$$\int d^4 p \oint_{\Sigma_x} \mathcal{F} u_\mu \, d\sigma_\mu = 0 \qquad (2.7)$$

See Fig. 6.4.

Gauss's theorem in four dimensions permits (2.7) to be written

$$\int d^4 p \int d^4 x \, \frac{\partial}{\partial x_\mu} (\mathcal{F} u_\mu) = 0 \qquad (2.8)$$

A similar argument stemming from the definition (2.3b) gives

$$\int d^4 x \oint_{\Sigma_p} \mathcal{F} \dot{p}_\mu \, d\tilde{\sigma}_\mu = \int d^4 x \int d^4 p \, \frac{\partial}{\partial p_\mu} (\mathcal{F} \dot{p}_\mu) = 0 \qquad (2.9)$$

Since $\mathcal{F} = \mathcal{F}(x_\mu, p_\mu)$, we may write

$$\frac{d\mathcal{F}}{d\tau} = \dot{x}_\mu \frac{\partial \mathcal{F}}{\partial x_\mu} + \dot{p}_\mu \frac{\partial \mathcal{F}}{\partial p_\mu} \qquad (2.10)$$

With Hamilton's equations (1.24), the preceding equation becomes

$$\frac{d\mathcal{F}}{d\tau} = \frac{\partial \mathcal{F} \dot{x}_\mu}{\partial x_\mu} + \frac{\partial \mathcal{F} \dot{p}_\mu}{\partial p_\mu} \qquad (2.11)$$

Integration of this equation gives

$$\int \frac{d\mathcal{F}}{d\tau} d^4x\, d^4p = \int \frac{\partial \mathcal{F} u_\mu}{\partial x_\mu} d^4x\, d^4p + \int \frac{\partial \mathcal{F} \dot{p}_\mu}{\partial p_\mu} d^4x\, d^4p \qquad (2.12)$$

With (2.8) and (2.9), the right side of (2.12) vanishes, and we obtain

$$\int \frac{d\mathcal{F}}{d\tau} d^4x\, d^4p = 0$$

Passing to zero volume gives

$$\frac{d\mathcal{F}}{d\tau} = 0 \qquad (2.13)$$

or, equivalently, with (2.10)

$$u_\mu \frac{\partial \mathcal{F}}{\partial x_\mu} + \dot{p}_\mu \frac{\partial \mathcal{F}}{\partial p_\mu} = 0 \qquad (2.14)$$

This equation represents a covariant one-particle Liouville equation. The term following $\dot{p}_\mu$ will be related to electrodynamic field variables, which in turn will be related to the distribution through the current. With these substitutions, the preceding equation is more properly termed the Vlasov equation.

6.2.3 Covariant Electrodynamics

To apply the above finding to a plasma, we must first write Maxwell's equations in covariant form. To these ends, we introduce the electromagnetic field tensor $F_{\nu\lambda}$. For l, j running from 1 to 3,

$$F_{lj} = \varepsilon_{ljk} B_k \qquad (2.15a)$$

where B_k represents the kth component of magnetic field and ε_{ijk} is the Levi-Civita symbol [see beneath (A.10′)]. With $\mathcal{E}_j$ written for the jth component of the electric field, we write

$$F_{4j} = -F_{j4} = i\mathcal{E}_j \qquad (2.15b)$$

These terms correspond to the matrix

$$F_{\mu\nu} = \begin{pmatrix} 0 & B_3 & -B_2 & -i\mathcal{E}_1 \\ -B_3 & 0 & B_1 & -i\mathcal{E}_2 \\ B_2 & -B_1 & 0 & -i\mathcal{E}_3 \\ i\mathcal{E}_1 & i\mathcal{E}_2 & i\mathcal{E}_3 & 0 \end{pmatrix} \qquad (2.16)$$

The covariant form of (2.15) is given

$$F_{\mu\nu} = \frac{\partial A_\nu}{\partial x_\mu} - \frac{\partial A_\mu}{\partial x_\nu} \qquad (2.17)$$

where A_μ is the four-vector potential (1.9b). In these variables, Maxwell's equations assume the form (cgs)

$$\frac{\partial F_{\mu\nu}}{\partial x_\mu} = \frac{4\pi}{c} J_\nu \tag{2.18}$$

where J_ν is the four-current (1.9a). The preceding covariant relation gives the two Maxwell equations:

$$\nabla \cdot \boldsymbol{\mathcal{E}} = 4\pi\rho, \qquad \nabla \times \mathbf{B} - \frac{1}{c}\frac{\partial \boldsymbol{\mathcal{E}}}{\partial t} = \frac{4\pi}{c}\mathbf{J} \tag{2.18a}$$

The covariant structure of the remaining Maxwell equations is

$$\frac{\partial F_{\mu\nu}}{\partial x_\lambda} + \frac{\partial F_{\lambda\mu}}{\partial x_\nu} + \frac{\partial F_{\nu\lambda}}{\partial x_\mu} = 0 \tag{2.19}$$

which gives

$$\nabla \cdot \mathbf{B} = 0, \qquad \nabla \times \boldsymbol{\mathcal{E}} + \frac{1}{c}\frac{\partial \mathbf{B}}{\partial t} = 0 \tag{2.19a}$$

The covariant form of the Lorentz force law is

$$\dot{p}_\mu = \frac{e}{mc} F_{\mu\nu} p_\nu \tag{2.20}$$

(see Problem 6.4).

6.2.4 Vlasov Equation

Combining the Lorentz force law (2.20) with the equation of motion (2.14) gives

$$u_\mu \frac{\partial \mathcal{F}}{\partial x_\mu} + \frac{e}{mc} F_{\mu\nu} p_\nu \frac{\partial \mathcal{F}}{\partial p_\mu} = 0$$

or equivalently,

$$p_\mu \frac{\partial \mathcal{F}}{\partial x_\mu} + \frac{e}{c} F_{\mu\nu} p_\nu \frac{\partial \mathcal{F}}{\partial p_\mu} = 0 \tag{2.21}$$

This scalar equation has the explicit form

$$\frac{\partial \mathcal{F}}{\partial t} + \mathbf{v}\cdot\frac{\partial \mathcal{F}}{\partial \mathbf{x}} + \left(e\boldsymbol{\mathcal{E}} + \frac{e}{c}\mathbf{v}\times\mathbf{B}\right)\cdot\frac{\partial \mathcal{F}}{\partial \mathbf{p}} + \frac{e}{c}(\mathbf{v}\cdot\boldsymbol{\mathcal{E}})\frac{\partial \mathcal{F}}{\partial p_4} = 0 \tag{2.21a}$$

Integrating over p_4 and recalling the connecting relation (2.5) gives the desired form (with components of $\tilde{\mathbf{p}}$ taken to be independent):

$$\frac{\partial f}{\partial t} + \mathbf{v}\cdot\frac{\partial}{\partial \mathbf{x}} f + \left(e\boldsymbol{\mathcal{E}} + \frac{e}{c}\mathbf{v}\times\mathbf{B}\right)\cdot\frac{\partial f}{\partial \mathbf{p}} = 0 \tag{2.22}$$

(see Problem 6.5). Note that the sole difference between the form of (2.22) and the classical Vlasov equation (2.2.30) is contained in the presence of **p** in (2.22), which is the relativistic momentum (1.5).

One further note is relevant to (2.2). For this equation to be properly termed the Vlasov equation, the fields $\mathcal{E}$ and **B** must be self-consistently defined. That this is so in the present case follows from the fact that (2.22) includes the force equation (2.20), with field variables coupled to the current density through (2.18). The picture is made complete if current and charge densities are expressible as functionals over the distribution function. These relations are given by

$$\mathbf{J}(\mathbf{x}, t) = eN \int \frac{\mathbf{p}}{m\gamma} f(\bar{\mathbf{x}}, \mathbf{p}) d\mathbf{p} \tag{2.23a}$$

$$ic\rho(\mathbf{x}, t) = eN \int \frac{p_4}{m\gamma} f(\bar{\mathbf{x}}, \mathbf{p}) d\mathbf{p} \tag{2.23b}$$

where N is the total number of particles in the system. Recall that $\mathbf{p}/m\gamma = \mathbf{v}$, the velocity in the lab frame. It will be shown that $f(x_\mu, \mathbf{p})$ and $d\mathbf{p}/\gamma$ are both relativistic invariants.

6.2.5   Covariant Drude Formulation of Ohm's Law

A monoenergetic beam of electrons of number density n and velocity **v** with respect to the lab frame has the current density

$$\mathbf{J} = en\mathbf{v} \tag{2.24a}$$

If n_0 is density measured in a frame moving with the beam, then

$$n = \gamma n_0 \tag{2.24b}$$

due to length contraction along the direction of motion as given by (1.21). With (1.7), we may write

$$\mathbf{u} = \gamma \mathbf{v}$$

and the current (2.24a) may be rewritten

$$\mathbf{J} = en_0\mathbf{u} \tag{2.25}$$

With (1.7), the preceding expression for current may be written

$$\mathbf{J} = \frac{en_0}{m} \mathbf{p} \tag{2.26}$$

which is the vector component of the four-vector [see (1.9a)]:

$$J_\mu = \frac{en_0}{m} p_\mu \tag{2.27}$$

This form gives the invariant

$$J_\mu J_\mu = -e^2 n_0^2 c^2 \tag{2.27a}$$

We may write down a covariant form of Rayleigh dissipation (4.2.32) in terms of *proper frequency* ν_0. This is frequency measured on a clock moving with the beam. The equation appears as

$$\frac{dp_\mu}{d\tau} = -\nu_0 p_\mu \tag{2.28}$$

Combining this relation with Lorentz force law (2.20) gives

$$\frac{dp_\mu}{d\tau} = \frac{e}{mc} F_{\mu\alpha} p_\alpha - \nu_0 p_\mu \tag{2.29}$$

The fourth component of this equation reads

$$\frac{dE}{d\tau} = e\mathbf{u} \cdot \boldsymbol{\mathcal{E}} - \nu_0 E \tag{2.30}$$

That is, in a frame moving with the beam, energy increases at a rate at which the electric field does work on the particles and decreases due to collisions at a rate proportional to the energy.

To obtain the covariant form of Ohm's law, first we assume steady state and set $dp_\mu/d\tau = 0$. Combining (2.29) and (2.27) then gives the desired relation:

$$J_\mu = \sigma F_{\alpha\mu} \frac{u_\alpha}{c} \tag{2.31}$$

where we have written

$$\sigma = \frac{e^2 n_0}{\nu_0 m} \tag{2.31a}$$

The vector component of (2.31) returns the standard form of Ohm's law,

$$\mathbf{J} = \sigma \left(\boldsymbol{\mathcal{E}} + \frac{\mathbf{v} \times \mathbf{B}}{c} \right) \tag{2.32}$$

which, we see, is a relativistically valid equation. Note in particular that $\mathbf{J}$, $\boldsymbol{\mathcal{E}}$, and $\mathbf{B}$ in this equation are lab-frame values and that σ contains rest-frame parameters.

The fourth component of (2.31), with J_4 given by (2.27), returns the energy balance equation [right side of (2.30)]

$$\nu_0 E = e\mathbf{u} \cdot \boldsymbol{\mathcal{E}} \tag{2.33}$$

6.2.6 Lorentz Invariants in Kinetic Theory

First we note

$$F(\mathbf{p})d^3p = d^3p \int_V F(\mathbf{x}, \mathbf{p})d^3x \tag{2.34}$$

represents the number of particles in the spatial volume V and in the momentum volume d^3p about the point $\mathbf{p}$. [Note that $F(\mathbf{x}, \mathbf{p})$ is normalized to total number of particles.] As this is a pure number, it is the same in all reference frames and thus may be termed a Lorentz invariant.

The same argument leads us to conclude that

$$F(\mathbf{x})d^3x = d^3x \int F(\mathbf{x}, \mathbf{p})d^3p \qquad (2.35)$$

is likewise a Lorentz invariant.

Consider next how the volume element d^3p transforms under a Lorentz transformation. To answer this question, we note a general rule of invariance:[10] if C_μ and D_μ are two four-vectors, then the ratio of two parallel components, C_l/D_l is a Lorentz invariant.

With (1.15a), we see that d^3p is an element of surface on the hypersurface

$$p_\mu p_\mu = p^2 - \left(\frac{E}{c}\right)^2 = -m^2c^2$$

Thus we may write

$$dp_x\, dp_y\, dp_z = d\sigma_4 \qquad (2.36)$$

where $d\sigma_4$ is a differential of hypersurface in four-dimensional $\bar{\mathbf{p}}$-space. With the above-stated rule of invariance, we conclude that

$$\frac{d\sigma_4}{p_4} = \frac{d^3p}{p_4} = \frac{d^3p}{iE/c} \qquad (2.37)$$

is a Lorentz invariant. That is, under a Lorentz transformation we find

$$\frac{d^2p}{iE/c} = \frac{d^3p'}{iE'/c} \qquad (2.38a)$$

or, equivalently,

$$d^3p' = \frac{\gamma'}{\gamma} d^3p \qquad (2.38b)$$

The invariant d^3p/E may be cast in four-dimensional form as follows. First we write (see Problem 6.8)

$$\frac{d^3p}{iE/c} = \int_{p_4} \delta(p_\mu^2 + m^2c^2)d^4p \qquad (2.39a)$$

which is evidently a Lorentz invariant. In this expression we are treating the four components of p_μ as independent. As in (2.22), the delta function in (2.39a) maintains the constraint (1.15a). As the integration in (2.39a) is only

---

[10]L. D. Landau and E. M. Lifshitz, *Classical Theory of Fields*, 4th ed., Pergamon Press, Elmsford, N.Y. (1975).

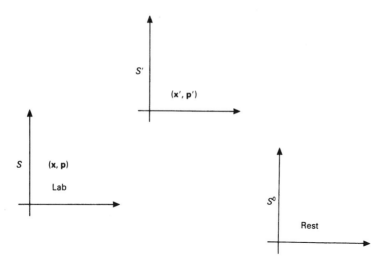

FIGURE 6.5. Particles described by (2.41) are at rest in S^0.

over p_4, we obtain the alternative form

$$\frac{d^3 p}{iE/c} = d^3 p \int \delta(p_\mu^2 + m^2 c^2) dp_4$$

With the invariants (2.34) and (2.37) at hand, we write

$$F(\mathbf{p}) d^3 p = E F(\mathbf{p}) \frac{d^2 p}{E}$$

This equality indicates that $E F(\mathbf{p})$ is a Lorentz invariant. That is, under a Lorentz transformation

$$F'(\mathbf{p}') E' = F(\mathbf{p}) E \qquad (2.40)$$

In addition to the distribution invariants (2.34) and (2.35), it is evident that

$$F(\mathbf{x}, \mathbf{p}) d^3 x \, d^3 p \qquad (2.41)$$

is likewise a Lorentz invariant. To discover the manner in which $F(\mathbf{x}, \mathbf{p})$ transforms, we must examine the manner in which $d^3 x$ transforms in kinetic theory.

To these ends, we note that in relativistic kinetic theory it is necessary to discern between three frames of reference. The independent variables of $F(\mathbf{x}, \mathbf{p})$ are defined with respect to the lab frame, S. An inertial frame, S', moves relative to the lab frame, where measured variables have values $\mathbf{x}'$ and $\mathbf{p}'$. An additional frame, S^0, is introduced such that particles with momenta defined by (2.41) are at rest (see Fig. 6.5). The *proper volume* $d^3 x^0$ lies in this frame. Recalling the description leading to (1.21) indicates that the primed frame of that expression is the frame of the proper volume in the present discussion. Since both S and S' move relative to S^0, where the differential of particles

(2.41) are at rest, our previous result (1.21) gives the relations

$$d^3x = \frac{1}{\gamma} d^3 x^0 \tag{2.42a}$$

$$d^3x' = \frac{1}{\gamma'} d^3 x^0 \tag{2.42b}$$

Thus we find

$$\gamma d^3 x = \gamma' d^3 x' \tag{2.43}$$

so that $\gamma d^3 x$ is a Lorentz invariant. With (2.41) and (2.43), we obtain

$$F(\mathbf{x}, \mathbf{p}) d^3 x \, d^3 p = F'(\mathbf{x}', \mathbf{p}') d^3 x' \, d^3 p' = F'(\mathbf{x}', \mathbf{p}') d^3 x \, d^3 p \tag{2.44}$$

We may conclude that both the distribution

$$F(\mathbf{x}, \mathbf{p})$$

as well as the phase volume

$$d^3 x \, d^3 p$$

are Lorentz invariants. A compilation of these Lorentz invariants is given in Table 6.1. The time dependence of these distributions is tacitly assumed. Recall also that F is normalized to total number of particles in the system at hand. The volume elements $d^4 x$ and $d^4 p$ are invariant due to the orthogonal property of the Lorentz transformation, $\det L = 1$ [see (1.13)].

6.2.7   Relativistic Electron Gas and Darwin Lagrangian

Let us consider a collection of N charges interacting with each other through the Coulomb interaction. Charge and current densities due to the jth particle are given by

$$\rho^{(j)}(\mathbf{x}, t) = q_j \delta[\mathbf{x} - \mathbf{x}_j(t)] \tag{2.45}$$

$$\mathbf{J}^{(j)}(\mathbf{x}, t) = q_j \mathbf{v}_j(t) \delta[\mathbf{x} - \mathbf{x}_j(t)] \tag{2.46}$$

where $\mathbf{x}_j(t)$ and $\mathbf{v}_j(t)$ are written for the coordinate and velocity of the jth particle of charge q_j.

We work in the Coulomb gauge[11] defined by the condition

$$\nabla \cdot \mathbf{A}^{(j)} = 0$$

where $\mathbf{A}^{(j)}$ is the vector potential due to the jth particle. In this gauge, potentials are given by

$$\phi^{(j)}(\mathbf{x}, t) = \int d\mathbf{x}' \frac{\rho^{(j)}(\mathbf{x}', t)}{|\mathbf{x} - \mathbf{x}'|} \tag{2.47}$$

---

[11]The Lorentz gauge was defined in Problem 1.22.

TABLE 6.1. Lorentz Invariants in Kinetic Theory

A. Distribution Functions and Volume Elements

$$F(\mathbf{x})d^3x; \quad F(\mathbf{p})d^3p; \quad F(\mathbf{x}, \mathbf{p})d^3x\, d^3p$$

$$F(\mathbf{p})E; \quad F(\mathbf{x}, \mathbf{p})$$

$$d^4x; \quad d^4p; \quad d^3x\, d^3p; \quad \frac{d^3p}{E}; \quad E\, d^3x$$

$$\frac{d^3p}{iE/c} = \int_{p_4} \delta(p_\mu^2 + m^2c^2)d^4p$$

B. Invariants for N-body Systems

$$\sum_{i=1}^{N}(\mathbf{p}_i^2 + m_i^2c^2) = \sum_{i=1}^{N}\frac{E_i^2}{c^2}$$

or, equivalently,[a]

$$\sum_{i=1}^{N} p_i^\mu p_{i\mu} = -\sum_{i=1}^{N} m_i^2 c^2$$

With

$$\mathbf{P}_{\text{total}} = \sum_{i=1}^{N}\mathbf{p}_i$$

$$E_{\text{total}} = \sum_{i=1}^{N}E_i$$

we may write

$$\left[\sum_{i=1}^{N} p_i^\mu\right]\left[\sum_{j=1}^{N} p_{j\mu}\right] = \mathbf{P}_{\text{total}}^2 - \frac{E_{\text{total}}^2}{c^2} \equiv -M^2c^2$$

where M is called the *invariant mass*.

C. Collisional Invariants

$$\sum_{i=1}^{N} p_{i\mu} = \sum_{i=1}^{N} p'_{i\mu}$$

With $N = 2$, we find[b]

$$(p_{1\mu} + p_{2\mu})^2 = (p'_{1\mu} + p'_{2\mu})^2 \equiv -s$$

$$(p_{1\mu} - p'_{1\mu})^2 = (p'_{2\mu} - p'_{2\mu})^2 \equiv -t$$

$$(p_{1\mu} - p'_{2\mu})^2 = (p'_{1\mu} - p_{2\mu})^2 \equiv -u$$

---

[a] Contra and covariant notation is described in Section 6.4.
[b] The terms s, t and u are called Mandelstam variables. Note in particular that only two of these variables are independent since $s + t + u = 2c^2(m_1^2 + m_2^2)$ (see Problem 6.24).

$$\mathbf{A}^{(j)}(\mathbf{x}, t) = \int d\mathbf{x}' \, dt' \, \frac{\delta(t - t' - (1/c)|\mathbf{x} - \mathbf{x}'|)}{|\mathbf{x} - \mathbf{x}'|}$$

$$\times \left[\frac{\mathbf{J}^{(j)}(\mathbf{x}', t')}{c} - \frac{1}{4\pi c} \frac{\partial}{\partial t'} \nabla' \phi^{(j)}(\mathbf{x}', t') \right] \qquad (2.48)$$

With these parameters at hand, the relativistic Lagrangian is written

$$L(\mathbf{x}^N, \mathbf{v}^N, t) = \sum_{i=1}^{N} \left\{ -mc^2 \sqrt{1 - \frac{v_i^2}{c^2}} - \sum_{i<j} \right.$$

$$\left. \times \left[q_i \phi^{(j)}(\mathbf{x}_i, t) - \frac{q_i}{c} \mathbf{A}^{(j)}(\mathbf{x}_i, t) \cdot \mathbf{v}_i(t) \right] \right\} \quad (2.49)$$

We wish to obtain the form (2.49) to $O(1/c^2)$. Toward this end, as is evident from (2.49), it is necessary only to obtain the vector potential to $O(1/c)$, which in turn is obtained from (2.48). With $\mathbf{x}$ written for $\mathbf{x}_i$, we write

$$\mathbf{A}^{(j)}(\mathbf{x}, t) = \int d\mathbf{x}' \, dt' \, \frac{\delta(t - t')}{c|\mathbf{x} - \mathbf{x}'|} \left[\mathbf{J}^{(j)}(\mathbf{x}', t') - \frac{1}{4\pi} \frac{\partial}{\partial t'} \nabla' \phi^{(j)}(\mathbf{x}', t') \right] + O\left(\frac{1}{c^2}\right)$$

Performing the t' integration, we find

$$\mathbf{A}^{(j)}(\mathbf{x}, t) = \int d\mathbf{x}' \, \frac{1}{c|\mathbf{x} - \mathbf{x}'|} \left[\mathbf{J}^{(j)}(\mathbf{x}', t) - \frac{1}{4\pi} \frac{\partial}{\partial t} \nabla' \phi^{(j)}(\mathbf{x}', t) \right] + O\left(\frac{1}{c^2}\right) \tag{2.50}$$

Let us evaluate $\phi(\mathbf{x}, t)$. Inserting (2.45) into (2.47) gives

$$\phi^{(j)}(\mathbf{x}, t) = \frac{q_j}{|\mathbf{x} - \mathbf{x}_j(t)|} \tag{2.51}$$

Substituting this expression together with (2.46) for $\mathbf{J}^{(j)}$ into (2.50), we obtain

$$\mathbf{A}^{(j)}(\mathbf{x}, t) = \frac{q_j \mathbf{v}_j(t)}{c|\mathbf{x} - \mathbf{x}_j(t)|} - \frac{1}{4\pi c} \frac{\partial}{\partial t} \int d\mathbf{x}' \frac{1}{|\mathbf{x} - \mathbf{x}'|} \nabla' \left(\frac{q_j}{|\mathbf{x}' - \mathbf{x}_j(t)|} \right) + O\left(\frac{1}{c^2}\right) \tag{2.52}$$

Now note that

$$\frac{\partial}{\partial t} \left(\frac{1}{|\mathbf{x}' - \mathbf{x}_j(t)|} \right) = \frac{\mathbf{v}_j(t) \cdot [\mathbf{x}' - \mathbf{x}_j(t)]}{|\mathbf{x}' - \mathbf{x}_j(t)|^3}$$

Substituting this result into the integral term of (2.52) (labeled **I**) gives

$$\mathbf{I} = -\frac{1}{4\pi c} \int d\mathbf{x}' \frac{q_j}{|\mathbf{x} - \mathbf{x}'|} \nabla' \left(\frac{\mathbf{v}_j(t) \cdot [\mathbf{x}' - \mathbf{x}_j(t)]}{|\mathbf{x}' - \mathbf{x}_j(t)|^3} \right)$$

Changing variables,

$$\mathbf{y} = \mathbf{x}' - \mathbf{x}_j(t)$$

and integrating by parts, we obtain

$$I = -\frac{1}{4\pi c} \nabla \int dy \, \frac{\mathbf{v}_j(t) \cdot \mathbf{y}}{y^3} \frac{q_j}{|\mathbf{r}_j - \mathbf{y}|}$$

Here we have labeled

$$\mathbf{r}_j = \mathbf{x} - \mathbf{x}_j(t) \tag{2.52a}$$

Integrating the preceding expression gives

$$I = -\frac{q_j}{2c} \nabla \left(\frac{\mathbf{v}_j(t) \cdot \mathbf{r}_j}{r_j} \right)$$

which when differentiated gives

$$I = -\frac{q_j}{2c} \left[\frac{\mathbf{v}_j(t)}{r_j} - \frac{\mathbf{r}_j(\mathbf{v}_j \cdot \mathbf{r}_j)}{r_j^3} \right] \tag{2.53}$$

Substituting this result into (2.52), we obtain

$$A^{(j)}(\mathbf{x}, t) = \frac{q_j}{2c} \left[\frac{\mathbf{v}_j(t)}{r_j} + \frac{\mathbf{r}_j(\mathbf{v}_j \cdot \mathbf{r}_j)}{r_j^3} \right] \tag{2.54}$$

Combining this result with (2.51) and substituting into the Lagrangian (2.49) gives

$$L(\mathbf{x}^N, \mathbf{v}^N, t) = \sum_{i=1}^{N} \left[-mc^2 \sqrt{1 - \frac{v_i^2}{c^2}} \right] \tag{2.55}$$

$$+ \sum_{i<j} \left\{ -\frac{q_i q_j}{r_{ij}} + \frac{q_i q_j}{2c^2} \left[\frac{\mathbf{v}_i \cdot \mathbf{v}_j}{r_{ij}} + \frac{(\mathbf{v}_i \cdot \mathbf{r}_{ij})(\mathbf{v}_j \cdot \mathbf{r}_{ij})}{r_{ij}^3} \right] \right\}$$

where $\mathbf{v}_i \equiv \mathbf{v}_i(t)$ and r_{ij} is given by (2.52a) with $\mathbf{x}$ rewritten $\mathbf{x}_i$ [as in (2.49)]. The preceding expression is the desired structure of the Lagrangian of an N-charged particle gas valid to $O(1/c^2)$. It was first obtained by C. G. Darwin.[12]

6.3 The Relativistic Maxwellian[13]

6.3.1 Normalization

With (1.1.19), we write

$$H = c\sqrt{m^2 c^2 + p^2}$$

---

[12]C. G. Darwin, *Phil. Mag.* 39, 537 (1920).
[13]J. L. Synge, *The Relativistic Gas*, North Holland, Amsterdam (1957).

$$= mc^2\sqrt{1 + \frac{p^2}{m^2c^2}} \equiv mc^2\sqrt{1 + \mu^2} \qquad (3.1)$$

where

$$\mu^2 \equiv \frac{p^2}{m^2c^2} = \frac{\beta^2}{1 - \beta^2} \qquad (3.2)$$

With (3.1), the Maxwellian may be written

$$f_0(\mu) = A e^{-H(\mu)/k_BT}$$

$$= A \exp\left[-\frac{mc^2}{k_BT}\sqrt{1 + \mu^2}\right] \qquad (3.3)$$

The normalization of f_0 may be written

$$\int f_0 \, d^3\mu = 1 \qquad (3.4a)$$

or, equivalently,

$$4\pi A \int_0^\infty d\mu \, \mu^2 \exp\left[-\frac{mc^2}{k_BT}\sqrt{1 + \mu^2}\right] = 1 \qquad (3.4b)$$

We wish to evaluate the normalization parameter A. Let us define

$$s^2 \equiv 1 + \mu^2$$

$$s \, ds = \mu \, d\mu$$

so that (3.4b) becomes

$$4\pi A \int_1^\infty \left(\exp -\frac{mc^2 s}{k_BT}\right)(s^2 - 1)^{1/2} s \, ds = 1 \qquad (3.5)$$

Noting that

$$\int (s^2 - 1)^{1/2} s \, ds = \frac{1}{3}(s^2 - 1)^{3/2}$$

permits (3.5) to be integrated by parts. There results

$$\frac{4\pi A mc^2}{3k_BT} \int_1^\infty \left(\exp -\frac{mc^2 s}{k_BT}\right)(s^2 - 1)^{3/2} ds = 1 \qquad (3.6)$$

We recall the integral expression for the Bessel function $K_\nu(z)$[14]

$$K_\nu(z) = \frac{(z/2)^\nu \Gamma(1/2)}{\Gamma[\nu + (1/2)]} \int_1^\infty e^{-zt}(t^2 - 1)^{\nu - (1/2)} dt$$

$$\left[\mathrm{Re}\left(\nu + \frac{1}{2}\right) > 0, \qquad |\arg z| < \frac{\pi}{2}\right] \qquad (3.7)$$

---

[14]I. S. Gradshteyn and I. M. Ryzhik, *Tables of Integrals, Series and Products*, Academic Press, New York (1965), Eq. (3), section 8.432.

The asymptotic value of $K_\nu(z)$ for large z is given by[15]

$$K_\nu(z) \sim \sqrt{\frac{\pi}{2z}} \, e^{-z} \left[1 + \frac{1}{2z} + \cdots \right] \tag{3.8}$$

Employing (3.7) permits the normalization (3.6) to be written

$$A = \frac{mc^2}{k_B T} \left[\frac{1}{4\pi K_2(mc^2/k_B T)} \right] \tag{3.9}$$

Our complete relativistic Maxwellian then appears as

$$f_0(\mu) = \frac{mc^2/k_B T}{4\pi K_2(mc^2/k_B T)} \exp\left(-\frac{mc^2}{k_B T} \sqrt{1 + \mu^2} \right) \tag{3.10}$$

6.3.2 The Nonrelativistic Domain

With the definition (3.2), we see that

$$1 + \mu^2 = 1 + \frac{\beta^2}{1 - \beta^2} = \frac{1}{1 - \beta^2}$$

which, in the nonrelativistic domain, gives

$$1 + \mu^2 \simeq 1 + \beta^2$$
$$\mu^2 \simeq \beta^2 \tag{3.11}$$
$$\sqrt{1 + \mu^2} \simeq \sqrt{1 + \beta^2} \simeq 1 + \frac{\beta^2}{2}$$

Setting

$$z = \frac{mc^2}{k_B T}$$

permits (3.10) to be rewritten

$$f_0(\mu) = \frac{z}{4\pi K_2(z)} e^{-z\sqrt{1+\mu^2}} \tag{3.12}$$

Inserting the asymptotic formulas (3.8) and (3.11) [corresponding to $k_B T \ll mc^2$] into the preceding result gives

$$f_0(\mu) \sim \frac{z}{4\pi \sqrt{\pi/2z} \, e^{-z}} e^{-z} e^{-z\beta^2/2}$$
$$= \frac{c^3}{(2\pi RT)^{3/2}} e^{-mv^2/2k_B T} \tag{3.13}$$

---

[15]Ibid., Eq. (6), section 8.451.

In this same limit,

$$d^3\mu \to \frac{d^2v}{c^3}$$

and $d^3\mu f_0(\mu) \to d^3v f_0(v)$, where $f_0(v)$ is the classical Maxwellian (3.3.58).

The relativistic Maxwellian (3.10) may be employed with velocity differential through the transformation

$$(mc)^3 d^3\mu = d^3 p = m^3\gamma^5 d^3v$$

(see Problem 6.10).

6.4 Non-Cartesian Coordinates

We close this chapter with a brief introduction to relativity in non-Cartesian coordinates.

6.4.1 Covariant and Contravariant Vectors

Vectors that transform as the gradient are called *covariant vectors* and are written with a subscript, for example, B_α. Vectors that transform as displacement are called *contravariant vectors* and are written with a superscript.[16] Thus, for example, coordinate components are written x^α.

Consider the transformations of coordinates

$$x'^\mu = x'^\mu(x^\alpha) \tag{4.1}$$

For differential of displacement, we write

$$dx'^\mu = \frac{\partial x'^\mu}{\partial x^\alpha} dx^\alpha \tag{4.2}$$

Thus *contravariant vectors* transform as

$$A'^\mu = \frac{\partial x'^\mu}{\partial x^\alpha} A^\alpha \tag{4.3}$$

For the gradient of a scalar function ϕ, we obtain

$$\frac{\partial \phi}{\partial x'^\mu} = \frac{\partial \phi}{\partial x^\alpha} \frac{\partial x^\alpha}{\partial x'^\mu} \tag{4.4}$$

Thus *covariant vectors* transform as

$$B'_\mu = \frac{\partial x^\alpha}{\partial x'^\mu} B_\alpha \tag{4.5}$$

---

[16]In the following section, we find that a contravariant vector may be transformed to a covariant vector, and vice versa. A more basic definition of these vectors is given in terms of the transformation equations (4.3) and (4.5).

A *contravariant second-rank tensor* $A^{\alpha\beta}$ (with 16 components) transforms as

$$A'^{\alpha\beta} = \frac{\partial x'^{\alpha}}{\partial x^{\mu}} \frac{\partial x'^{\beta}}{\partial x^{\nu}} A^{\mu\nu} \tag{4.6}$$

A *contravariant second-rank tensor* transforms according to

$$B'_{\alpha\beta} = \frac{\partial x'^{\mu}}{\partial x^{\alpha}} \frac{\partial x'^{\nu}}{\partial x^{\beta}} B_{\mu\nu} \tag{4.7}$$

whereas the mixed second-rank tensor K^{α}_{β} transforms as

$$K'^{\alpha}_{\beta} = \frac{\partial x'^{\alpha}}{\partial x^{\mu}} \frac{\partial x^{\nu}}{\partial x'^{\beta}} K^{\mu}_{\nu} \tag{4.8}$$

We take the inner product of two vectors to be defined as the product of their contravariant and covariant forms

$$\mathbf{B} \cdot \mathbf{A} \equiv B_{\mu} A^{\mu} \tag{4.9}$$

Note in particular that the inner product

$$\begin{aligned}
\mathbf{B}' \cdot \mathbf{A}' &= \frac{\partial x^{\alpha}}{\partial x'^{\mu}} \frac{\partial x'^{\mu}}{\partial x^{\nu}} B_{\alpha} A^{\nu} \\
&= \frac{\partial x^{\alpha}}{\partial x^{\nu}} B_{\alpha} A^{\nu} = \delta_{\alpha\nu} B_{\alpha} A^{\nu} \\
&= B_{\nu} A^{\nu} = \mathbf{B} \cdot \mathbf{A}
\end{aligned} \tag{4.10}$$

is a Lorentz invariant.

6.4.2   *Metric Tensor*

In (1.16), we wrote the invariant internal

$$(ds)^2 = (dx)^2 - c^2 (dt)^2 \tag{4.11}$$

This expression is a special case of the more general relation

$$(ds)^2 = g_{\alpha\beta} dx^{\alpha} dx^{\beta} \tag{4.12}$$

Here we have introduced the *metric tensor* $g_{\alpha\beta} = g_{\beta\alpha}$. In Cartesian space, such as employed in the earlier components of this chapter, $g_{\alpha\beta}$ is diagonal with (no summation)

$$g_{\alpha\beta} = g_{\alpha\alpha} \delta_{\alpha\beta} \tag{4.13a}$$

where

$$g_{11} = g_{22} = g_{33} = 1, \qquad g_{44} = -1 \tag{4.13b}$$

The contravariant four-vector event is then given by (x, ct), whereas the covariant event four-vector is given by $(x, -ct)$. Thus

$$x^{\mu} x_{\mu} = x^2 - c^2 t^2$$

An additional property of the metric tensor for flat space is given by

$$g^{\alpha\beta} = g_{\alpha\beta}$$

More generally,

$$g_{\alpha\nu}g^{\nu\beta} = \delta^{\beta}_{\alpha} \tag{4.14}$$

where δ^{β}_{α} is the four-dimensional Kronecker delta symbol.

The procedure for changing an index on a tensor from contravariant to covariant, or vice versa, is obtained by contraction with the metric tensor. Thus, for example,

$$A^{\alpha} = g^{\alpha\beta} A_{\beta}$$
$$F^{\alpha}_{\beta} = g^{\alpha\nu} A_{\nu\beta} \tag{4.15}$$
$$F^{\alpha\beta} = g^{\alpha\nu} g^{\beta\gamma} A_{\nu\gamma}$$

6.4.3   Lagrange's Equations

With the preceding preliminaries at hand, we return to the relativistic action integral [recall (1.1.4)]. In arbitrary geometry, it is given by

$$S = \int -mc\sqrt{-g_{\alpha\beta}(\bar{\mathbf{x}})\dot{x}^{\alpha}\dot{x}^{\beta}} \, d\lambda \tag{4.16}$$

where λ parameterizes the curve

$$x^{\mu} = x^{\mu}(\lambda) \tag{4.16a}$$

and

$$\dot{x}^{\mu} \equiv \frac{dx^{\mu}}{d\lambda} \tag{4.16b}$$

The effective covariant form implied by (4.16) is given by (see Problem 6.23)

$$\bar{L} = -mc\sqrt{-g_{\alpha\beta}\dot{x}^{\alpha}\dot{x}^{\beta}} \tag{4.17}$$

As noted previously, any function of L (satisfying conditions described in Problem 1.46), when substituted into the action integral returns Lagrange's equations. Thus, in place of (4.17), we work with

$$\tilde{L} \equiv \frac{m}{2} g_{\alpha\beta}\dot{x}^{\alpha}\dot{x}^{\beta} \tag{4.18}$$

Elements of Lagrange's equations are then given by

$$\frac{\partial \tilde{L}}{\partial \dot{x}^{\alpha}} = mg_{\alpha\beta}\dot{x}^{\beta} \tag{4.19a}$$

$$\frac{\partial \tilde{L}}{\partial x^{\alpha}} = \frac{m}{2}\left(\frac{\partial}{\partial x^{\alpha}} g_{\beta\mu}\right)\dot{x}^{\beta}\dot{x}^{\mu} \tag{4.19b}$$

$$\frac{d}{d\lambda}\left(\frac{\partial \tilde{L}}{\partial \dot{x}^{\alpha}}\right) = m g_{\alpha\beta}\ddot{x}^{\beta} + m \frac{\partial g_{\alpha\beta}}{\partial x^{\mu}}\dot{x}^{\mu}\dot{x}^{\beta} \tag{4.19c}$$

Inserting these into Lagrange's equations gives

$$m g_{\alpha\beta}\ddot{x}^{\beta} = m\left(\frac{1}{2}\frac{\partial g_{\beta\mu}}{\partial x^{\alpha}} - \frac{\partial g_{\alpha\beta}}{\partial x^{\mu}}\right)\dot{x}^{\mu}\dot{x}^{\beta} \tag{4.20}$$

This is the equation of motion for a free particle in a geometry described by the metric tensor $g_{\alpha\beta}$. In flat space, components of $g_{\alpha\beta}$ are given by (4.13), and (4.20) reduces to

$$m g_{\alpha\beta}\ddot{x}^{\beta} = m g_{\alpha\alpha}\ddot{x}^{\alpha} = 0$$

(no sum over α). We obtain

$$m\ddot{x}_1 = m\ddot{x}_2 = m\ddot{x}_3 = 0, \qquad -m\ddot{x}_4 = 0$$

If in the last equation we choose the constant of integration to be mc, then we find $t = \lambda$, and the first three equations return conservation of momentum.

6.4.4 The Christoffel Symbol

The equation of motion (4.20) may be cost in a more symmetric form. First note that it may be rewritten

$$m g_{\alpha\beta}\ddot{x}^{\beta} = \frac{m}{c}\left(\frac{\partial g_{\beta\mu}}{\partial x^{\alpha}} - \frac{\partial g_{\alpha\beta}}{\partial x^{\mu}} - \frac{\partial g_{\alpha\mu}}{\partial x^{\beta}}\right)\dot{x}^{\beta}\dot{x}^{\mu}$$

Multiplying this equation by $g^{\alpha\nu}$ gives

$$g^{\alpha\nu}g_{\alpha\beta}\ddot{x}^{\beta} = \delta^{\nu}_{\beta}\ddot{x}^{\beta} = \frac{1}{2}g^{\alpha\nu}\left(\frac{\partial g_{\beta\mu}}{\partial x^{\alpha}} - \frac{\partial g_{\alpha\beta}}{\partial x^{\mu}} - \frac{\partial g_{\alpha\mu}}{\partial x^{\beta}}\right)\dot{x}^{\beta}\dot{x}^{\mu}$$

There results

$$\ddot{x}^{\nu} + \Gamma^{\nu}_{\beta\mu}\dot{x}^{\beta}\dot{x}^{\mu} = 0 \tag{4.21}$$

where $\Gamma^{\nu}_{\beta\mu}$ is the *Christoffel symbol*:

$$\Gamma^{\nu}_{\beta\mu} \equiv \frac{1}{2}g^{\alpha\nu}\left(\frac{\partial g_{\alpha\beta}}{\partial x^{\mu}} + \frac{\partial g_{\alpha\mu}}{\partial x^{\beta}} - \frac{\partial g_{\beta\mu}}{\partial x^{\alpha}}\right) \tag{4.21a}$$

The relation (4.21) is the equation for a "free" particle in a geometry described by the Christoffel symbol. We may also view it as an equation for the *geodesic* in the given geometrical environment. A geodesic is the shortest internal between two points in the space-time continuum. This may be seen from the fact that (4.21) follows from rendering the action (4.16) stationary.

6.4.5 Liouville Equation

With the effective Lagrangian (4.18), we may construct the corresponding Hamiltonian. First note that canonical momenta are given by

$$p_\alpha = \frac{\partial \tilde{L}}{\partial \dot{q}_\alpha} = m g_{\alpha\beta} \dot{x}^\beta \qquad (4.22)$$

where, we recall, a dot represents differentiation with respect to the parameter λ. Thus we obtain

$$H = p_\alpha \dot{x}^\alpha - \tilde{L} = \frac{p_\alpha g^{\alpha\beta} p_\beta}{2m} \qquad (4.23)$$

[Note that the preceding reduces to (1.25) in flat space.]

The one-particle Liouville equation is then written

$$\frac{\partial}{\partial \lambda} f(x^\alpha, p_\beta, \lambda) = [H, f]$$

$$= \frac{\partial H}{\partial x^\alpha} \frac{\partial f}{\partial p_\alpha} - \frac{\partial f}{\partial x^\alpha} \frac{\partial H}{\partial p_\alpha} \qquad (4.24)$$

Inserting the Hamiltonian (4.23) into this expression gives

$$\frac{\partial f}{\partial \lambda} + \frac{g^{\alpha\beta} p_\beta}{m} \frac{\partial f}{\partial x^\alpha} - \frac{p_\alpha p_\beta}{2m} \frac{\partial g^{\alpha\beta}}{\partial x^\gamma} \frac{\partial f}{\partial p_\gamma} = 0 \qquad (4.25)$$

which is the one-particle Liouville equation appropriate to non-Cartesian space. As noted in Section 1.4.5, the most general solution of the preceding equation is an arbitrary function of solution to its characteristic equations. These are given by

$$\dot{x}^\alpha = \frac{g^{\alpha\beta} p_\beta}{m} = \frac{\partial H}{\partial p_\alpha} \qquad (4.25a)$$

$$\dot{p}_\gamma = -\frac{p_\alpha p_\beta}{2m} \frac{\partial g^{\alpha\beta}}{\partial x^\gamma} = -\frac{\partial H}{\partial x^\gamma} \qquad (4.25b)$$

As the inserts on the right of these equations indicate, they are Hamilton's equations derived from the Hamiltonian (4.23).

The formalism introduced in this section finds application in general relativity, where space–time curvature derives from large energy–mass concentration.[17]

Problems

6.1. Show that the determinant of the matrix $\hat{L}$ of a Lorentz transformation is unity.

---

[17]For further discussion, see C. W. Misner, K. S. Thorne, and J. A. Wheeler, *Gravitation*, W. H. Freeman and Co., San Francisco (1973).

6.2. If $\hat{L}(\beta_1)$ and $\hat{L}(\beta_2)$ represent two Lorentz transformations with respect to a common reference frame, then show that

$$\hat{L}(\beta_1)\hat{L}(\beta_2) = L(\beta_{12})$$

$$\beta_{12} = \frac{\beta_1 + \beta_2}{1 + \beta_1\beta_2}$$

Note that this example establishes the group property of Lorentz transformations.

6.3. With the relativistic Hamiltonian for a charged particle in an electromagnetic field given by (1.26), show that Hamilton's equations give

$$\frac{d\mathbf{p}}{dt} = e\mathcal{E} + \frac{e}{c}\mathbf{v} \times \mathbf{B}$$

$$\frac{d\mathbf{x}}{dt} = \frac{1}{\gamma m}\left(\mathbf{p} - \frac{e}{c}\mathbf{A}\right) = \mathbf{v}$$

6.4. Show that the vector components of the covariant form (2.20)

$$\frac{dp_\mu}{d\tau} = \frac{e}{mc}F_{\mu\nu}p_\nu$$

return the Lorentz force law.

6.5. (a) Show that (2.21) and (2.21a) are equivalent.
   (b) Integrate (2.21a) over p_4 to obtain the relativistic Vlasov equation (2.22).

6.6. Show that J_4 as given by (2.31) is the same as that given by (1.9a).

6.7. What is the relation between the conductivity parameter with proper values,

$$\sigma_0 = \frac{e^2 n_0}{v_0 m}$$

and σ with lab-frame values, $\sigma = e^2 n/vm$?

6.8. Show that

$$\frac{d^3 p}{E} = i\frac{d^3 p}{c}\int \delta(p_\mu^2 + m^2c^2)dp_4$$

Answer

First write

$$p_\mu^2 + m^2c^2 = p^2 + m^2c^2 + p_4^2 = \left(\frac{E}{c}\right)^2 + p_4^2$$

With the expression

$$\delta(y^2 - a^2) = \frac{1}{2a}[\delta(y - a) + \delta(y + a)]$$

the preceding integral over p_4 gives

$$i\int \delta\left[\left(\frac{E}{c}\right)^2 - (ip_4)^2\right]dp_4 = \frac{1}{E/c}$$

6.9. Obtain an expression for the relativistic Maxwellian for a gas of noninteracting particles of rest mass m confined to a volume V that is immersed in an externally supported potential field $\Phi(x)$.

6.10. Show that

$$d^3p = m^3\gamma^5 d^3v$$

Answer[18]

With

$$p_i = \gamma m v_i$$

we obtain

$$\frac{\partial p_i}{\partial v_j} = \gamma m \left[\delta_{ij} + \gamma^2 \frac{v_i v_j}{c^2}\right]$$

$$= \gamma m \left[\delta_{ij} + (\gamma^2 - 1)\frac{v_i v_j}{c^2}\right] \equiv R_{ij}$$

Introduce the matrix

$$\hat{Q} = \frac{\overline{\overline{\mathbf{vv}}}}{c^2}$$

which has the properties

$$\hat{Q}^n = \hat{Q}, \qquad \operatorname{Tr}\hat{Q} = 1$$

We may then write

$$\hat{R} = \gamma n[\hat{I} + (\gamma^2 - 1)\hat{Q}]$$

Thus

$$\det R = \gamma^3 m^3 \det[\hat{I} + (\gamma^2 - 1)\hat{Q}]$$

Now define

$$\hat{K} \equiv \ln[\hat{I} + (\gamma^2 - 1)\hat{Q}]$$

so that

$$\hat{K} = \sum_{n=1}^{\infty}(-1)^{n-1}\frac{(\gamma^2 - 1)^n}{n}\,\hat{Q}^n$$

$$= \hat{Q}\sum_{n=1}^{\infty}(-1)^{n-1}\frac{(\gamma^2 - 1)^n}{n} = \hat{Q}\ln\gamma^2$$

Note that

$$\det \hat{R} = \gamma^3 m^3 \det e^{\hat{K}}$$

---

[18]Due to C. Litwin.

Now use the identity

$$\det e^{\hat{K}} = e^{\operatorname{Tr} \hat{K}}$$

But

$$\operatorname{Tr} \hat{K} = \ln \gamma^2 \operatorname{Tr} \hat{Q} = \ln \gamma^2$$

We conclude that

$$\det \hat{R} = \gamma^3 m^3 e^{\ln \gamma^2} = m^3 \gamma^5$$

6.11. An electron plasma exits at temperature of $10^6 K$. Is it necessary to use relativistic kinetic theory for this plasma?

6.12. A student argues that relativistic kinetic theory for a many-particle system is inconsistent because proper time can only be defined with respect to one particle. Thus the notion of a relativistic distribution function is not well defined. Is the student's argument valid? If not, why not?

Answer

This problem is obviated if we work with distribution functions with variables defined with respect to the lab frame. The time that enters this distribution is the single lab time t. The distributions f and $F = Nf$ encountered in this chapter were so defined.

6.13. For the monoenergetic beam described in Section 2.5, show that $J_\mu J_\mu = -e^2 n_0^2 c^2$.

6.14. If A_μ is a four-vector, then show that $d^4 A$ is a Lorentz invariant.

6.15. (a) What is the value of the relativistic Maxwellian $f_0(\mu)$ at $v = c$.
   (b) Why is $f_0(\mu)$ physically unrealistic at $v > c$?

6.16. (a) Argue that an electromagnetic wave that is a plane wave in a given frame is a plane wave in all inertial frames.
   (b) With property (a), argue the existence of the four-wave vector (1.9c).
   (c) Let the frame S' move with velocity $\mathbf{v}$ relative to S. If a stationary oscillator in S emits a wave of frequency ω, then show that the observed frequency in S' is

$$\omega' = \frac{[1 - \beta \cos \theta]\omega}{\gamma}$$

where $\cos \theta = \mathbf{k} \cdot \mathbf{v}/kv$. The preceding formula for ω' is the relativistic expression of the *Doppler effect*. At $\theta = \pi/2$, this phenomenon is referred to as the *transverse Doppler shift*.

6.17. Show that the Lorentz transformation is orthogonal. That is, establish (1.12).

6.18. Derive the form that the relativistic Maxwellian (3.10) assumes in the nonrelativistic limit, $\beta \ll 1$.

6.19. The relativistic Lagrangian for a neutral particle of rest mass m in a potential field, $V(\mathbf{r})$ is given by

$$L = -\gamma^{-1} mc^2 - V(\mathbf{r})$$

(a) Show that in the nonrelativistic limit this form reduces to the classical expression (1.1.3).

(b) Employing Lagrange's equations, show that the given relativistic Lagrangian leads to

$$\frac{d}{dt}(m\mathbf{u}) = -\nabla V$$

6.20. Show that Lagrange's equations applied to the N-body Lagrangian (2.49) return the Lorentz force law.

6.21. In the vicinity of a *black hole*, the differential interval in spherical coordinates is given by

$$(ds)^2 = \alpha^{-1}(dr)^2 + r^2[(d\theta)^2 + (\sin\theta \, d\phi)^2] - \alpha(c \, dt)^2$$

$$\alpha \equiv 1 - \frac{2mG}{c^2 r}$$

where M is the mass of the black hole, G is the gravitational constant, and c is the speed of light. From this expression, write down the *Schwarzchild* metric $g_{\alpha\beta}$.

6.22. What equations do the fourth components of the left equalities in (4.25) correspond to for the case that the parameter $\lambda = \tau$, proper time?

6.23. In flat space, the single free-particle Lagrangian is given by (see Problem 6.19)

$$L = -\gamma^{-1}mc^2$$

(a) Does the effective Lagrangian $\tilde{L}$ given by (4.17) reduce to the preceding form in the flat-space limit? Explain your answer.

(b) How can your finding to part (a) be corrected?

Answer

(a) In the flat space limit we find

$$\tilde{L} = -mc\sqrt{-\left(\frac{dx}{d\lambda}\right)^2 + c^2\left(\frac{dt}{d\lambda}\right)^2}$$

which is not the desired form.

(b) The canonical action integral is written $S = \int L \, dt$. So, to bring $\tilde{L}$ to canonical form, the integral (4.16) must be multiplied by dt/dt, resulting in the flat-space canonical Lagrangian

$$L = \tilde{L}\frac{d\lambda}{dt} = -mc^2\gamma^{-1}$$

6.24. Establish the relation

$$-s - t - u = -2m_1^2 c^2 - 2m_2 c^2$$

where s, t and u are Mandelstam variables discussed in Table 6.1.

Answer

With given definitions, we write

$$-s - t - u = 3(p_{1\mu})^2 + 1(p'_{1\mu})^2 + (p_{2\mu})^2 + (p'_{2\mu})^2$$
$$+2p_{1\mu}p_{2\mu} - 2p_{1\mu}p'_{1\mu} - 2p_{1\mu}p'_{2\mu}$$

with

$$(p_{1\mu})^2 = (p'_{1\mu})^2 = -m^2c^2$$
$$(p_{2\mu})^2 = (p'_{2\mu})^2 = -m^2c^2$$

and

$$2p_{1\mu}(p'_{1\mu} + p'_{2\mu}) = 2p_{1\mu}(p_{1\mu} + p_{2\mu})$$

the result follows.

CHAPTER 7

Kinetic Properties of Metals and Amorphous Media

Introduction

Solid state matter falls into the categories: metals, semi-metals, semiconductors and insulators. Structures of solids may be separated into the groups: single crystal, polycrystal, and amorphous. A sample of single-crystal material comprises a periodic array of molecules (simple cubic, body-centered cubic, hexagonal, etc.). A sample of polycrystalline material is composed of single-crystal "grains" of material separated by "grain boundaries." An amorphous ("without form") material does not have a regular array of molecules. The single-crystal phase of alumina, Al_2O_3, with Fe impurities, is the gem sapphire, and with Cr impurities, it is the gem ruby. In polycrystalline form, this material is employed as a high-temperature resistant ceramic. Window glass is amorphous SiO_2 (fused silica). A common crystalline form of silica is quartz (the dominant component of sand).

A graphical means of distinguishing between these solid-state phases is given in terms of the radial distribution function defined in Section 2.2.C. The radial distribution function for a single-crystal sample is sketched in Fig. 7.1a, that of a polycrystalline sample in Fig. 7.1b, and that of an amorphous material in Fig. 7.1c.

A homogeneous mixture of two or more metals is called an *alloy*. The symbol for a "binary" alloy comprising metals A and B, is written $A_{1-x}B_x$ which indicates that in the alloy, x mole fraction of B atoms replace x mole fraction A atoms. Alloys are grouped into two classes: ordered and disordered. For example, the alloy brass, Cu–Zn, has an ordered phase, "β brass," which has a simple cubic lattice with a cesium chloride crystal structure. The alloy bronze is a multicomponent alloy composed of copper and a number of other elements, including, tin, aluminum, silicon, and nickel. Other alloys, such

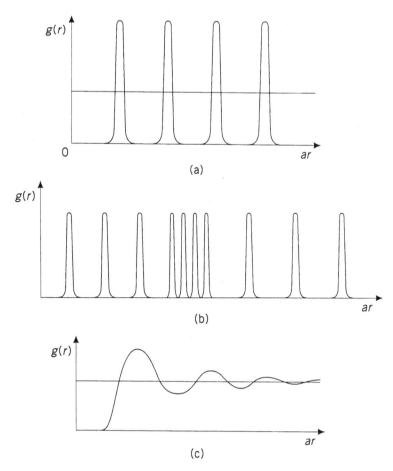

FIGURE 7.1. Sketches of the radial distribution for: (a) single-crystal, (long-range order) (b) polycrystal (partial order) and (c) amorphous material (short-range order). Characteristic intermolecular displacement a, is defined with respect to the displacement, r. The value $g(r) = 1$ is relevant to a perfect gas. Note that case (b) is dependent on the choice of origin-site. As $g(r)$ is an average entity, one may expect distortion of this curve, beyond a few intermolecular displacement.

as Cu−Pb and Zn−Cd are heterogeneous mixtures of grains of nearly pure single-metal components.

In the present chapter we are concerned with kinetic properties of metals and amorphous media. In the first of these descriptions, expressions for temperature-dependent electrical resistivity and thermal conductivity are derived for a class of metals with spherical-like Fermi surfaces. In the second part of the chapter, concepts are discussed relevant to amorphous media. Such notions include hopping, percolation, localization, and the related concept of the mobility edge. The chapter begins with a general discussion of the phenomenon of thermopower, which relates to the electric field developed in a metal due to a temperature gradient across the material.

7.1   Metallic Electrical and Thermal Conduction

7.1.1   Background

A number of processes contribute to metallic resistivity: electron scattering from impurities and lattice imperfections, electron–electron collisions, and electron–phonon collisions. This latter interaction has both *umklapp* and *normal* contributions.[1] In this section of the present chapter, metallic resistivity of a pure metal will be described, for which electron–phonon collisions are dominant. Umklapp processes are negligible at low temperature and are neglected in the discourse. Expressions for electrical and thermal conductivity in a metal were described briefly in Section 5.3.3. In the present discussion we are concerned with temperature dependence of these parameters.

Domains of thermal and electrical properties of a metal divide according to the Debye temperature of the metal, Θ_D. The Debye temperature partitions classical from quantum behavior of a given material. Classical properties occur in the domain $T \gg \Theta_D$, and quantum properties in the $T \lesssim \Theta_D$ interval. As was first derived by Bloch,[2] low-temperature electrical resistivity, ρ, of a pure metal (with a spherical, or nearly spherical Fermi surface) at $T \ll \Theta_D$, varies as T^5. A rough derivation of this dependence may be obtained employing mean-free-path estimates. At low temperature, electron dynamics is restricted to the Fermi surface. Let l_k denote mean-free-path displacement in k-space on the Fermi surface and let ν denote the electron–phonon collision frequency so that one may define an effective speed, $u \propto \nu/l_k$. One may then write, $k_B T \propto \hbar u l_k$. Recalling (3.4.18) for the diffusion coefficient, we write $D_k \propto l_k u \propto l_k^2 \nu$ relevant to diffusion in k-space on the Fermi surface. At lower temperature it may be shown that[3] electron–phonon collision frequency, $\nu \propto T^3$, which gives $D_k \propto T^5$. For electrical conductivity we recall (3.4.14), $\sigma = e^2 n/\nu m \propto e^2 n \nu_F \tau/p_F$ where τ is the time for an electron to diffuse a distance p_F on the Fermi surface. In diffusion motion, the mean square displacement is proportional to time (1.8.36) and we may write, $p_F^2 \propto D_k \tau$. There results $\sigma \propto n e^2 p_F^2/m D_k$, so that low-temperature metallic resistivity, $\rho = 1/\sigma \propto T^5$.

In the high-temperature domain, $T \gg \Theta_D$, it is generally observed that ρ varies as T. This increase in resistivity follows from the behavior of the Bose–Einstein distribution for phonons at high temperature. As note in

---

[1]W. A. Harrison, *Solid State Theory*, Dover, New York (1979).

[2]F. Bloch, *Z. Phys.* **59**, 208 (1930).

[3]E. M. Lifshitz and L. P. Pitaevskii, *Physical Kinetics*, Pergamon, New York (1981).

(5.3.68) one finds that the number of phonons at high temperature is given by, $n_q \propto k_B T / \hbar \omega(q)$. Since this number of scatterers increases with T so does the resistivity.

Here is a summary of these findings.

$$T \ll \Theta_D \quad \rho \propto T^5$$
$$T \gg \Theta_D \quad \rho \propto T \tag{1.1}$$

Whereas electrical conduction in a metal is due only to electron transport, both electrons and phonons contribute to thermal conduction. However, a simplifying property comes into effect. Heat transfer in a metal is due primarily to electron transport, owing to relativity high electron speeds (Fermi speed) compared to phonon speeds. Likewise, at low temperature, specific heat due to electrons exceeds that due to phonons. For pure metals, thermal and electrical conductivity are related according to the Weidemann–Franz law (5.3.43)

$$\frac{\kappa}{\sigma T} = \frac{\pi^2}{3} \left(\frac{k_B}{e} \right)^2 \tag{1.2a}$$

where κ denotes coefficient of thermal conductivity. This relation is valid over a wide temperature domain,[4] whereas in the low-temperature domain, $T \ll \Theta_D$, the Lorenz equation[5]

$$\frac{\kappa}{\sigma T} = A \left(\frac{T}{\Theta_D} \right)^2 \tag{1.2b}$$

is appropriate, where A is a constant. With (1.1) and the preceding relations we note the following characteristic values of κ.

$$T \ll \Theta_D \quad \kappa \propto 1/T^2$$
$$T \gg \Theta_D \quad \kappa = \text{constant} \tag{1.2c}$$

Measurements of κ indicate that at low temperature, κ behaves linearly in T and reduces to zero at 0K. This behavior is attributed to impurity scattering.

The main thrust of this component of the present chapter is to derive an expression for electrical conductivity in a metal relevant to the entire temperature range. Before doing so, we discuss a closely related phenomenon in metals for the situation when a gradient of temperature is present.

For future reference we note the following properties of the Fermi–Dirac distribution (5.3.1). We define

$$G_n \equiv \frac{1}{n!} \int_0^\infty (E - \mu)^n \left(-\frac{\partial f_0}{\partial E} \right) dE \tag{1.3a}$$

---

[4]N. W. Ashcroft and N. D. Mermin, *Solid State Physics*, Holt, Rinehart and Winston, New York (1976).

[5]J. M. Ziman, *Principles of the Theory of Solids*, Cambridge, New York, (1964), Section 7.8.

Then

$$G_n = \frac{(k_B T)^n}{n!} \int_{-\infty}^{\infty} \frac{z^n dz}{(e^z + 1)(1 - e^{-z})} \tag{1.3b}$$

$$G_n = \left(\begin{array}{cc} 2b_n (k_B T)^n, & n \text{ even} \\ 0, & n \text{ odd} \end{array} \right) \tag{1.3c}$$

Values of the constant b_n are readily calculated, e.g., $2b_0 = 1$, $2b_2 = \pi^2/6$. [See Problem 7.1 and recall (5.3.39b).] The following integrals are also noted.

$$\mathbf{B}_n \equiv \frac{1}{4\pi^3} \frac{1}{\hbar} \int \int \tau \overline{\overline{\mathbf{v}\mathbf{v}}} (E - \mu)^n \left(-\frac{\partial f_0}{\partial E} \right) \frac{dS}{v} dE \tag{1.3d}$$

where $\mathbf{B}_n$ is a tensor and $\int dS$ represents an integral over the Fermi surface. With (1.3) we obtain,

$$\int \Phi(E) \left(-\frac{\partial f_0}{\partial E} \right) dE = \Phi(E_F) + \frac{\pi^2}{6} (k_B T)^2 \left[\frac{\partial^2 \Phi}{\partial E^2} \right]_{E_F} + \cdots \tag{1.3e}$$

Employing (1.3e) we find

$$\mathbf{B}_0 = \frac{1}{4\pi^3} \frac{1}{\hbar} \int \tau \overline{\overline{\mathbf{v}\mathbf{v}}} \frac{dS}{v} \tag{1.3f}$$

$$\mathbf{B}_1 = \frac{\pi^2}{3} (k_B T)^2 \left[\frac{\partial \mathbf{B}_0}{\partial E} \right]_{E_F} \tag{1.3g}$$

$$\mathbf{B}_2 = \frac{\pi^3}{3} (k_B T)^2 \mathbf{B}_0 \tag{1.3h}$$

In the preceding we have recalled the relation (5.3.39a)

$$d\mathbf{k} = \frac{dS \, dE}{|\nabla_k E|} = \frac{dS \, dE}{\hbar v} \tag{1.3i}$$

where the right equality is relevant to a spherical energy surface and, we recall, dS is a surface element on the Fermi surface in $\mathbf{k}$-space.

7.1.2 Thermopower

Thermopower refers to the electric field developed in a metal due to a temperature gradient present in the material. (A closely allied phenomenon, electrochemical potential, was discussed previously in Section 3.4. We recall that in this phenomenon current flow occurs in a medium in which a density gradient exists in addition to an electric field.) Consider a long, thin metal bar along which there is a temperature gradient. Due to resulting charge flow, an electric field is generated in the direction opposite to the temperature gradient. At equilibrium there is an electric field with no charge flow. This electric field is called the *thermoelectric field*. One writes

$$\mathcal{E} = Q \nabla T \tag{1.3j}$$

where Q is labeled the thermopower. In typical metals, observed values of Q are of order $\mu V/K$.

Relaxation-time approximation

We wish to study thermopower in the relaxation-time approximation (5.3.35). The equilibrium distribution for conduction electrons in a metal in which there is a temperature gradient and an electric field is given by the $(k, \mathbf{r})$-dependent Fermi-Dirac distribution

$$f_0[E(k), T(\mathbf{r}), \mu(\mathbf{r})] = \frac{1}{\exp[(E - \mu)/k_B T] + 1} \tag{1.4}$$

In this model one is concerned with the change in the distribution due to diffusion effects for which one writes, $(\partial f_0/\partial t)_{\text{Diff}} = -\mathbf{v} \cdot \nabla f$, etc.[5] The corresponding steady-state version of (5.3.35) is then given by

$$-\left(\mathbf{v} \cdot \nabla f + \frac{e\mathcal{E}}{\hbar} \cdot \frac{\partial t}{\partial \mathbf{k}}\right) = \frac{1}{r}[f_0 - f] \tag{1.5}$$

where $\mathcal{E}$ denotes electric field. We introduce the perturbation distribution

$$f(\mathbf{r}, \mathbf{k}) = f_0 + g(\mathbf{r}, \mathbf{k}) \tag{1.6}$$

and the relations [see equation above (5.3.32)]

$$\mathbf{v}(\mathbf{k}) = \frac{1}{\hbar} \frac{\partial E}{\partial \mathbf{k}}, \qquad \frac{\partial f_0}{\partial \mathbf{k}} = \hbar \mathbf{v} \frac{\partial f_0}{\partial E} \tag{1.7a}$$

where $\mathbf{v}$ is group velocity. We note the derivative

$$\nabla f_0 = \frac{\partial f_0}{\partial T} \nabla T + \frac{\partial f_0}{\partial \mu} \nabla \mu \tag{1.7b}$$

Substituting the latter three equations in (1.5), and keeping $O(1)$ terms, one obtains

$$\mathbf{v} \cdot \left(\frac{\partial f_0}{\partial T} \nabla T + \frac{\partial f_0}{\partial \mu} \nabla \mu\right) + e\mathcal{E} \cdot \mathbf{v} \frac{\partial f_0}{\partial E} = \frac{1}{\tau(E)} g(\mathbf{k}, \mathbf{r}) \tag{1.8}$$

With

$$\frac{\partial f_0}{\partial E} = -\frac{f_0^2 \exp[(E - \mu)/(k_B T)]}{k_B T} \tag{1.9a}$$

we find

$$\frac{\partial f_0}{\partial T} = -\left(\frac{E - \mu}{T}\right) \frac{\partial f_0}{\partial E} \tag{1.9b}$$

$$\frac{\partial f_0}{\partial \mu} = -\frac{\partial f_0}{\partial E} \tag{1.9c}$$

---

[5]J. M. Ziman, *Principles of the Theory of Solids*, Cambridge, New York, (1964), Section 7.8.

Substituting these findings into (1.8) gives

$$\frac{\partial f_0}{\partial E} \mathbf{v} \cdot \left[-\left(\frac{E - \mu}{T} \right) \nabla T + e \left(\boldsymbol{\mathcal{E}} - \frac{1}{e} \nabla \mu \right) \right] = \frac{1}{\tau(E)} g(\mathbf{k}, \mathbf{r}) \qquad (1.10a)$$

For future reference this equation is rewritten as

$$g(\mathbf{k}, \mathbf{r}) = \alpha \mathbf{v} \cdot \left[\frac{1}{T} \Gamma(-\nabla T) + e \left(\boldsymbol{\mathcal{E}} - \frac{1}{e} \nabla \mu \right) \right] \qquad (1.10b)$$

$$\alpha \equiv \tau \frac{df_0}{dE}, \qquad \Gamma \equiv E - \mu \qquad (1.10c)$$

To calculate thermal conductivity in a solid one imagines a small region over which temperature is effectively constant. Recalling the thermodynamic relation between a differential of heat, dQ and a differential of entropy, dS

$$dQ = T \, dS$$

permits one to write the corresponding relation between thermal and entropy current densities, $\mathbf{J}_Q$ and $\mathbf{J}_S$

$$\mathbf{J} = T \mathbf{J}_S \qquad (1.11a)$$

($\mathbf{J}_Q$ was previously labeled $\mathbf{Q}$). With the entropy relation

$$T \, dS = dE - \mu \, dN$$

one writes

$$T \mathbf{J}_S = \mathbf{J}_E - \mu \mathbf{J}_N \qquad (1.11b)$$

where $\mathbf{J}_N$ is particle current density. With (1.11a) it follows that

$$\mathbf{J}_Q = \mathbf{J}_E - \mu \mathbf{J}_N \qquad (1.11c)$$

Local energy and particle current-density are given by [with $\mathbf{r}$-dependence of f tacitly implied]

$$\begin{pmatrix} \mathbf{J}_E \\ \mathbf{J}_N \end{pmatrix} = \int \frac{d\mathbf{k}}{4\pi^3} \begin{pmatrix} E(k) \\ 1 \end{pmatrix} \mathbf{v}(\mathbf{k}) f(\mathbf{k}) \qquad (1.12)$$

With (1.11c) we may then write

$$\mathbf{J}_Q = \int \frac{d\mathbf{k}}{4\pi^3} \Gamma(E) \mathbf{v}(\mathbf{k}) f(\mathbf{k}) \qquad (1.13)$$

As f_0 (1.4) is an even function of k, only the perturbation distribution g enters in the preceding integral. With (1.10b) we obtain

$$\mathbf{J}_Q = \int \frac{d\mathbf{k}}{4\pi^3} \Gamma \alpha(E) \overline{\mathbf{v}(\mathbf{k})\mathbf{v}(\mathbf{k})} \cdot \left[\frac{1}{T} \Gamma(-\nabla T) + e \left(\boldsymbol{\mathcal{E}} - \frac{1}{e} \nabla \mu \right) \right] \qquad (1.14)$$

Electric current density is given by [recall equation preceding (5.3.39)]

$$\mathbf{J} = e \int \frac{d\mathbf{k}}{4\pi^3} \mathbf{v}(\mathbf{k}) f(\mathbf{k}) \qquad (1.15)$$

Again with reference to (1.10b) we find

$$\mathbf{J} = e \int \frac{d\mathbf{k}}{4\pi^3} \alpha(E) \overline{\overline{\mathbf{v(k)v(k)}}} \cdot \left[\frac{1}{T} \Gamma(-\nabla T) + e \left(\mathcal{E} - \frac{1}{e} \nabla \mu \right) \right] \quad (1.16)$$

It may be assumed that in measurement of electric field in a metal, gradients in chemical potential are subsumed in the field. Consequently this term is ignored in the following relations. With (1.14) and (1.16) we write

$$\mathbf{J} = \mathbf{L}_{11}\mathcal{E} + \mathbf{L}_{12}(-\nabla T) \quad (1.17a)$$
$$\mathbf{J}_Q = \mathbf{L}_{21}\mathcal{E} + \mathbf{L}_{22}(-\nabla T) \quad (1.17b)$$

where the $\mathbf{L}_{ij}$ coefficients are as implied and are, in general, tensor forms. In the absence of a gradient of temperature, $\mathbf{L}_{11}$ is electrical conductivity, whereas in the absence of an electric current, $\mathbf{L}_{22}$ is the coefficient of thermal conductivity. Here one assumes, $|\mathbf{L}_{22}| \gg |\mathbf{L}_{21}\mathbf{L}_{11}^{-1}\mathbf{L}_{12}|$. The off-diagonal elements relate to the interplay of these two effects when both fields are present. Note that in general, $\mathbf{L}_{21} = T\mathbf{L}_{12}$.

For $\mathbf{L}_{11}$ we find

$$\mathbf{L}_{11} = \frac{e^2}{4\pi^3} \frac{1}{\hbar} \int \int d\mathbf{k}\tau \frac{\partial f_0}{\partial E} \overline{\overline{\mathbf{vv}}} \quad (1.18a)$$

which with (5.3.39b) and (1.3i), may be written

$$\mathbf{L}_{11} = \frac{e^2}{4\pi^3} \frac{1}{\hbar} \int \int \tau \overline{\overline{\mathbf{vv}}} \frac{dS}{v} = e^2\mathbf{B}_0 \quad (1.18b)$$

so that

$$\sigma = e^2\mathbf{B}_0 \quad (1.18c)$$

which is in accord with our previous finding (5.3.41).

For $\mathbf{L}_{22}$ we find [with (1.3d) and (1.3i)]

$$\mathbf{L}_{22} = \kappa = \frac{1}{T}\mathbf{B}_2 = T\pi^2 k_B^2 \mathbf{B}_0 \quad (1.18d)$$

which, with (1.18c) returns the Wiedemann-Franz law, (1.2a).

7.1.3 Electron–Phonon Scattering Matrix Elements

The analysis of the following sections addresses temperature-dependent electrical resistivity in monovalent metals. These include a subset of the alkali metals: Na, K, Rb, Cs, which each have bcc structures, and the noble metals: Cu, Ag, Au, which have have fcc structures. Alkali metal atoms include a single electron outside a closed shell noble-atom electron configuration. Thus, for example, potassium has the electronic configuration, [Ar] $4s^1$. Noble metal atoms include a single electron exterior to closed shell composed of a corresponding noble atom electronic configuration and a closed d^{10} shell. Thus, for example, copper has the electronic configuration, [Ar] $3d^{10}4s^1$. The metallic

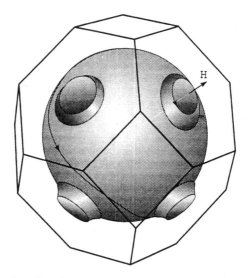

FIGURE 7.2. Fermi surface for a noble metal showing necks at contact with Bragg planes of the first Brillouin zone (truncated octahedron). Also shown are de Hass–van Alphen electron orbits due to an imposed magnetic field, H, in the [111] direction. (Reproduced with permission of N. W. Ashcroft and N. D. Mermin.)

state of both classes of elements is characterized by tightly bounded bands that lie significantly below conduction electron levels of the metal.

The Fermi surfaces for the alkali metals are spheres and lie, respectively, within the first Brillouin zone (regular rhombic dodecahedron). The Fermi surfaces for the noble metals are spherical except for "necks," which extend to the Bragg planes of the first Brillouin zone (truncated octahedron) in the [111] directions (Fig. 7.2). As noted earlier, temperature dependence of phenomena in solid materials separates according to weather temperature in less than or greater than the Debye temperature. Our analysis will rediscover this property as well as a residual resistivity, which comes into play near 0K.[6,7] Characteristic low-temperature alkali-metal resistance is shown in Fig. 7.3.

Interaction Hamiltonian

Our study begins with the derivation of an expression for the interaction Hamiltonian between electrons and ions in the lattice, in which ion displacement from equilibrium is incorporated in the analysis. Working in second quantization,

[6]Components of this analysis appeared previously in R. L. Liboff and G. K. Schenter, *Phys. Rev. B54*, 16591 (1996).

[7]For a review of these topics see: J. Bass, W. P. Pratt, Jr. and P. A. Schroeder, *Revs. Mod. Phys.* 62, 645 (1990).

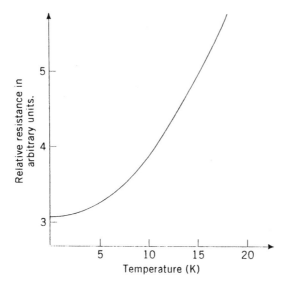

FIGURE 7.3. Characteristic relative resistance as a function of temperature for an alkali metal in the low-temperature domain at relatively small impurity concentration. For further discussion see D. K. C. MacDonald and K. Mendelssohn, *Proc. Roy. Soc.* (London) 202, 103 (1950); J. S. Dugdale, *Electrical Properties of Metals and Alloys*, Edward Arnold, London, 1977 Chap. 10; J. L. Olsen, *Electron Transport in Metals*, Interscience, New York (1962), Chap. 2.

the potential Hamiltonian component for the system may be written

$$\hat{H} = \int d\mathbf{x}\, \hat{\psi}^\dagger(\mathbf{x})\hat{\psi}(\mathbf{x}) \sum_{\mathbf{y}} \Phi[\mathbf{x} - \mathbf{y} - \hat{\mathbf{z}}(\mathbf{y})] \qquad (1.19)$$

where Φ is electron–ion interaction potential, $\hat{\psi}^\dagger$ is electron creation operator, $\mathbf{x}$ denotes electron position, $\mathbf{y}$ denotes ion equilibrium position and $\hat{\mathbf{z}}(\mathbf{y})$, ion displacement from equilibrium. For small ion displacements we write

$$\hat{H}_{\mathrm{INT}} = \hat{H} - \hat{H}(\hat{\mathbf{z}} = 0) = \int d\mathbf{x}\, \hat{\psi}^\dagger(\mathbf{x})\hat{\psi}(\mathbf{x}) \sum_{\mathbf{y}} \hat{\mathbf{z}}(\mathbf{y}) \cdot \frac{\partial}{\partial \mathbf{y}} \Phi(\mathbf{x} - \mathbf{y}) \quad (1.20)$$

Fourier expanding electron operators

$$\hat{\psi}^\dagger(\mathbf{x}) = \int \frac{d\mathbf{k}_2}{(2\pi)^3}\, e^{-i\mathbf{k}_2 \cdot \mathbf{x}}\hat{\psi}^\dagger(\mathbf{k}_2)$$

$$\hat{\psi}(\mathbf{x}) = \int \frac{d\mathbf{k}_1}{(2\pi)^3}\, e^{i\mathbf{k}_1 \cdot \mathbf{x}}\hat{\psi}(\mathbf{k}_1) \qquad (1.21a)$$

and quantizing ion displacements

$$\hat{\mathbf{z}}(\mathbf{y}) = \sum_{\mathbf{q},\mu} \sqrt{\frac{\hbar}{2MN\omega}} \left[\hat{\mathbf{e}}_\mu(\mathbf{q})e^{i(\mathbf{q}\cdot\mathbf{y})}\hat{a}_\mu(\mathbf{q}) + \hat{\mathbf{e}}_\mu(-\mathbf{q})e^{-i\mathbf{q}\cdot\mathbf{y}}\hat{a}_\mu^\dagger(\mathbf{q}) \right] \qquad (1.21b)$$

gives

$$
\hat{H}_{\text{INT}} = \int d\mathbf{x} \int \frac{d\mathbf{k}_1 d\mathbf{k}_2}{(2\pi)^6} \sum_{\mathbf{y}} \sum_{\pm} \sum_{\mathbf{q},\mu} \sqrt{\frac{\hbar}{2MN\omega}} \, \hat{\mathbf{e}}_\mu(\pm\mathbf{q}) \left[\psi^\dagger(\mathbf{k}_2)\psi(\mathbf{k}_1) \right]
$$

$$
\times \begin{bmatrix} \hat{a}_\mu(\mathbf{q}) \\ \hat{a}_\mu^\dagger(\mathbf{q}) \end{bmatrix} (\mp i\mathbf{q} \exp i[\pm\mathbf{q} \cdot \mathbf{y} - \mathbf{k}_2 \cdot \mathbf{x} + \mathbf{k}_1 \cdot \mathbf{x}])\Phi(\mathbf{x} - \mathbf{y}) \quad (1.22)
$$

where $\hat{\mathbf{e}}_\mu$ is a phonon polarization unit vector, $\mathbf{q}$ denotes phonon wave vector, $\omega = uq$, N represents total number of free electrons and M is ion mass. The column vector notation in (1.22) is such that the upper $\hat{a}_\mu(\mathbf{q})$ term corresponds to the $(+)$ phonon absorption mode and the lower $\hat{a}_\mu^\dagger(\mathbf{q})$ term corresponds to the $(-)$ phonon emission mode. In these expressions,

$$
\rho_M = Mn_M, \qquad n_M = \frac{n}{Z}, \qquad n = \frac{N}{V} \quad (1.22a)
$$

where ρ_M is ion mass density, n_M is ion number density, n is electron number density, Z is atomic valence, μ is polarization index and ω is phonon frequency. Now we note that, with $\mathbf{u} \equiv \mathbf{x} - \mathbf{y}$,

$$
\sum_{\mathbf{y}} e^{\pm i\mathbf{q}\cdot\mathbf{y}} \Phi(\mathbf{x} - \mathbf{y}) = -e^{\pm i\mathbf{q}\cdot\mathbf{x}} \sum_{\mathbf{u}} e^{\mp i\mathbf{q}\cdot\mathbf{u}} \Phi(\mathbf{u}) = \frac{N}{V} e^{\pm i\mathbf{q}\cdot\mathbf{x}} \tilde{\Phi}(\pm\mathbf{q}) \quad (1.23)
$$

where V is total crystal volume and

$$
\tilde{\Phi}(\mathbf{q}) = \int d\mathbf{x} \, e^{-i\mathbf{q}\cdot\mathbf{x}} \Phi(\mathbf{x}) \quad (1.24)
$$

is the Fourier transform of the potential. The preceding combines with the exponentials (1.23) to give

$$
H_{\text{INT}} = \int \frac{d\mathbf{k}_1 d\mathbf{k}_2}{(2\pi)^6} \sum_{\pm} \sum_{\mathbf{q}} \delta(\pm\mathbf{q} - \mathbf{k}_2 + \mathbf{k}_1) \sqrt{\frac{\hbar}{2MN\omega}} \quad (1.25)
$$

$$
\times \left\{ \mp i\mathbf{q} \cdot \hat{\mathbf{e}}_\mu(\pm\mathbf{q}) \frac{N}{V} \Phi(\pm\mathbf{q}) \left[\hat{\psi}^\dagger(\mathbf{k}_2)\hat{\psi}(\mathbf{k}_1) \right] \begin{bmatrix} \hat{a}_\mu(\mathbf{q}) \\ \hat{a}_\mu^\dagger(\mathbf{q}) \end{bmatrix} \right\}
$$

With this equation we see that only longitudinal modes contribute so we may set

$$
\mp i\mathbf{q} \cdot \hat{\mathbf{e}}_\mu(\pm\mathbf{q}) = \mp iq
$$

Thus, rewriting (1.25) in the form

$$
H_{\text{INT}} = \int \int \frac{d\mathbf{k}_1 d\mathbf{k}_2}{(2\pi)^6} \sum_{\pm} C_q \delta(\pm\mathbf{q} - \mathbf{k}_2 + \mathbf{k}_1) \hat{\psi}^\dagger(\mathbf{k}_2)\hat{\psi}(\mathbf{k}_1) \begin{bmatrix} \hat{a}_\mu(\mathbf{q}) \\ \hat{a}_\mu^\dagger(\mathbf{q}) \end{bmatrix} \quad (1.26)
$$

gives the coefficient

$$
C_q = \mp\sqrt{\frac{\hbar}{2MN\omega}} \, iq\frac{N}{V}\omega, \, \tilde{\Phi}(\pm q) \quad (1.27)
$$

For the shielded Coulomb potential one writes

$$\tilde{\Phi}(q) = \frac{4\pi Z e^2}{q^2 + q_{TF}^2} \tag{1.28}$$

where q_{TF} is the Thomas–Fermi shielding wavenumber

$$q_{TF}^2 = \left(\frac{4}{a_0}\right)\left(\frac{3n}{\pi}\right)^{1/3} \tag{1.28a}$$

and a_0 is the Bohr radius. With $\omega = uq$ we obtain

$$|C_q|^2 = \frac{\hbar(M\Omega^2)^2}{2Z^2\rho_M u} \frac{q}{(q^2 + q_{TF}^2)^2} \tag{1.29}$$

where Ω is the ion plasma frequency

$$\Omega^2 = \frac{4\pi n_M (Ze)^2}{M} \tag{1.29a}$$

As we will see below, the expression (1.29) comes into play in electron–phonon scattering matrix elements.

Scattering-rate matrix elements

Electron–phonon scattering-rate matrix elements are written $S_{k'k}^{(\alpha)}$ (with dimensions of inverse time) and are given by

$$S_{kk'}^{(\alpha)} = |\langle k', n'|H_{INT}|k, n\rangle|^2 \frac{2\pi}{\hbar} \delta(\Delta E) \tag{1.30a}$$

$$\delta(\Delta E) = \delta(E' - E - \alpha\hbar\omega) \tag{1.30b}$$

where E is electron energy, H_{INT} is given by (1.26). [$S_{kk'}^{(\alpha)}$ was previously labeled $w(k, k)$ in (5.3.59).] The equality

$$S_{kk'}^{(\alpha)} = S_{k'k}^{(-\alpha)} \tag{1.30c}$$

corresponds to symmetry of the electron–phonon interaction under time reversal and $\alpha = (+, -)$ corresponds to phonon (absorption, emission) in an electron–phonon scattering event. Momentum conservation in a collision is given by

$$k' = k + \alpha q \tag{1.31}$$

where $\hbar q$ is phonon momentum.

In the relation (1.30a), $|n\rangle$ denotes the many-phonon state

$$|n\rangle = |n_q, n_{q'}, \ldots\rangle \tag{1.32}$$

where

$$n_q = \frac{1}{e^{\hbar\omega/k_B T} - 1} \tag{1.33}$$

is the Bose–Einstein distribution. For the dispersion relation for phonons we write

$$\omega = uq \tag{1.34}$$

An estimate of the phonon speed u is given by the Bohm–Staver relation

$$u^2 = \frac{2}{3} \frac{ZE_F}{M} \tag{1.34a}$$

Fast relaxation of phonons to the distribution (1.33) is assumed in the analysis (the so-called "Bloch condition").

For metals, the matrix elements (1.30a) have the value[8]

$$S_{\mathbf{kk'}}^{(\alpha)} = |C_q|^2 \left(n_q + \frac{1}{2} - \alpha \frac{1}{2} \right) \frac{2\pi}{\hbar} \delta(\Delta E) \tag{1.35}$$

where with (1.29)

$$|C_q|^2 = RG(q) \tag{1.36a}$$

$$G(q) \equiv \frac{q/V}{(q^2 + q_{\mathrm{TF}}^2)^2} \tag{1.36b}$$

$$R \equiv \frac{\hbar (M\Omega^2)^2}{2\rho_M u Z^2} \tag{1.36c}$$

Note that R has dimensions of (energy)^2, G, is dimensionless and C_q has the units of energy. The quantity $\hbar\Omega/2$ may be identified with ion zero-point energy.

With (1.33) we see that phonon occupation numbers, $n_q = 0$, at $T = 0$K. Nevertheless, from (1.35) we note that the phonon emission ($\alpha = -1$) matrix element persists at this temperature. Thus, inelastic electron–phonon scattering maintains at $T = 0$K. The fact that phonon absorption scattering matrix elements vanish in this limit is consistent with the 0K limit. That is, as the ground phonon state at $T = 0$K is the state of lowest energy, no energy may be extracted from it in any interaction. The phonon emission at $T = 0$K is due to zero-point fluctuations which induce an electron to decay, consistent with (1.30b). As will be noted below, this phenomenon gives rise to a residual resistivity in a pure metal at $T = 0$K.

Assumptions

Here is a brief review of assumptions included in following analysis. In the absence of an electric field, it assumed that conduction electrons in the metal are in the Fermi–Dirac distribution

$$f_0(k) = \frac{1}{1 + \exp[(E - E_F)/k_B T]} \tag{1.37}$$

---

[8]F. Sietz, *The Modern Theory of Solids*, Dover, New York (1987); J. M. Ziman, *Electrons and Phonons*, Oxford, New York (1960).

with normalization

$$\int f_0(k)\frac{d\mathbf{k}}{(2\pi)^3} = n \tag{1.38}$$

where E_F is the Fermi energy.

For metals with a spherical or nearly spherical energy surface we may write

$$d\mathbf{k} = 4\pi k^2 dk = \frac{2\pi \sqrt{E}\, dE}{(\hbar^2/2m)^{3/2}} \tag{1.39}$$

and (1.39) may be rewritten

$$\int f_0(E)\sqrt{E}\, dE = n(2\pi)^2(\hbar^2/2m)^{3/2} \tag{1.40}$$

With ω written for phonon frequency, the following relations are assumed

$$\hbar\omega < \hbar\omega_D \ll E_F \simeq E \tag{1.41}$$

where ω_D is the Debye frequency. Furthermore, as electron wave vectors lie predominantly on the Fermi surface, we also conclude that: $q \ll k_F$ and electron scattering is predominantly small angle, or, equivalently, $\mathbf{k} \cdot \mathbf{q} \ll kq$.

Scattering matrix elements (1.35), (1.36), are employed in construction of the quantum Boltzmann equation presented in the following section, relevant to derivation of the distribution function for conduction electrons in a metal.

7.1.4 Quantum Boltzmann Equation

Our starting equation is the quantum Boltzmann equation

$$\frac{\partial f}{\partial t} + \frac{e\mathcal{E}}{\hbar} \cdot \frac{\partial f}{\partial \mathbf{k}} = \hat{J}(f) \tag{1.42}$$

$$\hat{J}(f) = \sum_\alpha \int \frac{d\mathbf{k}'}{(2\pi)^3} [f'(1-f)S_{\mathbf{k}'\mathbf{k}}^{(\alpha)} - f(1-f')S_{\mathbf{k}\mathbf{k}'}^{(\alpha)}] \tag{1.43}$$

which is noted to have the generic form of (5.4.5). As noted above, the sum over $\alpha = +1, -1$, corresponds to emission and absorption of a phonon, respectively, in an electron-phonon scattering event. The electric field is $\mathcal{E}$ and

$$f' \equiv f(\mathbf{k}', t), \qquad \text{etc.} \tag{1.44}$$

represents the electron distribution function where $\mathbf{k}'$ corresponds to "after" collision. We recall that the term, $(1 - f)$ in the "birth" term on the right side of (1.43) is a quantum inhibiting factor relevant to fermions (Section 5.4).

Substituting (1.35), (1.36) into (1.43) gives

$$\hat{J}(f) = \sum_\alpha \hat{I}_\alpha \left[f'(1-f)\left(n_q + \frac{1}{2} + \alpha\frac{1}{2}\right) - f(1-f')\left(n_q + \frac{1}{2} - \alpha\frac{1}{2}\right)\right] \tag{1.45}$$

where

$$\hat{I}_\alpha[\varphi_\alpha(q)] \equiv V \int \frac{d\mathbf{q}}{(2\pi)^3} |C_q|^2 \delta(\Delta E)\varphi_\alpha(q) \tag{1.45a}$$

In this expression, with (1.31), we have set $d\mathbf{k}' = d\mathbf{q}$ and

$$f'(\mathbf{k}) \equiv f(\mathbf{k}') = f(\mathbf{k} + \alpha\mathbf{q}) \tag{1.45b}$$

so that $f'(\mathbf{k})$ is α-dependent.

Lorentz–Legendre expansion[9]

To account for anisotropy of the distribution function due to the imposed $\mathcal{E}$-field, we employ the Lorentz expansion

$$f(\mathbf{k}) = f_0(k) + \mu f_1(k) + \cdots \tag{1.46a}$$

$$\mu = \hat{\mathbf{k}} \cdot \hat{\mathcal{E}} = \cos\theta \tag{1.46b}$$

$$f(\mathbf{k}') = f_0(\mathbf{k}') + \mu' f_1(\mathbf{k}') + \cdots \tag{1.46c}$$

$$\mu' = \hat{\mathbf{k}}' \cdot \hat{\mathcal{E}} \tag{1.46d}$$

where hatted variables are unit vectors.

Keeping terms to $O(\mu)$ in (1.46) and substituting the resulting form into the collision integral (1.45) gives

$$\hat{J}[f(\mathbf{k})] = \hat{J}_0(f_0) + \hat{J}_1(f_0, f_1) \tag{1.47}$$

where

$$\hat{J}_0(f_0) = \sum_\alpha \hat{I}_\alpha \left[(f_0' - f_0)\left(n_q + \frac{1}{2}\right) + \alpha\frac{1}{2}(f_0' + f_0) - \alpha f_0' f_0 \right] \tag{1.47a}$$

$$\hat{J}_1(f_0, f_1) = \mu \sum_\alpha I_\alpha \left[\left(\frac{\mu'}{\mu} f_1' - f_1\right)\left(n_q + \frac{1}{2}\right) + \alpha\frac{1}{2}\left(\frac{\mu'}{\mu} f_1' + f_1\right) \right.$$
$$\left. -\alpha\left(\frac{\mu'}{\mu} f_1' f_0 + f_0' f_1\right) \right]$$

$$\tag{1.47b}$$

Note that

$$f'(E) \equiv f(E') = f\left(E + \alpha\frac{\hbar\omega}{k_B T}\right) \tag{1.47c}$$

Substituting these expressions into (1.43) and passing to the steady-state limit, with the orthogonality of Legendre polynomials, we obtain the two equations (see Appendix D)

$$\frac{2e\mathcal{E}}{3\hbar k} \frac{\partial}{\partial E}(Ef_1) = \hat{J}_0(f_0) \tag{1.48a}$$

---

[9]H. A. Lorentz, *The Theory of Electrons*, B. G. Teubner, Leipzig (1909).

$$\mu \frac{2e\mathcal{E}}{\hbar k} E \frac{\partial}{\partial E} f_0 = \hat{J}_1(f_0, f_1) \tag{1.48b}$$

It is assumed that $f_0(E)$ is the Fermi-Dirac distribution (1.38). In that $\hat{J}_0(f_0)$ vanishes for this choice of f_0, (1.48a) corroborates the fact that the Fermi-Dirac distribution is relevant to the zero-field situation, $\mathcal{E} = 0$. The relation (1.48b) suffices to determine the correction f_1.

Reduction to integrals

To reduce the integral (1.47b) occurring in the right side of (1.48b), we first note that

$$\int d\mathbf{q} = \int_0^{q_D} dq\, q^2 \int_0^{2\pi} d\beta \int_{-1}^{1} d\cos\gamma \tag{1.50a}$$

where

$$\hat{\mathbf{q}} \cdot \hat{\mathbf{k}} = \cos\gamma \tag{1.50b}$$

Furthermore, we note that

$$\delta(E' - E - \alpha\hbar\omega) = \frac{k}{2qE}\,\delta\left[\cos\gamma - \left(-\alpha\frac{q}{2k} + \varepsilon\frac{k}{2q}\right)\right] \tag{1.51}$$

where we have set

$$\varepsilon \equiv \frac{\hbar\omega}{E} \tag{1.51a}$$

The delta function in $\hat{I}_\alpha$ restricts the domain of integration. With (1.51) we write

$$\cos\gamma = -\alpha\frac{q}{2k} + \varepsilon\frac{k}{2q} \tag{1.52}$$

Noting that

$$-1 < \cos\gamma < +1$$

and that ε is infinitesimally small, we obtain the upper (U) and lower (L) limits on the q integration:

$$q_U = 2k + \alpha\frac{\varepsilon k}{2} + O(\varepsilon^2) \tag{1.53}$$

$$q_L = 0 + O(\varepsilon^2)$$

Thus, (1.45a) reduces to

$$\hat{I}_\alpha(\varphi_\alpha) = \frac{Vk}{4\pi\hbar E}\int_0^{q_D} dq\, q|C_q|^2\varphi_\alpha \tag{1.54}$$

where q_D represents the Debye wavenumber,

$$q_D = \frac{\omega_D}{u} = [6\pi^2 n_M]^{1/3} \tag{1.55}$$

Inserting the preceding expression into (1.47b) and summing over α, we obtain

$$\hat{J}_1(f_0, f_1) = \hat{I} n_q \left\{ \left\langle \mu' f_1^{(+)} \left[e^y (1 - f_0) + f_0 \right] - \mu f_1 \left[e^y f_0^{(+)} + (1 - f_0^{(+)}) \right] \right\rangle_+ \right.$$
$$\left. + \left\langle \mu' f_1^{(-)} \left[e^u f_0 + (1 - f_0) \right] - \mu f_1 \left[e^y (1 - f_0^{(-)}) + f_0^{(-)} \right] \right\rangle_- \right\}$$

(1.56)

where $\langle \ \rangle_\pm$ correspond to $\alpha = \pm 1$ and

$$y \equiv \frac{\hbar \omega}{k_B T} = \frac{\hbar u q}{k_B T} = \frac{q}{Q}$$

(1.56a)

$$Q \equiv \frac{k_B T}{\hbar u}$$

(1.56b)

and, with (1.47c)

$$f^{(\pm)} = f(E \pm \hbar \omega)$$

(1.56c)

Note further that Q has the dimensions of wavenumber.

In (1.56) we introduced

$$\hat{I} \equiv \frac{Vk}{4\pi \hbar E} \int_0^{q_D} dq\, q |C_q|^2$$

(1.57)

With (1.36), the preceding is written

$$\hat{I} = \frac{kRV}{4\pi \hbar E} \int_0^{q_D} dq\, q G(q)$$

(1.58)

Since most of electron scattering occurs on the Fermi surface, we may write $k' \simeq k$. With (1.31) we then obtain

$$\mu' \approx \mu \left(1 - \frac{q^2}{2k^2} \right) = \mu \left[1 - \frac{Q^2 y^2}{2k^2} \right]$$

(1.59)

Substituting this expression into (1.56) permits the starting equation (1.48b) to be written

$$e\mathcal{E} \frac{\partial f_0}{\partial E} = \frac{V R m Q^2}{4\pi E \hbar^2} \int_0^{y_D} \frac{dy\, y G(y)}{e^y - 1} \left(L_1 - \frac{Q^2 y^2}{2k^2} L_2 \right)$$

(1.60)

where

$$L_1 \equiv f_1^{(+)} [e^y (1 - f_0) + f_0] - f_1 [e^y (1 - f_0^{(-)}) + f_0^{(-)}]$$
$$+ f_1^{(-)} [e^y f_0 + (1 - f_0)] - f_1 [e^y f_0^{(+)} + (1 - f_0^{(+)})]$$

(1.61)

$$L_2 \equiv f_1^{(+)} [e^y (1 - f_0) + f_0] + f_1^{(-)} [e^y f_0 + (1 - f_0)]$$

(1.62)

7.1.5 Perturbation Distribution

The relations (1.60), (1.62) comprise a self-contained integro-difference equation for the perturbation distribution, $f_1(E)$. When written in terms of nondimensional variables (x, y) the definition (1.56c) is given by

$$f^{(\pm)}(x) = f(x \pm y) \tag{1.63}$$

$$x \equiv E/k_B T, \qquad y \equiv \hbar\omega/k_B T$$

With (1.41) we note

$$x \gg y \tag{1.63b}$$

Thus, in this same limit, (1.56c) becomes

$$f^{(\pm)}(x) \simeq f(x) \tag{1.64}$$

(both for f_0 and f_1).

Substituting (1.64) and (1.61), (1.61), reduces to L_1 to zero. For L_2 there results

$$L_2 = f_1(e^y + 1) \tag{1.65}$$

Defining

$$F(E) \equiv -\frac{f_1(E)/E}{\partial f_0/\partial E} \tag{1.66}$$

and

$$B \equiv \frac{Rm}{4\pi\hbar^2} \tag{1.67}$$

(1.60) becomes

$$\frac{e\mathcal{E}}{B} = \frac{QW}{2k^2} F(E) \tag{1.68}$$

where W is the dimensionless integral

$$W(T) \equiv Q^3 V \int_0^{y_D} dy\, y^3 G(y) \left(\frac{e^y + 1}{e^y - 1}\right) \tag{1.69}$$

or, equivalently,

$$W(T) = \int_0^{y_D} \frac{dy\, y^4}{(y^2 + y_{TF}^2)^2} \left(\frac{e^y + 1}{e^y - 1}\right) \tag{1.69a}$$

$$y_{TF} \equiv \frac{\Theta_{TF}}{T}, \qquad y_D \equiv \frac{\Theta_D}{T} \tag{1.69b}$$

$$k_B \Theta_{TF} = \hbar u q_{TF}, \qquad k_B \Theta_D = \hbar u q_D \tag{1.69c}$$

It follows that

$$f_1(E) = \frac{-16\pi e\mathcal{E} E^2 \partial f_0/\partial E}{RK(T)} \tag{1.70a}$$

where the temperature-dependent term

$$K(T) \equiv QW(T) = \bar{Q}TW(T) \tag{1.70b}$$

$$\bar{Q} \equiv k_B/\hbar u \tag{1.70c}$$

Note that $K(T)$ has dimensions of wavenumber.[10] With (1.38) and (1.46), the expression (1.70a) gives the corrected electron distribution to the given order in μ.

7.1.6 Electrical Resistivity

Current density is given by [recall (1.15)]

$$\mathbf{J} = \int \frac{d\mathbf{k}}{(2\pi)^3} \frac{e\hbar\mathbf{k}}{m} f(\mathbf{k}) \tag{1.71}$$

Substituting (1.69) into the preceding we obtain

$$\begin{aligned}
\mathbf{J} &= \int \frac{d\mathbf{k}}{(2\pi)^3} \frac{e\hbar k}{m} \hat{\mathbf{k}} \cdot \hat{\boldsymbol{\mathcal{E}}} f_1(k) \\
&= \frac{e\hbar}{m(2\pi)^3} \hat{\boldsymbol{\mathcal{E}}} \cdot \int d\mathbf{k}\, \hat{\mathbf{k}}\hat{\mathbf{k}}\, f_1(k)k \\
&= \frac{1}{3} \frac{e\hbar}{m(2\pi)^3} \hat{\boldsymbol{\mathcal{E}}} \cdot \bar{\bar{I}} \int d\mathbf{k}\, f_1(k)k
\end{aligned} \tag{1.72}$$

where $\bar{\bar{I}}$ is the unit matrix. A double-barred variable represents a dyad. [See (B.1.1) et seq.] There results

$$\mathbf{J} = \frac{1}{3} \frac{e\hbar}{m(2\pi)^3} \hat{\boldsymbol{\mathcal{E}}} \int d\mathbf{k}\, f_1(k)k \tag{1.73}$$

With (1.39) we write

$$k\, d\mathbf{k} = 4\pi k^3 dk = 2\pi \left(\frac{2m}{\hbar^2}\right) E\, dE$$

It follows that

$$\mathbf{J} = \frac{em}{3\pi^2\hbar^3} \hat{\boldsymbol{\mathcal{E}}} \int dE\, E f_1(E) \tag{1.74}$$

In estimating f_1 it is further assumed that

$$\frac{\partial f_0}{\partial E} = -\delta(E - E_F) \tag{1.74a}$$

Since $k_B T \ll E_F$, f_0 is sharply peaked in the temperature range of interest (0K $\leq T \leq$ 300K) and (1.74a) remains a good approximation. Substituting

---

[10]The relation (1.70a) contradicts Bloch's principal assumption, $f_1(E) = \alpha \partial f_0(E)/\partial E$, where α is a constant. See: F. Block, Ref. 2.

the preceding relations in (1.70) gives the desired solution for the perturbations distribution $f_1(E)$. When substituted into (1.74) this solution gives the
conductivity (σ); resistivity (ρ) expression

$$\sigma = \frac{1}{\rho} = \frac{16}{3\pi} \frac{e^2 m E_F^3}{\hbar^3 R K(T)} = \left(\frac{e^2 m E_F^3}{3\pi \hbar^3 R} \right) \left(\frac{16 \hbar u}{k_B T W(T)} \right) \tag{1.75}$$

whose temperature dependence is contained entirely in $TW(T)$ [see (1.70b)].
As will be shown below, the preceding expression for ρ gives both Bloch's T^5
dependence at $(T/\Theta_D) \ll 1$ as well as canonical T dependence at $(T/\Theta_D) \gg$
1 in addition to a residual resistivity at $T = 0$K.

Properties of $W(T)$ and $S_1(\lambda)$

The function $W(T)$ is singular at $T = 0$K. To expose this singularity first we
note the relation

$$\frac{e^y + 1}{e^y - 1} = 1 + \frac{2}{e^y - 1}$$

so that [recall (1.69)]

$$W(T) = \int_0^{y_D} \frac{dy \, y^4}{(y^2 + y_{TF}^2)^2} \left(1 + \frac{2}{e^y - 1} \right)$$
$$\equiv W_1(T) + W_2(T) \tag{1.76}$$

The $W_2(T)$ contribution corresponds to the exponential term and is finite at
$T = 0$K. The singularity of $W(T)$ lies in $W_1(T)$. To obtain the T-dependence
of this singularity we introduce the variable

$$z \equiv Ty$$

There results (relabeling $z_D \equiv \Theta_D$, etc.)

$$W_1 = \frac{1}{T} \int_0^{\Theta_D} \frac{z^4 \, dz}{[z^2 + \Theta_{TF}^2]^2} \equiv \frac{\Theta_D S_1}{T} \tag{1.77}$$

where S_1 is the implied nondimensional temperature-independent integral. The
relation (1.77) indicates that $W(T)$ has a simple pole at $T = 0$K. Evaluating
the integral S_1 gives

$$S_1(\lambda) = 1 + \frac{1}{2(1 + \lambda^2)} - \frac{3}{2\lambda} \tan^{-1} \lambda \tag{1.77a}$$

$$\lambda \equiv \frac{\Theta_D}{\Theta_{TF}} = \frac{\omega_D}{\omega_{TF}} = \frac{q_D}{q_{TF}} = \frac{y_D}{y_{TF}} \tag{1.77b}$$

The parameter q_{TF} is given by (1.28a), q_D by (1.68c) and we have set $\tan^{-1}(0) =$
0.

The function $S_1(\lambda)$ is a positive monotonic function with properties

$$S_1(0) = S_1'(0) = 0$$
$$S_1(\lambda) \sim 1, \qquad S_1'(\lambda) \sim 0, \qquad \lambda \gg 1 \tag{1.77c}$$

For $\lambda \ll 1$, one obtains

$$S_1(\lambda) = \frac{\lambda^4}{5} + O(\lambda^6) \tag{1.77d}$$

Values of $S_1(\lambda)$ pertinent to the problem at hand are obtained as follows: First we note

$$\lambda^2 = \left(\frac{3\pi^5}{16}\right)^{1/3} a_0 \left(\frac{n}{Z^2}\right)^{1/3} \tag{1.78a}$$

or, equivalently (with $Z = 1$)

$$\lambda = 1.43 \times 10^{-4} n^{1/6} \tag{1.78b}$$

where n is electron density in cm^{-3}.

Among the alkali and noble metals, n is maximum for Cu, for which we obtain $\lambda_{\text{Cu}} = 0.96$. In this group n is minimum for Cs, for which we obtain $\lambda_{\text{Cs}} = 0.66$. We may conclude that the expansion (1.77d) is appropriate to the metals addressed in this analysis. A more accurate description of $S_1(\lambda)$ is obtained by curve fitting this function to a parabola in the λ-domain of interest. There results

$$S_1(\lambda) = -0.042\lambda + 0.11\lambda^2, \qquad 0.60 \leq \lambda \leq 1.00 \tag{1.78c}$$

Thus

$$\begin{aligned} S_1(\lambda_{\text{Cs}}) &\simeq 0.020 \\ S_1(\lambda_{\text{Cu}}) &\simeq 0.061 \end{aligned} \tag{1.78d}$$

Combining (1.77b) and (1.77c) we obtain

$$10^4 S_1(\lambda) = -0.06 n^{1/6} + 0.16 \times 10^{-4} n^{1/3} \tag{1.78e}$$

values from which are seen to agree with (1.78d).

We note that W_1 as given by (1.76), with (1.70b) gives

$$K_1(T) = \bar{Q}\Theta_D S_1 = \frac{k_B \Theta_D}{\hbar u} S_1(\lambda) \tag{1.78f}$$

which is independent of temperature. The temperature dependence of the distribution f_1, resides entire in W_2.

Temperature dependence of the W_2 integral

To examine the finite integral W_2 we revert to y dependence and write

$$W_2 = \int_0^{\Theta_D/T} \frac{2 dy\, y^4}{[y^2 + (\Theta_{\text{TF}}/T)^2]^2} \frac{1}{(e^y - 1)} \tag{1.79a}$$

With these results at hand, we consider first the high temperature limit.

Case (a) $T \gg \Theta_D$: In this limit, expanding the integral (1.78a) about $y_D = 0$ we obtain

$$W_2 = -\frac{\lambda^2}{1+\lambda^2} + \log(1+\lambda^2) + O(y_D) \tag{1.79b}$$

In the limit of $\lambda \ll 1$,

$$W_2 \to \frac{1}{2}\lambda^4 + O(\lambda^6) \tag{1.79c}$$

With the preceding, in the said limit, (1.75) gives the result

$$\rho = \frac{3\pi}{16}\left(\frac{\hbar^3 R}{e^2 m E_F^3}\right)\left(\frac{k_B T}{\hbar u}\right)\left(\frac{\lambda^4}{2} + \frac{\Theta_D S_1}{T}\right) \to \frac{\pi\lambda^4}{32}\left(\frac{\hbar R}{e^2 m E_F^3}\right)\left(\frac{k_B T}{\hbar u}\right) \tag{1.80}$$

which is noted to have the canonical form $\rho \propto T$.

Case (b) $T \ll \Theta_D$: In this limit we obtain

$$W_2 = \left(\frac{T}{\Theta_{TF}}\right)^4 \int_0^\infty \frac{2dy\, y^4}{e^y - 1} \equiv \left(\frac{T}{\Theta_{TH}}\right)^4 S_2 \tag{1.81}$$

where S_2 is the implied non-dimensional, temperature-independent integral with the value

$$\frac{1}{2}S_2 = \Gamma(5)\zeta(5) = 24.886 \tag{1.81a}$$

and Γ and ζ are Gamma and Zeta functions, respectively (see Appendix B).

General resistivity expressions

Employing results from the preceding sections we now obtain explicit expressions for the residual and Bloch components of resistivity. With (1.69a), (1.75) we write

$$\rho = \frac{K(T)}{A} = \frac{\bar{Q}TW}{A} \tag{1.82}$$

Note the relations

$$\bar{Q}\Theta_D = \frac{\omega_D}{u} = q_D$$

$$\bar{Q}\Theta_{TF} = \frac{\omega_{TF}}{u} = q_{TF} \tag{1.83}$$

$$A \equiv \frac{16 e^2 m E_F^3}{3\pi\hbar^3 R} \tag{1.84}$$

The parameter A has dimensions of wavenumber, so that K/A has the correct resistivity dimensions (cgs): time. The parameter R is defined in (1.36c).

Collecting results we write

$$K(T) = Q \left[\frac{\Theta_D S_1}{T} + \left(\frac{T}{\Theta_{TF}} \right)^4 S_2 \right] \qquad (1.85a)$$

$$K(T) \equiv K_0 + K_B(T) \qquad (1.85b)$$

where K_0 is independent of T and $K_B(T)$ heads to the Bloch result. Inserting this finding into (1.53) gives

$$\rho = \frac{\bar{Q}}{A} \left[\Theta_D S_1 + \Theta_{TF} \left(\frac{T}{\Theta_{TF}} \right)^5 S_2 \right] \qquad (1.86a)$$

$$\equiv \rho_0 + \rho_B(T) \qquad (1.86b)$$

where

$$\rho_0 = \frac{q_D S_1(\lambda)}{A} \qquad (1.86c)$$

is the component of resistivity due to electron–phonon scattering that survives at 0K and ρ_B is the Bloch contribution. We note that ρ_B may be written in the more canonical form

$$\rho_B = \frac{\omega_{TF}}{uA} \left(\frac{\Theta_D}{\Theta_{TF}} \right)^5 \left(\frac{T}{\Theta_D} \right)^5 \qquad (1.86d)$$

where, with (1.28a), (1.55) one notes that

$$(\Theta_D/\Theta_{TF})^6 = \left(\frac{q_D}{q_{TF}} \right)^6 = (3\pi^5/16)(a_0^3 n_M^2/n) \qquad (1.86e)$$

The relations (1.57) indicate that ρ_0 dominates over ρ_B for temperatures

$$\left(\frac{T}{\Theta_{TF}} \right)^5 \ll \frac{\lambda S_1(\lambda)}{S_2} \simeq \frac{\lambda S_1(\lambda)}{250} \qquad (1.87a)$$

or, equivalently

$$\frac{T}{\Theta_D} \ll \left[\frac{S_1(\lambda)}{250\lambda^4} \right]^{1/5} \equiv \tau(\lambda) \qquad (1.87b)$$

For Cs we find $\tau = 0.25$. For Cu we find $\tau = 0.20$. Thus, one expects ρ_0 to come into play at

$$y_D \gg 5 \qquad (1.87c)$$

for the class of metals considered.

7.1.7   Scale Parameters of ρ_0

We wish to obtain the manner in which ρ_0 scales with basic metallic parameters. To these ends we write

$$\rho_0 = \bar{\rho}_0 S_1(\lambda) \tag{1.88}$$

First, consider the $\bar{\rho}_0$ factor. With (1.86c) we write

$$\bar{\rho}_0 = \frac{3\pi^2}{8} \frac{k_B \Theta_D}{mu^2} \frac{\hbar(\hbar\Omega)^2}{E_F^3} \tag{1.89a}$$

To find the manner in which $\bar{\rho}_0$ scales with metallic parameters, in (1.89a) we set all parameters that are constant with respect to change of metallic samples $(e, m, k_B, \text{etc.})$ equal to one. There results

$$\bar{\rho}_0 \propto \frac{n_M \Theta_D Z}{E_F^4} \tag{1.89b}$$

To further reduce this relation recall (1.22a) and note that

$$E_F \propto n^{2/3}$$
$$\Theta_D \propto n^{2/3} Z^{1/6} / M^{1/2}$$

It follows that

$$\bar{\rho}_0 \propto Z^{1/6}/nM^{1/2} \tag{1.89c}$$

Large-mass consistency limit

As ion mass grows large, $\Omega \to 0$, and electrons do not interact with the lattice [see (1.29)]. It follows that in this limit one should find that $\rho_0 \to 0$. To explore this situation we examine the limit, $M \to \infty$, at otherwise fixed ion parameters: n_M and Z. With these constraints we note that

$$\bar{\rho}_0 \propto \frac{\Theta_D}{u^2} \Omega^2 = \frac{q_D^2 \Omega^2}{u} \tag{1.90a}$$

We recall that $u \propto \sqrt{Z/M}$, $q_D^2 \propto n_M^{2/3}$, $\Omega^2 \propto n_M Z^2/M$, and that

$$\lambda = \frac{\Theta_D}{\Theta_{TF}} = \frac{q_D}{q_{TF}} \propto n_M^{1/3} \tag{1.90b}$$

which together with $S_1(\lambda)$ are constant under the said constraints. There results

$$\bar{\rho}_0 \propto \frac{\text{constant}}{M^{1/2}} \to 0 \tag{1.90c}$$

This property agrees with the preceding observation that in the given limit, electrons do not interact with the lattice so that $\rho_0 \to 0$. In this same limit, $\rho_B \to 0$, as $T \to 0$, providing $T/M < 1$ [as follows from (1.86)].

7.1.8   Electron Distribution Function

Returning to the expansion (1.46) and inserting the solution (1.70) gives the electron distribution

$$f(E, \mu) = f_0(E) + \frac{D\mu}{K(T)} E^2 \frac{\partial f_0}{\partial E} \tag{1.91a}$$

$$D \equiv \frac{32e\mathcal{E}}{R} \tag{1.91b}$$

where $f_0(E)$ is the Fermi-Dirac distribution (1.37) and R is given by (1.36c). Consider the function $K(T)$ as given by (1.85). Let us suppose that there is no residual term and set $K_0 = 0$. Then as $T \to 0\text{K}$, $K(T) \to 0$ and the perturbation term in (1.91a) grows singular at all E thereby violating the Lorentz expansion (1.46). At $T = 0\text{K}$, $\partial f_0 / \partial E$ is zero except at $E = E_F$. However, with $K_0 = 0$, this zero is divided by $K_B(0) = 0$ and the distribution (1.85) is indeterminate. For the case $K_0 > 0$, as found in the present analysis, this pathological behavior of $f(E)$ is circumvented and, save for the singular point $E = E_F$ at $T = 0\text{K}$, a well-defined distribution results for all E.

Reduced resistivity

We note that the reciprocal of the right side of (1.75) may be written

$$\rho = H(\Theta_D, \Theta_{\text{TF}}) \frac{W(y_D, \lambda)}{y_D} \tag{1.92a}$$

where the coefficient $H(\Theta_D, \Theta_{\text{TF}})$ is an implied and dimensionless "reduced" resistivity is given by

$$\tilde{\rho} \equiv \frac{\rho}{H(\Theta_D, \Theta_{\text{TF}})} = \frac{W(y_D, \lambda)}{y_D} = \frac{W_1(y_D, \lambda)}{y_D} + \frac{W_2(y_D, \lambda)}{y_D} \tag{1.92b}$$

and we have recalled that

$$(\Theta_{\text{TF}}/T)^2 = y_D^2/\lambda^2$$

The term W_1 includes the residual resistivity result whereas W_2 includes both Bloch's T^5 result as well as the linear T high-temperature dependence. When $\tilde{\rho}$ is plotted vs temperature $(1/y)$, the preceding result (1.92) exhibits the characteristic behavior depicted in Fig. 7.3.

It should be borne in mind that the resistivity results presented above are relevant to ideal monovalent metals free of defects, dislocations and impurities.

7.1.9   Thermal Conductivity

It has been noted that the thermionic properties of metals are complex and not well explained by standard analyses.[8] With the preceding results at hand we return to the problem of metallic thermal conductivity and describe a model which gives agreement with characteristics of the coefficient of thermal conductivity for a metal over three basic ranges of temperature. We recall that the Weidemann–Franz law for $\kappa/\sigma T$ as given by (1.2a) is valid over a wide temperature range. Having found a residual resistivity at $T = 0K$, we revert to this law, which when taken with the preceding results (1.86c) returns the experimentally correct low-temperature thermal conductivity linear-temperature dependence[10]

$$\kappa = \frac{\pi^2}{3\rho_0} \left(\frac{k_B}{e}\right)^2 T \tag{1.93}$$

Application of this relation to specific metals is derived from the λ-dependence contained in ρ_0. At higher temperatures but still beneath the Debye temperature, where W_2 [see (1.92b)] comes into play, one reverts to the augmented Lorenz relation (1.2b) which returns the correct T^{-2} thermal conductivity behavior. Above the Debye temperature, we return the Weidemann–Franz relation and write

$$\kappa = \frac{\pi^2}{3} \frac{T}{\rho(T)} \left(\frac{k_B}{e}\right)^2 \tag{1.94}$$

where $\rho(T)$ again includes the $W_2(T)$ component but is now relevant to the limit, $T \gg \Theta_D$. At these conditions (1.94) gives the correct behavior (1.2c), $\kappa = $ constant.

The preceding results imply a thermal conductivity vs. temperature curve which rises linearly from the origin and then decays as T^{-2} from a maximum value and levels off to a constant at $T \lesssim \Theta_D$, in agreement with characteristics of measured values (Fig. 7.4).

To assist the reader in this analysis, a list of key parameter-relations and definitions is presented in Table 7.1.

7.2   Amorphous Media

7.2.1   Background

As noted in the introduction to this chapter, the amorphous structure is a common form of material. We recall that window glass is a fused alloy of Na_2O and SiO_2. Other types of glass that are of interest because of their semiconducting properties are compounds of S, Se, and Te with elements such as As and Ge. Such compounds are called *chalcogenide glasses* (compounds of any of the group 6 elements of the periodic chart, excluding oxides). Common liquids are amorphous. Liquid crystals are composed of rodlike molecules (length $\gg$ diameter). In the *smectic* phase, liquid crystals have long-range order along the molecular axis, as well as molecular orientational order, but are disordered in the plane normal to the molecular axis. At higher temperature, in the *nematic* phase, the liquid crystal remains only with molecular orientational order.

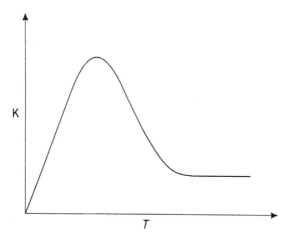

FIGURE 7.4. Sketch of observed temperature dependence of the coefficient of thermal conductivity of a metal. At low temperature $\kappa \approx T$ whereas at $T \gg \Theta_D$, $\kappa \approx$ constant. At intermediate temperature $\kappa \approx T^{-2}$. The central peak of the graph is observed to rise with increased purity of the metal. Fur further discussion, see J. M. Ziman, *Electrons and Phonons*, ibid., Chapter 9; H. M. Rosenberg, *The Solid State*, 3rd ed., Oxford, New York (1992), Chapter 8; G. K. White, *Proc. Phys. Soc. (London) A66*, 559 (1953).

As shown in Fig. 7.1, the radial distribution function of ions in an amorphous material has short range order, and is void of long-range periodicity. In what manner may one expect to find properties of crystal structures in an amorphous sample? As noted in Fig. 5.5, the energy spectrum of crystalline material has a band structure. The fact that window glass is transparent to visible light is indicative of a quasi band spectrum with an effective band gap of the order of several eV. Liquid mercury, as well as other molten metals, have high electrical conductivity, again indicative of a quasi band spectrum with a partially filled conduction band.

An assortment of phenomena are relevant to amorphous materials. These include the nature of the band structure and related concepts of: localization, mobility edge, hopping, percolation, and the metal-insulator transition.

Extended states

The notions of extended and localized states play an important role in the study of electrical properties of amorphous material. Here is a brief review of these concepts. Consider the Schrödinger equation for a particle of mass m moving in a periodic potential, $V(x) = V(x + a)$, where a is the lattice constant of the array.

$$\varphi''(x) - W\varphi(x) + K^2\varphi(x) = 0 \tag{2.1}$$

TABLE 7.1. Key Parameter-Relations and Definitions for Section 1.3 et seq.

$$\rho_M = M n_M, \qquad n_M = \frac{n}{Z}, \qquad n = \frac{N}{V} \tag{1.22a}$$

$$q_{TF}^2 = \left(\frac{4}{a_0}\right)\left(\frac{3n}{\pi}\right)^{1/3} \tag{1.28a}$$

$$\Omega^2 = \frac{4\pi n_M (Ze)^2}{M} \tag{1.29a}$$

$$u^2 = \frac{2}{3}\frac{Z E_F}{M} \tag{1.34a}$$

$$G(q) \equiv \frac{q/V}{(q^2 + q_{TF}^2)^2} \tag{1.36b}$$

$$R \equiv \frac{\hbar (M\Omega^2)^2}{2\rho_M u Z^2} \tag{1.36c}$$

$$q_D = \frac{\omega_D}{u} = [6\pi^2 n_M]^{1/3} \tag{1.55}$$

$$y \equiv \frac{\hbar\omega}{k_B T} = \frac{\hbar u q}{k_B T} = \frac{q}{Q} \tag{1.56a}$$

$$Q \equiv \frac{k_B T}{\hbar u} \tag{1.56b}$$

$$x \equiv E/k_B T, \qquad y \equiv \hbar\omega/k_B T$$

$$B \equiv \frac{Rm}{4\pi\hbar^2} \tag{1.67}$$

$$y_{TF} \equiv \frac{\Theta_{TF}}{T}, \qquad y_D \equiv \frac{\Theta_D}{T} \tag{1.69b}$$

$$k_B \Theta_{TF} = \hbar u q_{TF}, \qquad k_B \Theta_\hbar = \hbar u q_D \tag{1.69c}$$

$$\lambda \equiv \frac{\Theta_D}{\Theta_{TF}} = \frac{\omega_D}{\omega_{TF}} = \frac{q_D}{q_{TF}} = \frac{y_D}{y_{TF}} \tag{1.77b}$$

$$\lambda^2 = \left(\frac{3\pi^5}{16}\right)^{1/3} a_0 \left(\frac{n}{Z^2}\right)^{1/3} \tag{1.78a}$$

$$\bar{Q}\Theta_D = \frac{\omega_D}{u} = q_D \tag{1.83}$$

$$\bar{Q}\Theta_{TF} = \frac{\omega_{TF}}{u} = q_{TF}$$

$$A \equiv \frac{16 e^2 m E_F^3}{3\pi\hbar^3 R} \tag{1.84}$$

$$(\Theta_D/\Theta_{TF})^6 = \left(\frac{q_D}{q_{TF}}\right)^6 = (3\pi^5/16)(a_0^3 n_M^2/n) \tag{1.86e}$$

$$(\Theta_{TF}/T)^2 = y_D^2/\lambda^2$$

$$W \equiv \frac{2m}{\hbar^2} V, \qquad E \equiv \frac{\hbar^2 K^2}{2m}$$

Substituting the periodic function $\varphi(x) = \varphi(x+a)$, into (2.1) returns the original equation, indicating that any periodic function with period a is a solution to this equation. To gain further information on the eigenvalue problem we set

$$\varphi(x) = e^{ikx}u(x) \tag{2.2a}$$

The following equation results for $u(x)$.

$$u'' + 2iku' - (k^2 + W - K^2)u = 0 \tag{2.2b}$$

It follows that $u(x)$ is also periodic and dependent on k.

The solutions (2.2a) are called *Bloch waves*. Each such wavefunction is present throughout the periodic array and is an example of an *extended state*. As depicted in Fig. 5.5, allowed k-values comprise a band structure. Electron wavefunctions in valence or conduction bands are Bloch waves. An example of localized wavefunctions is given by electronic states of unionized n-type impurities in a host semiconductor, such as, for example, antimony atoms in crystalline silicon. After bonding to the silicon host, one electron remains weakly bound to the antimony atom, and is in a localized state. Energy levels of this electron are discrete and hydrogen like. On promotion to the conduction band, this electron goes into an extended state. If disorder is introduced into the regular crystal, localization of the wavefunction occurs and k is not a good quantum number.

As first discovered by P. W. Anderson,[11] the wavefunction of an electron in a lattice of sufficient disorder, is localized. The real part of an extended wavefunction is shown in Fig. 7.5a and that of a localized wavefunction is shown in Fig. 7.5b. Note that the envelope of the localized wavefunction decays exponentially. The exponential coefficient α is called the inverse localization length. As will be denoted below, Anderson localization is relevant to intrinsic type material.

7.2.2   Localization

Consider the array of quantum wells shown in Fig. 7.6. For the regular array, (Fig. 7.6a) the band width is labeled B, whereas for the disordered array, (Fig. 7.6b), the related width of the spread of states is labeled W. We consider the parameter

$$\Gamma \equiv W/B$$

In the limit $\Gamma \ll 1$, extended states prevail. We work in the tight-binding approximation in which the overlap of atomic wavefunctions is considered negligible except for nearest neighbors. This approximation is consistent with Fig. 7.1c, which indicates short-range order in an amorphous material. Approximate solutions of the time-independent Schrödinger equation, relevant

[11]P. W. Anderson, *Phys. Rev. 109*, 1492 (1958). See also: J. M. Ziman, *J. Phys. C2*, 1230 (1969); D. C. Herbert and R. Jones, *J. Phys. C4*, 1145 (1971); N. F. Mott and E. A. Davis, *Electronic Properties in Non-Crystalline Materials*, Oxford, New York (1979); P. A. Lee and T. .V. Ramakrishnan, *Rev. Mod. Phys. 57*, 287 (1985).

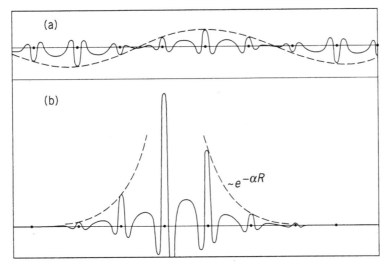

FIGURE 7.5. (a) Real part of a Bloch extended-state wavefunction of a single-particle in a periodic potential. (b) Real part of a localized wavefunction of a particle in a disordered potential array, exhibiting an exponentially decaying envelope. From: R. Zallen (1983). Reprinted by permission of John Wiley and Sons, Inc.[12]

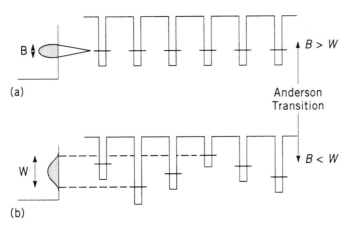

FIGURE 7.6. Energy of states of a particle in a one-dimensional quantum-well array. (a) For the periodic array the band width is labeled B. (b) For the disordered array the width of the spread of states is labeled W. Localization occurs in the limit $W/B \geq 1$. From: R. Zallen (1983). Reprinted by permission of John Wiley and Sons, Inc.

to electrons in a given band, at $\Gamma < 1$, are given by

$$\Psi = \sum_n \exp(i\mathbf{k} \cdot \mathbf{a}_n)\psi(|\mathbf{r} - \mathbf{a}_n|) \qquad (2.3)$$

where $\mathbf{a}_n$ are atomic sites and $\psi(\mathbf{r})$ wavefunctions are centered at these points. The width B is given by

$$B = 2zI \tag{2.4a}$$

where z is coordination number and I is the transfer integral

$$I = \int \psi(|\mathbf{r} - \mathbf{a}_n|) H \psi(|\mathbf{r} - \mathbf{a}_{n+1}|) \tag{2.4b}$$

where H is the Hamiltonian of the system. For a lattice composed of hydrogen atoms, for S-like wavefunctions, one obtains

$$I = 2[1 + (\rho/a_0)] \exp(-\rho/a_0) \tag{2.4c}$$

where I is measured in Rydberg units ($R_y = e^2/2a_0$), a_0 is the Bohr radius and ρ is atomic separation. This expression exhibits a sharp decay of the overlap integral with increase in separation, which is consistent with the tight-binding approximation.

To examine the disordered case for which $\Gamma \gtrsim 1$, as noted above, one introduces a random distribution of potential depths at each site, with variance W. In this limit, the wavefunction (2.3) loses phase coherence in passing from one atomic site to the next and assumes the form

$$\Psi = \sum_n c_n \exp(i\phi_n) \psi(|\mathbf{r} - \mathbf{a}_n|) \tag{2.5}$$

where ϕ_n are random phases and c_n are constants.

The primary finding of Anderson is that as Γ is increased, at some critical value, wavefunctions become localized. Let us call this critical value, Γ_c. For coordination number 6, Anderson found $\Gamma_c = 5$. Edwards and Thouless[13] obtained the value $\Gamma_c = 2$. More recently, Elyutin et al.,[14] found the value $\Gamma_c = 1.7$.

Localized wavefunctions have the form

$$\Psi = \exp(-|\mathbf{r} - \mathbf{r}_0|/\xi) \sum_n c_n \exp(i\phi_n) \psi(|\mathbf{r} - \mathbf{a}_n|) \tag{2.6}$$

Each such wavefunction is localized about a point $\mathbf{r}_0$ in space. The parameter ξ is called the *localization length* and decreases with increasing disorder, W, thereby increasing the rate of decay of Ψ away from atomic sites.

Metal-insulator transition

As was shown by Mott,[15] if the value of Γ is insufficient to give localization throughout the band, at band edges, energies of localized and extended

---

[12]R. Zallen, *The Physics of Amorphous Solids*, Wiley, New York (1983).
[13]J. T. Edwards and D. Thouless, *J. Phys.* C5, 807 (1972).
[14]P. V. Elyutin, et al., *Phys. Status Solidi* B124, 279 (1984).
[15]N. F. Mott, *Adv. Phys. 16*, 49C (1967).

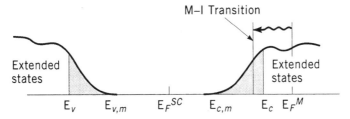

FIGURE 7.7. Density of states vs. energy depicting the metal-insulator transition for an amorphous material. Shaded areas represent localized states. Valence and conduction mobility-edges are labeled E_v and E_c, respectively and E_{vm}, E_{cm} denote respective tail-edges of localized states of valence and conduction band-edges. Fermi energies of respective semiconductor and metal phases are written E_F^{SC}, E_F^M. Transition of the Fermi energy from E_F^M to the domain of localized states corresponds to the metal–insulator (M–I) transition. As noted in the text, widths of tail edges of localized states are greatly exaggerated in this figure.

states are separated at critical respective energies, E_v and E_c, labeled "mobility edges." For crystalline metals the Fermi energy, E_F, lies in the conduction band. For crystalline semiconductors (in the "nondegenerate limit"), E_F lies in the band gap. These properties are roughly maintained in amorphous materials, so that again, for an amorphous metal E_F lies in the conduction band. However, in an amorphous metal, the band-edges have tails, composed of localized states (Fig. 7.7) and properties of the sample are critically dependent on the location of E_F relative to E_c. The region between E_v and and E_c is called the "mobility gap." In typical amorphous materials, tail edges extend to a very small fraction of the mobility gap.

If $E_F > E_c$, the material acts as a metal. If $E_F < E_c$, electrical conductivity of the material decreases, with transport of charge carriers limited either to hopping or excitation of carriers to E_c. In this case the material may be considered an insulator. Consider that the value of E_F can be varied, (Fig. 7.7) and be made to fall from a value in excess of E_c to a value below E_c. In this event there is a sudden change of the material from a metal to an insulator. This transition is called the metal-insulator transition. The value of E_F may be made to change in a number of ways. Thus, for example, E_F will change value by altering the allow composition, by applying external stress to the sample, or by the application of electric or magnetic fields. In Fig. 7.7 we have recalled that the Fermi energy of an intrinsic semiconductor lies near the midpoint of the energy gap. As noted previously, this type of localization is relevant to intrinsic conduction materials.

7.2.3   Conduction Mechanisms and Hopping

A number of experiments indicate that at low temperatures, conductivity of amorphous alloys does not fall beneath a temperature-independent minimum value of $5 \times 10^5 \ \Omega^{-1} \ m^{-1}$. With (5.6.30) and Problem 5.52, we write

$$\sigma = \frac{e^2 l}{12\pi^3 \hbar} S_F \tag{2.7a}$$

where S_F is a measure of the Fermi surface. The preceding relation is relevant to extended states in a metal sample. In this event, the mean free path, $l \gtrsim a$, where a is atomic spacing, or the distance over which the wavefunction loses phase memory. As l reduces to a, states become localized. At the Ioffe-Regel minimum,[16] $kl \cong 1$ and $1 \cong a$ where k denotes wavevector. Substituting these relations into (2.7a) gives the conductivity

$$\sigma_{\min} = \frac{e^2}{3\pi^2 a \hbar} \tag{2.7b}$$

This value of σ is the minimum on the delocalized side of E_c.[17]

Mechanisms for conductivity in amorphous media separate according to temperature. Three basic intervals are: high, lower and very low temperature, relevant to the respective degrees to which thermal energy can excite electrons to domains within or above regions of localized states. For conductivity due to motion of electrons within domains of localized states, the primary mechanism is that of hopping. This mode of conductivity depends to a large degree on thermal excitation. In the event that thermal fluctuations are insufficient to supply hopping activation energy, a fourth mechanism for conductivity may be available which depends on electrons tunneling to states at nearby or distant sites of similar energy. (We recall that *activation energy* refers to the energy required to carry a particle above a given potential barrier.)

High temperature

For temperature sufficiently high, thermal energy excites electrons above E_c to extended states in the conduction band and standard ohmic current flows in response to a potential gradient. In this event conductivity is given by

$$\sigma = \sigma_0 \exp\{-(E_c - E_F)/k_B T\} \tag{2.8}$$

where σ_0 is a constant. Such conductivity applies to Ge and Si at room temperature.

Lower temperature

At lower temperatures, thermal energy is sufficient only to excite electrons if the interval of localized states $(E_{c,m}, E_c)$ to the conduction-extended states

---

[16] A. F. Ioffe and A. R. Regel, *Progress in Semiconductors*, Heywood, London (1960).

[17] For further discussion see, N. F. Mott and E. A. Davis, *Electronic Properties in Non-Crystalline Materials, ibid*; S. R. Elliott, *Physics of Amorphous Materials*, Longman, New York (1983).

domain $E > E_c$. For conductivity to occur, electrons must absorb sufficient energy, ΔW_1, to enable hopping from one localized state to the next. Activation energy is, $E_{c,m} + \Delta W_1$, with related conductivity given by

$$\sigma = \sigma_1 \exp\{-(E_{c,m} + \Delta W_1 - E_F)/k_B T\} \tag{2.9}$$

The constant σ_1 is approximately one thousand times smaller than σ_0. This mode of conductivity is called, thermally assisted hopping.

Very low temperature

In semiconducting materials the Fermi energy typically lies in the band gap. In an amorphous semiconductor the same is true and it is argued that there is a high concentration of localized states in the vicinity of the Fermi energy. Thermally assisted hopping may still occur for transitions to states close to the Fermi energy with resulting conductivity

$$\sigma = \sigma_2 \exp(-\Delta W_2/k_B T) \tag{2.10}$$

where ΔW_2 is approximately half the spread of the distribution of localized states about E_F. The constants σ_1 and σ_2 both depend on the hopping frequency of electrons.

Variable-range hopping

At still lower temperature, for which thermal fluctuations are insufficient to supply hopping activation energy, ($k_B T \ll \Delta W_2$), there is an additional mechanism which permits electrical conductivity. In this mechanism, electrons in a given energy state tunnel to nearby, or possibly distant sites, to states of similar energy. This process is called, *variable-range hopping*. The temperature dependence of this mechanism was first derived by N. F. Mott.[18]

Since electrons are in localized states, their wavefunction varies with displacement R, as $\exp(-\alpha R)$, where α is a constant. It follows that the probability of hopping to a site at R, is given by $\exp(-2\alpha R)$. Again it is assumed that states are concentrated at E_F. With the density of states written $g(E_F)$ the number of states in the interval dE about E_F is $V g(E_F)dE$, where V is volume. To obtain the energy of hopping, we assume that V contains one state, and write

$$V g(E_F)dE = 1 \tag{2.11}$$

in which case dE is the spread between one state and the next. It follows that the average separation (in energy) between neighboring states, ΔW_3, is

$$\Delta W_3 = 1/[V g(E_F)] \tag{2.12}$$

[18]N. F. Mott, *Conduction in Non-Crystalline Materials*, Oxford, New York, (1987).

If we concentrate on states in the volume $V = 4\pi R^3/3$, the latter expression becomes

$$\Delta W_3 = 3/[4\pi R^3 g(E_F)] \tag{2.13}$$

This expression indicates that the farther an electron tunnels, the greater is the probability that it finds a site with small ΔW_3. The probability of a given value of ΔW_3 is $\exp\{-\Delta W_3/k_B T\}$. It follows that the probability of hopping through the displacement R to a nearby site at ΔW_3 is

$$P(R, \Delta W_3) = \exp\{-2\alpha R - (\Delta W_3/k_B T)\} \tag{2.13a}$$

Substituting the value ΔW_3 from (2.13) into the preceding relation gives the probability

$$P(R) = \exp\{-[2\alpha R + (3/4\pi R^3 g(E_F)k_B T)]\} \tag{2.13b}$$

This probability is maximum at the minimum of the exponent. There results

$$R = (9/8\pi \alpha g(E_F)k_B T)^{1/4} \cong (1/\pi \alpha g(E_F)k_B T)^{1/4} \tag{2.14a}$$

On substituting this value of R into (2.13) and then inserting the resulting value of ΔW_3 into the generic form

$$\sigma = \sigma_3 \exp(-\Delta W_3/k_B T) \tag{2.14b}$$

gives

$$\sigma = \sigma_3 \exp(-B/T^{1/4}) \tag{2.14c}$$

$$B = 2 \left(\frac{3}{2\pi}\right)^{1/4} \left(\frac{\alpha^3}{k_B g(E_F)}\right)^{1/4} \tag{2.14d}$$

This $T^{-1/4}$ behavior of conductivity has been observed in amorphous Ge film.[19]

With the preceding discussion it is noted that a plot of $\log \sigma$ versus T^{-1} should reveal four distinct regions, three of which should display straight-line behavior with respective slopes $(E_c - E_F)$ [see (2.8)], $(E_a + \Delta W_1 - E_F)$ [see (2.9)], and ΔW_2 [see (2.10)]. This behavior has likewise been observed[20] for conductivity in the amorphous alloy, $(As_2 Se_3)_{0.95} Tl_{0.05}$ with slight curvature at large T^{-1} due to variable-range hopping.

7.2.4 Percolation Phenomena

Consider a network comprising a random array of independent resistors. Each component either has a finite resistance (probability, p) or an open-circuit infinite resistance (probability, $1 - p$). A voltage source is placed across the network. Current flows providing there is a connecting path of finite resistance elements between terminals, which occurs at some critical value of p.

[19] M. H. Gilbert and C. J. Adkins, *Phil. Mag* **34**, 143 (1976).
[20] M. F. Kotakata, et al., *Semicond. Sci. Tech.* **1**, 313 (1986).

TABLE 7.2. Applications of Percolation Theory

System	Critical Phenomenon
Mobility edge in amorphous semiconductors	Localized to extended states
Dilute magnetic allows	Para-ferromagnetic transition
Resistive-open circuit array	No current-current transition
Liquid flow in porous media	Static to continuous flow
Supercooling of molten glass	Glass transition
Polymer gelatin	Liquid-gel transition
Disease spread in a population	Containment-epidemic transition
Composite superconductor-metal materials	Normal-superconducting transition
Variable-range hopping	No current-current transition
Hydrofracturing of a rock	Generation of a crack
Neural networks	No current-current transition

The event of current flow at the critical value of p is an example of the phenomenon of percolation. Another example of this phenomenon is that of the phase transition of a dilute magnetic alloy from a paramagnetic to a ferromagnetic state, with p now denoting the fraction of magnetic atoms in the alloy. The occurrence of a phase transition to the ferromagnetic state depends on the existence of a sufficiently large connected cluster of interacting magnetic atoms. This transition configuration occurs at a critical value of p. The phrase *percolation* was first used by J. M. Hemmersly[21] in the study of the passage of a fluid through a network of channels, some of which are blocked, and some of which are open. Numerous other examples of percolation are described in the literature.[21,22] (See Table 7.2.)

Bond and site percolation; clusters

Percolation divides into two categories: *bond* and *site* percolation. In site percolation sites are either occupied or unoccupied. In bond percolation, bonds in an array are either connected or disconnected.

Occupied sites fall into a *cluster* if any two sites belonging to that cluster have a chain of occupied states connecting them. A measure of the size of a cluster is given by the number of occupied sites in the cluster. Consider a uniform infinite square array of lattice points with site occupation probability, p. Three finite rectangular subsections of this array are shown in Fig. 7.8, with respective p-

[21] J. M. Hemmersly, *Proc. Cambridge Phil. Soc. 54*, 642 (1957).

[22] D. Stauffer, *Introduction to Percolation Theory*, Taylor and Francis, London (1985); J. W. Essam, *Percolation and Cluster Size*, in *Phase Transitions and Critical Phenomena*, Vol. 2, ed., C. Dumb and M. S. Green, Academic, New York (1972); H. L. Frisch and J. M. Hemmersly, *J. Soc. Indust. Appl. Math. 11*, 894 (1963); G. Grimmett, *Percolation*, Springer, New York (1989); M. F. Thorpe and M. I. Mitkova, eds. *Amorphous Insulators and Semiconductors*, Kluwer Academic, Dordrecht (1997).

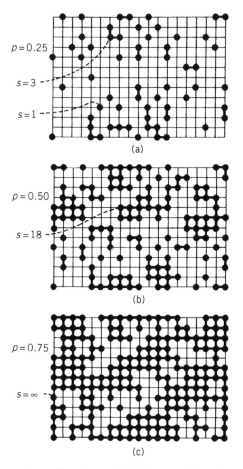

$p = 0.25$

$s = 3$

$s = 1$

(a)

$p = 0.50$

$s = 18$

(b)

$p = 0.75$

$s = \infty$

(c)

FIGURE 7.8. Rectangular subsections of infinite-two-dimension arrays illustrating site percolation on the square lattice. Cluster size (s) is shown for three clusters, and occupation probability (p) is for the three lattices. For the square lattice, percolation ($s = \infty$) occurs at $p = 0.59$. From: R. Zallen (1983). Reprinted by permission of John Wiley and Sons, Inc.

values: 0.25, 0.50, 0.75. It is evident that the cluster size for case (c) is infinite and percolation has occurred. One may conclude that percolation for the two-dimensional square array occurs at p-value, $0.50 < p_c \leq 0.75$. The critical value if $p_c = 0.59$.[21] If p_c^B and p_c^S represent, respectively, critical p-values for bond and site percolation, respectively, then it may be shown that $p_c^B \leq p_c^S$.[23] Examples of this property are shown for four periodic two-dimensional configurations and the Bethe lattice in Fig. 7.9. (At every branching of a Bethe lattice a stem bifurcates into two branches.)

[23] J. W. Essam, *Percolation and Cluster Size* (ibid).

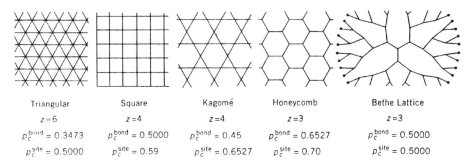

Triangular	Square	Kagomé	Honeycomb	Bethe Lattice
$z=6$	$z=4$	$z=4$	$z=3$	$z=3$
$p_c^{bond} = 0.3473$	$p_c^{bond} = 0.5000$	$p_c^{bond} = 0.45$	$p_c^{bond} = 0.6527$	$p_c^{bond} = 0.5000$
$p_c^{site} = 0.5000$	$p_c^{site} = 0.59$	$p_c^{site} = 0.6527$	$p_c^{site} = 0.70$	$p_c^{site} = 0.5000$

FIGURE 7.9. Critical site and bond percolation p-values, and coordinate number (z) for four two-dimensional periodic arrays. Also shown in the Bethe lattice with effective infinite dimensionality (no closed loop exists for this lattice). From: R. Zallen (1983). Reprinted by permission of John Wiley and Sons, Inc.

We wish to obtain a relation for the mean size of a finite cluster, with site probability p. (For bond percolation, p represents edge probability.) To this end we introduce the number of clusters of size s, per lattice site, $n_s(p)$. It follows that the probability that a given site in a cluster of size s is occupied in $sn_s(p)$. (In a lattice of 100 sites, which at a given value of p has, say, three clusters of size s, $n_s(p) = 0.03$. If $s = 10$, then $sn_s(p) = 0.3$.) Let $P(p)$ represent the fraction of occupied sites for the infinite cluster. Note that $P(1) = 1$ and $P(p) = 0$, for $p < p_c$. The probability that a site is occupied for the infinite cluster is $pP(p)$. The probability p, that a site is occupied is the sum of occupation probabilities overall cluster sizes. That is,

$$\sum_s sn_s(p) + pP(p) = p \tag{2.15a}$$

Since $sn_s(p)$ represents the probability that a given site in a cluster of size s is occupied, it follows that the mean size of finite clusters is given by

$$S(p) = \frac{\sum_s s^2 n_s(p)}{\sum_s sn_s(p)} \tag{2.15b}$$

7.2.5 Pair-Connectedness Function and Scaling

Percolation is closely akin to critical phenomena in statistical physics which is characterized by the behavior of thermodynamic variables as a function of the temperature-displacement variable, $t = (T - T_c)/T_c$ where T_c is critical temperature.[24] A universality of such behavior is present in which such thermodynamic variables scale with t, near the critical point, independent of

---

[24]K. Huang, *Statistical Mechanics*, 2nd ed. Wiley, New York (1987); M. Plischke and B. Bergersen, *Equilibrium Statistical Physics*, Prentice Hall, Englewood Cliffs, NJ (1989); H. E. Stanley, *Introduction to Phase Transitions and Critical Phenomena*, Oxford, New York (1987).

material, but dependent on dimension, d. The exponents of these scaling terms are called *critical exponents* and are conventionally labeled: $(\alpha, \beta, \gamma, \eta, \nu, \sigma)$. Of these six critical exponents, only two are independent.

In a similar vein, percolation phenomena likewise exhibits critical behavior with related scaling relations.[24] To demonstrate these relations the following analogues with variables of statistical physics are noted. The total number of clusters per site is analogous to the free energy per site.

$$G(p) = \sum_s n_s(p) \tag{2.16}$$

A parameter relevant to a given cluster is the *pair connectedness* function, $C(p, r)$, which is the probability that occupied sites a distance r apart are in a common cluster. This function is analogous to the pair correlation function of a statistical physics.[25,26] The mean size of finite clusters is analogous to the susceptibility parameter in magnetic systems. It is then implied that these variables have the following power-law singularities at the percolation transition.[26]

$$G(p) \simeq |p - p_c|^{2-\alpha} \tag{2.17a}$$

$$P(p) \simeq (p - p_c)^{\beta} \tag{2.17b}$$

$$S(p) \simeq (p - p_c)^{-\gamma} \tag{2.17c}$$

$$C(p, r) \simeq \frac{\exp[-r/\xi(p)]}{r^{d-2+\eta}} \tag{2.17d}$$

where d is dimension and the correlation length ξ, which, with comparison to known physical systems, is expected to diverge as

$$\xi(p) \simeq (p - p_c)^{-\nu} \tag{2.17e}$$

Universality of critical exponents for the percolation phenomenon indicates that the percolation exponents α, β, etc., should depend only on the dimensionality of the system, not the lattice specifies nor whether the percolation is site or bond type.

Rushbrooke, Fisher, and Josephson scaling laws

The basic premise of percolation scaling theory[26] is that for given correlation length, ξ, for p near p_c, a cluster size s_ξ exists which is the dominant contribution to the percolation functions (2.17a), (2.17b), (2.17c). The cluster size s_ξ diverges as $p \to p_c$ and is assumed to have the power-law behavior

$$s_\xi \simeq |p - p_c|^{-1/\sigma} \tag{2.18}$$

which defines the critical exponent σ. It is further suggested that

$$n_s(p) \simeq n_s(p_c)\phi\left(\frac{s}{s_\xi}\right) \tag{2.19}$$

[25]R. L. Liboff, *Phys. Rev.* A39, 4098 (1989).
[26]D. Stauffer, *Phys. Rev.* 54, 1 (1979).

where the function $\phi(x) \to 0$, as $x \to \infty$, and $\phi(x) \to 1$, as $x \to 0$ but is otherwise unspecified. Simulation calculations indicate that for large cluster size, s, the probable number of occupied sites at the critical percolation state behaves as, $n_s(p_c) \simeq s^{-\tau}$, where the parameter τ depends on the dimension, d, of the system. Inserting this behavior of $n_s(p)$, together with (2.18) into (2.19) gives

$$n_s(p) \simeq s^{-\tau}\phi(s|p - p_c|^{1/\sigma}) \tag{2.20}$$

With (2.17a) and (2.20) we obtain for the total number of clusters per site,

$$G(p) \simeq \sum_s s^{-\tau}\phi(s|p - p_c|^{1/\sigma})$$

$$= |p - p_c|^{(\frac{\tau-1}{\sigma})} \int dx\, \chi^{-\tau}\phi(x) \tag{2.21a}$$

where

$$x = s/s_\xi \tag{2.21b}$$

and χ is as implied and is the analogue of zero-field susceptibility. Assuming a convergent integral, with (2.17a), we obtain

$$\alpha = 2 - \frac{\tau - 1}{\sigma} \tag{2.22a}$$

In like manner there results

$$\gamma = (3 - \tau)/\sigma, \qquad \beta = (\tau - 2)/\sigma \tag{2.22b}$$

These equations give Rushbrooke's scaling law[27]

$$\alpha + 2\beta + \gamma = 2 \tag{2.23}$$

Another relation among these exponents may be obtained as follows. Integration of the pair-connectedness function over the volume of the sample gives a measure of the mean cluster size, $S(p)$, in dimension, d

$$S(p) \simeq |p - p_c|^{-\gamma} \simeq \int dr\, r^{d-1} \left(\frac{\exp[r/\xi(p)]}{r^{d-2+\eta}}\right) \simeq |p - p_c|^{-\nu(2-\eta)} \tag{2.24}$$

which gives Fisher's scaling law[28]

$$\gamma = \nu(2 - \eta) \tag{2.25}$$

Lastly, assuming a slowly varying integrand (2.18) and (2.21) give for the concentration of clusters of dominant size,

$$n_{s_\xi}(p) \simeq |p - p_c|^{-\tau/\sigma}$$

---

[27]G. S. Rushbrook, *J. Chem. Phys. 29*, 842 (1963).
[28]M. E. Fisher, *Phys. Rev. 180*, 594 (1969).

Assuming that this concentration is inversely proportional to the volume, $\xi^d(d)$, (in dimension d) occupied by these clusters, gives

$$|p - p_c|^{\tau/\sigma} = |p - p_c|^{\nu d}$$

which gives Josephson's hyperscaling law[29]

$$\nu d = 2 - \alpha \tag{2.26}$$

Equations (2.22), (2.25), and (2.26) are coupled relations for the six critical exponents relevant to percolation scaling theory.

7.2.6 Localization in Second Quantization

Working in second quantization, the Hamiltonian of a one-dimensional potential array may be written[30]

$$H = -\sum_{i \neq j} t_{ij}(c_i^\dagger c_j + c_j^\dagger c_i) + \sum_j \varepsilon_j c_j^\dagger c_j \tag{2.27}$$

relevant to an electron in a given band. In the preceding, ε_j is the energy of an electron at j sites, with a corresponding Wannier wavefunction centered at these sites. The t_{ij} matrix couples the ith and jth sites, and $c_j^\dagger$ and c_j are operators which create and annihilate an electron at the jth site, respectively. [t_{ij} matrices are also called: the *transfer integral*, the *overlap energy integral* or the *hopping integral*, cf., (2.4b)] Sums in (2.27) are over the sites of a perfect lattice with disorder introduced by taking either or both of the ε_j and t_{ij} coefficients as random variables subject to a given probability distribution.

We recall the following fermion anticommutation relation rules:

$$[c_i^\dagger, c_j^\dagger]_+ = [c_i, c_j]_+ = 0$$
$$[c_i^\dagger, c_j]_+ = \delta_{ij} \tag{2.27a}$$

Note also that

$$c_i^\dagger |0\rangle = |i\rangle, \qquad c_i |i\rangle = |0\rangle \tag{2.27b}$$

where $|0\rangle$ includes the empty ith site and the ket vector, $|i\rangle$, describes an electron at the ith site.

In the tight-binding approximation of (2.27) one sets $t_{ij} = t$, for nearest neighbors and zero otherwise. Energies ε_j are set equal to ε_A with probability P_A, of ε_B with probability P_B independent of the site j. This labeling provides a one-dimensional model of a disordered binary allow.

---

[29]B. D. Josephson, *Proc. Phys. Soc* **92**, 269 (1967).
[30]A review of one-dimensional disordered systems is given by: K. Ishii, *Prog. Theor. Phys. Supplement, 53*, 77 (1973).

Extended states for the uniform case

Let us show that for the uniform case, for which either $P_A = 1$ or $P_B = 1$, and in the tight-binding approximation, one does not obtain localization. In this event the Hamiltonian (2.27) reduces to

$$H = -t \sum_m (c^\dagger_{m+1} c_m + c^\dagger_m c_{m+1}) + \varepsilon_{A,B} \sum_j c^\dagger_j c_j \qquad (2.28)$$

We consider a lattice of length Na, corresponding to a chain of N atoms and lattice constant a. The Hamiltonian (2.26) may then be diagonalized through the transformation

$$c_j = \frac{1}{\sqrt{N}} \sum_k b_k e^{-ikja} \qquad (2.29a)$$

$$c^\dagger_j = \frac{1}{\sqrt{N}} \sum_k b^\dagger_k e^{ikja} \qquad (2.29b)$$

$$k = \frac{2\pi n}{(Na)}$$
$$n = 0, \pm 1, \pm 2, \ldots, \pm[(N/2) - 1], \pm N/2 \qquad (2.29c)$$

Note that the latter sequence corresponds to periodic boundary conditions. There results

$$H = \sum_k \varepsilon(k) b^\dagger_k b_k = \sum_k \varepsilon(k) n_k \qquad (2.30)$$

where n_k is the occupation number of the kth site, 0 or 1. Eigenenergies are given by

$$\begin{aligned} \varepsilon(k) &= \varepsilon_A - 2t \cos ka, \qquad \text{for } P_A = 1 \\ (k) &= \varepsilon_B - 2t \cos ka, \qquad \text{for } P_B = 1 \end{aligned} \qquad (2.31a)$$

For either case energies lie in a band of width $4t$. The eigenfunction corresponding to $\varepsilon(k)$ is obtained by operating on the state $|0\rangle$ with the creation operator $b^\dagger_k$. There results

$$|k_\varepsilon\rangle = b^\dagger_k |0\rangle = \frac{1}{\sqrt{N}} \sum_j b^\dagger_j \exp(ik_\varepsilon ja) |0\rangle \qquad (2.31b)$$

The coordinate representation of this eigenstate is given by the Dirac product

$$\langle x | k_\varepsilon \rangle = \sum_m \langle x | m \rangle \langle m | k_\varepsilon \rangle \qquad (2.31c)$$

This relation permits one to identify the probability amplitude of finding the site m occupied as the product

$$\langle m | k_\varepsilon \rangle = \frac{1}{\sqrt{N}} \exp(imk_\varepsilon a) \qquad (2.31d)$$

It follows that the probability of finding an electron on the mth site is the constant $1/N$, independent of m or k_ε. We may therefore identify the wavefunction $\langle x|k_\varepsilon\rangle$ as an extended state.

Disordered One-Dimensional Lattice

Transfer-matrix method

As noted above, to describe the disordered situation one sets energies $\varepsilon_m = \varepsilon_A$ with probability P_A, and $\varepsilon_m = \varepsilon_B$ with probability P_B, for all m. We outline the transfer matrix method[31] for examining eigenfunctions and related eigenenergies, E, of the Hamiltonian (2.27) in the tight-binding approximation. Eigenfunctions are written in the form

$$|k_E\rangle = \sum_{n=1}^{N} A_n|n\rangle = \sum_{n=1}^{N} A_n c_n^\dagger|0\rangle \tag{2.32}$$

so that $\langle m|k_E\rangle = A_m$. With (2.27a) we obtain

$$\langle m|H|k_E\rangle = E A_m = \varepsilon_m A_m - t(A_{m+1} + A_{m-1}) \tag{2.33}$$

where $m = 1, 2, \ldots$. To satisfy periodic boundary conditions we set $A_{N+1} = A_1$ and $A_N = A_0$. Introducing the two-component column vector

$$\psi_m \equiv \begin{pmatrix} A_{m+1} \\ A_m \end{pmatrix} \tag{2.34a}$$

gives the recursion relation

$$\psi_m = \hat{T}_m \psi_{m-1} \tag{2.34b}$$

$$\hat{T}_m(\varepsilon_m, E) = \begin{pmatrix} \frac{\varepsilon_m - E}{t} & -1 \\ 1 & 0 \end{pmatrix} \tag{2.34c}$$

Note that $\det \hat{T}_j = 1$, so that eigenvalues, $\mu^{(\pm)}$, of $\hat{T}$ satisfy the relation

$$\mu_+ \mu_- = 1 \tag{2.35a}$$

which gives the forms

$$\mu_\pm = e^{\pm i\theta} \tag{2.35b}$$

In the transfer-matrix method, solution for eigenenergies E is reduced to finding solutions to the equation

$$\psi_N = [\hat{T}_N(\varepsilon_N, E)\hat{T}_{N-1}(\varepsilon_{N-1}, E)\ldots\hat{T}_1(\varepsilon_1, E)]\psi_N \tag{2.36a}$$

---

[31]J. Ziman, *Models of Disorder*, Cambridge, New York (1979), Section 8.2: M. Plischke and B. Bergersen, *Equilibrium Statistical Mechanics, ibid*; M. L. Mehta, *Random Matrices*, Academic, New York (1991).

or, equivalently,

$$\sum_{m=1}^{N} \hat{T}_m(\varepsilon_m, E) = \hat{I} \qquad (2.36b)$$

where $\hat{I}$ is the identity matrix.

Band gap for the uniform array

We wish to demonstrate the existence of a gap in the spectrum of energies for the case of a uniform array. In this event, $\varepsilon_j = \varepsilon_A$ for all m. With (2.34c) and (2.35) it follows that the eigenvalues of $\hat{T}$ are given by

$$\mu_+ = R + iK = e^{i\theta} \qquad (2.37a)$$

$$\mu_- = R - iK = e^{-i\theta}$$

$$R = \frac{\varepsilon_A - E}{2t}, \qquad K = \sqrt{1 - R^2} \qquad (2.37b)$$

Addition of equations (2.37a), with (2.37b) gives

$$E = \varepsilon_A - 2t \cos\theta \qquad (2.38)$$

Periodic boundary conditions ($\mu^N = 1$), imply the discrete θ-spectrum, $\theta_m = 2\pi m/N$.

The condition

$$|\varepsilon_A - E| > 2t \qquad (2.39)$$

implies that K is purely imaginary, in which case eigen μ-values are real. For K purely imaginary, $K = i|K|$ and $\mu_+\mu_- = -1$ which violates (2.35a). We may conclude that for real eigen μ-values, eigenenergies do not exist. This corresponds to the existence of an energy gap for the uniform potential array.

Disordered chain

For the disordered case, in the product (2.36b), transfer matrix components are either of the form $\hat{T}_m = \hat{T}_A$ or $\hat{T}_B$, depending respectively on whether site m is occupied by atom A or atom B. To examine this situation we recall the Saxon-Hunter theorem which states that any spectral region that is a spectral gap for both a pure A-type chain and a pure B-type chain, is also a gap for any mixed lattice of A- and B-type atoms.[32] Consider the case that an eigenenergy lies in such a forbidden domain. In this case, with real eigen μ-values, one again encounters a violation of (2.35a). However, let us consider for the moment that solutions for this situation do exist.

---

[32] J. M.Luttinger, *Phillips Res. Repts. 6*, 303 (1951).

Let $W_\pm$ denote eigenvectors of $\hat{T}$ corresponding respectively to the eigenvalues, $\mu_\pm$. The column wavefunction ψ_n may be written, in general, as a superposition of these eigenvectors and we write

$$\psi_n = \alpha_n^{(+)} W_+ + \alpha_n^{(-)} W_- \tag{2.40}$$

where $\alpha_n^{(\pm)}$ are n-dependent parameters. It follows that

$$\psi_{n+1} = \hat{T}\psi_n = \alpha_n^{(+)}\mu_+ W_+ + \alpha_n^{(-)}\mu_- W_- \tag{2.41a}$$

Returning to the situation of real eigen μ-values, we examine the case that $|\mu_+| > |\mu_-|$ so that

$$|\mu_+| = 1/|\mu_-| > 1 > |\mu_-| \tag{2.41b}$$

Consider that a wavefunction of the form (2.40) starts at the beginning of the chain and passes through p such forbidden cells. In this case

$$\psi_p = (\hat{T})^p \psi_0 = \alpha_n^{(+)}(\mu_+)^p W_+ + \alpha_n^{(-)}(\mu_-)^p W_- \tag{2.41c}$$

The larger eigenvalue quickly dominates and ψ_p grows as $(\exp \gamma p)$ where the growth coefficient $\gamma = \ln |\mu_+|$ This hypothetical analysis hints at the existence of localized states in forbidden domains of the energy spectrum.

Extensive numerical work of Ishii[30] indicates that solutions do in fact exist in such domains, and that corresponding solutions are localized. This conclusion is akin to our previous observation that in a disordered material, localized states occur at the edges of the forbidden band gap (mobility edge). In a closely allied, fundamental work by Matsuda and Ishii[33] it was demonstrated that all one-dimensional disordered systems are localized.

Problems

7.1. Show the equivalence between the integral relations (1.3b) and (1.3a). Hint: Introduce the parameter $\Gamma \equiv E - \mu$.

7.2. Derive the Fermi–Dirac integral series representation (1.3e).

7.3. Establish the Fermi–Dirac vector coefficients (1.3f)–(1.3h).

7.4. Establish validity of the time-reversal scattering-rate symmetry relation (1.30c).

7.5. Working in second quantization, write down the total Hamiltonian for an electron-phonon system, incorporating the potential Hamiltonian component, (1.19). Hint: See (5.6.6b).

7.6. For each of the cases listed in Table 7.1, indicate if the related transition site or bond percolation.

---

[33] H. Matsuda and K. Ishii, *Prog. Theor. Phys. Suppl* **45**, 56 (1970).

7.7. Employ the transformation equations (2.29) to obtain the diagonalized Hamiltonian (2.30) and related eigenenergies (2.31a).

7.8. Show that in the tight-binding approximation, the Hamiltonian (2.27) reduces to the Hamiltonian (2.28).

7.9. Derive the eigenenergy relation (2.33) from the wavefunction (2.32) and Hamiltonian (2.27) under the said conditions.

7.10. What is the width of the energy gap for the uniform array described in (2.37), et seq.?

7.11. Show that Grüneisen's[34] integral for metallic resistivity

$$\rho = A T^5 \int_0^{\Theta/T} \frac{x^5 \, dx}{(e^x - 1)(1 - e^{-x})}$$

reduces to Bloch's form at $T \ll \Theta$, and has the correct linear T dependence at $T \gg \Theta$ where Θ represents the Debye temperature.

7.12. **(a)** Describe conditions under which extended states exist in an amorphous material.

(b) Describe conditions under which an amorphous metal does not conduct. In both answers, define parameters.

7.13. A student argues that tearing a piece of paper is percolative. Is the student correct. Explain your answer.

Answer

Tearing a piece of paper is a local and not a global event. For this reason the process is not percolative.

---

[34]E. Grüneisen, *Ann. Physik 16*, 530 (1932).

Location of Key Equations

Name	*Location*
Balescu–Lenard equation	(4.2.71)
BBKGY equations	(2.1.20)
Boltzmann equation	(3.2.14)
Chapman–Kolmogorov equation	(1.7.4)
Conservation equations (in absolute variables)	(3.14, 3.18, 3.19)
Conservation equations (in relative variables)	(3.14, 3.30, 3.31)
Density matrix, N-body, equation of motion	(5.5.23)
Euler equations	(3.5.23)
Fock-space function, equation of motion	(5.5.29)
Fokker–Planck equation	(5.2.31)
Grad's second equation	(2.6.12)
Green's function equation, coupled one- and two-particle	(5.7.46)
Green's function, N-body, equation of motion	(5.7.59)
Hamilton's equations	(1.1.12)
Hierarchies of classical and quantum distributions	(Table 5.2)
Kinetic equations, connecting relations	(Figure 4.9)
Krook–Bhatnager–Gross equation	(4.2.7)
Kubo formula, classical	(3.4.66)
Kubo formula, quantum mechanical	(5.6.10)
Lagrange's equations	(1.1.7)
Landau equation	(4.2.51)
Landau equation, Fermi liquid	(5.4.13)
Liouville equation	(1.4.7)
Liouville equation, quantum mechanical	(5.2.9)
Liouville equation, relativistic	(6.4.25)
Master equation	(1.7.20, 5.2.28)

Navier–Stokes equation   (3.4.8)
Pauli equation   (5.2.28)
Photon kinetic equation   (5.3.18)
Plasma convergent kinetic equation   (4.2.75)
Relaxation-time model kinetic equation   (5.3.35)
Uehling–Uhlenbeck equation   (5.4.5)
Vlasov equation   (2.2.30)
Vlasov equation, relativistic   (6.2.22)
Wigner distribution, equation of motion   (5.2.54)
Wigner distribution, hierarchy   (5.5.33)
Wigner–Moyal equation   (5.2.71)

List of Symbols

(See also Table 7.1, page 487)

$\mathbf{A}, \mathcal{A}$	Vector and scalar functions in Chapman–Enskog expansion (3.5.36a).
$\mathbf{A}$	Vector potential.
$A(\mathbf{k}, \omega)$	Spectral density function (5.7.12, 5.8.23).
$(\hat{A})\text{W}$	Wigner counterpart of $\hat{A}$ (5.2.55).
$\hat{A}^{(s)}$	s-particle operator (5.5.10).
$\mathbf{a}$	Acceleration. Friction coefficient in Fokker–Planck equation (4.2.30).
$\hat{a}^{\dagger}, \hat{a}$	Creation and annihilation operators.
$\hat{a}_{\text{H}}, \hat{a}_{\text{H}}^{\dagger}$	Operators $\hat{a}$ and $\hat{a}^{\dagger}$ in the Heisenberg representation.
$a_{(\mathbf{k})}$	Fourier coefficient in Prigogine expansion (2.3.8).
$\overset{=}{B}, \mathcal{B}$	Tensor and scalar functions in Chapman–Enskog expansion (3.5.36b).
$\mathbf{B}$	Magnetic field.
BE	Boltzmann equation.
BL	Balescu–Lenard equation.
BY_s	sth equation in BBKGY hierarchy.
b	Nondimensional impact parameter.
$\overset{=}{\mathbf{b}}$	Friction coefficient in Fokker–Planck equation (4.2.30).
C	$mC^2 = k_B T$ in most of text. $mC^2 = 3k_B T$ when C is thermal speed (3.4.2).
$C_2(1, 2)$	Two-particle correlation function (2.2.18).
C_{12}	Effective two-particle speed (3.4.25).
$C(\mathbf{q})$	Coefficient of electron–phonon interaction (5.8.1).

$\bar{C}$	$\bar{C}^2 = 2RT$ (3.5.28).
CK	Chapman–Kolmogorov equation.
$\mathbf{c}$	$\mathbf{c} = \mathbf{v} - \langle \mathbf{v} \rangle$. Deviation from mean microscopic velocity.
$\bar{\mathbf{c}} = \mathbf{c}/C$	Nondimensional microscopic velocity (3.5.90).
c	Speed of light.
c_V	Specific heat per molecule (3.4.34).
$\bar{c}_V$	Specific heat per molecule per mass (3.4.35).
D	Self-diffusion coefficient (3.4.2). Density of points in phase space (Chapter 1).
$D(\xi)$	Variance of $rv\xi$.
D_{12}	Mutual diffusion coefficient (3.4.29).
D/Dt	Convective time derivative.
$\hat{D}$	Differential operator in Boltzmann equation (3.5.7a).
E	Energy.
E_F	Fermi energy.
$\mathbf{E}$	Perturbative electric field (4.1.4). Two-particle electric field (4.2.61).
$\bar{\mathbf{E}}$	Fourier transform of $\mathbf{E}$ (4.1.6b).
$\mathcal{E}$	Expectation (1.8.2a).
E_K	Relative kinetic energy density (3.3.26).
$\bar{E}(x)$	Elliptic integral (3.6.22).
e_K	Absolute kinetic energy density.
$F(t)$	Time-dependent component of perturbation Hamiltonian (3.4.59).
$F(\theta)$	Term in collision cross section (3.6.212).
$F(\omega)$	Number of photons per frequency interval (Chapter 5).
$F(\beta)$	Integral in dielectric constant (4.1.68).
$\bar{F}_s$	s-tuple distribution function (1.6.12).
F_s	$F_s = V_s f_s$. Bogoliubov s-particle distribution.
F	Distribution function normalized to total number of particles.
$\mathcal{F}$	Relativistic distribution (6.2.1).
f	Classical one-particle distribution function in most of text.
$\bar{f}$	Nondimensional distribution function (3.5.90).
f^0, F^0	Local Maxwellians.
f_0, F_0	Absolute Maxwellians.
f_0	Exponential component of Maxwellian (4.1.27).
f_1	Arbitrary single-peaked one-dimensional distribution (4.1.48). One-particle distribution function.
$\tilde{f}_1$	Velocity component of f_1 (4.1.48).
$f_0(E)$	Equilibrium occupation number per energy level (Chapter 5).
$f_0(\omega)$	Equilibrium distribution for number of photons per frequency mode (Chapter 5).

f_s	s-particle distribution function normalized to unity in most of text.
f_n^Δ	Truncated distribution in Grad's analysis (Chapter 2).
G_1, G_2	Generating functions (Chapter 1).
$\mathbf{G}_{ij}$	Two-particle force (2.1.7a).
$\mathbf{G}(\mathbf{x}, \mathbf{x}')$	Two-particle force (2.2.27).
$\mathbf{G}(\mathbf{x}, t)$	Mean force field (2.2.28b).
$G_{R,A}$	Green's function, retarded and advanced.
$\hat{\bar{G}}$	s-particle Fock space operator (5.5.7).
g_i	Constant of motion (Chapter 1).
$g(r)$	Radial distribution function (2.2.44).
$g(\omega)$	Number of normal modes per frequency interval (Chapter 5).
$g_{\mu\nu}$	Metric tensor (Chapter 6).
$\mathbf{g}$	Relative two-particle velocity.
H	Hamiltonian.
$H_{i_1,i_2,\dots}^{(n)}$	Tensor Hermite polynomial (3.5.92).
$\bar{H}$	Time-independent component of perturbation Hamiltonian (3.4.59).
$h(r)$	Total correlation function (2.2.47).
$h_s^{(N)}$	s-particle conditional distribution function (1.6.10).
h	Enthalpy per particle (Problem 3.48).
$\mathcal{H}$	Boltzmann $\mathcal{H}$ function (3.3.42).
$\mathfrak{H}$	Hilbert space [beneath (5.1.18)].
$\hat{I}_s$	Collision term in BY$_s$ (2.2.2).
$\hat{I}_s^n$	Collision term in BY$_s$ (2.2.25).
$\bar{\bar{I}}$	Identity operator.
$\mathbf{I}$	Integral [above (6.2.52a)]. Tensor term (3.5.100).
$\hat{J}(f)$	Collision integral.
$\mathbf{J}$	Current density.
$\mathbf{K}$	Externally supported force field (3.2.1). Acceleration (3.5.7a). Momentum of center of mass (3.3.68).
K	Constant in Coulomb interaction (4.2.7a). Constant in inverse radial potential (3.1.7).
$\hat{K}_s$	Kinetic energy operator of sth particle (2.1.9).
$\hat{K}(\mathbf{v}, \mathbf{v}_1)$	Collision-integral kernel operator (3.6.3).
$K(\mathbf{p}, \mathbf{p}')$	Quasi-particle interaction function (5.4.11).
$\bar{K}, \bar{K}_0$	Quasi-particle interaction function and constant (5.4.32a).
$\bar{K}(x)$	Elliptic integral (3.6.22).
k_0	Wave number of closest approach (4.2.73a).
k_d	Debye wave vector (5.3.52).
k_B	Boltzmann's constant.
k_{TF}	Thomas–Fermi wave number (5.3.46).

$\mathbf{k}_i$	Momentum of ith molecule (3.3.68).
$\mathbf{k}$	Momentum (5.2.73, and following).
KBG	Krook–Bhatnager–Gross equation.
$\hat{\mathcal{K}}$	Operator in Boltzmann collision integral (3.6.29).
l	Mean free path (3.2.2).
$\mathcal{L}_2$	Space of square integrable functions.
LIM	Limiting process [defined above (5.7 to 5.19)].
$L(q, \dot{q}, t)$	Lagrangian (1.1.3).
$\hat{L}_N$	N-particle Liouville operator (2.1.3).
$\delta\hat{L}$	Perturbation Liouville operator (2.3.4).
M	Mach number (Chapter 4).
$M_1(n), M_2(n)$	Moments of probability distribution (1.7.24).
M_{12}	Rate of momentum transport per unit volume (3.4.26).
$\mathcal{N}(\mathbf{z})$	Phase density (2.5.4).
$\mathcal{N}$	Number of ensemble points in phase space (1.4.1b).
N	Total number of particles.
N_z	Number of zeros (4.1.67).
N_p	Number of poles (4.1.67).
n	Particle number density.
$\hat{O}_{ij}$	Operator in Liouville equation (2.1.7).
$\bar{\bar{p}}$	Pressure tensor in lab frame (3.3.15).
p	Scalar pressure (3.4.4).
$\mathfrak{p}$	Probability of step to right (1.1.8 to 1.1.14). Momentum (3.4.23).
$\mathbf{p}^N$	$3N$-dimensional momentum vector relevant to N point particles (5.2.46).
$\mathbf{p}_F$	Fermi momentum (5.3.34a).
p_μ	Four-momentum (Chapter 6).
$\bar{\mathbf{p}}$	Momentum four-vector (6.1.2). Momentum three-vector, Problem 1.40.
$\mathbf{p}$	Momentum.
$\bar{\bar{P}}$	Relative pressure tensor (3.3.24).
P	Probability.
$\mathcal{P}$	Principal part (4.1.71).
$\mathfrak{q}$	Probability of step to left (1.8.14).
$\mathbf{q}$	Heat flow vector in lab frame (3.3.16).
$\mathbf{Q}$	Relative heat flow vector.
$\mathbf{Q}_K$	Kinetic energy flow vector (2.1.34a).
$\mathbf{Q}_\phi$	Potential energy flow vector (2.1.34b).
$\bar{\bar{Q}}_L$	Landau collision tensor (4.2.72).
$\bar{\bar{Q}}$	Collision form in derivation of the Balescu–Lenard equation (4.2.64).
$\bar{\bar{Q}}_{\mathrm{BL}}$	Balescu–Lenard collision term (4.2.69).

$\bar{\bar{Q}}_{\mathrm{L}}$	Landau collision term (4.2.72).
$\mathbf{r}$	Interparticle displacement vector. Radius vector.
r	Total number of events (1.8.21).
rv	Random variable.
R	k_B/m.
$\mathbf{R}$	Radius to center of mass (Problem 1.7).
$\hat{S}(\mathbf{k})$	Momentum operator (5.2.76b).
S	Action integral (1.1.4).
$S(k)$	Structure factor (2.2.51).
$S_m^{(n)}$	Sonine polynomial (3.5.54a).
$\hat{\mathbf{S}}$	Spin operator (Chapter 5).
$\hat{\hat{S}}$	Superoperator (5.6.11).
$\bar{\bar{S}}$	Response component of pressure tensor (3.4.4).
s	Impact parameter.
T	Temperature. Kinetic energy.
$\bar{\bar{T}}$	Tensor of products of relative velocity, $\mathbf{g}$ (Chapter 4).
$\hat{T}$	Integral-kernel operator (3.6.11). Time ordering operator (Chapter 5).
$U(T)$	Total radiant energy density (5.3.12).
U	Generalized potential (Chapter 1).
$U(1, 2)$	Single-time interaction potential (5.7.49a).
$\hat{U}$	Unitary operator for Schroedinger equation (5.1.39).
$\mathbf{u}(\mathbf{x}, t)$	Macroscopic fluid velocity (Chapter 3).
u_d	Downstream fluid speed (4.2.12).
u_p	Upstream fluid speed (4.2.12).
$\bar{V}$	Renormalized volume (2.3.9).
$\mathbf{V}$	Vector in Γ-space (1.4.16a).
$\mathbf{v}_F$	Fermi velocity [beneath (5.4.31)].
$\mathbf{v}$	Microscopic velocity (Chapter 3).
$w(\mathbf{v}', \mathbf{v})$	Probability scattering rate (4.3.1), (5.3.5).
$\tilde{w}(\mathbf{v}', \mathbf{v})$	Augmented scattering probability rate (4.3.3).
$\tilde{w}_{ij}$	Scattering rate (1.7.20).
$\tilde{w}_{ij}$	Transition probability (1.7.22).
w_{nk}	Transition probability (5.2.20).
$\mathbf{x}_k$	kth particle position vector (5.2.55).
$\mathbf{x}^N$	$3N$ particle position vector relevant to N point particles.
x_μ	Four-displacement (Chapter 5).
$\bar{\mathbf{x}}$	Displacement four-vector (6.1.1). Displacement three-vector (Problem 1.40).
Z_N	N-body partition function (Chapter 3).

Greek and Other Symbols

α, β, γ	Constants in homogeneous distribution (3.5.32).
α	Nondimensionalization parameter (2.2.11).
α_n	Matrix element in Chapman–Enskog expansion (3.5.56).
α	Apsidal vector (3.1.38).
$\alpha(\mathbf{n})$	Displacement of Fermi surface in direction of $\mathbf{n}$ (5.4.31).
α_x, etc.	Components of spin eigenvector (5.1.33).
β_x, etc.	Components of spin eigenvector (5.1.33).
β	Dimensionless frequency (4.1.42).
β_n	Matrix element in Chapman–Enskog expansion (3.5.64).
β_p	Dimensionless plasma frequency (4.1.50).
$\delta_{\mathbf{k}}$	Kronecker symbol (2.3.9).
δ_{ij}	Kronecker-delta symbol.
$\delta(\mathbf{r})$	Three-dimensional delta function.
ε	Dielectric constant (Chapters 4 and 5). Parameter of smallness.
$\varepsilon(x, p)$	Quasiparticle energy (5.4.10).
$\mathcal{E}$	Electric field.
$\Phi(\mathbf{x}_i, \mathbf{x}_j)$	Interaction potential (1.4.24).
Φ_K	Fourier transform of interaction potential (2.3.24).
$\hat{\phi}^{\dagger}$	Field creation operator (5.5.3).
$\hat{\phi}_{\mathrm{H}}$	Field operator in Heisenberg picture (5.5.7a).
$\varphi_E(\mathbf{x}^N)$	Stationary energy eigenstate (5.1.8).
γ	Relativistic parameter (Chapter 6). Nondimensionalization parameter (2.2.11).
Γ	Flux vector (No./cm^2-s) (3.3.11).
Γ	Maximum total scattering rate (4.3.4a).
$\Gamma^{\nu}_{\beta\mu}$	Christoffel symbol (6.4.21).
η	Coefficient of viscosity Chapter 3.
$\eta(t)$	Coarse-grained entropy (3.8.12).
κ	Coefficient of thermal conductivity (3.4.3). Constant in Coulomb cross section (4.2.21a). Absorption coefficient (5.3.28).
$\hat{\kappa}_s$	Kinetic energy operator (2.2.22a).
κ_{12}	Coefficient in mutual diffusion expression (3.4.26).
$\kappa_{\mu\nu}$	Coefficient of thermal conductivity tensor (5.3.42).
$\lambda(\mathbf{v})$	Total scattering rate (4.3.2).
$\tilde{\lambda}(\mathbf{v})$	Maximum scattering rate (3.3a).
λ_d	Thermal de Broglie wavelength (5.3.5). Debye distance (4.1.36).
λ_{TF}	Thomas–Fermi wavelength (5.3.46a).
Λ	Plasma parameter (Chapters 2 and 4).
$\hat{\Lambda}$	Liouville operator (1.4.20).
$\overset{=}{\Lambda}$	Symmetric strain tensor (3.4.6).

μ	Reduced mass. Mobility (Chapter 3). Chemical potential (5.3.1). Relativistic parameter (6.3.2).
$\mu(A)$	Measure of set A (3.8.3a).
ν	Collision frequency (3.5.2).
ν_{12}	Collision frequency for mutual diffusion (3.4.24).
$\Omega^{(n,q)}$	Element of collision integral in Chapman–Enskog expansion (3.5.62).
ω_p	Plasma frequency (4.1.18).
$\omega(c)$	Maxwellian (3.5.91)
$\Psi(\mathbf{x}^N, t)$	Wavefunction (5.1.1).
ψ_D	Electrochemical potential (3.4.15a).
$\prod_{ij}$	Momentum flux (5.4.23).
$\prod(z, t \mid z_0, t_0)$	Two-time distribution (1.7.2).
$\prod(\mathbf{z}, t)$	Coarse-grained entropy (3.8.13).
ρ	Mass density (3.3.18a).
$\hat{\rho}$	Density operator (5.2.1).
ρ_{nq}	Density matrix (5.2.3).
σ	Stefan–Boltzmann constant (5.3.15).
σ_{01}, σ_{02}	Molecular diameters (3.4.30).
σ_{12}	Sum of molecular radii (3.4.30).
$\sigma_{\mu\nu}$	Electrical conductivity tensor (5.3.41).
$\sigma(E, \theta)$	Scattering cross section (3.1.24).
$\sigma(\theta)$	Scattering cross section (3.1.31).
σ_{12}	Scattering cross section for rigid spheres (3.4.30).
σ_T	Total scattering cross section (3.1.34).
$\hat{\sigma}$	Pauli spin operator (5.2.39).
$d\sigma_\mu$	Element of hypersurface in $\bar{\mathbf{x}}$ space (6.2.2).
$d\tilde{\sigma}_\mu$	Element of hypersurface in $\bar{\mathbf{p}}$ space (6.2.3b).
$d\Sigma$	Element of surface area (3.8.7).
τ	Relaxation time (5.3.35a; 5.3.79). Proper time (6.1.23).
ξ	Dimensionless velocity (Chapter 3). Quantum parameter (5.4.6a).
Ξ	Time-ordering operator (5.8.7).
$\hat{\Box}$	Collision operator (3.6.1).
$\sum$	Symmetric sum of gradients (3.5.101).
Δ	Collision term in photon kinetic equation (5.3.19).

Vector Formulas and Tensor Notation

A.1   Definitions

The vector function of position $\mathbf{A}(\mathbf{r})$ is a set of three functions A_μ; $\mu = 1, 2, 3$. These represent the projection of the vector $\mathbf{A}$ onto the Cartesian axis, so that $A_1 \equiv A_x$, $A_2 \equiv A_y$, $A_3 \equiv A_z$. The *tensor notation* of the vector $\mathbf{A}$ is A_μ.

More generally, the set of 3^n functions

$$\psi_{i_1 i_2 \cdots i_n}(\mathbf{r})$$

in which each of the n indexes $\{i_p\}$ run from 1 to 3, is called an nth-rank tensor in three dimensions. A scalar carries no indexes and so is a zeroth-rank tensor. A vector is a first-rank tensor.

In the text the *vector notation* for a second-rank tensor $\phi_{\mu\nu}$ is

$$\bar{\bar{\phi}}$$

A popular second-rank tensor has as its nine components the products of the components to two vectors, $\mathbf{A}$ and $\mathbf{B}$ and is call a *dyad*.

$$\bar{\bar{D}} \equiv \mathbf{A}\mathbf{B} \tag{A.1}$$

$$D_{\mu\nu} \equiv A_\mu B_\nu \tag{A.2}$$

A second rank tensor $\psi_{\mu\nu}$ is symmetric if

$$\psi_{\mu\nu} = \psi_{\nu\mu} \tag{A.3}$$

It is antisymmetric (or *skew*) if

$$\psi_{\mu\nu} = -\psi_{\nu\mu} \tag{A.4}$$

a symmetric or antisymmetric tensor can always be constructed from an arbitrary tensor in the following manner:

$$\psi_{\mu\nu}^{(s)} = \frac{1}{2}(\psi_{mu\nu} + \psi_{\nu\mu}) \tag{A.5}$$

$$\psi_{\mu\nu}^{(a)} = \frac{1}{2}(\psi_{mu\nu} - \psi_{\nu\mu}) \tag{A.6}$$

The trace of a tensor is the sum of its diagonal components:

$$\mathrm{Tr}\,\phi_{\mu\nu} \equiv \sum_{\mu=1}^{3} \psi_{\mu\mu} \tag{A.7}$$

A tensor of zero trace can be made from $\bar{\bar{\psi}}$:

$$\psi_{\mu\nu}^{(0)} = \psi_{\mu\nu} - \frac{1}{3}(\mathrm{Tr}\,\bar{\bar{\phi}})\delta_{\mu\nu} \tag{A.8}$$

(Repeated indexes are summed.) The second-rank Kronecker delta symbol $\delta_{\mu\nu}$ is defined as

$$\begin{aligned} \delta_{\mu\nu} &= 1 \quad &\text{for } \mu = \nu \\ \delta_{\mu\nu} &= 0 \quad &\text{otherwise} \end{aligned} \tag{A.9}$$

A symmetric tensor of zero trace is

$$\psi_{\mu\nu}^{(s0)} = \psi_{\mu\nu}^{(s)} - \frac{1}{3}(\mathrm{Tr}\,\bar{\bar{\phi}})\delta_{\mu\nu} \tag{A.10}$$

In *vector notation*, equations (A.8) and (A.10) appear as

$$\bar{\bar{\phi}}^{(0)} = \bar{\bar{\phi}} - \frac{1}{3}(\mathrm{Tr}\,\bar{\bar{\phi}})\bar{\bar{I}} \tag{A.8'}$$

$$\bar{\bar{\phi}}^{(s0)} = \bar{\bar{\phi}}^{(s)} - \frac{1}{3}(\mathrm{Tr}\,\bar{\bar{\phi}})\bar{\bar{I}} \tag{A.10'}$$

The third rank Levi–Civita symbol $\varepsilon_{l\nu\mu}$ is defined as

$$\begin{aligned} \varepsilon_{l\nu\mu} &= +1 \text{ if } l\nu\mu \text{ is an even permutation of } 1, 2, 3 \\ &= -1 \text{ if } l\nu\mu \text{ is an odd permutation of } 1, 2, 3 \\ &= 0 \text{ if any two indexes are equal} \end{aligned}$$

The *vector* or "cross product to two vectors is defined by

$$\mathbf{A} \times \mathbf{B} = \begin{vmatrix} \mathbf{i} & \mathbf{j} & \mathbf{k} \\ A_x & A_y & A_z \\ B_x & B_y & B_z \end{vmatrix} \tag{A.11}$$

The triad of unit vectors $(\mathbf{i}, \mathbf{j}, \mathbf{k})$ lies along the Cartesian axis with $\mathbf{i}$ in the x direction, $\mathbf{j}$ in the y direction, and $\mathbf{k}$ in the z direction. In tensor notation the cross product appears as

$$(\mathbf{A} \times \mathbf{B})_\mu = \varepsilon_{\mu l p} A_l B_p \tag{A.12}$$

The *inner* or *dot* product of two vectors is defined by

$$\mathbf{A} \cdot \mathbf{B} = A_x B_x + A_y B_y + A_z B_z \tag{A.13a}$$

$$\mathbf{A} \cdot \mathbf{B} = A_\mu B_\mu \tag{A.13b}$$

The *del operator* ∇ is written for

$$\nabla \equiv \mathbf{i}\frac{\partial}{\partial x} + \mathbf{j}\frac{\partial}{\partial y} + \mathbf{k}\frac{\partial}{\partial z} \tag{A.14a}$$

$$\nabla = \frac{\partial}{\partial x_\mu} \tag{A.14b}$$

The gradient of a scalar function ϕ is defined by

$$\text{grad }\phi \equiv \nabla\phi (\text{grad }\phi)_\mu = \frac{\partial}{\partial x_\mu}\phi \tag{A.15b}$$

The curl of a vector is defined by

$$\text{curl }\mathbf{A} \equiv \nabla \times \mathbf{A} \tag{A.16a}$$

$$(\text{curl }\mathbf{A})_\mu = \varepsilon_{\mu l v}\frac{\partial}{\partial x_l}A_v \tag{A.16b}$$

The divergence of a vector is defined by

$$\text{div }\mathbf{A} \equiv \nabla \cdot \mathbf{A} \tag{A.17a}$$

$$\text{div }\mathbf{A} \equiv \frac{\partial}{\partial x_\mu}A_\mu \tag{A.17b}$$

A.2   Vector Formulas and Tensor Equivalents

Some useful vector formulas together with their tensor-notation equivalents are given below:

$$\nabla \cdot \phi\mathbf{A} = \phi(\nabla \cdot \mathbf{A}) + (\mathbf{A} \cdot \nabla)\phi \tag{A.18a}$$

$$\frac{\partial}{\partial x_l}\phi A_l = \phi\frac{\partial}{\partial x_l}A_l + A_l\frac{\partial}{\partial x_l}\phi \tag{A.18b}$$

$$\nabla \times \phi\mathbf{A} = \phi\nabla \times \mathbf{A} - \mathbf{A} \times \nabla\phi \tag{A.19a}$$

$$\varepsilon_{\alpha\beta v}\frac{\partial}{\partial x_\beta}\phi A_v = \varepsilon_{\alpha\beta v}\phi\frac{\partial}{\partial x_\beta}A_v + \varepsilon_{\alpha\beta v}A_v\frac{\partial}{\partial x_\beta}\phi \tag{A.19b}$$

$$\nabla \cdot \mathbf{A} \times \mathbf{B} = \mathbf{B} \cdot \nabla \times \mathbf{A} - \mathbf{A} \cdot \nabla \times \mathbf{B} \tag{A.20a}$$

$$\frac{\partial}{\partial x_\mu}\varepsilon_{\mu v\tau}A_v B_\tau = \varepsilon_{\mu v\tau}A_v\frac{\partial}{\partial x_\mu}B_\tau + \varepsilon_{\mu v\tau}B_\tau\frac{\partial}{\partial x_\mu}A_v \tag{A.20b}$$

$$\nabla(\mathbf{A} \cdot \mathbf{B}) = (\mathbf{A} \cdot \nabla)\mathbf{B} + (\mathbf{B} \cdot \nabla)\mathbf{A} + \mathbf{A} \times (\nabla \times \mathbf{B}) + \mathbf{B} \times (\nabla \times \mathbf{A}) \tag{A.21a}$$

$$\frac{\partial}{\partial x_l}A_\mu B_\mu = A_\mu\frac{\partial}{\partial x_l}B_\mu + B_\mu\frac{\partial}{\partial x_l}A_\mu \tag{A.21b}$$

$$\nabla \times (\nabla \phi) = 0 \tag{A.22a}$$

$$\varepsilon_{l\mu\nu} \frac{\partial}{\partial x_\mu} \frac{\partial}{\partial x_\nu} \phi = 0 \tag{A.22b}$$

$$\nabla \cdot (\nabla \times \mathbf{A}) = 0 \tag{A.23a}$$

$$\frac{\partial}{\partial x_\mu} \varepsilon_{\mu lp} \frac{\partial}{\partial x_l} A_p = 0 \tag{A.23b}$$

$$\nabla \times (\nabla \times \mathbf{A}) = \nabla(\nabla \cdot \mathbf{A}) - (\nabla \cdot \nabla)\mathbf{A} \tag{A.24a}$$

$$\varepsilon_{\mu lk} \frac{\partial}{\partial x_l} \varepsilon_{knp} \frac{\partial}{\partial x_n} A_p = \varepsilon_{\mu lk}\varepsilon_{knp} \frac{\partial}{\partial x_l} \frac{\partial}{\partial x_n} A_p \tag{A.24b}$$

$$\varepsilon_{\kappa\mu l}\varepsilon_{knp} = \delta_{\mu n}\delta_{lp} - \delta_{\mu p}\delta_{ln} \tag{A.25}$$

(Use equation A.25 in A.24b to obtain A.24a.)

$$\nabla \cdot \mathbf{r} = 3 \tag{A.26a}$$

$$\frac{\partial}{\partial x_\mu} x_\mu = 3 \tag{A.26b}$$

$$\nabla \times \mathbf{r} = 0 \tag{A.27a}$$

$$\varepsilon_{klp} \frac{\partial}{\partial x_l} x_p = \varepsilon_{klp}\delta_{lp} = 0 \tag{A.27b}$$

$$(\mathbf{A} \cdot \nabla)\mathbf{r} = \mathbf{A} \tag{A.28a}$$

$$A_\mu \frac{\partial}{\partial x_\mu} x_l = A_\mu \delta_{\mu l} = A_l \tag{A.28b}$$

$$\nabla \times (\mathbf{A} \times \mathbf{B}) = -\mathbf{B}\nabla \cdot \mathbf{A} - \mathbf{A} \cdot \nabla\mathbf{B} + \mathbf{A}\nabla \cdot \mathbf{B} + \mathbf{B} \cdot \nabla\mathbf{A} \tag{A.29a}$$

$$\varepsilon_{lnp} \frac{\partial}{\partial x_n} \varepsilon_{pkh} A_k B_h = \frac{\partial}{\partial x_a} B_a A_l - \frac{\partial}{\partial x_d} A_d B_l$$
$$= \frac{\partial}{\partial x_a}(B_a A_l - A_a B_l) \tag{A.29b}$$

$$\nabla \times (\mathbf{A} \times \mathbf{B}) = \nabla \overline{\overline{(\mathbf{BA} - \mathbf{AB})}} \tag{A.30a}$$

$$(\mathbf{A} \times \mathbf{B}) \times (\mathbf{C} \times \mathbf{D}) = (\mathbf{A} \cdot \mathbf{C})(\mathbf{B} \cdot \mathbf{D}) - (\mathbf{A} \cdot \mathbf{D})(\mathbf{B} \cdot \mathbf{C})$$

$$\varepsilon_{\mu kl} A_k B_l \varepsilon_{\mu dp} C_d D_p = A_\varphi C_\varphi B_k D_k - A_n D_n B_j C_j \tag{A.30b}$$

$$(\mathbf{A} \times \mathbf{B}) \times (\mathbf{C} \times \mathbf{D}) = [\mathbf{A} \cdot (\mathbf{B} \times \mathbf{D})]\mathbf{C} - [\mathbf{A} \cdot (\mathbf{B} \times \mathbf{C})]\mathbf{D} \tag{A.31a}$$

$$\varepsilon_{h\mu\varphi}\varepsilon_{\mu kl}\varepsilon_{\varphi np} A_k B_l C_n D_p = \varepsilon_{iba} A_i B_b D_a C_h - \varepsilon_{jde} A_j B_d C_e D_h \tag{A.31b}$$

The symmetric sum of gradients operator $\sum$ is defined by

$$\left(\sum \bar{\bar{\phi}} \right)_{k\mu\nu} \equiv \frac{\partial \phi_{\mu\nu}}{\partial x_k} + \frac{\partial \phi_{\nu k}}{\partial x_\mu} + \frac{\partial \phi_{k\mu}}{\partial x_\nu} \tag{A.32}$$

APPENDIX B

Mathematical Formulas

B.1 Tensor Integrals and Unit Vector Products

Let

$$A_{i_1 i_2 \cdots i_N} \equiv \int \hat{v}_{i_1} \hat{v}_{i_2} \cdots \hat{v}_{i_N} f(v^2) d\mathbf{v} \tag{B1.1}$$

where $\hat{v}_{i_k}$ are Cartesian components of the unit vector $\hat{\mathbf{v}}$ so that the index i_k runs from 1 to 3. Then

$$A_{i_1 i_2 \cdots i_N} = 0, \qquad \text{for } N \text{ odd}$$

whereas, for N even,

$$A_{i_1 i_2 \cdots i_N} = \frac{1}{(N+1)!!} \sum_P \delta_{i_1 i_2} \delta_{i_3 i_4} \cdots \delta_{i_{N-1} i_N} \int f(v^2) d\mathbf{v} \tag{B1.2}$$

where the summation is over the P permutations of the integers $i_1 \cdots i_N$ and $(N+1)!!$ denotes the skip factorial

$$(N+1)!! = (N+1)(N-1)(N-3)\cdots$$

With these formulas, the case $N = 2$ gives

$$A_{i_1 i_2} = \frac{1}{3} \delta_{i_1 i_2} \int f(v^2) d\mathbf{v} \tag{B1.3}$$

For $N = 4$, we obtain

$$A_{i_1 i_2 i_3 i_4} = \frac{1}{15} [\delta_{i_1 i_2} \delta_{i_3 i_4} + \delta_{i_1 i_3} \delta_{i_4 i_4} + \delta_{i_2 i_4} \delta_{i_2 i_3}] \int f(v^2) d\mathbf{v} \tag{B1.4}$$

B.2   Exponential Integral, Γ Function, ζ Function, and Error Function[1]

B.2.1   Exponential Integral

The definition of this function is

$$E_1(z) = \int_z^\infty \frac{e^{-t}}{t} dt \tag{B2.1}$$

where $|\arg z| < \pi$. $E_1(z)$ has the expansion

$$E_1(z) = -\gamma - \ln z - \sum_{n=1}^\infty \frac{(-1)^n z^n}{n n!} \tag{B2.2}$$

where γ is Euler's constant, which is given by

$$\gamma = \lim_{m \to \infty} \left(1 + \frac{1}{2} + \frac{1}{3} + \cdots + \frac{1}{m} - \ln m\right) = -\int_0^\infty dt\, e^{-t} \ln t$$
$$= 0.5772157$$

Here are a few other integral properties of $E_1(z)$:

$$\int_0^\infty \frac{e^{-at}}{b+t} dt = e^{ab} E_1(ab) \tag{B2.3}$$

$$\int_0^z \frac{1 - e^{-t}}{t} dt = E_1(z) + \ln z + \gamma \tag{B2.4}$$

$$\int_0^\infty E_1^2(t) dt = 2 \ln 2$$

$E_1(z)$ is a member of the set of integrals

$$E_n(z) = \int_1^\infty \frac{e^{-zt}}{t^n} dt, \qquad n = 0, 1, 2, \ldots \tag{B2.5}$$

with $\mathrm{Re}\, z > 0$. These functions have the recurrence relation

$$E_{n+1}(z) = \frac{1}{n}[e^{-z} - z E_n(z)], \qquad n = 1, 2, \ldots \tag{B2.6}$$

Furthermore,

$$\frac{d E_n(z)}{dz} = -E_{n-1}(z) \tag{B2.7}$$

---

[1]For additional properties, see M. Abramowitz and I. A. Stegun, *Handbook of Mathematical Functions*, Dover, New York (1970).

and

$$E_n(0) = \frac{1}{n-1}, \qquad n > 1$$
$$E_0(z) = \frac{e^{-z}}{z}$$

(B2.8)

and

$$\int_0^\infty e^{-at} E_n(t)\, dt = \frac{(-1)^{n-1}}{a^n} \left[\ln(1+a) + \sum_{k=1}^{n-1} \frac{(-1)^k a^k}{k} \right], \qquad a > -1$$

(B2.9)

The asymptotic expansion of $E_n(t)$ for large z is given by

$$E_n(z) \sim \frac{e^{-z}}{z} \left[1 - \frac{n}{z} + \frac{n(n+1)}{z^2} - \frac{n(n+1)(n+2)}{z^2} + \cdots \right]$$

(B2.10)

for ($|\arg z| < 3\pi/2$). Two functions closely related to the exponential integral are given by

$$Ei(x) = -\mathcal{P} \int_{-x}^\infty \frac{-e^{-t}\, dt}{t} = \mathcal{P} \int_{-\infty}^x \frac{e^t\, dt}{t}$$

(B2.11)

for $x > 0$, where $\mathcal{P}$ denotes principal part, and

$$li(x) = \mathcal{P} \int_0^x \frac{dt}{\ln t} = Ei(\ln x)$$

(B2.12)

for $x > 1$. Expansion of $Ei(x)$ is given by

$$Ei(x) = \gamma + \ln x + \sum_{n=1}^\infty \frac{x^n}{nn!}$$

(B2.13)

for $x > 0$. We note also that

$$\int_0^x \frac{e^t - 1}{t}\, dt = Ei(x) - \ln x - \gamma$$

(B2.14)

$$\int_0^\infty \frac{e^{iat}\, dt}{t \mp ib} = e^{\pm ab} \left[\begin{array}{c} E_1(ab) \\ -Ei(ab) + i\pi \end{array} \right]$$

(B2.15)

In the latter relation the $-$ sign on the left side corresponds to the $+$ exponential and top member of the column and $a > 0, b > 0$.

Γ-Function

The Γ is defined by

$$\Gamma(z) = \int_0^\infty t^{z-1} e^{-t}\, dt, \qquad \operatorname{Re} z > 0$$

(B2.16)

For $z = n$, an integer,

$$\Gamma(n) = (n - 1)! \tag{B2.17}$$

so that

$$\Gamma(n + 1) = n\Gamma(n) \tag{B2.18}$$

Here are some typical values:

$$\Gamma\left(\frac{1}{2}\right) = \sqrt{\pi}, \qquad \Gamma(1) = 1$$
$$\Gamma\left(\frac{3}{2}\right) = \frac{1}{2}\Gamma\left(\frac{1}{2}\right) = \frac{\sqrt{\pi}}{2} \tag{B2.19}$$

Stirling's asymptotic expression for $z \to \infty$ in the domain $|\arg z| < \pi$ is given by

$$\Gamma(z) \sim e^{-z} z^{z+(1/2)} (2\pi)^{1/2} \left[1 + \frac{1}{12z} + \frac{1}{288z^2} + \cdots \right] \tag{B2.20}$$

Riemann ζ Function

This function is defined by

$$\zeta(z) = \sum_{k=1}^{\infty} \frac{1}{k^z} \tag{B2.21}$$

for $\mathrm{Re}\, z > 1$. First we note that

$$\zeta(z) = \prod_{p} \left(\frac{p^z}{p^z - 1} \right) \tag{B2.22}$$

where the product is over all primes p. We may also write

$$\zeta(z) = \frac{1}{\Gamma(z)} \int_0^{\infty} \frac{x^{z-1}}{e^x - 1} \, dx \tag{B2.23}$$

Here are some special values:

$$\zeta(1) = \infty \qquad\qquad \zeta(2) = \frac{\pi^2}{6}$$
$$\zeta\left(\frac{3}{2}\right) = 2.612 \qquad \zeta(4) = \frac{\pi^4}{90}$$
$$\zeta\left(\frac{5}{2}\right) = 1.341 \qquad \zeta(6) = \frac{\pi^6}{945}$$

Error Function

The error function is defined by

$$\text{erf } z = \frac{2}{\sqrt{\pi}} \int_0^z e^{-t^2} dt \tag{B2.24}$$

We also define

$$\text{erfc } z = 1 - \text{erf } z = 2\sqrt{\pi} \int_z^{\infty} e^{-t^2} dt \tag{B2.25}$$

The error function has the series expansions

$$\text{erf } z = \frac{2}{\sqrt{\pi}} \sum_{n=0}^{\infty} \frac{(-1)^n z^{2n+1}}{n!(2n+1)} \tag{B2.26}$$

Asymptotic expansions are given by

$$x \ll 1 : \text{erf } x \sim \frac{2}{\sqrt{\pi}} \left(x - \frac{1}{3}x^3 + \frac{1}{10}x^5 + \cdots \right) \tag{B2.27}$$

$$x \gg 1 : \text{erf } x \sim 1 - \frac{e^{-x^2}}{x\sqrt{\pi}} \left(x - \frac{1}{2x^2} + \cdots \right) \tag{B2.28}$$

Some integral relations for this function are as follows:

$$\int_0^{\infty} e^{-at} \text{erf } bt \, dt = \frac{1}{a} \exp \left(\frac{a^2}{4b^2} \right) \text{erfc} \left(\frac{a}{2b} \right) \tag{B2.29}$$

$$\int_0^{\infty} \sin(2at) \text{erfc } bt \, dt = \frac{1}{2a} \left[1 - \exp \left(\frac{-a^2}{b^2} \right) \right] \tag{B2.30}$$

$$\int_0^{\infty} e^{-at} \text{erf } \sqrt{bt} \, dt = \frac{1}{a} \sqrt{\frac{b}{a+b}} \tag{B2.31}$$

$$\int_0^{\infty} e^{-at} \text{erfc} \sqrt{\frac{b}{t}} \, dt = \frac{1}{a} e^{-2\sqrt{ab}} \tag{B2.32}$$

B.3 Other Useful Integrals

$$\int_{-\infty}^{\infty} e^{-y^2} dy = \sqrt{\pi} \tag{B3.1}$$

$$E(n) \equiv \int_0^{\infty} e^{-ax^2} x^n \, dx, \qquad n \geq 0, a > 0 \tag{B3.2}$$

$$E(n) = \frac{1}{2} \Gamma \left(\frac{n+1}{2} \right) a^{-(n+1)/2} \tag{B3.3}$$

$$E(0) = \frac{1}{2} \sqrt{\pi} a^{-1/2} \qquad E(3) = \frac{1}{2} a^{-2}$$

$$E(1) = \frac{1}{2}a^{-1} \qquad E(4) = \frac{3}{8}\sqrt{\pi}a^{-5/2} \tag{B3.4}$$

$$E(2) = \frac{1}{4}\sqrt{\pi}a^{-3/2} \qquad E(5) = a^{-3}$$

$$\int_0^\infty \frac{x^{\frac{n}{2}-1}\,dx}{e^x + 1} = \Gamma(n)\lambda(n) \tag{B3.5}$$

$$\lambda(n) \equiv \sum_{k=1}^\infty \frac{(-1)^{k-1}}{k^n} \tag{B3.6}$$

$$\lambda(1) = \ln 2, \qquad \lambda(4) = \frac{7\pi^4}{720}$$

$$\lambda(2) = \frac{\pi^2}{12} \qquad \lambda(6) = \frac{31\pi^6}{30,240} \tag{B3.7}$$

Green's Identities

Green's First Identity

$$\int_V \left(\phi\nabla^2\psi + \nabla\phi \cdot \nabla\psi\right) dV = \int_S (\phi\nabla\psi) \cdot d\mathbf{S} \tag{B3.8}$$

Green's Second Identity

$$\int_V \left(\phi\nabla^2\psi - \psi\nabla\phi\right) dV = \int_S (\phi\nabla\psi - \psi\nabla\phi) \cdot d\mathbf{S} \tag{B3.9}$$

Other Identities

$$\int_V \nabla \times \mathbf{A}\,dV = \int_S d\mathbf{S} \times \mathbf{A}$$

$$\int_C \phi\,d\mathbf{l} = \int_S d\mathbf{S} \times \nabla\phi \tag{B3.10}$$

B.4 Hermite and Laguerre Polynomials and the Hypergeometric Function

Hermite polynomials

These polynomials are defined by the relation

$$H_n(z) = (-1)^n e^{z^2}\left(\frac{d^n}{dz^n}e^{-z^2}\right), \qquad n = 1, 2, 3, \ldots \tag{B4.1}$$

They satisfy the differential equation

$$\left(\frac{d^2}{dz^2} - 2z\frac{d}{dz} + 2n \right) H_n(z) = 0 \tag{B4.2}$$

and have the generating function

$$\exp\left(-s^2 + 2sz\right) = \sum \frac{s^n}{n!} H_n(z) \tag{B4.3}$$

Recurrence relations are given by

$$\frac{d}{dz} H_n = 2n H_{n-1}$$

$$\left(2z - \frac{d}{dz} \right) H_n = H_{n-1} \tag{B4.4}$$

$$2z H_n = H_{n+1} + 2n H_{n-1}$$

Here is a list of the first few polynomials

$$\begin{aligned}
H_0 &= 1 & H_1 &= 2z \\
H_2 &= 4z^2 - 2 & H_3 &= 8z^3 - 12z \\
H_4 &= 16z^4 - 48z^2 + 12 & H_5 &= 32z^5 - 160z^3 + 120z
\end{aligned} \tag{B4.5}$$

Tensor Hermite polynomials are discussed in Section 3.5.10.

Laguerre Polynomials

These polynomials are defined by the relation

$$L_n^0 = e^z \frac{d^n}{dz^n} \left(e^{-z} z^n \right)$$

$$L_n^k = (-1)^k \frac{d^k}{dz^n} L_{n+k}^0 \qquad k, n = 0, 1, 2, \ldots \tag{B4.6}$$

These functions satisfy the differential equation

$$\left[z\frac{d^2}{dz^2} + (k + 1 - z)\frac{d}{dz} + n \right] L_n^k = 0 \tag{B4.7}$$

and have the generating function

$$\frac{e^{-zt/(1-s)}}{(1 - s)^{k+1}} = \sum_{n=0}^{\infty} \frac{s^n}{(n+k)!} L_n^k(z), \qquad |s| < 1 \tag{B4.8}$$

Orthogonality relations for these polynomials are given by

$$\int_0^{\infty} e^{-z} z^k L_n^k L_m^k \, dz = \frac{[(n+k)!]^3}{n!} \delta_{nm} \tag{B4.9}$$

Hypergeometric Function

These functions are defined by the series (for $|z| < 1$)

$$F(a, b, c; z) = 1 + \frac{ab}{c} \frac{z}{1!} + \frac{a(a + 1)b(b + 1)}{c(c + 1)} \frac{z^2}{2!}$$

$$+ \frac{a(a + 1)(a + 2)b(b + 1)(b + 2)}{c(c + 1)(c + 2)} \frac{z^3}{3!} + \cdots \quad (B4.10)$$

and are related to the differential equation

$$\left[z(1 - z)\frac{d^2}{dz^2} + (c - (a + b + 1)z)\frac{d}{dz} - ab \right] \varphi = 0 \quad (B4.11)$$

where $c \neq -n$, $(n = 1, 2, \ldots)$. The function $F(a, b, c; z)$ is the solution of (B4.11) for $\varphi(0) = 1$.

The general solution of the differential equation (B4.11) is given by

$$\varphi = A_1 F(a, b, c; z) + A_2 z^{1-c} F(a + 1 - c, b + 1 - c, 2 - c; z) \quad (B4.12)$$

for $|z| < 1$ and $c \neq 0, \pm 1, \pm 2, \ldots$.

For large z,

$$F(a, b, c; z) \sim \frac{\Gamma(c)\Gamma(b - a)}{\Gamma(b)\Gamma(c - a)}(-z)^{-a} + \frac{\Gamma(c)\Gamma(a - b)}{\Gamma(a)\Gamma(c - b)}(-z)^{-b} \quad (B4.13)$$

The relation of these functions to associated Legendre polynomials is given by [for $|\arg(z \pm 1)| < \pi$]

$$P_l^m(z) = \frac{\Gamma(l + m + 1)}{\Gamma(l - m + 1)} \cdot \frac{(z^2 - 1)^{m/2}}{2^m \Gamma(1 + m)} F\left(m - l, m + l + 1, m + 1, \frac{1 - z}{2} \right)$$
$$(B4.14)$$

whereas the relation to the Legendre polynomials is given by

$$P_l(z) = F\left(-l, l + 1, 1; \frac{1 - z}{2} \right) \quad (B4.15)$$

Confluent Hypergeometric Function

These functions are defined by the series (convergent in the whole complex plane)

$$\Phi(a, c; z) = 1 + \frac{a}{c}\frac{z}{1!} + \frac{a(a + 1)}{c(c + 1)}\frac{z^2}{2!} + \frac{a(a + 1)(a + 2)}{c(c + 1)(c + 2)}\frac{z^3}{3!} + \cdots \quad (B4.16)$$

and stem from the differential equation

$$\left[z\frac{d^2}{dz^2} + (c - z)\frac{d}{dz} - a \right] \varphi = 0 \quad (B4.17)$$

The general solution of (B4.17) is given by (for c noninteger)

$$\varphi = A_1 \Phi(a, c; z) + A_2 x^{1-c} \Phi(a - c + 1, 2 - c; z) \quad (B4.18)$$

The relation of these functions to the hypergeometric series is given by

$$\varphi(a, c; z) = \lim_{b \to \infty} F\left(a, b, c; \frac{z}{b}\right) \tag{B4.19}$$

For large z

$$\Phi(a, c; z) \sim e^{-i\pi a} \frac{\Gamma(c)}{\Gamma(c - a)} z^{-a} + \frac{\Gamma(c)}{\Gamma(a)} e^{z} z^{a-c} \tag{B4.20}$$

Physical Constants

Velocity of light in vacuum	c	2.9979×10^8 m/s
		2.9979×10^{10} cm/s
Planck's constant	h	6.6261×10^{-34} J s
		6.6261×10^{-27} erg sec
		4.1357×10^{-15} eV s
	$\hbar$	1.0546×10^{-34} j S
		1.0546×10^{-27} erg s
		6.5821×10^{-16} eV s
Avogadro's number	N_0	6.0221×10^{23} atoms/mol
Boltzmann's constant	k_B	1.3807×10^{-23} J/K
		1.3807×10^{-16} erg/K
		8.6174×10^{-5} eV/K
Gas constant	$R = N_0 k_b$	8.3145 J/mol K
		8.3145×10^7 erg/mol K
		1.9870 cal/mol K
Volume of 1 mole of perfect gas, at normal temperature and pressure		22.241 liters
Electron charge	e	1.6022×10^{19} C
		4.8032×10^{-10} esu
Electron rest mass	m	9.1094×10^{-31} kg
		9.1094×10^{-28} g
		0.511 MeV
Proton rest mass	M_p	1.6726×10^{-27} kg
		1.6726×10^{-24} g
		1.0073 amu
		938.27 MeV

Neutron rest mass	M_n	1.675×10^{-27} kg
		1.675×10^{-24} g
		939.57 MeV
Bohr magneton	$\mu_B = \frac{e\hbar}{2mc}$	2.273×10^{-21} erg gauss^{-1}
Ratio of proton mass to electron mass	$\frac{M_p}{m}$	1836.1
Charge-to-mass ratio of electron	$\frac{e}{m}$	1.7588×10^{11} C/kg
		5.2730×10^{17} esu/g
Stephan-Boltzmann constant	σ	5.6697×10^{-5} erg cm^{-2} s^{-1} K^{-4}
Rydberg constant	R_y	109,737.32 cm^{-1}
Bohr radius	a_0	0.52918 Å
Triple point of water		273.16 K
Atomic mass unit	1 amu	1.6605×10^{-24} g
		931.5 MeV

Useful Conversion Constants and Units

Constants, mks units	ε_0	8.8542×10^{12} F/m
	μ_0	$4\pi \times 10^{-7}$ H/m
1 electron volt	eV	1.6022×10^{-12} erg
		1.6022×10^{-19} J
		3.829×10^{-20} cal
		11.605 K
1 coulomb		0.1 abcoulomb (emu)
		2.9979×10^9 statcoulomb (esu)
1 weber per square meter		10^4 gauss
1 eV/molecule		96.485 kJ/mol
		23.06 kcal/mol
1 entropy unit $\equiv$ 1 eu		$N_0 k_B = 2$ cal/K
1 poise (unit of viscosity)		1 dyne-s/cm^2
1 bar (unit of pressure)		10^6 dynes/cm$^2 = 0.9869$ atm.

APPENDIX   D

Lorentz–Legendre Expansion

In this appendix we wish to outline the precursor relations leading to (7.1.48). The expansion (7.1.47) is formally written

$$f(E, \mu, \phi) = \sum_{l=1}^{\infty} \varepsilon^l f_l(E, \phi) P_l(\mu) \tag{D1.1a}$$

$$\mu = \hat{\mathcal{E}} \cdot \hat{\mathbf{k}} = \cos \theta \tag{D1.1b}$$

where hatted variables are unit vectors and $P_l(\mu)$ are Legendre polynomials. The variables (k, θ, ϕ) in (D1.1) are spherical coordinates of the $\mathbf{k}$ vector (Fig. D.1). These unit vectors have the following Cartesian components:

$$\hat{\mathbf{k}} = [\cos \phi \sqrt{1 - \mu^2}, \sin \phi \sqrt{1 - \mu^2}, \mu] \tag{D1.2a}$$

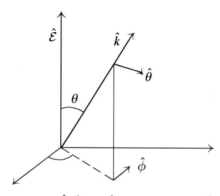

FIGURE D.1. In this diagram, $\hat{\mathcal{E}}$, $\hat{\mathbf{k}}$, and $\hat{\theta}$ lie in a plane and $\hat{\phi}$ lies in a plane normal to $\hat{\mathcal{E}}$.

$$\hat{\theta} = [\mu \cos \phi, \mu \sin \phi, -\sqrt{1 - \mu^2}] \tag{D1.2b}$$

$$\hat{\phi} = [-\sin \phi, \cos \phi, 0] \tag{D1.2c}$$

Furthermore,

$$\frac{\partial}{\partial \mathbf{k}} = \left[\hat{\mathbf{k}} \frac{\partial}{\partial k} + \hat{\theta} \frac{\partial}{k \partial \theta} + \hat{\phi} \frac{\partial}{(k \sin \theta) \partial \phi} \right] \tag{D1.3a}$$

We choose $\mathcal{E}$ to lie in the z direction so that

$$\mathcal{E} \cdot \frac{d}{d\mathbf{k}} = \mathcal{E} \left[\hat{\mathbf{z}} \cdot \hat{\mathbf{k}} \frac{\partial}{\partial k} + \hat{\mathbf{z}} \cdot \hat{\theta} \frac{\partial}{k \partial \theta} + \hat{\mathbf{z}} \cdot \hat{\phi} \frac{\partial}{(k \sin \theta) \partial \phi} \right] \tag{D1.3b}$$

With (D1.2c) the third term in (D1.3b) vanishes and one is left with

$$\mathcal{E} \cdot \frac{d}{d\mathbf{k}} = \mathcal{E} \left[\mu \frac{\partial}{\partial k} - \sqrt{1 - \mu^2} \frac{\partial}{k \partial \theta} \right] \tag{D1.3c}$$

With the given orientation of $\mathcal{E}$ one may take $f_l = f_l(E)$ in the expansion (D1.1a). With

$$\frac{\partial}{\partial k} = \frac{2E}{k} \frac{\partial}{\partial E} \tag{D1.4a}$$

$$\frac{\partial}{\partial \theta} = -\sqrt{1 - \mu^2} \frac{\partial}{\partial \mu} \tag{D1.4b}$$

one obtains

$$\mathcal{E} \cdot \frac{d}{d\mathbf{k}} = \frac{2\mathcal{E}}{k} \left[\mu E \frac{\partial}{\partial E} + \left(\frac{1 - \mu^2}{2} \right) \frac{\partial}{\partial \mu} \right] \tag{D1.5a}$$

It follows that

$$\mathcal{E} \cdot \frac{d}{d\mathbf{k}} (f_0 + \mu f_1) = \frac{2\mathcal{E}}{k} \left[\mu E \frac{\partial}{\partial E} + \left(\frac{1 - \mu^2}{2} \right) \frac{\partial}{\partial \mu} \right] (f_0 + \mu f_1) \tag{D1.5b}$$

or equivalently,

$$\mathcal{E} \cdot \frac{d}{d\mathbf{k}} (f_0 + \mu f_1) = \frac{2\mathcal{E}}{k} \left[\mu E \frac{\partial f_0}{\partial E} + \mu^2 E \frac{\partial f_1}{\partial E} + \left(\frac{1 - \mu^2}{2} \right) f_1 \right] \tag{D1.5c}$$

where it was noted that $\partial \mu f_1(E) / \partial \mu = f_1(E)$ and that $\partial f_0 / \partial \mu = 0$. The first few Legendre polynomials are given by

$$P_0(\mu) = 1, \quad P_1(\mu) = \mu, \quad P_2(\mu) = \frac{1}{2}(3\mu^2 - 1)$$

$$u^2 = \frac{2}{3} P_2(\mu) + \frac{1}{3} P_0(\mu) \tag{D1.5d}$$

Keeping Legendre polynomials up to order $l = 1$, we note that $\mu^2 = \frac{1}{3} P_0(\mu)$. The relation (D1.5c) reduces to

$$\mathcal{E} \cdot \frac{d}{d\mathbf{k}} f = \frac{2\mathcal{E}}{k} \left[\frac{1}{3} \frac{\partial (E f_1)}{\partial E} P_0 + E \frac{\partial f_0}{\partial E} P_1 \right] + \cdots \tag{D1.6}$$

This relation gives rise to (7.1.48).

APPENDIX E

Additional References

E.1 Early Works

1. Analytical Theory of Dynamics

Hamilton, Sir W. R.
  (a) "Problem of Three Bodies by My Characteristic Function" (1833).
  (b) "On a General Method in Dynamics, by which the Study of the Motions of all Free Systems of Attracting or Repelling Points Is Reduced to the Search and Differentiation of One Central Relation, or Characteristic Function," Phil. *Trans. Roy. Soc.*, Part 11, 247–308 (1834).
  (c) "Second Essay on a General Method in Dynamics," *Phil. Trans. Roy. Soc.*, Part I, 95–144 (1835).

Jacobi, K. G. J.
  Vorlesungen über Dynamik, Berlin, G. Reimer, 1866, second edition, 1884.

Lagrange, J. L.
  (a) *Mecanique Analytique* (1788), *Oeuvres, 6*, 335 (1773).
  (b) *Mém. de L'Institut de France* (1808).

Liouville, J.
  "Note sur la théorie de la variation des constantes arbitraries," *J. de Math., 3*, 342 (1838).

Poincaré, H.
  (a) *Oeuvres*, Paris, Gauthier-Villars, 1951–1954.
  (b) *Methodés Nouvelles de la Mechanique Céleste* (1892). Reprinted by Dover, New York, 1957.
  (c) "Sur le probléme des trois corps et les équations de la dynamique," *Acta Math., 13*, 1 (1890); *Oeuvres, 7*, 272.

Poisson, S. D.
- (a) "Sur la chaleur des gaz et des vapeurs," *Ann. Chim., 23*, 337 (1823), *Ann. Physik, 76*, 269 (1824); English translation with notes by J. Herapath, *Phil. Mag., 62*, 328 (1823).
- (b) "Mémoire sur la théorie du son," *Nouv. Bull. Sci. Soc. Philomat. (Paris) 1*, 19 (1807); *J. École Polyt. (Paris), 7*, 14° Cah., 319 (1808).
- (c) *Nouvelle théorie de l'action capillaire*, Paris, Bachelier, 1831.
- (d) *Théorie mathematique de la chaleur*, Paris, Bachelier, 1835.

2. Kinetic Theory of Matter

Boltzmann, L.
- (a) *Lectures on Gas Theory*, translated by S. G. Brush, University of California Press, Berkeley, 1964. This text contains an excellent history of the theory and a very extensive bibliography.
- (b) "Further Studies on the Thermal Equilibrium Among Gas-molecules," *Wien. Ber., 66*, 275 (1872).
- (c) "On the Thermal Equilibrium of Gases Subject to External Forces," *Wien. Ber., 72*, 427 (1875).
- (d) "On the Formulation and Integration of Equations Which Determine the Molecular Motion in Gases," *Wien. Ber., 74*, 503 (1876).
- (e) *Wien. Ber., 81*, 117 (1880).
- (f) *Wien. Ber., 84*, 40 (1881).
- (g) *Wien. Ber., 84*, 1230 (1881).
- (h) *Wien. Ber., 86*, 63 (1882).
- (i) *Wien. Ber., 88*, 835 (1883).
- (j) *Jahresb, d. D. Math. Verein, 6*, 130 (1889).

Maxwell, J. C.
- (a) *The Scientific Papers of James Clerk Maxwell*, Cambridge University Press, 1890; reprinted by Hermann, Paris, 1927, and by Dover, New York, 1952.
- (b) "Illustrations of the Dynamical Theory of Gases. I. On the Motions and Collisions of Perfectly Elastic Spheres. II. On the Process of Diffusion of Two or More Kinds of Moving Particles Among One Another. III. On the Collision of Perfectly Elastic Bodies of Any Form." *Phil. Mag.* [4], *19*, 19; *20*, 21, 33 (1860); *Brit. Assoc. Rept. 29* (2), 9 (1859); Athenaeum, p. 468 (Oct. 8, 1859); L'Institut *364* (1859).
- (c) "Viscosity or Internal Friction of Air and Other Gases," *Phil. Trans., 156*, 249 (1866); *Proc. Roy. Soc. (London), 15*, 14 (1867).
- (d) "On the Dynamical Theory of Gases," *Phil. Trans., 157*, 49 (1867); *Proc. Roy. Soc. (London), 15*, 146 (1867).
- (e) "A Discourse on Molecules," *Nature, 8*, 437 (1873); *Phil Mag.* [4], *46*, 453 (1873); *Les Mondes, 32*, 311, 409 (1873); *Pharmaceut, J., 4*, 404, 492, 511 (1874).

(f) "On the Dynamical Evidence of the Molecular Constitution of Bodies" (lecture). *Nature, 11*, 357, 374 (1875); *J. Chem. Soc., 13*, 493 (1875); *Gazz. Chim. Ital., 5*, 190 (1875).

(g) "On Stresses in Rarified Gases Arising from Inequalities of Temperature," *Phil. Trans., 170*, 231 (1880); *Proc. Roy. Soc. (London), 27*, 304 (1878).

(h) "On Boltzmann's Theorem on the Average Distribution of Energy in a System of Material Points," *Trans. Cambridge Phil. Soc., 12*, 547 (1879); *Phil. Mag.* [5], *14*, 299 (1882); *Ann. Physik Beibl., 5*, 403 (1881).

3. Theory of Rare Gases and Solution to the Boltzmann Equation

Burnett, D.

(a) "The Distribution of Velocities in a Slightly Non-Uniform Gas, *Proc. Lond. Math. Soc., 39*, 385 (1935).

(b) "The Distribution of Molecular Velocities and the Mean Motion in a Non-Uniform Gas," *Proc. Lond. Math. Soc., 40*, 382 (1935).

Chapman, S.

(a) *Phil. Trans. Roy. Soc. A, 211*, 433 (1912).

(b) On the Law of Distribution of Velocities, and on the Theory of Viscosity and Thermal Conduction, in a Non-Uniform Simple Monatomic Gas," *Phil. Trans. Roy. Soc. A., 216*, 279 (1916).

(c) *Phil. Trans. Roy. Soc. A, 217*, 115 (1917).

(d) *Phil. Mag., 34*, 146 (1917).

(e) *Phil. Mag., 34*, 182 (1919).

(f) "On Certain Integrals Occurring in the Kinetic Theory of Gases," *Manchester Mem., 66*, 1 (1922).

(g) *Manchester Mem., 7*, 1 (1929).

(h) "On Approximate Theories of Diffusion," *Phil. Mag., 5*, 630 (1928).

(i) "On the Convergence of the Infinite Determinants in the Lorentz Case," *J. Lond. Math. Soc., 8*, 266 (1933).

Chapman, S., and Hainsworth, W.

"Some Notes on the Kinetic Theory of Viscosity, Conduction, and Diffusion," *Phil. Mag., 48*, 593 (1924).

Enskog, D.

(a) *Phys. Zeri., 12*, 56 and 633 (1911).

(b) *Ann. der Phys., 38*, 731 (1912).

(c) The Kinetic Theory of Phenomena in Fairly Rare Gases, Dissertation, Uppsala, 1917.

(d) "The Numerical Calculation of Phenomena in Fairly Rare Gases," *Svensk. Vet. Akad.* (Arkiv. f. Mat., Ast. och. Fys.), *16*, 1 (1921).

(e) "Kinetic Theory of Thermal Conduction, Viscosity, and Self-Diffusion in Certain Dense Gases and Liquids," *Svensk. Akad. Handl., 63*, No. 4 (1922).

(f) *Svensk. Akad.* (Arkiv. f. Mat., Ast. och. Fys.), *21A*, No. 13 (1928).

Hasse, H. R.
 Phil. Mag., 1, 139 (1926).

Hasse, H. R., and Cook, W. R.
 (a) *Phil. Mag., 3*, 977 (1927).
 (b) *Proc. Roy. Soc.*, 196 (1929).
 (c) *Phil. Mag., 12*, 554 (1931).

Hilbert, D.
 Math. Ann., 72, 562 (1912).

James, C. G. F.
 Proc. Camb. Phil. Soc., 20, 477 (1921).

Jeans, J.
 (a) *An Introduction to the Kinetic Theory of Gases*, Cambridge University Press, New York, 1952.
 (b) *Kinetic Theory of Gases*, Cambridge University Press, New York, 1946.
 (c) Phil. Trans. Roy. Soc. A, 196, 399 (1901).
 (d) *Quart. J. Math., 25*, 224 (1904).

Kennard, E. H.
 Kinetic Theory of Gases, McGraw-Hill, New York, 1938.

Loeb, L. B.
 The Kinetic Theory of Gases, 2d ed., McGraw-Hill, New York, 1934.

Lorentz, H. A.
 (a) "On the Equilibrium of Kinetic Energy Among Gas-Molecules," *Wien. Ber., 95*, 115 (1887).
 (b) "The Motions of Electrons in Metallic Bodies," *Proc. Amsterdam Acad., 7*, 438, 585, 684 (1905).
 (c) *The Theory of Electrons*, B. G. Teubner, Leipzig, 1909.

Massey, H. S. W., and Mohr, C. B. O.
 (a) "On the Rigid Sphere Model," *Proc. Roy. Soc. A, 141*, 434 (1933).
 (b) "On the Determination of the Laws of Force between Atoms and Molecules," Proc. *Roy. Soc., 144*, 188 (1934).

Pidduck, F. B.
 (a) "The Kinetic Theory of the Motions of Ions in Gases," *Proc. Lond. Math. Soc., 15*, 89 (1916).
 (b) "The Kinetic Theory of a Special Type of Rigid Molecule," *Proc. Roy. Soc. A., 101*, 101 (1922).

E.2   Recent Contributions to Kinetic Theory and Allied Topics

Abrikosov, A. A., Gorkov, L. P., and Dzyaloshinskii, I. E., *Methods of Quantum Field Theory in Statistical Physics*, Prentice-Hall, Englewood Cliffs, N.J. (1963). Reprinted by Dover, New York (1975).

Akhiezer, A. I., and Peletminskii, S. V., *Methods of Statistical Physics*, Pergamon, Elmsford, N.Y. (1981).

Arnold, V. I., *Mathematical Methods of Classical Mechanics*, Springer-Verlag, New York (1978).

Allis, W. P., "Motion of Ions and Electrons," *Hand. Physik*, vol. XXI, Springer-Verlag, Berlin (1956).

Balescu, R., *Statistical Mechanics of Charged Particles*, Wiley, New York (1967).

Barut, A. O., *Electrodynamics and Classical Theory of Fields and Particles*, Dover, New York (1980).

Bellman, R., G. Birkhoff and Abu-Shumays, I., eds., *Transport Theory*, American Mathematical Society, Providence, R.I. (1969).

Bernstein, J., *Kinetic Theory in the Expanding Universe*, Cambridge, New York (1988).

Brenig, W., *Statistical Theory of Heat: Non-Equilibrium Phenomena*, Springer, New York (1989).

Burgers, J. M., *The Nonlinear Diffusion Equation*, D. Reidel, Dordrecht, The Netherlands (1974).

Burshtein, A. I., *Introduction to Thermodynamics and Kinetic Theory of Matter*, Wiley, New York (1996).

Carruthers, P., and Zachariasen, F., "Quantum Collision Theory with Phase-Space Distributions," *Rev. Mod. Phys.* 55, 245 (1983).

Cercignani, C., *Mathematical Methods in Kinetic Theory*, Plenum, New York (1969).

——, *Theory and Application of the Boltzmann Equation*, Elsevier, New York (1975).

Case, K. M., and Zweifel, P. F., *Linear Transport Theory*, Addison-Wesley, Reading, Mass. (1967).

Chapman, S., and Cowling, T. G., *The Mathematical Theory of Non-Uniform Gases*, 3rd ed., Cambridge University Press, New York (1974).

Cohen, E. D. G., *Fundamental Problems in Statistical Mechanics*, Wiley, New York (1962).

Cohen, E. G. D., and Thiring, W. (eds.), *The Boltzmann Equation: Theory and Application*, Springer-Verlag, New York (1973).

Curtiss, C. F., Hirschfelder, J. O., and Bird, R. B., *The Molecular Theory of Gases and Liquids*, Wiley, New York (1969).

Davidson, R. C., *Methods in Nonlinear Plasma Theory*, Academic Press, New York (1972).

de Boer, J., and Uhlenbeck, G. E., *Studies in Statistical Mechanics*, North-Holland, Amsterdam (1970).

de Groot, S. R., and Mazur, P., *Non-Equilibrium Thermodynamics*, North Holland, Amsterdam (1962).

Doniach, S., and Sondheimer, E. H., *Green's Functions for Solid State Physicists*, Benjamin-Cummings, Menlo Park, Calif. (1974).

Dresden, M., "Recent Developments in the Quantum Theory of Transport and Galvonomagnetic Phenomena" *Rev. Mod. Phys. 33*, 265 (1961).

Duderstadt, J. J., and Martin, W. R., *Transport Theory*, Wiley, New York (1979).

Ebeling, W., *Transport Properties of Dense Plasma*, Birkhauser, Boston, (1984).

Ecker, G., *Theory of Fully Ionized Plasmas*, Academic Press, New York (1972).

Eu, B. C., *Kinetic Theory and Irreversible Thermodynamics*, Wiley, New York (1992).

Family, F., and Landau, D. P., eds. *Kinetics of Aggregation and Gelation*, North Holland, Amsterdam (1984).

Farquhar, I. E., *Ergodic Theory in Statistical Mechanics*, Wiley-Interscience, New York (1964).

Ferziger, J. H., and Kaper, H. G., *Mathematical Theory of Transport Processes in Gases*, North Holland, Amsterdam (1962).

Fetter, A. L., and Walecka, J. D., *Quantum Theory of Many-Particle Systems*, McGraw-Hill, New York (1971).

Finkelstein, R. J., *Nonrelativistic Mechanics*, Benjamin, Menlo Park, Calif. (1973).

Frenkel, J., *Kinetic Theory of Liquids*, Oxford, New York (1946).

Fujita, S., *Introduction to Non-Equilibrium Statistical Mechanics*, W. B. Saunders, Philadelphia (1966).

Goldstein, H., *Classical Mechanics*, 2nd ed., Addison-Wesley, Reading, Mass. (1973).

Gombosi, T. I., *Gas Kinetic Theory*, Cambridge, Now York (1994).

Grad, H., "On the Kinetic Theory of Rarefied Gases," *Comm. Pure and Appl. Math. 2*, 331 (1949).

——, "Principles of the Kinetic Theory of Gases," *Hand. Physik*, vol. XII, Springer-Verlag, Berlin (1958).

——, "The Many Faces of Entropy," *Comm. Pure Appl. Math. 14*, 323 (1961).

Grandy, W. T., *Foundations of Statistical Mechanics, Vol. 2, Nonequilibrium Phenomena*, D. Reidel, Boston (1988).

Groot, S. R. de, and Leeuwen, W. A. Van., *Relativistic Kinetic Theory: Principles and Applications*, North Holland, New York (1980).

Haken, H., *Quantum Field Theory of Solids*, North-Holland, Amsterdam (1976).

Harris, S., *Introduction to the Theory of the Boltzmann Equation*, Holt, Rinehart and Winston, New York (1971).

Harrison, L. G., *Kinetic Theory of Living Patterns*, Cambridge, New York (1993).

Haug, A., *Theoretical Solid State Physics*, H. S. H. Massey (trans.), Pergamon, Elmsford, N.Y. (1972).

Hillery, M., O'Connell, R. F., Scully, M. O., and Wigner, E. P., "Distribution Functions in Physics: Fundamentals," *Phys. Repts. 106*, 122 (1984).

Kac, M., *Probability and Related Topics in Physical Sciences*, Wiley Interscience, New York (1959).

Kadanoff, L. P., and Baym, G., *Quantum Statistical Mechanics*, Benjamin, Menlo Park, Calif. (1962).

Keldysh, L. V., "Diagram Technique for Nonequilibrium Processes," *Sov. Phys. JETP 20*, 1018 (1965).

Klimontovich, Y., *Statistical Theory of Non-Equilibrium Processes in Plasma*, MIT Press, Cambridge, Mass. (1969).

Koga, T., *Introduction to Kinetic Theory. Stochastic Processes in Gaseous Systems*, Pergamon Press, Elmsford, N.Y. (1970).

Kogan, M. N., *Rarefied Gas Dynamics*, Plenum, New York (1969).

Kohn, W., and Luttinger, J. M., "Quantum Theory of Electron Transport Phenomena," *Phys. Rev. 108*, 570 (1975).

Kubo, R., *Statistical Mechanics*, Wiley, New York (1965).

Liboff, R. L., *Introduction to the Theory of Kinetic Equations*, Wiley, New York (1969); Mir, Moscow (1974), Krieger, Melbourne, Fla. (1979).

——, *Introductory Quantum Mechanics*, 4th ed., Addison-Wesley, San Francisco (2002).

——, and Rostoker, N. (eds.), *Kinetic Equations*, Gordon and Breach, New York (1971).

Lie, T-J., and Liboff, R. L., "Consideration of Particle Exchange in Quantum Kinetic Theory," *Ann. Phys. 67*, 349 (1971).

Lifshitz, E. M., and Pitaevskii, L. P., *Physical Kinetics*, Pergamon, Elmsford, N.Y. (1981).

——, and ——, *Statistical Physics, Part Two*, Pergamon, Elmsford, N.Y. (1980).

Mahan, G. D., "Quantum Transport Equation for Electric and Magnetic Fields," *Phys. Repts. 145*, 251 (1987).

——, *Many-Particle Physics*, Plenum, New York (1981).

Mitchner, M., and Kruger, C. H., Jr., *Partially Ionized Gases*, Wiley, New York (1973).

McQuarrie, D. A., *Statistical Mechanics*, Harper & Row, New York (1976).

Montgomery, D. C., and Tidman, D. A., *Plasma Kinetic Theory*, McGraw-Hill, New York (1964).

Negele, J. W., and Orland, H., *Quantum Many-Particle Systems*, Addison-Wesley, Reading, Mass. (1988).

Nicholson, D. R., *Introduction to Plasma Theory*, Wiley, New York (1983).

O'Raifeartaigh, L., *General Relativity. Papers in Honor of J. L. Synge*, Oxford University Press, New York (1972).

Pathria, R. K., *Statistical Mechanics*, Pergamon, Elmsford, N. Y. (1972).

Peletminskii, S., and Vatsenko, A., "Contribution to the Quantum Theory of Kinetic and Relaxation Processes," *Sov. Phys. JETP, 26*, 773 (1968).

Pines, D., *The Many-Body Problem*, Benjamin-Cummings, Menlo Park, Calif., (1962).

Pomraning, G. C., *The Equations of Radiation Hydrodynamics*, Pergamon, Elmsford, N.Y. (1973).

Pozhar, L.A., *Transport of Inhomogeneous Fluids*, World Scientific, Singapore (1994).

Present, R. D., *Kinetic Theory of Gases*, McGraw-Hill, New York (1958).

Prigogine, I., *Non-Equilibrium Statistical Mechanics*, Wiley, New York (1962).

——, *From Being to Becoming*, W. A. Freeman, San Francisco (1980).

Rammer, J., and Smith, H., "Quantum Field-Theory Methods in Transport Theory of Metals," *Rev. Mod. Phys., 58*, 323 (1986).

Reed, T. M., and Gubbins, K. E., *Applied Statistical Mechanics*, McGraw-Hill, New York (1973).

Reichl, L., *A Modern Course in Statistical Physics*, University of Texas Press, Austin (1986).

Resibois, P., and de Leener, M., *Classical Kinetic Theory of Fluids*, Wiley, New York (1977).

Rickayzen, G., *Green's Functions and Condensed Matter*, Academic Press, New York (1980).

Riskin, H., *The Fokker-Planck Equation*, 2nd ed., Springer-Verlag, New York (1989).

Roos, B. W., *Analytic Functions and Distributions in Physics and Engineering*, Wiley, New York (1969).

Sampson, D. H., *Radiative Contributions to Energy and Momentum Transport in a Gas*, Wiley-Interscience, New York (1965).

Schenter, K. G., *A Unified View of Classical and Quantum Kinetic Theory with Application to Charge-Carrier Transport in Semiconductors*, Thesis, Cornell (1988).

Sinai, Ya. G., *Introduction to Ergodic Theory*, Princeton University Press, Princeton, N.J. (1976).

Stewart, J. M., *Non-Equilibrium Relativistic Kinetic Theory*, Springer-Verlag, New York (1971).

Sudarshan, E. C. G., and Mukunda, N., *Classical Dynamics*, Wiley, New York (1974).

Synge, J. L., *Relativity, the Special Theory*, 2nd ed., North-Holland, Amsterdam (1965).

——, *Relativity, the General Theory*, North-Holland, Amsterdam (1966).

Thornber, K. K., and Feynman, R. P., "Velocity Acquired by an Electron in a Finite Crystal in a Polar Field," *Phys. Rev. B1*, 4900 (1970).

Thouless, D. J., *Quantum Mechanics of Many Body Systems*, Academic Press, New York (1961).

Uehling, E. A., and Uhlenbeck, G. E., "Transport Phenomena in Einstein-Bose and Fermi-Dirac Gases. I," *Phys. Rev., 43*, 552 (1933).

Uhlenbeck, G. E., and Ford, G. W., *Lectures in Statistical Mechanics*, American Mathematical Society, Providence, R.I. (1963).

van Leewen, W. A., van Weert, Ch. G., and de Groot, S. R., *Relativistic Kinetic Theory*, North-Holland, Amsterdam (1980).

Vincenti, W. G., and Kruger, C. H., *Introduction to Physical Gas Dynamics*, Wiley, New York (1965).

Wing G. M, *An Introduction to Transport Theory*, Wiley, New York (1962).

Wolfram, S., "Cellular Automaton Fluids 1: Basic Theory," *J. Stat Phys.*, 4, 471 (1986).

Wu, T. Y., *Kinetic Equations of Gases and Plasmas*, Addison-Wesley Reading, Mass. (1966).

Zaslavsky, G. M., *Chaos in Dynamical Systems*, Harwood Academic, New York (1984).

Ziff, R. M., New Class of Solvable-Model Boltzmann Equations, Phys. Rev. Letts, **45** 306 (1980).

Zimon, J. M., *Electrons and Phonons*, Oxford University Press, New York (1960).

——, *Elements of Advanced Quantum Theory*, Cambridge University Press, New York (1969).

Zubarev, D. N., *Nonequilibrium Statistical Thermodynamics*, P. J. Shepard (trans.), Consultants Bureau, New York (1974).

——, and Kalashnikov, V. P., "Derivation of the Nonequilibrium Statistical Operator from the Extremum of the Information Entropy," *Physica, 46*, 550 (1970).

APPENDIX F

SCIENCE AND SOCIETY IN THE CLASSICAL GREEK AND ROMAN ERAS

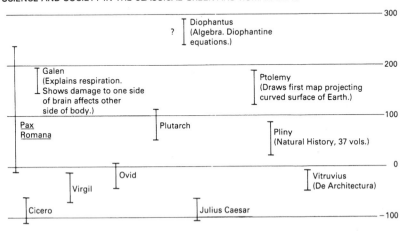

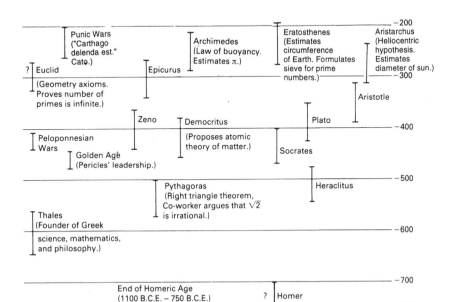

Index

Acoustic phonons, 379
Action-angle variables, 245
Action integral, 2, 478
Activation energy, 512
Adiabatic law, 198, 265
Alkali metals, 487
Alloy, 480
    binary, 480
    ordered and disordered, 480, 520
Alumina, 480
Amorphous media, 506ff
Anderson, P. W., 508
Apsidal vector, 142, 302
Atmospheres, law of (*See* Barometer
    formula)
Autocorrelation function, 59, 189,
    312, 327
    and Green's function, 441
Averages (*See also* Expectation)
    ensemble, 343
    phase, 37, 241
    phase density, 125
    time, 242

Balazs, N. L., 354
Balescu-Lenard equation, 313, 316
Barometer formula, 171
Barut, A. O., 455
BBKGY hierarchy, 78

Bergman, P. G., 449
Bethe lattice, 517
Birkhoff's theorem, 242
Black hole, 478
Bloch
    condition, 492
    waves, 508
Bogoliubov distribution, 117
Bogoliubov hypothesis, 115
Boltzmann equation, 112, 124, 131,
    152
    assumptions in derivation, 152
    quasi-classical, 385, 493
Bose-Einstein distribution, 382, 387
        (*See also* Planck distribution)
Boson, composite, 439
β-Brass, 480
Brass, 480
Brillouin zone, 488
Bronze, 480
Bruns's theorem, 253
Burnett equations, 199

Canonical distribution, 133, 344
Canonical invariants, 15
Canonical transformation, 11, 247
Carbon dioxide, viscosity, 212
Central limit theorem, 54, 172, 270
Chapman-Enskog expansion, 194,
    265

Chapman-Kolmogorov equation, 41, 317
Characteristic curves, 25
Characteristic function, 48, 69
Characteristics, method of, 24, 94, 304
Chebyshev's inequality, 71
Christoffel symbol, 473
Coarse graining, 243
Collision:
    inverse, 147
    reverse, 148
Collisional invariants, 155
Collision frequency, 194, 297
Collision operator, linear, 217
    negative eigenvalues, 218
    spectrum for Maxwell molecules, 220
    symmetry, 217
Completely continuous operator, 224
Conductivity, 177, 209, 238
Conservation equations:
    absolute, 159
    from BY and BY, 83
    relative, 160
Constants of motion, 4, 243, 254
Contravariant vectors, 470
Convergent kinetic equation, 315
Convolution integral, 191
Coordinates, generalized, 1
Correlation expansion, 89, 310
Correlation function, 59, 89
    total, 96, 134
Coulomb gauge, 464
Covariance, 59, 452
Covariant vectors, 470
Cross section, scattering:
    Coulomb, 145, 234
    differential, 143
    rigid sphere, 145
    total, 144
Current density
    entropy, 486
    particle, 486
    thermal, 486

Darwin Lagrangian, 467
deBroglie wavelength, 364
Debye

distance, 95, 113, 326
frequency, 493
shielding potential, 284
temperature, 482, 498
wave number, 283, 314, 495
Deformation potential, 379
Degeneracy:
    classical, 250
    quantum, 377
Degenerate plasma, 377
Degrees of freedom, 1
de Groot, S. R., 449
Density expansion, 117
Density matrix, 395, 430
    equation of motion, 399
Density operator, 331
Deuterium, liquid, 259
Diagrammatic analysis (See Prigogine analysis)
Dielectric function, plasma (See Plasma)
Dielectric time, 113
Diffusion, 175
    coefficient, 483
    mutual, 181, 208
    self, 180, 188
Diffusion coefficient, 304, 327
Diffusion equation, 67, 187
Dispersion relation:
    degenerate plasma, 377
    warm plasma, 286, 377
Distribution functions, classical (See also Quantum distributions)
    conditional, 37
    joint probability, 36
    reduced, 37
    s-tuple, 38
Doppler effect, 477
Drift-diffusion equation, 177
Drude conductivity, 177, 239, 407
    covariant, 460
Druyvesteyn distribution, 236
    absolute, 237

Edwards, J. T., 510
Electrical conductivity, 487
    temperature domains, 512
Electrochemical potential, 178, 484
Electrodynamics, covariant, 458

Electron gas, 464
Electron mobility (*See* Mobility, electrical)
Electron-phonon scattering rate (*See* Scattering rate)
Elliptic integrals, complete, 221
Elyutin, P. V., 510
Energy bands, 369
Energy equation, 97
Energy shell, 35, 240
Ensemble, 20
  average, 189, 343
Entropy:
  Boltzmann, 165
  change, 275
  Gibbs, 164, 275
Equal a priori probabilities, 345
Equation of state, 97, 134
Equipartition of energy, 179, 382
Ergodic hypothesis, 241
Ergodic motion, 240, 264
Error function, 545
Euler equations, 198
Euler's constant, 542
Event, relativistic, 450
Exchange transformation, 13
Expectation:
  of a classical function, 37
  of a random variable, 47
  and wave function, 330
  and Wigner distribution, 353
Exponential integral, 542
Exponents, critical, 518
Extended states, 506, 521

Fermi-Dirac distribution, 371, 387, 483, 492
Fermi energy, 371, 4343
Fermi liquid, 387
Fermi momentum, 388
Fermi sphere, 374
Field tensor, electromagnetic, 458
Flow chart, 333
Fluid dynamical variables (*See* Macroscopic variables)
Fock space, 395, 433
  Hamiltonian, 433
  operator, 396, 438

Fokker-Planck equation, 304, 308, 315, 317, 327
Four-current, 453
Four-vector, 451
  covariant and contravariant, 471
Four-vector potential, 453, 459
Four-wave vector, 453
Free energy, 518
Friction coefficient, 304, 327
Friedel oscillations, 377

Gamma function, 543
Gauss' equation, 280, 281
Gaussian distribution, 56, 59, 173
Generalized coordinates, 1
Generalized potential (*See* Potential)
Generating function, 11
Goldstein, H., 455
Grad's first and second equations, 129, 262
Grad's method of moments, 213
Grains, 480
Grazing collisions, 301, 303
Green's function equations, coupled, 415, 422
Green's functions:
  and averages, 410
  diagrammatic representation, 416, 440
  retarded and advanced, 415
  s-body, 410, 438
  and Schroedinger equation, 408
Group property:
  canonical transformations, 15
  Lorentz transformations, 475

Hamiltonian, 4, 61, 66, 139, 254, 262, 272, 274, 396, 399
  relativistic, 5, 455, 467
Hamilton-Jacobi equation, 248
Hamilton's equations, 4
  covariant, 455
Hamilton's principle, 2, 76
Hard potential (*See* Potential)
Heisenberg picture, 339, 396, 422, 429
Hénon-Heiles Hamiltonian, 254
  anti, 274
Hermite polynomials, tensor, 213, 546

Hermiticity, 27
Hessian, 257
Heteroclinic orbit, 74
Hierarchies, quantum and classical,
    table of, 402
Hillery, R. F., 354
Histogram, bin storage, 323
Hole mobility, 369
Homoclinic orbit, 74
Honeycomb lattice, 517
Hopping, 506
    thermally assisted, 513
    variable range, 513
H-theorem, 166, 392, 446
Hydrofracturing of a rock, 515
Hypergeometric function, 548
    confluent, 548
Hypersonic flow, 298

Ideal gas, 30, 270
Impact parameter, 142
Incompressible phase density, 24, 136
Inelastic scattering, 319
Insulators, 480
Integrable motion, 254
Integral invariants of Poincaré, 68
Integral of motion (*See* Constants of
    motion)
Inverse scattering, 147
Ioffe-Regel minimum, 512
Irreversibility, 162, 167, 240

Jacobian, 17
Jacobi's identity, 11
Jennings, B. K., 354
Joint probability distribution (*See*
    Distribution functions)

Kagomé lattice, 517
KAM theorem, 257, 274
KBG equation (*See* Krook-
    Bhatnager-Gross
    equation)
Kinetic equation:
    convergent, 315
    definition, 82
    photon, 366
    table of, 296
Klimontovich picture, 124, 136

Kronicker-delta symbol, four
    dimensional, 472
Krook-Bhatnager-Gross equation,
    96, 194, 295, 328
Kubo formula:
    classical, 192
    quantum, 404
Kursunoglu, B., 456

Lagrange's equations, 3
    covariant, 472
Lagrangian, 2, 60, 62, 132
    relativistic, 466, 472
Laguerre polynomials, 204, 547
Landau damping, 288, 326, 377
Landau equation, 309, 315
Laplace-Runge-Lenz vector
Large-mass consistency limit, 503
Lasing criterion, 368
Lattice:
    Bethe, 517
    Honeycomb, 517
    Kagomé, 517
    one-dimensional, 522
Legendre expansion, 227, 261
Legendre polynomials, 229
Legendre transformation, 14
Length contraction, 454
Levi-Civita symbol, 538
Liapunov exponent, 257
Lie derivative, 135
Light cone, 450
Liouville equation, 22, 78, 344, 399
    in non-Cartesian coordinates, 474
    one-particle, 93
    relativistic, 457
Liouville operator, 27, 98
    resolvent, 33
Liouville theorem, 17, 276
Localization, 508
    length, 510
    in second quantization, 520*ff*
Localized states, 506
Logarithmic singularity, 295, 303
Long-time limit, 110
Long-time tails, 189, 277
Lorentz
    expansion, 494
    force, 5, 62

gauge, 66, 464
  invariants, 453
  in kinetic theory, table of, 465
  transformation, 452, 475
Lorentzian form (*See* Spectral
    function)
Lorenz
  number, 373, 376
  relation, 483

Mach number, 298
Macroscopic variables:
  absolute, 159
  relative, 160
Magnetic alloys, dilute, 515
Mandelstam variables, 465, 478
Markov process, 41
Master equation, 43, 319, 446
Maxwellian, 96, 225, 275, 306, 327
  absolute, 169
  local, 169
  relativistic, 467, 475
Maxwell interaction, 144, 219, 221,
    271
Maxwell molecules (*See* Maxwell
    interaction)
Maxwell's equations, 66, 459
Mean-free-path, 179
  acoustic phonon, 384
  estimates, 178
Measurement:
  and commutators, 332
  and operators, 331
Measure of a set, 241
Metal-insulator transition, 506, 510
Metals, 480
  electron transport in, 369
Method of moments, Grad's, 213
Metrically indecomposable sets, 242
Metrically transitive transformations,
    242
Metric space, 133
Metric tensor, 471
Microcanonical distribution, 344
Mixed states, 341, 434
Mixing flow, 243
Mobility
  edge, 506, 511
  electrical, 177, 192

electron, 369
gap, 511
hole, 369
Monte Carlo analysis, 319, 328
Mott, N. F., 510
Mott transition, 378

Navier-Stokes equations, 199
Neural networks, 515
Neutron, magnetic moment, 430
Nitrogen, viscosity, 212
Noble metals, 487
Non-Cartesian coordinates, 474
Nyquist criterion, 292, 327

Occupation numbers, 395
O'Connell, M. O., 354
Optical phonons, 379

Pair connectedness, 518
Partition function, 432
  grand, 433
Pauli principle, 339, 394
Percolation, 506, 514*ff*
  bond, 515
  cluster, 515
  site, 515
Periodic motion, conditional, 249
Phase average (*See* Averages)
Phase density averages, 125
Phonon polarization, 490
Phonon scattering (*See* Relaxation
    time)
Planck distribution, 364, 379
Plasma:
  dielectric function, 282
  frequency, 113, 282, 326
  parameter, 113, 304, 314
Plasma waves, 285
  stable modes, 286
  unstable modes, 291
Plemelj's formula, 362
Poincaré map, 254
Poincaré recurrence theorem, 163
Poisson brackets, 10, 64
Poisson distribution, 52, 58, 69
Polycrystalline material, 480
Polymer gelatin, 515
Potential:

Potential: (*continued*)
  generalized, 66
  hard, 220
  soft, 220
Prandtl number, 185
Pratt, W. P., 488
Pressure tensor, 175
Prigogine analysis, 97
Principal part, 326
Projection representation, 350, 438
Proper frequency, 461
Proper time, 454
Proper volume, 463
Pure state, 341, 397, 398, 432

Quantum distributions (*See*
    Bose-Einstein distribution;
    Fermi-Dirac distribution;
    Planck distribution)
Quantum-modified Boltzmann
    equation (*See* Boltzmann
    equation)
Quantum statistics, 386
Quartz, 480
Quasi-classical kinetic equation,
    433 (*See also* Boltzmann
    equation)
Quasi-classical limit, 386, 392
Quasi-free particle, 413
  energy, 428
  lifetime, 428

Radial distribution function, 96, 135,
    448, 480
Radiation field, 366
Random numbers, 323
Random phases, 345, 510
Random variables:
  definition, 47
  sums of, 49
Random walk, 44, 50, 55, 187
Range of interaction, 86
Rayleigh dissipation, 304
Rayleigh's formula, 365
Reichl, L. E., 394
Relativistic Maxwellian (*See*
    Maxwellian)
Relativistic Vlasov equation (*See*
    Vlasov equation)

Relativity, postulates, 450
Relaxation time approximation, 485
  phonon scattering, 379
Representations:
  coordinate, 335
  momentum, 335
Resistivity, metallic, 481
  Bloch component of, 501
  electrical, 498, 501
  low-temperature, 483
  reduced, 504
  residual, 488, 501
Resolvent of Liouville operator (*See*
    Liouville operator)
Riemann zeta function, 544
Rigid spheres:
  collision integral, 153
  transport coefficients, 210
Rosenbluth-Rostoker limit, 96
Rough spheres, 273

Sample space, 47
Saxon-Hunter theorem, 523
Scattering:
  angle, 140
  inelastic, 319
  matrix, 142, 258, 488, 493
  rate, 362
Scattering cross section (*See* Cross
    section)
Schenter, G. K., 394, 488
Schmidt number, 185
Schroeder, P. A., 488
Schwartzchild metric, 478
Scully, M. O., 354
Second quantization, 395
Self-consistent solution, 92
Self-scattering mechanism, 320
Semiconductor, 369, 480
Semimetals, 480
Shock front, 298
Shock waves, 298
Sine-Gordon equation, 70, 71
Single-crystal material, 480
Small shot noise, 53
Smooth spheres, 273
Sodium, band structure, 370
Soft potential (*See* Potential)
Solids: categories of, 480

Sonine polynomials, 204, 221, 224, 259, 266
Sound speed, 268
Spectral function, 409, 425
    Lorentzian form, 426
Spectral theorem, 224
Spherical harmonics, 220
Spin, 338
    and density matrix, 348, 430
    and density of states, 371
Spinor, 430
Spontaneous decay, 366
Stable modes (*See* Plasma)
Statistical balance, 167, 381, 387
Stefan-Boltzmann law, 366
Stimulated decay, 366
Stirling's approximation, 57, 544
Stosszahlansatz, 152
Strain, 176
Strain-acoustic interaction, 379
Strain-optical interaction, 379
Strain tensor, symmetric, 176
Stress tensor, 175, 203, 265, 266
Structure factor, 97, 135
Subsonic flow, 298
Superposition principle, 41, 337
Sutherland model, 211
Symmetry properties of distribution functions, 39, 66, 73, 75
Synge, J. L., 467

Temperature, 161
Tensor:
    covariant and contravariant, 472
    equivalents, 539
    integrals, 471
    metric, 471
Thermal conductivity, 175, 185, 202
    coefficient of, 483, 487, 505
Thermal speed, 169, 179, 275
Thermopower, 484
Thomas-Fermi potential, 377, 378
Thomas-Fermi screening:
    distance, 376, 377
    wave number, 376, 377
Thouless, D., 510
Three-body problem, 253
Tight-binding approximation, 508, 520

Time dilation, 454
Tori, invariant, 252, 257
Transfer integral, 510
Transfer matrix method, 522
Transonic flow, 298
Transport coefficients, 174
    Chapman-Enskog estimates, 210
    mean-free-path estimates, 186

Uehling-Uhlenbeck equation, 388
Uhlenbeck, G. E., 204, 220
Umklapp process, 482
Uncertainty, in quantum mechanics, 333, 334
Unstable modes (*See* Plasma waves)

Van Leewen, W. A., 449
van Weert, Ch. G., 449
Variance, 48, 161
Variational technique, 442
Vector potential, 66, 452, 466, 475
Velocity autocorrelation function, 441
Viscosity coefficient, 175, 184, 204, 211
    observed values, 212
Vlasov equation, 92
    for a plasma, 279
    relativistic, 459, 475
Vlasov fluid, 92

Walker, C. H., 257
Wang Chang, C. S., 204, 220
Wave number of closest approach, 314
Weidemann-Franz Law, 483
Weyl correspondence, 354, 435, 438
Wigner distribution, 351, 396, 435, 436, 437, 438
    equation of motion, 353, 396
Wigner-Moyal equation, 358, 399, 431
World line, 450

Zallen, R., 510
Zero-point energy, ion, 492
Zero sound, 377, 391
Ziman, J. M., 508
Zwanzig, R., 188

Graduate Texts in Contemporary Physics

B.M. Smirnov: **Physics of Atoms and Ions**

M. Stone: **The Physics of Quantum Fields**

F.T. Vasko and A.V. Kuznetsov: **Electronic States and Optical Transitions in Semiconductor Heterostructures**

A.M. Zagoskin: **Quantum Theory of Many-Body Systems: Techniques and Applications**